3rd International Conference on Artificial Intelligence Applications and Technologies (AIAAT 2019)

IOP Conference Series: Materials Science and Engineering
Volume 646

Beijing, China
1 – 3 August 2019

ISBN: 978-1-5108-9845-5
ISSN: 1757-8981

Printed from e-media with permission by:

Curran Associates, Inc.
57 Morehouse Lane
Red Hook, NY 12571

Some format issues inherent in the e-media version may also appear in this print version.

Additional copies of this publication are available from:

Curran Associates, Inc.
57 Morehouse Lane
Red Hook, NY 12571 USA
Phone: 845-758-0400
Fax: 845-758-2633
Email: curran@proceedings.com
Web: www.proceedings.com

Table of contents

Volume 646

2019 3rd International Conference on Artificial Intelligence Applications and Technologies (AIAAT 2019)

1–3 August 2019, Beijing, China

Accepted papers received: 13 September 2019
Published online: 17 October 2019

Preface

Papers

system modeling

neural network

natural language processing

computational intelligence

computer vision

automation and control

IOP Conf. Series: Materials Science and Engineering **646** (2019) 011001 doi:10.1088/1757-899X/646/1/011001

PREFACE

2019 3rd International Conference on Artificial Intelligence Applications and Technologies (AIAAT 2019) was held in Beijing, China from Aug.1 to 3, 2019. AIAAT 2019 was organized by Hong Kong Society of Mechanical Engineers (HKSME), sponsored by Xidian University, York Univeristy, Donghua University, Fudan University. The conference provides a useful and wide platform both for display the latest research and for exchange of research results and thoughts in Artificial Intelligence Applications and Technologies and other topics. The participants of the conference were from almost every part of the world, with background of either academia or industry, even well-known enterprise. The success and prosperity of the conference is reflected high level of the papers received.

The proceedings are a compilation of the accepted papers and represent an interesting outcome of the conference. This book covers 6 chapters: system modeling; neural network; computational intelligence; pattern recognition; computer vision; automation and control.

We would like to acknowledge all of those who supported AIAAT 2019. Each individual and institutional help were very important for the success of this conference. Especially we would like to thank the organizing committee for their valuable advices in the organization and helpful peer review of the papers.

We sincerely hope that AIAAT 2019 will be a forum for excellent discussions that will put forward new ideas and promote collaborative researches. We are sure that the proceedings will serve as an important research source of references and the knowledge, which will lead to not only scientific and engineering progress but also other new products and processes.

Prof. Dan Zhang
York University, Canada

IOP Conf. Series: Materials Science and Engineering **646** (2019) 011002 doi:10.1088/1757-899X/646/1/011002

COMMITTEES

Conference Chairman

Prof. Dan Zhang, York University, Canada

Program Committee

Prof. Xiaochuang Xu, Fudan University, China (Chair)

Prof. Jie Zhang, Donghua University, China

Prof. Wilson Q. Wang, Lakehead University, Canada

Advisory Committee

Prof. Ge Cheng, Donghua University, China

Organizing Committee

Prof. Zidong Wang, Chiba Institute of Technology, Japan

Local Committee

Prof. Chunzhan Yu, Beijing Forestry University, China

Prof. Jinsong Bao, Donghua University, China

Prof. Sheng Guo, Beijing Jiaotong University, China

Prof. Hesheng Wang, Shanghai Jiaotong University, China

Prof. Xuechao Duan, Xidian University, China

Technical Program Committee

Prof. Xiao-Zheng Jin, HeFei University of Technology, China

Prof. Spilios Fassois, the University of Patras, Greece

Prof. Shinn-Liang Chang, National Formosa University, Taiwan

Prof. Shao Cheng, Dalian University of Technology, China

Prof. Robert Babuska, Delft University of Technology, Netherlands

Prof. Qin Jiahu, University of Science and Technology of China, China

IOP Conf. Series: Materials Science and Engineering **646** (2019) 011002 doi:10.1088/1757-899X/646/1/011002

IOP Conf. Series: Materials Science and Engineering **646** (2019) 011003 doi:10.1088/1757-899X/646/1/011003

Peer review statement

All papers published in this volume of *IOP Conference Series: Materials Science and Engineering* have been peer reviewed through processes administered by the proceedings Editors. Reviews were conducted by expert referees to the professional and scientific standards expected of a proceedings journal published by IOP Publishing.

IOP Conf. Series: Materials Science and Engineering **646** (2019) 012001 doi:10.1088/1757-899X/646/1/012001

Semi-supervised Soft Sensor Modeling Method for Crude Furfuryl Alcohol Distillation Process Based on Cosine Similarity-Discount

L H Li[1,*] and Y S Dai[1]

[1]College of Information and Control Engineering, China University of Petroleum (East China), Shandong Qingdao, 266580, China

[*]Lilonghao_upc@qq.com

Abstract. In view of the dynamic time variation and multirate data problems in crude furfuryl alcohol distillation process, too few samples of modeling data caused poor generalization ability of soft sensor model and reduced the accuracy of soft sensor model prediction. To solve this problem, this paper proposes a semi-supervised soft sensor modeling method based on cosine similarity-discount(SSSMI-COSD). By calculating the cosine similarity between labeled and unlabeled samples in the same time interval, and combining with the proposed constraints, the data clustering between labeled and unlabeled samples is realized. In addition, in order to avoid the ill-conditioned problem caused by high-dimensional input variables, the clustered data are fused by discount factor value (DFV). An actual data of a crude furfuryl alcohol distillation process (CFADP) simulation experiment were carried out. The results show that the proposed SSSMI-COSD method can effectively improve the soft sensor model prediction accuracy for a crude furfuryl alcohol distillation process.

1. Introduction

The distillation process is a process of separating a complex mixture into a single chemical product
The distillation process is a process of separating a complex mixture into a single chemical product
with relatively high purity through heat and mass transfer. The principle is to make use of the different relative volatility of each component of the mixture, through evaporation and condensation at a certain temperature and pressure, so that the light components in the liquid phase and the recombinant components in the gas phase can be transferred to each other, thus realizing the separation of components [1]. For distillation column, its characteristics are affected not only by its own internal factors, but also by external factors, resulting in dynamic time-varying distillation process[2]. Because the labeled sample data need to be obtained by utilizing people's professional knowledge and experience, it leads to the difficulty of obtaining valuable labeled sample information. Thus, the classifier with supervised classification based on only a small number of labeled samples does not have good generalization ability[3]. Unlabeled samples are usually collected by various sensor transducers on DCS system, which is characterized by real-time acquisition. Comparing with manual analysis, it is easier to get and larger in quantity[4], and include more sample information[5]. Semi-supervised ensemble learning can further improve the performance of soft sensor model on the basis of ensemble learning by utilizing a large number of unlabeled samples[6].

Aiming at the problem of semi-supervised soft sensing, generous number of scientific researchers have studied it. Yuan[7] proposes a nonlinear semi-active soft sensing modeling method based on semi-supervised weighted probabilistic principal component regression, to disposes unequal data sets with only a small amount of tagged data. Luo[8] aims at the low prediction accuracy of reservoir parameters, combining LSSVM with semi-supervised learning, establishing a semi-supervised regression SLSSVM model. Ferreira[9] applies a semi-supervised soft sensor model to CO training, and improves the quality of unlabeled data by eliminating irrelevant features by feature ranking method. Jin[10] aims at most of the current chemical process soft sensor models do not use labeled samples, proposes a semi-supervised limit learning machine applied to the modeling of chemical process soft sensor.

To solve the problem of few labeled samples and poor generalization ability of soft sensor model in crude furfuryl alcohol distillation process, this paper proposes a semi-supervised soft sensor modeling method based on cosine similarity-discount. This method aiming at the problem of data multiplicity in crude furfuryl alcohol distillation process, by calculating cosine similarity between labeled and unlabeled samples in the same time interval, analyzes the semblance between labeled and unlabeled samples. Based on the definition of constraints, the data clustering of unlabeled samples is realized. The generalization ability of soft sensor model has been improved. In order to avoid the ill-conditioned problem caused by high-dimensional input variables, the clustered data are fused by discount factor value. A new data set of soft sensor modeling has been established. An actual data of a crude furfuryl alcohol distillation process (CFADP) simulation experiment were carried out. The results show that compared with the traditional soft sensor modeling method, the SSSMI-COSD method has better fitting performance and effectively improves the soft sensor model prediction accuracy for a crude furfuryl alcohol distillation process.

2. Data Sample Marking and Fusion Method Based on Cosine Similarity-Discount

Cosine similarity[11,12] is a method of calculating relevance. This method calculates the angle cosine between two vectors as a measure of similarity. The greater the cosine similarity, the greater the similarity between the two vectors; the smaller the similarity between the two vectors, the smaller the similarity between them[13].

The cosine similarity formula between variable $X=(x_1,x_2,\cdots,x_n)$ and $Y=(y_1,y_2,\cdots,y_n)$ is:

$$\cos\langle X,Y\rangle=\frac{X\bullet Y}{\|X\|\|Y\|}=\frac{\sum_{i=1}^{n}x_i y_i}{\left(\sum_{i=1}^{n}x_i^2\right)^{1/2}\left(\sum_{i=1}^{n}y_i^2\right)^{1/2}} \tag{1}$$

Where x_i and y_i represent the ith element values of variables X and Y.

The cosine similarity values are $C_j, 1\le j\le p$.

In order to improve the data quality of samples of soft sensor modeling and eliminate the influence of abnormal data on the data quality, this paper proposes the numerical constraints of cosine similarity of clustering data to achieve high-quality clustering of unlabeled sample data with constraints.

If the cosine similarity between unlabeled sample data and labeled sample data is greater than the threshold value ξ, it is considered that the current unlabeled sample data can be labeled. Otherwise, the unlabeled sample data can't be labeled. The formula is expressed as follows:

$$\begin{cases} C_j\ge\xi; & u_{ij}, 1\le i\le n; 1\le j\le p \text{ labeled} \\ C_j<\xi; & u_{ij}, 1\le i\le n; 1\le j\le p \text{ unlabeled} \end{cases} \tag{2}$$

Because the length of clustered data can't be determined, the value of the discount factor in the discount method[14] can be adjusted dynamically. At the same time, the function of new data can be enhanced, the influence of old data can be reduced[15], and the quality of input sample data with data multiplicity problem can be effectively improved.

IOP Conf. Series: Materials Science and Engineering **646** (2019) 012001 doi:10.1088/1757-899X/646/1/012001

At the same time, in order to ensure the integrity of clustering sample data in the same time period, a numerical calculation method of discount factor and constraints are proposed in this paper.

Assuming that the sample size is P in the same time period after clustering, discount factor values are $\lambda_1, \lambda_2, \cdots, \lambda_p$.

Then the formula for calculating the discount factor values is:

$$\lambda_1 + \lambda_2 + \cdots + \lambda_p = 1 \tag{3}$$

The constraint is as follows:

$$\lambda_1 > \lambda_2 > \cdots > \lambda_p \tag{4}$$

The time span between the new labeled sample data with discount factor value and the original labeled sample data is the largest.

The sample set S of semi-supervised soft sensor modeling data is:

$$S = \begin{Bmatrix} v_j(t_k) | k = 1, 2, \cdots, M; j = 1, 2, \cdots, N \\ y(t_k) | k = 1, 2, \cdots, M \end{Bmatrix} \tag{5}$$

Where $v_j(t_k)$ is the jth input variables of time t_k, $y(t_k)$ is the output variables of time t_k, $t_k(k = 1, 2, \cdots, M)$ represents the sampling time of M sample points output by the system.

Soft sensor model is established based on modeling sample data set S.

3. Soft Sensor Modeling Based on LSSVM

LSSVM[16] has better calculation speed, convergence accuracy and generalization performance. It is more suitable for crude furfuryl alcohol distillation process with small sample size.

The model of LSSVM[17] is:

$$y(x) = \omega^T \phi(x) + b \tag{6}$$

Where $\phi(\bullet)$ is a nonlinear transformation function. ω is an adjustable weight vector. b is an offset. The objective function[18] is:

$$\min J(\omega, \xi) = \frac{1}{2}\omega^T \omega + \frac{C}{2}\sum_{i=1}^{l} \xi^T \xi \tag{7}$$

$$s.t. \quad y_i = \omega^T \varphi(x_i) + b + \xi_i \quad (i = 1, 2, \cdots, l) \tag{8}$$

Where x_i, y_i are the corresponding input and output vectors. ξ_i is the difference between the system output value and actual value. C is a regularization parameter.

Because of the good universality of radial basis kernel function[19], the radial basis function is chosen.

$$k(x, x') = \exp\left\{-\frac{\|x - x'\|^2}{2\sigma^2}\right\}, \quad \sigma > 0 \tag{9}$$

4. Simulation and Analysis of Crude Furfuryl Alcohol Distillation Process

The principle diagram of distillation process in a crude furfuryl alcohol production enterprise is shown in Figure 1.

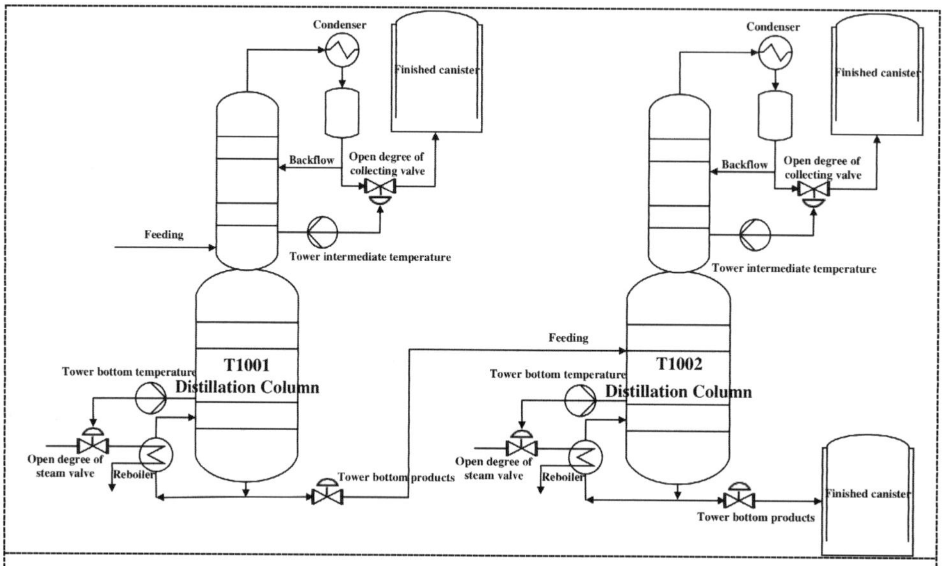

Figure 1. The working principle of a twin-column DP.

The two-column distillation process shown in Figure 1. Because of the purity quality index value of the product needs to be obtained through laboratory tests, and has a certain lag, it is necessary to establish a soft-sensing model. Soft sensor modeling parameters are shown in Table 1:

Table 1. Descriptions of modeling parameters in DP SSM.

Modeling parameter	Description
u_1	T1001 Distillation column bottom temperature
u_2	T1001 Distillation column middle temperature
u_3	T1002 Distillation column bottom temperature
u_4	T1002 Distillation column middle temperature
y	Product purity

The simulation experiment data used in this paper are from DCS system of a furfuryl alcohol production enterprise. The auxiliary variable sampling period is 1 h, and the dominant variable sampling period is 12 h.

The soft sensor model based on traditional labeled sample data adopts the following model structure:

$$\hat{y}_L(t_k) = f_{LDP}\left(u_j(t_k)\right), j = 1,2,3,4 \tag{10}$$

According to the characteristics of crude furfuryl alcohol distillation process and related expert knowledge, the model structure based on semi-supervised clustering fusion data is as follows:

$$\hat{y}_S(t_k) = f_{SDP}\left(v_j(t_k)\right), j = 1,2,3,4 \tag{11}$$

$$v_j(t_k) = f_{Dis}\begin{bmatrix} u_j(t_{k-n}) \\ \vdots \\ u_j(t_k) \end{bmatrix} \tag{12}$$

Where $y(t_k)$ is the purity obtained in the laboratory at time t_k, $u_j(t_{k-n})$ represents the sampling value at the time t_{k-n} of the jth auxiliary variable. $v_j(t_k)$ represents a discounted fusion value of $u_j(t_{k-n})$. $\hat{y}(t_k)$ is the predicted value of soft sensor model.

According to the cosine similarity value of unlabeled sample data, the clustering constraint $\xi = 0.98$ is setted in this paper.

The parameters of LSSVM soft sensor modeling method are set to: $C = 30, \sigma^2 = 3$.

In this paper, Root Mean Square Error (RMSE) and Mean Absolute Error (MAE) are used to evaluate the fitting and prediction performance of regression analysis.

The formula of RMSE is:

$$RMSE=\left(\frac{1}{N}\sum_{i=1}^{N}\left|y_i - y_i\right|^2\right)^{1/2} \tag{13}$$

The formula of MAE is:

$$MAE=\frac{1}{N}\sum_{i=1}^{N}\left|y_i - y_i\right| \tag{14}$$

Where N is the total number of samples, y_i is the predicted output value, and y_i is the actual value.

The training results of soft sensor model based on labeled sample data and semi-supervised fusion data are shown in Figure 2 and Figure 3.

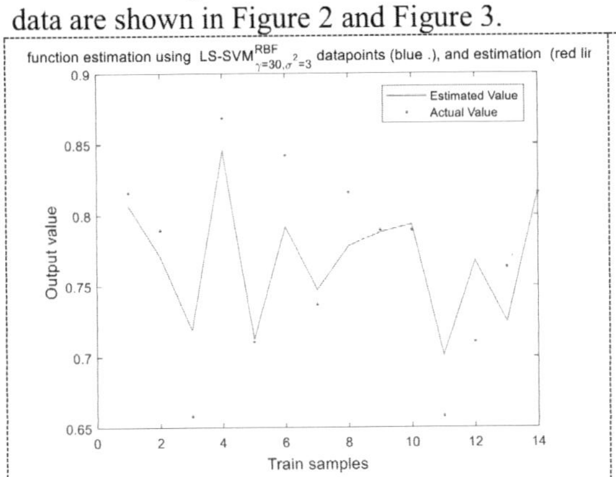

Figure 2. LSSVM data training graph based on labeled sample data.

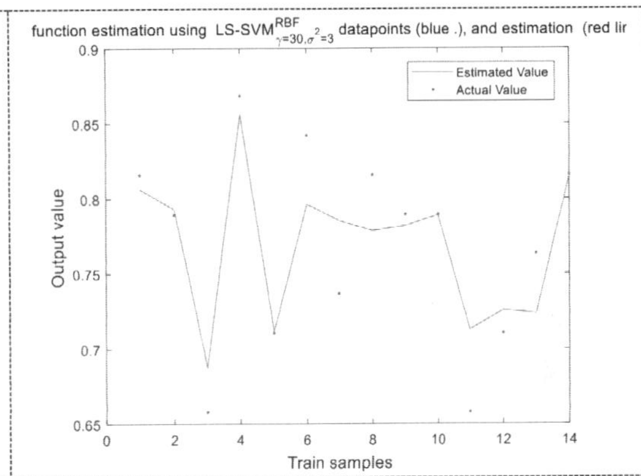

Figure 3. LSSVM data training graph based on semi-supervised fusion data.

Compared with Figure 2 and Figure 3, the training curve shows some deviations. However, the training curve based on semi-supervised fusion data is closer to the actual data and better than the training curve based on labeled sample data.

Table 2. Comparison of fitting performance of different training samples.

Training samples	*RMSE*	*MAE*
Labeled Samples	0.0331	0.0256
Semi-supervised Fusion Samples	0.0291	0.0219

The prediction results of soft sensor model test data based on labeled sample data and semi-supervised fusion data are shown in Figure 4 and Figure 5.

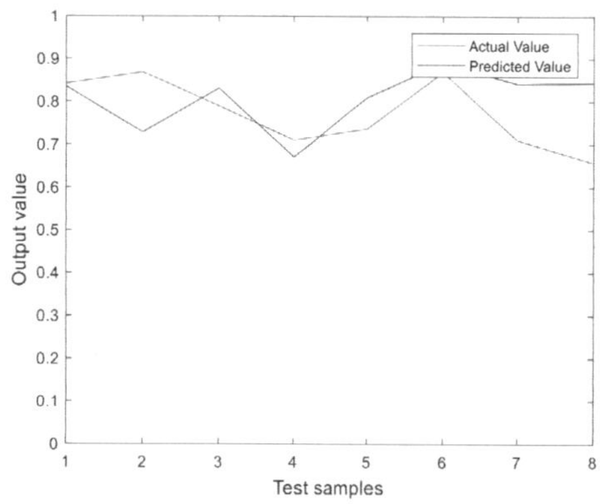

Figure 4. LSSVM model data prediction chart based on labeled sample data.

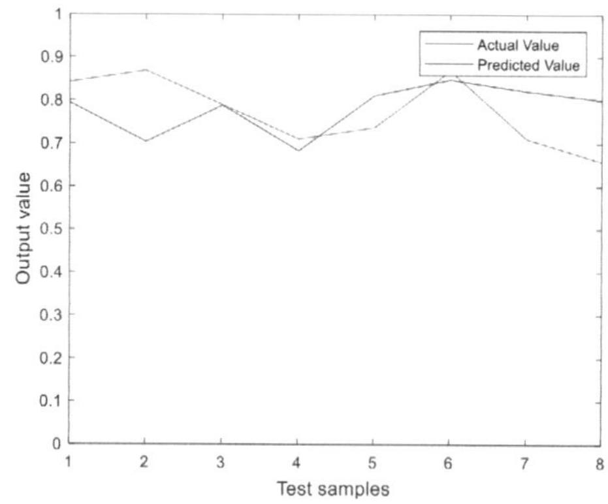

Figure 5. LSSVM model data prediction chart based on semi-supervised fusion data.

Compared with Figure 4 and Figure 5, the data prediction curves show some deviations, but the prediction curves based on semi-supervised fusion data are more compact and uniform than those based on labeled sample data. The soft sensor model based on semi-supervised fusion data improved RMSE and MAE by 7.48% and 51.94% respectively compared with labeled sample data.

Table 3. Comparison of predictive performance of different test samples.

Training samples	*RMSE*	*MAE*
Labeled Samples	0.1003	0.0799
Semi-supervised Fusion Samples	0.0928	0.0384

Comparing the LSSVM fitting curve of Figure 2 and Figure 3 with the RMSE and MAE evaluation index values of Table 2, the following conclusions can be drawn: although the training curve has some deviation, the semi-supervised dynamic fusion sample data fitting curve is better than the labeled sample data fitting curve, and RMSE, MAE evaluation index values have been improved.

Comparing the LSSVM prediction curve of Figure 4 and Figure 5 with the RMSE and MAE evaluation index values of Table 3, the following conclusions can be drawn: Although the prediction curve has some deviation, the semi-supervised dynamic fusion sample data prediction curve is better than the labeled sample data prediction curve, and RMSE, MAE evaluation index values have improved.

Through the simulation modeling process and test data prediction based on the actual industrial data of crude furfurfuryl alcohol distillation process, and comparing the corresponding evaluation index values, combined with the above analysis, the following conclusions can be drawn:

The soft sensor model based on cosine similarity and discount weighted semi-supervised clustering fusion data is superior to the soft sensor model based on labeled sample data in fitting degree and prediction accuracy.

5. Conclusion

In this paper, the crude furfuryl alcohol distillation process is taken as the research background. The simulation and modeling of the actual data of the distillation process show that the proposed semi-supervised soft sensing modeling method based on cosine similarity-discount can realize semi-supervised clustering and fusion of the unlabeled sample data of crude furfurfuryl alcohol distillation process. The semi-supervised data fusion can effectively improve the fitting degree of the soft sensor model and the accuracy of data prediction through the simulation experiment of establishing the soft sensor model.

Acknowledgments
Authors wishing to acknowledge assistance or encouragement from colleagues, special work by technical staff or financial support from organizations should do so in an unnumbered Acknowledgments section immediately following the last numbered section of the paper.

References
[1] Wu X C and Li J T 2005 *J. Kun. Uni. Sci. Tec.* 51-4
[2] Li D 2016 *Liaoning. Uni. Tec.*
[3] Shang N L, Wang X L, Shen K X, Lu Y L, Ma X P and Lan Y H 2015 *Com. APP. Soft.* 162-6+179
[4] Zhang J and Yang H Z 2013 *Com. App. Che.* 1453-6
[5] Wang W, Xi J, Chong A and Li L 2017 *IEEE. Tra. Hum. Sys.* 650-60
[6] Ge Z, Huang B and Song Z 2015 *J. Che.* 793-804
[7] Yuan X, Ge Z, Huang B, Song Z and Wang Y 2016 *IEEE. Tra. Ind. Inf.* 532-41
[8] Luo W P, Li H Q and Shi N 2016 *App. Geo.* 406-15
[9] Ferreira V, Souza F and Rui A 2017 *IEEE. Int. Con. Ind. Inf.* 257-62
[10] Jin X 2015 *Zhejiang. Uni. Tec.*
[11] Kalhori H, Alamdari M and Lin Y 2018 *Mea.* 648-57
[12] Hernandez A and Garcia N 2016 *IEEE. Lat. Ame. Tra.* 2857-61
[13] Geng Z Q, Hu H X and Han Y M 2017 *J. Bei. Uni. Che. Tec.* 87-94
[14] Fan J and Wang G 2017 *Opt. Meth. Soft.* 139-49
[15] Song K, Wang H and Li P 2004 *J. Che. Ind. Eng.* 942-6
[16] Suykens J and Johan A K 2002 *Int. J. Circ. Theor. Appl.* 605-15
[17] Li C X, Ding X D and Zheng X F 2017 *J. Vib. Sho.* 52-8
[18] Suykens J, Brabanter J and Lukas L 2002 *Neurocom.* 85-105
[19] Roushangar K, Saghebian S M and Mouaze D 2017 *Int. J. Sed. Rese.* 51-4

IOP Conf. Series: Materials Science and Engineering **646** (2019) 012002 doi:10.1088/1757-899X/646/1/012002

A Trend Forecast of Import and Export Trade Total Volume based on LSTM

Qinglin Qu[1], Zhao Li[2,3,4,*], Juanjuan Tang[4], Shiwei Wu[4] and Ruishuang Wang[4]

[1]Shandong Business Information Center
[2]Qilu University of Technology (Shandong Academy of Sciences), Shandong Computer Science Center (National Supercomputer Center in Jinan)
[3]Shandong Provincial Key Laboratory of Computer Networks
[4]Big Data Engineering Technology Research Center of E-Government, Shandong

[*]Corresponding author's e-mail address: liz@sdas.org

Abstract. The monthly import and export data are usually with the challenging characteristics including large scale, nonlinear and hard to fit, leading to their development trends can be hardly predicted. To tackle this problem, we propose to leverage the LSTM-based recurrent neural network to forecast the development trend in this paper. We demonstrate the effectiveness of our approach based on the monthly import and export data of Shandong Province from January 2001 to June 2018. In particular, we achieve the MSE score of 124.39, which outperforms the traditional time series model less than 12.2%.

1. Introduction

Import and export trade reflects the opening up of a country (region) to a certain extent. At present, the development of foreign trade is in the key stage of accelerating the pace of structural adjustment and the transformation between the new and old energy in our province. Generally, the environment of development is beneficial to the import and export trade. However, the trade frictions have been obvious recently. Influenced by the thought of "anti-globalization" in western countries, the trend of protectionism is on the rise and still spreading. More and more countries abused the emergency measures of the trade and have launched an investigation into the anti-dumping and anti-subsidy of the imported products as well as imposed high import tariffs. Therefore, we need to pay attention to the development situation of import and export trade and adjust the variable parameters of import and export monitoring system which is related to the macro economy. It is necessary for us to make a right prediction for the trend of total import and export volume in a certain period of time in the future and to make dynamic information, which has important reference significance for management departments and enterprises to know about the dynamic information of import and export trade.

In recent years, some domestic scholars have also done a lot of research on the prediction of total volume of import and export. Luo Wenbin[1] et al. predicted the monthly growth rate of China's total imports and exports from January 1990 to May 2006 by using the small data sets arithmetic, chaotic characteristic analysis of time series and prediction model based on radial basis function neural network. Li Jumeip[2] et al. forecast the annual import and export data of China from 1984 to 2005 by using the time series analysis and established the corresponding model of time series ,then got the conclusion that the effect of short-term prediction of ARMA is better than the effect of long-term prediction .

Duan Peng [3] judged the single order and the integral order of monthly import and export data by using the method proposed by Beaulieu & Miron based on monthly data of import and export trade in 2009. He used the SARIMA model to forecast the data of import and export trade in the next two years. He concluded that the growth rate of export and trade surplus will slow down and the growth rate of import and the total volume of trade will accelerate in 2008. The growth rate of export and the trade surplus will increase and the growth rate of import and the total volume of trade will increase in 2009.Bai Shuang [4] studied the related indicators of Shenzhen's import and export trade in 2013.Firstly,she extracted the four main factors with 99% resolution by adopting Factor Analysis. Then, ARMA, BP and SVM are used to predict according to the prediction results of sub-predictors, the weight of the three sub-predictors in the combined predictor is optimized by the combined prediction method, and the comprehensive prediction results with higher accuracy are obtained.

Wu Xin [5] et al. selected VAR model which is related to the import and export to analyze the structural characteristics of import and export trade and forecast the total volume of import and export according to the causal relationship and the lag relationship between the variables and the total volume of import and export. Ma Ming [6] used GM (1,1) model to make a gray forecast to the trade deficit of maize farming industry in China in 2016. Duan Lei [7] et al. used BP neural network and discrete GM combination forecasting model to forecast the annual data of Jiangsu's total volume of import and export from 2013 to 2015, and got some better results in 2017.

In summary, the models used by scholars for forecasting the date of import and export can be divided into three types: 1. Traditional time series models include auto-regressive (AR), moving average (MA), auto-regressive moving average (ARMA), auto-regressive integrated moving average (ARIMA) or SARIMA with seasonal characteristics. This kind of model mainly considers the fluctuation and trend of quantity of import and export. 2. Regression analysis: Screening other economic variables which are related to total volume of import and export in order to analyze the most suitable influence factors. Using multiple regression, stepwise regression or VAR models or simultaneous equation models. This kind of model analyzed the influence of other factors on import and export in order to deduce the change direction of import and export. But the influence factors of imports and exports are not well determined. 3. Nonlinear prediction: This kind of method mainly includes the gray prediction, SVM model, BP neural net, wavelet transform and so on. The strategies used in forecasting can be divided into two types: 1. single prediction; 2. combination prediction of two or more methods.

But most scholars use the annual data of import and export in the statistical yearbook. According to the actual situation of data, the annual data is much smoother than the monthly data, and the fluctuation is small. It is easy to get a better trend of fitting. But it is difficult to fit if the fluctuation of monthly data is great and quick. However, for the monitoring and early warning of import and export, the use of forecasting cycle of the annual data is longer. In recent years, in-depth learning has developed rapidly, and significant progress has been made in the accuracy of time series prediction. We are very good at dealing with nonlinear data with relatively long intervals and delays. Therefore, this paper forecast the trend of monthly data of import and export over the years by using Long Short-Term Memory (LSTM) network and compares the prediction results with those traditional analysis.

2. The prediction of trend based on LSTM

2.1. The introduction of LSTM

Before introducing LSTM, I will introduce recurrent neural network(RNN) first. The results are from the input layer to the hidden layer and then to the output layer in the process of being fully connected with neural network. Layers and layers are fully or partially connected, and the nodes between each layer are not connected. But the nodes between hidden layers of RNN are connected. The input of the hidden layer includes not only the input of the input layer, but also the output of the hidden layer of the previous moment. It not only learns the information of the current moment, but also depends on the previous sequence information. Therefore, it can describe the relationship between the current output and the previous information, and it has unique advantages in processing the time series data.

The typical structure of RNN is shown on the left of the Figure 1. From the figure, we can see that the main structure, S of RNN not only comes from the input layer X, but also has a circular edge to provide the state at the current moment. At the same time, the state of S will be transferred from the current step to the next step. On the right of the Figure 1, the loop is expanded so that we can see the transformation of the information between the hidden layers clearly.

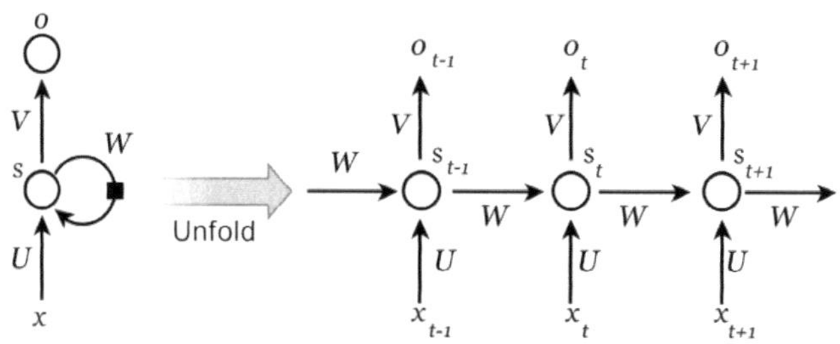

Figure 1. The structure of RNN.

But RNN can only memorize several parts of the sequence, so its performance in long sequence is much worse than that in short sequence. The effect of RNN becomes worse when facing the phenomenon of "long-distance dependence" . Because in the process of gradient return, every return must be multiplied by a rectangular with W matrix and calculated repeatedly, which will bring some problems in training, such as When the weightiness W is less than 1, the error of back propagation will be smaller and smaller, which may cause gradient dispersion. When the weightiness W is larger than 1slightly, the error of back propagation will become larger and larger, which will cause gradient explosion. In this case, no matter what numerical operation is performed on the gradient, the parameters cannot be updated by using the gradient and then LSTM is born at that right moment.

LSTM [8-10] is a special net structure with three “gates”: “forgetting gate”, “input gate” and “output gate”. The forgetting gate is mainly used to remember the state of memory cells and forget the information in memory cells selectively. Input Gates also act on memory cell states, selectively recording new information into new cell states. The output gate acts on the input and the output of the hidden layer, so that the final output includes both the state of cell and the input, and the result is updated to the next hidden layer. Through these three gates, LSTM can determine which information should be forgotten and which information should be retained more effectively. The cell structure of LSTM is shown in Figure 2, in which C is the cell state, X is the input and “a” is the output of each layer.

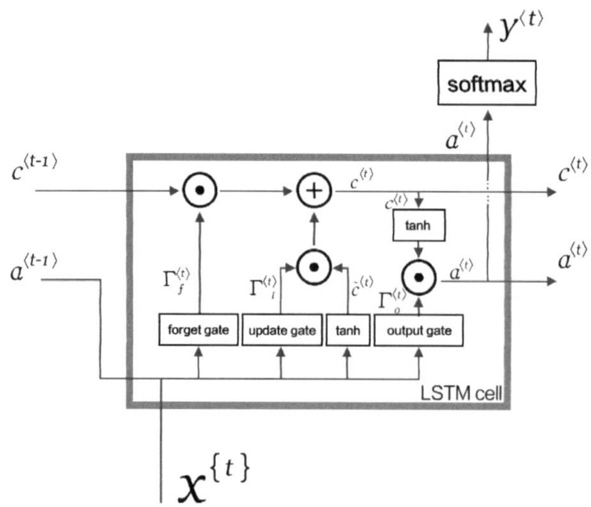

Figure 2. The structure of the cell LSTM.

According to the structure of LSTM , the calculating formula of each LSTM cell is as follows: $\Gamma_f^{\langle t\rangle}$represents the forgetting threshold, $\Gamma_u^{\langle t\rangle}$ represents the input threshold, $\tilde{c}^{\langle t\rangle}$ represents the cell state at the previous moment, $c^{\langle t\rangle}$ represents the cell state (where the cycle occurs), $\Gamma_0^{\langle t\rangle}$ represents the output threshold, $a^{\langle t\rangle}$ represents the output of the current cell and$a^{\langle t-1\rangle}$ represents the output of the previous cell.

$$\Gamma_f^{\langle t\rangle} = \sigma\left(W_f\left[a^{\langle t-1\rangle}, x^{\langle t\rangle}\right] + b_f\right) \tag{1}$$

$$\Gamma_u^{\langle t\rangle} = \sigma\left(W_u\left[a^{\langle t-1\rangle}, x^{\langle t\rangle}\right] + b_u\right) \tag{2}$$

$$\tilde{c}^{\langle t\rangle} = tanh\left(W_c\left[a^{\langle t-1\rangle}, x^{\langle t\rangle}\right] + b_c\right) \tag{3}$$

$$c^{\langle t\rangle} = \Gamma_f^{\langle t\rangle} \circ c^{\langle t-1\rangle} + \Gamma_u^{\langle t\rangle} \circ \tilde{c}^{\langle t\rangle} \tag{4}$$

$$\Gamma_0^{\langle t\rangle} = \sigma\left(W_0\left[a^{\langle t-1\rangle}, x^{\langle t\rangle}\right] + b_0\right) \tag{5}$$

$$a^{\langle t\rangle} = \Gamma_o^{\langle t\rangle} \circ tanh\left(c^{\langle t\rangle}\right) \tag{6}$$

LSTM trained with Back Propagation (BPTT) . To calculate the errors between the output value and the real value of each LSTM neuron. The gradient of each weight is calculated according to the corresponding errors, and the gradient descent algorithm is used to update the weight.

2.2. The empirical analysis of LSTM

2.2.1. Description and Pre-process of Data. The data which is used in this paper comes from the monthly import and export data of China Customs of Provincial Department of Commerce in Shandong Province. The data spans from January 2001 to June 2018, there are 210 samples with a unit of 100 million yuan in total.

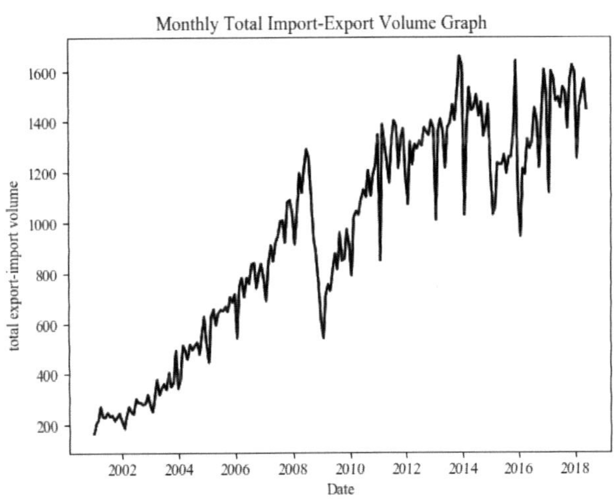

Figure 3. Trend for monthly volume of import and export.

We can see that the fluctuation of monthly data of import and export is great in the Figure 3, and it shows an overall growth trend, which is not a stable but volatile time sequence. Before 2008, imports and exports grew rapidly, showing an exponential growth trend. The financial crisis in 2008 had a great impact on imports and exports, the total volume is decreased by 50%. The total imports and exports in January 2009 were only half of those in August 2008. It slowed down considerably after 2010.

The difference of raw data. The advantage of LSTM is to deal with time sequence with certain correlation. The auto-correlation of the original sequence is not strong. LSTM carries out differential processing, It can be seen the sequence after difference has 12-order correlation from the autocorrelation and partial autocorrelation plots of the difference sequence, which means the current quantitative value is related to the value of previous 12-period.

Normalization of input data. Because LSTM is sensitive to the size of the input data, especially when using sigmoid (default) or tanh activation functions, the data need to be normalized first. In this paper, we will scale the data between the maximum and minimum by standardizing the maximum and minimum.To transform the sequence , , ...Transformation formula is

$$y_i = \frac{x_i - \min_{1\le j\le n}\{x_j\}}{\max_{1 j\le n}\{x_j\} - \min_{1\le j\le n}\{x_j\}} \tag{7}$$

After transformation, the original sequence is scaled between [0,1].

To transform the time sequence into the problem of supervising learning . According to the auto-correlation, we predict the data of the 13th month by using the data of every 12 months. There are 197 samples after processing the original date.

To divide training set and test set. The first 80% of the samples are taken as training set and the remaining 20% as test set without disturbing the original order of the data. Thus, there are 157 samples in the training set and 40 samples in the test set.

To build input data that can be put into the LSTM. The data format required by LSTM is tensor format. The predictive variable (X) must be a three-dimensional array, its dimensions are samples, timesteps, and features. The first dimension represents the number of samples; the second dimension is the time step (lag order); and the third dimension represents the number of predictor variable (1 is a single variable, n is multivariate). The input form of training set is (157, 12, 1), and the input form of test set is (40, 12, 1).

2.2.2. The construction and prediction of model. The LSTM model used in this paper includes an input layer, two LSTM layers which are as a hidden layer and an output layer, and each hidden layer contains a Dropout layer to prevent over-fitting. After several adjustments, the input dimension of the input layer is selected to be 1. There are 128 units contained in the first hidden layer, the output dimension of the second hidden layer is 256, and the output dimension of the output layer is 1. The loss function used in compiling is mean square error (MSE), the optimizer is Adam, and epochs is 200. Keras framework is used to train the model.

To train the model on the training set and to predict on the test set. The predicted values of the training set and the test set are output respectively. We obtained the mean square errors of the training set and the prediction set by processing the normalization and inverse difference. It is as shown in Figure 4. The mean square error of training set is 84.49 and test set is 124.39.

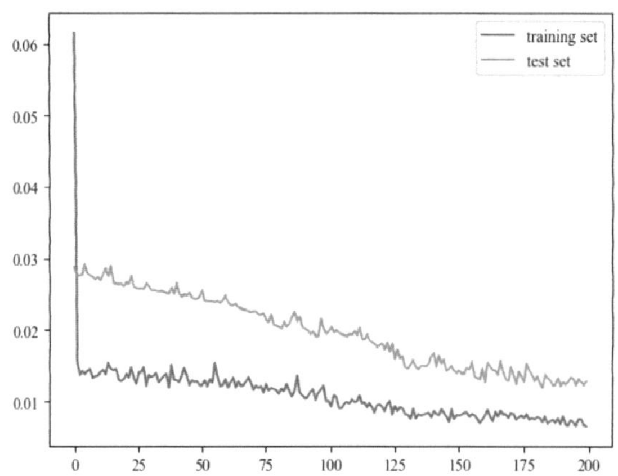

Figure 4. The MSE of the training set and test set.

To combine the prediction results of training set with that of test set and to observe the fitting effect of LSTM in the same graph with the original data set. The effect is shown in Figure 5. We can see that the model fits the trend change of import and export data very well.

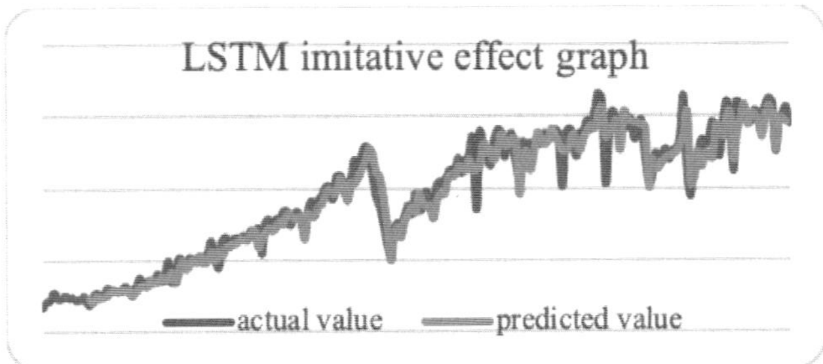

Figure 5. The fitting effect of the import and export of LSTM.

2.2.3. The cubic smoothing prediction of index. Classical time sequence model [11-14] is used to forecast the data of import and export. We choose the cubic exponential smoothing method here. Although the method is simple, the effect of prediction is better in practice. The cubic exponential smoothing algorithm can predict time series with both trend and seasonality. It retains seasonal information on the basis of quadratic exponential smoothing, so that it can predict time sequence with seasonality. A new parameter P is added in order to express the trend after smoothing. It has two methods, the accumulation and the multiplication.

After training, the α is 0.6023, it means that about 60% of the current level depends on historical observations; β is 0; γ is 0.6163, which means that the estimated 61.6% of the trend is dependent on historical data. The mean square error of training set is 83.19, and that of test set is141.69.

To combine the prediction results of training set with that of test set and to observe the fitting effect in the same graph with the original data set. The effect is shown in Figure 6.

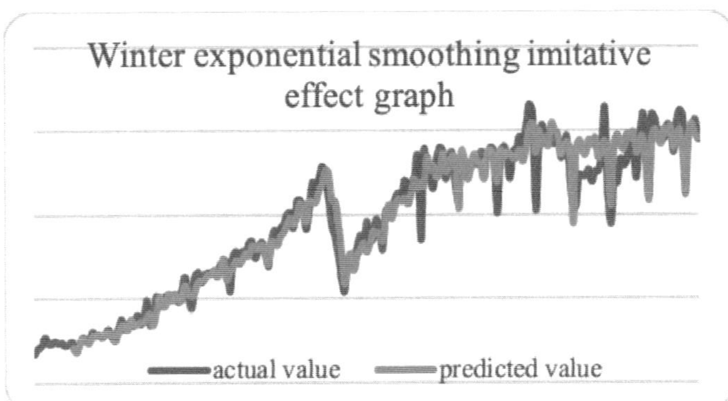

Figure 6. The fitting effect of import and export data of the cubic exponential smoothhing method.

3. Conclusions

We adopt LSTM recurrent neural network and cubic exponential smoothing method of classical time sequence to forecast the trend of import and export data monthly in this paper. From the training error and test error of this two training models, we can see that the training error of the cubic exponential smoothing method is lower than that of LSTM, but the prediction error of the test set is higher than that of LSTM model. This phenomenon illustrate that the generalization ability of LSTM is stronger, we can also see it from the picture of fitting effect. Both of them can better fit the trend and fluctuation of data in training set, but from the date after January 2015, we can see that LSTM model is better than cubic exponential smoothing method to fit the trend and fluctuation of data . Therefore, LSTM provides a better method to fit monthly data of import and export.

Acknowledgments
This work has been supported by Key Research and Development Plan in Shandong Province, Grant/Award Number: 2018GGX101012.

References

[1] Luo Wenbin,Wang Haiyan, Zhao Lindu.Analysis and Prediction of Chaotic Characteristic of Time Sequence of Total Import and Export Volume [J]. Industrial Technology Economy, 2007, 26 (1): 115-119.
[2] Li Jumei, Fan Liang and Cui Junfeng. Forecasting the total volume of China's import and export by time sequence analysis method [J]. Chongqing Institute of Technology (Natural Science), 2008, 22 (12): 136-138.
[3] Duan Peng. Import and Export of Commodities in China: Prediction Based on SARIMA [J]. Journal of Central China Normal University (Humanities and Social Sciences Edition), 2009, 48 (1): 60-66.
[4] Bai Shuang. Research on Forecasting Model of Foreign Trade of Import and Export in Shenzhen Based on Combination Prediction [D]. Tianjin: Tianjin University, 2013.
[5] Wu Xin, Wang Wenjie. Prediction of Potential of China's Import and Export Trade [J]. Reference for Economic Research, 2016 (33): 34-39.
[6] Ma Ming. Analysis and Scale Forecast of Import and Export Trade of Maize Farming Industry in China [D]. Hangzhou: Zhejiang University, 2016.
[7] Duan Lei, Hu Mingli. The research on forecasting of total foreign trade imports and exports based on BP neural network and discrete GM combined model [J]. China Management Science, 2017, 25 (11): 650-655.
[8] Wang Xudong, Yan Ke, Lu Huijuan, Ye Minchao. Test of LSTM Univariate Demand of Short-term Household Electricity [J]. Journal of International Trade, 2018, 29 (2): 142-148.
[9] Lu Zenan, Shang Yulin. The prediction of Steel price based on LSTM neural network model [J]. Science and technology horizon, 2017:116-117.
[10] Sun Ruiqi. Research on the forecasting model of the trend of US stock index price based on LSTM neural network [D]. Beijing: Capital University of Economics and Trade, 2015.
[11] Luo Hongben. Research on Intelligent Hybrid Forecasting of Financial Time Series Based on Grey-ARIMA [J]. Financial Theory and Practice (Bimonthly), 2014, 35 (188): 27-34.
[12] Wang Shuhua. Study on Combination Forecasting Model Based on Time Sequence [D]. Qinhuangdao: Yanshan University, 2011.
[13] Li Su. Prediction of the development trend of total volume of China's import and export trade based on gray system [J]. Business Research, 2008 (371): 113-115.
[14] Shen Hanxi, Lin Jian. The prediction of import and export of China's foreign trade based on ARIMA l: 2006-2010 [J]. International trade issues, 2007 (6): 112-114.24-2.

Research on the Prediction of Popular Opinion Trend of Web News based on BP neural Network and LSTM

Jiale Wu*, **Lulin Zhang and Fulian Yin**

Communication University of China, Department of Science and Technology

*E-mail:wujiale0713@163.com

Abstract. Current prediction methods on public opinion trend of network news are often carried out according to human experience or traditional time series prediction, but they sometimes lack scientific principle and immediacy. This paper researches on the prediction of public opinion trend of network news based on deep learning. Firstly, we use the BP neural network and LSTM to predict public sentiment trend, then combine word2vec with LSTM to classify the text sentiment and finally use doc2vec algorithm and k-means to cluster the text. Experiment result shows that the mean square error with good performance, which reveals the beneficial effect of our prediction method.

Keywords. Network news, Time series predicting, Deep learning, LSTM, Emotional classification.

1. Introduction

The rapid spread of online news in the era of self-media has brought an important influence on the spread of public opinion. However, the progress of existing public opinion monitoring is often not as fast as the speed of Internet information dissemination. However, with the development of deep learning technology, it has become the mainstream research technology in the field of natural language processing, which can capture the more comprehensive and in-depth implicit text features of public sentiment texts of social platforms.

The application of deep learning technology in the field of online news sensation mainly focuses on the sentiment analysis of popular opinion texts. In 2002, the emotion classification method based on the naive Bayes model, maximum entropy model and support vector machine model SVM was first proposed by Pang B at the EMNLP conference[1]. After that, the exploration of text sentiment classification centering on the traditional machine learning model began, and this method was also applied to the sentiment classification of network news and papers. In 2006, the sentiment dictionary method used word matching scores to calculate the sentiment of texts and applied it to the analysis of public opinion in network news[2]. However, this method cannot avoid the flaws out of context. In 2011, Latent Semantic Indexing (LSI), which aims to perform dimensionality reduction[3] on raw text data, but semantics still cannot be extracted by this method. In 2014,taking advantage of the CNN to classify text sentences and using single-layer CNN for text modeling[4] was proposed, which achieved good results under the condition of good quality of word vectors, but if there is a semantic noise in the original network news lyric text word vector, it would make the classification effect worse. In order to solve this problem, Luo Fan et al. proposed a Hierarchical RNN-CNN (H-RNN-CNN) model combining RNN and CNN[5], including double-layer RNN and top-level CNN, which contributes to the long-text emotional orientation and cross-sentence information judgment[6]. Then, using CNN to capture local information, a network of attention models incorporating LSTM and CNN was proposed by Hu Zhaoju et al. to extract textual context information, and the AT-BL&C model was obtained by replacing the LSTM with the BILSTM that implements the attention mechanism, and then the better classification effect of the online news public opinion text was achieved [7-8].

Based on the above technical achievements, this paper proposes a method for further integrating the application of network news public opinion predicting and deep learning technologies.

2. Research on predicting Trend of Internet News Development

2.1 BP Neural Network and LSTM

Back Propagation Neural Network is a multilayer feed-forward network trained by error back propagation algorithm. The original default identity activation function is shown in Eq. (1):

$$f(x)=x \tag{1}$$

The Rectified Linear Unit (ReLU), also known as a modified linear unit, is a commonly used activation function in an artificial neural network[9]. The ReLU function is shown in Eq. (2):

$$f(x)=\max(0,x) \tag{2}$$

The ReLU function converges much faster than the sigmoid function or the tanh function. The disadvantage is that when a large gradient flows through the ReLU neuron and the parameters are updated, the neuron loses any data activation capability. The Leaky-ReLU function after improving ReLU is shown in Eq. (3):

$$f(x)=x, x\geq 0 \quad f(x)=a\,x, x<0 \tag{3}$$

The LSTM neural network model[10] is based on the RNN model to design the memory unit in the neural network node, which is equivalent to the "processor" for judging the usefulness of information. It is used to control the flow of information in the form of a gate structure. Only the information that meets the algorithm's authentication remain and the non-compliant information was deleted through the Oblivion Gate.

This design solves the problem of information disappearing or exploding when the RNN model transmits information over long distances, which can learn the characteristics of long-distance dependence and is suitable for network news public opinion prediction experiments.

2.2 Experiments and Comparative Analysis

The experiment of this paper takes the case of Jiangge, a network news event, as an example. It collected a total of 63 days of micro-index and Baidu index hot word search volume data before and after the fermentation of the Jiangge case and then obtained an integrated search index after the addition calculation.

This paper used the time series sliding window method to process the comprehensive search index into a form that can be used as a neural network input. For example, suppose that the BP neural network structure is 6-X-1, that is, there are 6 input nodes and 1 output node, which uses the 6-day time series as the prediction basis to predict the 7th day. This paper successively uses 6 consecutive days of data as input data for the prediction model of the development trend of network news events, and the 7th-day data is output as the predicted target data. At the same time, we used the first two-thirds of the collected network news event data as a training sample that constitutes a prediction for the development trend of network news events, and the latter one-third of the data as a test sample. Next, this paper compared the experiments of three layers of BP neural network structure in a single hidden layer and four layers of BP neural network structure in a double hidden layer, and we found that the mean squared error of the four-layer BP neural network structure 6-5-3-1 was the lowest, equal to 0.017396. In the case of consistent data conditions, the experimental results are shown in Table 1.

Table 1. Comparison of different transfer functions in BPthree-layer four-layer neural network output layer

	Identity activation function MSE	Leaky-ReLU MSE
three-layer BP 6-5-1	0.019322	0.015920
three-layer BP 6-5-3-1	0.017396	0.015938

Then we choose LSTM time recursive neural network for predictive analysis of data series and uses Python Keras to construct and train the LSTM model. Based on the LSTM time-recursive neural network algorithm, this experiment compares the prediction effect using different related input layer node time-steps and iterative times epochs. According to the comprehensive experimental results, it can be concluded that the optimal result of the prediction effect is the mean square error 0.0030. Figure 1 shows the trends of the original search data and prediction data of the comprehensive search index in the Jiangge case.

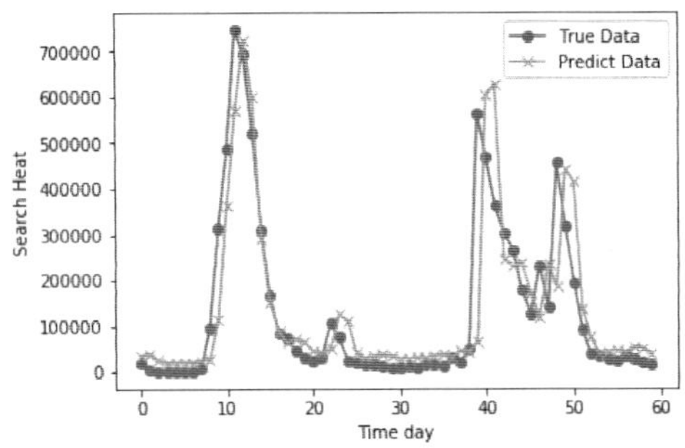

Figure 1. The developmental raw data and prediction data trends of LSTM experimental jiangge case

After a comprehensive comparison of the BP neural network algorithm and the LSTM deep learning network for the data set of the jiangge case, we can see that LSTM has a better prediction effect. Moreover, the mean square error that can be achieved after training is significantly reduced, and the fitting effect between the original data curve and the predicted data curve is better.

3. Research on the Prediction of Public Sentiment Trend

3.1 Sentiment Classification based on LSTM and sentiment index

The LSTM time recursive neural network model is good at processing sequence data and can learn long-distance data dependencies. Before putting network news lyric text data into the LSTM model, we need to first perform word vectorization of Word2vec. In 2013, using Word2vec to learn the vector representation of words was proposed by Mikolov et al. It is mainly divided into two methods, which are the CBOW (continuous bag of words) that predicts the target word with the context and the Skip-gram [11-12] that predicts the context with the center word. LSTM deep learning classification requires training and learning through training sets. Therefore, the field and quality of training set corpus in text sentiment classification influenced the training of the model to a great extent, and thus affect the classification effect. In order to fit the theme of this paper's sentiment analysis based on social news media in this paper, we introduced the Chinese microblog sentiment analysis of the 2012 CCF Natural Language Processing and Chinese Computing Conference under the e-commerce sentiment training data of various fields with a large amount of test data. We add the parsed emotional text data of social media comments to the original Chinese sentiment analysis corpus, and then iteratively train the LSTM model with the merged corpus to obtain the sentiment classification model. Finally, we divide the overall Weibo lyric corpus data into daily text collections by date and then put it into a model based on deep learning, so we can get the network news sensational daily positive emotional text quantity *posnum* and negative emotional text quantity *negnum*. This paper defines the daily emotional index to reflect the proportion of positive sentiment texts and the calculation method is shown in Eq. (3):

$$x = \frac{posnum}{posnum + negnum} \tag{4}$$

3.2 Algorithm for text vectorization

The categorization of web news lyric text topics requires first processing the text as a vector based on Word2vec. In 2014, doc2vec, or paragraph2vec, was proposed by Le, Mikolov or some others. This is an unsupervised framework, which can get text vectors and expand Word2vec. The learned vector can calculate the similarity between texts by calculating the distance. Doc2vec's two models: DBOW (distributed bag of words) and DM (distributed memory) correspond to CBOW and skip-gram in Word2vec respectively. The DBOW model is the probability of predicting a set of random words in a document given a document vector, while the DM model is the probability of predicting words given a context and a document vector.

This paper vectorizes the online news lyrics and then perform K-means clustering. The K-means algorithm is the most classical clustering method. It sets the K points in the space as the cluster center, and classifies the closest objects, and then iteratively calculates and updates the values of each cluster center. Finally, until the center point position becomes stable, the number of custom classifications is Clustering is complete.

In order to determine the appropriate number of clusters for the network news public opinion topic, we set up different clustering numbers for comparison and then perform text processing on the classified network news comment subsets: text segmentation, text preprocessing and word frequency calculation. The specific calculation method is: We select the top 30 words, and then calculate the topic overlapping high-frequency words again and again. After calculating the average of all combinations, the obtained value is the average high-frequency word repetition rate. This paper proposed the average high-frequency word within the suitable range with the lowest repetition rate, where the topic difference is relatively largest, and the clustering value is clustered as the topic classification number.

3.3. Experiments and results

Set the number of nodes in each hidden layer unchanged to change the number of input layer nodes, which is the number of days related to the prediction. Finally, we find that the mean square error of the four-layer BP neural network structure 6-5-3-1 is the lowest, equal to 0.026184, under the condition that other data conditions are consistent. Comparing the experimental results of the different nodes in the hidden layer, this paper can conclude that the 6-2-1 mean-squared error of the three-layer BP neural network structure is relatively lowest, which is approximately equal to 0.024151.

By comprehensively comparing the BP neural network algorithm and LSTM deep learning network experiment results, we can conclude that when the set value time-steps is 5 in the LSTM network experiment, which means the LSTM set prediction output data is related to the data of the first 5 days, and the mean square error is the lowest, which is 0.0200.

When the number of network news media clustering topics is specified as 2, 3, 4, and 5 respectively, text data vectorization and clustering experiments are performed and the average high-frequency word repetition rate is calculated. According to the experimental results, this paper selected the results with a relatively large degree of thematic differences, which means the number of cluster classifications defined at this time is 2. The experimental results are shown in Table 2.

Table 2. Different cluster classification number average high-frequency word repetition rate comparison results

Number of clusters	2	3	4	5
The average repetition rate of high-frequency words	0.60	0.69	0.63	0.62

4. Conclusion

This paper introduces deep learning technology into the research of network news dissemination effect predicting, and analyzes and studies the development trend of network news events and the prediction model of online newsgroup sentiment trends. This paper takes jiangge case, a popular online news incident, as an example. Firstly, we use the BP neural network and LSTM to predict public sentiment trend, then combine word2vec with LSTM to classify the text sentiment and finally use doc2vec algorithm and k-means to cluster the text. Based on the Python language, we conduct prediction experiments on various modules of network news events. Among them, the best result of the network news event trend predicting experiment is the mean square error of 0.0030, and the best result of the network news public opinion sentiment trend predicting experiment is the mean square error of 0.02, which has achieved a good experimental effect.

ACKNOWLEDGMENT

This work is supported by the National Natural Science Foundation of China (No. 61801440), the Fundamental Research Funds for the Central Universities and the Communication University of China's state-of-the-art training research project (No. CUC18A015-1).

References

[1] Pang B, Lee L, Vaithyanathan S. 2002 Sentiment Classification Using Machine Learning Techniques[C] *Proc of EMNLP* **79-86**.
[2] Kim SM, Hovy E. 2006 Extracting Opinions, Opinion Holders, and Topics Expressed in Online News Media Text[C] *Proc. of Wksp.* on Sentiment and Subjectivity in Text at Proc. of the Int'l Conf. on Computational Linguistics the, Ann. *Meeting of the Assoc. for Computational Linguistics*:**1-8**.
[3] Bader B W, Kegelmeyer WP, Chew P A. 2011 Multilingual Sentiment Analysis Using Latent Semantic Indexing and Machine Learning[C] *IEEE, International Conference on Data Mining Workshops. IEEE Computer Society,***45-52**.
[4] Kim Y. 2014 Convolutional Neural Networks for Sentence Classification[J]. *Eprint Arxiv*, 31(1):**132-164**.
[5] Luo Fan, Wang Houfeng. 2018 Chinese Text Sentiment Classification Based on RNN and CNN Hierarchical Networks[J]. Journal of Peking University(*Science and Technology*), (4):**1-6**.
[6] Hu Zhaoju, Zhao Xiaowei. 2018 Sentiment Analysis Based on Word Vector Technology and Hybrid Neural Network[J]. *Research on Application of Computer*, (12):**1-6.**
[7] Huang Jianguo, Luo Hang, Wang Houjun, et al. 2009 Application of GA-BP Neural Network to Study Time Series Prediction [J]. *Journal of University of Electronic Science and Technology*, 38(5): **687-692**.
[8] Ding S, Jia W, Su C, et al. 2010 An Improved BP Neural Network Algorithm Based on Factor Analysis[J]. *Journal of Convergence Information Technology*, 5(4):**103-108**.
[9] Jiang Anbo, Wang Weiwei. 2018 Research on Optimization of ReLU Activation Function[J].*Transducer and Microsystem*,37(02):**50-52**.
[10] Hochreiter S, Schmidhuber J. 1997 Long Short-Term Memory[J]. *Neural Computation,* 9(8):**1735-1780**.
[11] Mikolov T, Chen K, Corrado G, et al. 2013 Efficient Estimation of Word Representations in Vector Space[J]. *Computer Science,* 31(1):**132-164**.
[12] Pan Bo, Zhang Qingchuan, Yu Zhongzhong, Cao Shuai. 2018 The Application of Doc2vec in Salary Forecast[J].*Application Research of Computers*, 35(01):**155-157**.

IOP Conf. Series: Materials Science and Engineering **646** (2019) 012004 doi:10.1088/1757-899X/646/1/012004

Design and Analysis of a Novel Planar Translational Parallel Robotic Mechanism with Three Limbs

Z H Chen, Y B Zhang and K M Wang

School of Mechatronics Engineering, Henan University of Science and Technology, Luoyang 471003, China.

Corresponding author Email: yanbin_zh@163.com

Abstract. A novel planar translational parallel robotic mechanism with three limbs is designed. The degree of freedom (DOF) and output characteristics of mechanism are studied based on the screw theory. The mathematic models including position, velocity and acceleration are derived according to different forms of the actuated inputs. When the linear displacements of two cylindrical joints are selected as the actuated inputs, the velocity Jacobian is an identity matrix and the mechanism has fully isotropic kinematics characteristics. Singularity of the mechanism is discussed and all singular configurations are explored. Finally, kinematics simulation of the mechanism is carried out by ADMAS and MATLAB software, and the kinematic curves are drawn.

1. Introduction

Compared with the traditional serial mechanism, parallel mechanism has the advantages of compact structure, strong bearing capacity and small synthetic error, and becomes one of the hot topics in the field of mechanism and robotics [1-2]. Because of its low manufacturing cost and relatively simple motion control, less degree of freedom (DOF) parallel mechanism has received more attention and research in recent years [3-4]. Several typical low DOF parallel mechanisms have been successfully applied in the world, such as Delta mechanism [5], Hunt mechanism [6], Diamond mechanism [7], and so on.

As a common mechanism in the mechanical processing industry, planar translational parallel mechanism has attracted wide attention from scholars at home and abroad [8-9]. Most of the planar translational parallel mechanisms have strong kinematic coupling. It is undeniable that strong motion coupling can improve the stiffness and load-carrying capacity of the mechanism. However, it also brings difficulties in solving kinematics and dynamics of the mechanism, complex control design and reduced workspace. In some applications for requiring high accuracy and low bearing capacity, decoupled parallel mechanisms show their superior performance. Carricato [10] studied the type synthesis method of decoupled translational parallel mechanism and designed many new mechanisms. Zhang [11] established the type synthesis method of uncoupled parallel mechanism based on the actuation screw theory and designed many new mechanisms.

In this paper, a novel planar translational robotic parallel mechanism with three limbs is proposed. Mobility of the mechanism is analyzed based on the position and orientation characteristic method. Kinematic equations, including position, velocity and acceleration, are derived. Singularities of the mechanism are discussed in detail. Finally, kinematic simulations are performed as well.

2. Structural design and mobility analysis

The diagram of the novel planar translational parallel robotic mechanism is shown in Figure 1. It consists of a moving platform (*MP*) connected to a fixed base (*FB*) and three kinematic chains, including two active chains L_1 and L_2 and a pure constrained chain L_3. Two active chains have the same structure composed of a cylindrical joint (C), a prismatic joint (P) and a revolute joint (R) from the platform to the base in serial. Axes of C joint and R joint are parallel to each other and perpendicular to P joint. So this chain can be recorded as $SOC\{-C_1(\perp P_1)//R_1-\}$. The chain L_3 only consists of three R joints with parallel axes and is recorded as $SOC\{-R_{31}//R_{32}//R_{33}-\}$. Axes of three joints mounted on the base are orthogonal. Similarly, the axes of three R joints installed on the platform are also perpendicular to each other.

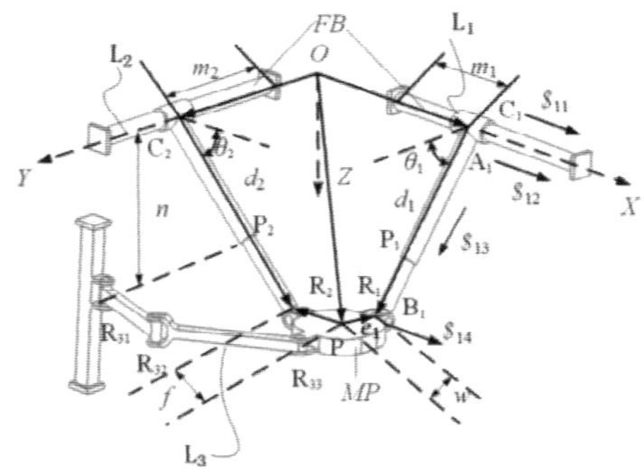

Figure 1. Diagram of the novel parallel robotic mechanism

According to the configuration relationship of mechanism joints, the motion characteristic matrix of three chains can be written as follows

$$M_1 = \begin{Bmatrix} t^3 \\ r^1(//R_1) \end{Bmatrix} \tag{1}$$

$$M_2 = \begin{Bmatrix} t^3 \\ r^1(//R_2) \end{Bmatrix} \tag{2}$$

$$M_3 = \begin{Bmatrix} t^2(\perp R_{31}) \\ r^1(//R_{31}) \end{Bmatrix} \tag{3}$$

where, the right superscript number denotes the DOF of the limb, *t* the translational DOF and *r* the rotary DOF.

From equations (1), (2) and (3), the motion characteristic matrix of the mechanism can be obtained, and have

$$M_P = M_1 \cap M_2 \cap M_3 = \begin{Bmatrix} t^2(\perp R_{31}) \\ r^0 \end{Bmatrix} \tag{4}$$

Equation (4) shows the limb only has two translational degrees of freedom.

Therefore, DOF of parallel mechanism can be calculated by the following formula

$$F = \sum_{i=1}^{g} f_i - \min\left\{\sum_{i=1}^{\delta} \lambda_i\right\} + \Omega \tag{5}$$

where, F is DOF of the mechanism, g the total number of the joints in the mechanism, f_i DOF of the *i*th joint, λ_i the number of independent displacement equations in the *i*th loop, δ the number of independent loop, $\min\{\bullet\}$ the minimum number of independent displacement equations in a limb,

Ω the virtual constraint number of mechanism. Therefore, DOF of the mechanism is $F = 10 - 10 + 2 = 2$. This result is consistent with the previous analysis.

3. Kinematics analysis of mechanism

The static coordinate *O-XYZ* is attached on the base and its origin point *O* falls at the intersection of two C joint axes. X-axis and Y-axis coincide with the corresponding C joint axis, respectively. Z-axis conforms to the right-hand principle. C joints in two active limbs are selected as the active joints of the mechanism. In Figure 1, point A_1 is the intersection of C_1 joint and P_1 joint axes, point B_1 is the intersection of R_1 joint and P_1 joint axes, point P is the centroid of the platform. If the coordinate of point P is (x, y, n), so the vectors $\boldsymbol{OP} = \boldsymbol{r} = (x, y, n)$ and $\boldsymbol{PB}_1 = \boldsymbol{e}_1 = (0, w, 0)$.

In terms of the screw theory, the kinematic screw of each joint in chain L_1 can be written as

$$\begin{aligned} \$_{11} &= (0 \quad 0 \quad 0; \quad 1 \quad 0 \quad 0) \\ \$_{12} &= (1 \quad 0 \quad 0; \quad 0 \quad 0 \quad 0) \\ \$_{13} &= (0 \quad 0 \quad 0; \quad 0 \quad \cos\theta_1 \quad \sin\theta_1) \\ \$_{14} &= (1 \quad 0 \quad 0; \quad 0 \quad d_1\sin\theta_1 \quad -d_1\cos\theta_1) \end{aligned} \tag{6}$$

Where, d_1 is distance between joints C_1 and R_1, θ_1 is angular displacement of joint C_1.

According to the reciprocal screw theory, we can get the reciprocal screw of limb L_1, and yields

$$\begin{aligned} \$^r_{11} &= (0 \quad 0 \quad 0; \quad 0 \quad 1 \quad 0) \\ \$^r_{12} &= (0 \quad 0 \quad 0; \quad 0 \quad 0 \quad 1) \end{aligned} \tag{7}$$

According to the selection principle for the actuated joints of parallel mechanism, it can be proved that the linear displacements or angular displacements of two cylindrical joints can be used as the actuated inputs. Next, the kinematics of the mechanism will be analyzed in terms of two different input forms.

3.1 Angular displacements as the actuated inputs

3.1.1 Position analysis. Based on the spatial polygon enclosed by points *O*, A_1, B_1 and P (seeing Fig.1), the following vector equation can be written, and have

$$\boldsymbol{OP} + \boldsymbol{PB}_1 = \boldsymbol{OA}_1 + \boldsymbol{A}_1\boldsymbol{B}_1 \tag{8}$$

Substituting the relative parameters into equation (9), we get

$$\boldsymbol{r} + \boldsymbol{e}_1 = m_1\boldsymbol{s}_{12} + d_1\boldsymbol{s}_{13} \tag{9}$$

Where, $\boldsymbol{s}_{12}$ is the direction vector of the actuated screw $\$_{12}$ in chain L_1, $\boldsymbol{s}_{13}$ is the direction vector of P_1 joint, m_1 is the distance from points *O* to A_1.

Based on equations (6) and (7), the actuation screw $\$'_{a1}$ applied to the moving platform by the actuated joint C_1 through chain L_1 can be calculated, yields

$$\$'_{a1} = (0 \quad -sin\theta_1 \quad cos\theta_1; \quad d_1 \quad 0 \quad 0) \tag{10}$$

Equation (10) shows that the actuation screw $\$'_{a1}$ is a line vector and it direction vector is $\boldsymbol{s}'_{a1} = (0 \quad -\sin\theta_1 \quad \cos\theta_1)$.

Making the dot products with vector $\boldsymbol{s}'_{a1}$ on both sides of equation (9), and have

$$(\boldsymbol{r} + \boldsymbol{e}_1)\boldsymbol{s}'_{a1} = (m_1\boldsymbol{s}_{12} + d_1\boldsymbol{s}_{13})\boldsymbol{s}'_{a1} \tag{11}$$

Because the actuation screw $\$'_{a1}$ is perpendicular to the axes of C_1 and P_1 joints, it does not work in either direction. Then equation (11) can be simplified as follows

$$-(y + w)\sin\theta_1 + n\cos\theta_1 = 0 \tag{12}$$

Where, n is the distance between the platform and XOY plane and it is constant, w is the distance between points P to B_1.

Similarly, by using the vector polygon enclosed by limb L_2, points O and P, the following equation can be obtained, and

$$-(x+f)\sin\theta_2 + n\cos\theta_2 = 0 \tag{13}$$

where, f is the distance between points P to B_2.

Then, the coordinate of the point P can be derived from equations (12) and (13), i.e.,

$$y = n\cot\theta_1 - w \tag{14}$$

$$x = n\cot\theta_2 - f \tag{15}$$

Equations (14) and (15) are the forward kinematics of the mechanism when the angular displacements of C joints are selected as the actuated inputs.

3.1.2 Velocity analysis. According to the position equations, the velocity equation is achieved by the first-order kinematic influence coefficient method, and

$$\mathbf{V} = \mathbf{J}\dot{\boldsymbol{\theta}} \tag{16}$$

Where, $\boldsymbol{V} = [v_x, v_y]^{\mathrm{T}}$ is the output velocity vector of the moving platform, $\dot{\boldsymbol{\theta}} = [\dot{\theta}_2, \dot{\theta}_1]^{\mathrm{T}}$ is the input velocity vector of the actuated joints, $\boldsymbol{J}$ is the velocity Jacobian matrix, and

$$\boldsymbol{J} = \begin{bmatrix} -n\csc^2\theta_2 & 0 \\ 0 & -n\csc^2\theta_1 \end{bmatrix} \tag{17}$$

Equation (17) shows that the Jacobian is a diagonal matrix, which implies that each output motion of the platform is only controlled by one actuated input. Therefore, the mechanism has uncoupled kinematic characteristics under this kind of input forms.

3.1.3 Acceleration analysis. Acceleration equation of the mechanism can be obtained by using the second-order kinematic influence coefficient method, and have

$$\boldsymbol{a} = \boldsymbol{H}\dot{\theta}^2 + \boldsymbol{J}\ddot{\theta} \tag{18}$$

Where, $\boldsymbol{a} = [a_x, a_y]^{\mathrm{T}}$ is the output acceleration vector of the platform, $\ddot{\boldsymbol{\theta}} = [\ddot{\theta}_2, \ddot{\theta}_1]^{\mathrm{T}}$ is the input acceleration vector of actuated joints, $\boldsymbol{H}$ is

$$\boldsymbol{H} = \begin{bmatrix} 2n\csc\theta_2\cot\theta_2 & 0 \\ 0 & 2n\csc\theta_1\cot\theta_1 \end{bmatrix} \tag{19}$$

3.2 Linear displacements as the actuated inputs

3.2.1 Position analysis. When the linear displacement of the joint C_1 is taken as the actuated input, the actuation screw $\$_{a1}$ of limb L_1 can be calculated, and

$$\$_{a1} = (1 \quad 0 \quad 0; \quad 0 \quad 0 \quad 0) \tag{20}$$

Making dot products with the direction vector $s_{a1} = (1 \quad 0 \quad 0)$ of $\$_{a1}$ on both sides of equation (8), yields

$$(\boldsymbol{r} + \boldsymbol{e}_1)\boldsymbol{s}_{a1} = (m_1\boldsymbol{s}_{11} + d_1\boldsymbol{s}_{13})\boldsymbol{s}_{a1} \tag{21}$$

Equation (20) shows that the actuation screw $\$_{a1}$ applied to the platform by joint C_1 is a linear force vector parallel to X-axis and perpendicular to the direction of the kinematic screw $\$_{13}$. Consequently, the screw $\$_{a1}$ does not work in the direction of the joint P_1. Then, equation (21) can be simplified as follows

$$x = m_1 \tag{22}$$

Similarly, based on the chain L_2 we get

$$y = m_2 \tag{23}$$

Equations (22) and (23) are the forward kinematics when the liner displacements of C joints is selected as the actuated inputs.

3.2.2 Velocity analysis. Differentiating equations (22) and (23) with respect to time and rewriting them in matrix form, we have

$$\boldsymbol{V} = \boldsymbol{J}\dot{m} \tag{24}$$

where, $\dot{\mathbf{m}} = [\dot{m}_1, \dot{m}_2]^{\mathrm{T}}$ is the output velocity vector of the platform, $\boldsymbol{J}$ is the velocity Jacobian matrix of mechanisms and it is a 2×2 identity matrix. Therefore, when the linear displacements of two C joints are used as the inputs, not only it is a one-to-one control relationship between the inputs and the outputs, but also the determinant of the Jacobian is equal to one. So the mechanism shows fully isotropic characteristics in the whole workspace.

3.2.3 Acceleration analysis. The acceleration equation of the mechanism can be obtained by differentiating equation (24) with respect to time, and have

$$\boldsymbol{a} = \boldsymbol{J}\ddot{m} \tag{25}$$

where, $\ddot{\mathbf{m}} = [\ddot{m}_1, \ddot{m}_2]^{\mathrm{T}}$ is the input acceleration vector of actuated joint.

4. Singularity analysis

In this paper, the singular configurations of the mechanism are analyzed by the singularity of the Jacobian matrix $\boldsymbol{J}$. When $|\boldsymbol{J}| = 0$ is occurrence, the mechanism is stuck. If $|\boldsymbol{J}| \to \infty$ is occurrence, the mechanism will lose its bearing capacity

When the linear displacements of two C joints are taken as the actuated inputs, the Jacobian matrix is an unit one and its determinant is always equal to 1. Therefore, there is no any singular configuration in the whole workspace for the mechanism.

When the angular displacements of two C joints are selected as the inputs, the value of the Jacobian matrix determinant by means of equation (18), and have

$$|\boldsymbol{J}| = n^2 \csc^2 \theta_1 \csc^2 \theta_2 \tag{26}$$

According to equation (26), we know that when $\theta_1 = k\pi$ or $\theta_2 = k\pi$ (k=0, 1), the value of $|\boldsymbol{J}|$ will attend infinite. At these configurations, the mechanism is out of control, in other words, it shows singular. According to the geometric form of the mechanism, we can get $\theta_i = \arcsin(d_i / n)$. If structural size satisfies $d_i \neq 0$, θ_i will be always not zero, which means these singular configurations can be avoided.

In addition, if the axes of three revolute joints in the pure constrained chain are located in the same plane, the kinematic screws corresponding to these revolute joints will be linear dependent. In a result, mechanism will be boundary singularity. However, the singular configuration can be avoided by adjusting the structural length of the chain L_3.

In summary, whatever the angular displacements or linear displacements of the cylindrical joints are chosen as the actuated inputs, the Jacobian matrix keeps diagonal, which means there is an one-to-one mapping relationship between the inputs and the outputs. So the mechanism has well kinematic decoupling. Especially, when the linear displacements are used as the inputs, the Jacobian of the mechanism is an identity matrix. The mechanism has the same kinematics transferring performance along all directions. Therefore, the linear driving mode is more suitable for practical application.

5. Example simulation

Three-dimensional model prototype of the mechanism is set up by Solidworks software, as shown in Figure 1. Then it is imported into ADAMS software to redefine the constraints of each joint. With the linear displacement of the joint in the limb as input, the input displacement equations of the limb L_1 and L_2 are $X = 10\cos(\pi t / 10)$ and $Y = 5\cos(\pi t / 5 + \pi / 2)$. The kinematic curves including position, velocity and acceleration are plotted based on the virtual prototype, seeing Figure 2. Furthermore, simulation curves are also given by using MATLAB software in terms of the theoretical kinematic equations established here, shown in Figure 3. Comparing two simulation results, it is obvious that the corresponding curves are identical, which verifies the correctness of the kinematical models.

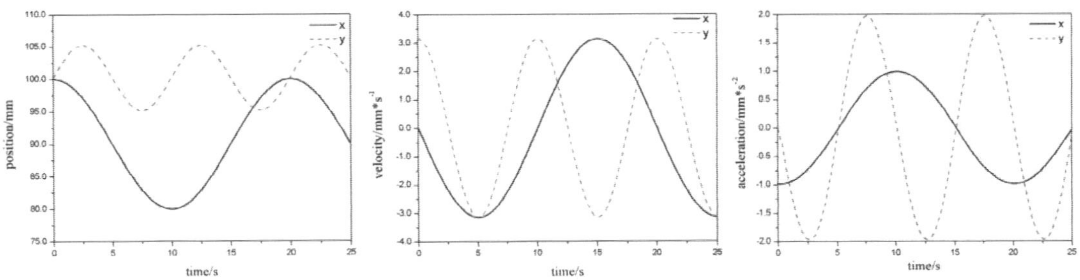

(a) Position curves (b) Velocity curves (c) Acceleration curves

Figure 2. Virtual prototype simulation curves

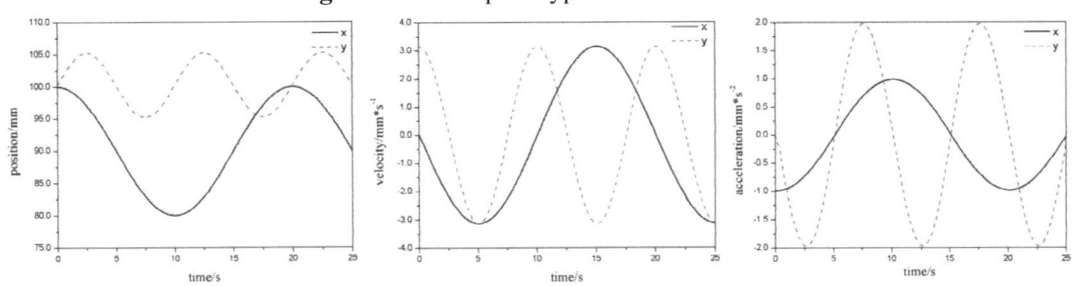

(a) Position curves (b) Velocity curves (c) Acceleration curves

Figure 3. Numerical simulation curves

6. Conclusions

A novel planar translational parallel robotic mechanism with three limbs is proposed in this paper. The output characteristics of the mechanism are analyzed and the kinematic equations of the mechanism under two different input modes are derived. Then, kinematical simulations of the mechanism are completed by ADAMS and MATLAB. The results verify the correctness of the theoretical analysis. It is more interesting that the mechanism has the fully-isotropic characteristics when the linear displacements of two cylindrical joints are selected as the actuated inputs.

Acknowledge

The authors would like to thank the financial support from the Fundamental Project of Key Scientific Research of Henan Advanced Education (18A460001), and Scientific and Technological Project of Henan Province(192102210221).

References

[1] Sato S, Takagi M and Miyoshi T 2018 Development of a novel 6-DOF parallel mechanism moving platform for investigating the effects of multimodal sensory feedback information regarding postural stability in humans *J. Journal of Medical Engineering & Technology* **42**(2) 105-112

[2] Lu Y, Li X P, Zhang C G and Liu Y 2016 Analysis of kinematics and statics for a novel 6-DoF

parallel mechanism with three planar mechanism limbs *J. Robotic* **34**(04) 957-972

[3] Kim J S, Jeong Y H and Park J H 2016 A geometric approach for forward kinematics analysis of a 3-SPS/S redundant motion manipulator with an extra sensor using conformal geometric algebra *J Meccanica* **51**(10)2289- 2304.

[4] Bi W Y, Xie F G, Liu X J and Luo X 2019 Optimal design of a novel 4-degree-of-freedom parallel mechanism with flexible orientation capability *J. proceedings of the institution of mechanical engineers part B-journal of engineering manufacture* **233**(2)623-642.

[5] Xu D T 2018 Kinematic Reliability and Sensitivity Analysis of Delta Parallel Mechanism *J. Advanced Materials Research* **15**(1)1-8.

[6] Wang Y, Yu J J, Pei X 2018 Fast forward kinematics algorithm for real-time and high-precision control of the 3-RPS parallel mechanism *J. Frontiers of Mechanical Engineering***13**(3)368-375.

[7] Huang T, Zhang X-L, Li M, Chetwynd D G and Gosselin C M 2004 Conceptual design and dimensional synthesis of a novel 2-DOF translational parallel robot for pick-and-place operations *J. ASME Journal of Mechanical Design* **126**(3)449-455.

[8] Yao J, Li T M, Wang L P and Chen F F 2017 Systematic design method and experimental validation of a 2-DOF compliant parallel mechanism with excellent input and output decoupling performances *J Applied Sciences- Basel* **7**(6)1-28.

[9] Zhang J F, Chen Q H, Wu C Y and Li Q C 2014 Kinematic calibration of a 2-DOF translational parallel manipulator *J. Advanced Robotics* **28**(10)1-8.

[10] Carricato M and Parenti-Castelli V. 2002 Singularity-free fully- isotropic translational parallel mechanisms *J. International Journal of Robotics Research* **21**(2)161-176.

[11] Zhang Y B, Zhao Y F, Jing X L and Li X P. 2018 Type synthesis of uncoupled translational parallel manipulators based on actuation wrench screw theory *J. Advances in Mechanical Engineering* **10**(1)1-10.

Calibration Method and Experiment of in-Situ Radioactivity Measurement in Seawater Based on Monte Carlo Simulation

Y Y Zhang, B W Wu, X D Feng, D Y Liu and Y Zhang

Institute of Oceanographic Instrumentation, Qilu University of Technology (Shandong Academy of Sciences), Shandong Provincial Key Laboratory of Ocean Environmental Monitoring Techno1ogy,National Engineering and Technological Research Center of Marine Monitoring Equipment, No 37 Miaoling Road, 266061 Qingdao, China

triciayyz@163.com

Abstract. Some research on the in-situ radioactivity measurement in seawater and development of underwater spectrometer using NaI(Tl) scintillation crystal are being carried out for the automatic and continuous monitoring in the marine environment. The gamma spectrometer must be efficiency calibrated for the detection of radionuclides in the seawater. For the problem of traditional experiments, a simulation model of in-situ radioactivity measurement in the seawater was established by using Monte Carlo statistical method. The model simulates the in-situ measurement environment of seawater and the characteristics of the underwater spectrometer developed by ourselves. By calculating the interaction between gamma-ray photons emitted by radionuclides in seawater and various atoms in seawater and the internal structures of the spectrometer, the marine detection efficiency of the underwater spectrometer developed was calculated and verified by the field experiments. The results show that this kind of non-experimental calibration method is effective and feasible for the underwater spectrometer and further research and experiments will continue.

1. Introduction

The radioactivity measurement in seawater has been using the method of in-situ sampling and returning to the laboratory for processing and analysis. Using the traditional method, it is impossible to monitor the marine radioactivity environment effectively, and even more difficult to supply the pollution warning timely. It is necessary to carry out research on the in-situ monitoring technology of marine radioactivity, which can guide sampling, and provide the important technical support for the continuous radioactivity monitoring in the marine environment, as well as efficient response to the emergency accidents.

The spectrometer using NaI(Tl) scintillation crystal is always the main method of studying the in-situ radioactivity monitoring in the marine environment, because of high detection efficiency, wide temperature range, stable performance, low power consumption and low cost. The Fukushima accident has once again attracted the attention of international maritime countries to this technology, and related research is still continuing [1-8].

The efficiency calibration of the spectrometer is very important for the quantitative determination of radionuclides in the seawater. But it is very difficult for the traditional method using some standard sources because few radionuclides can be dissolved in the seawater, and more importantly, the large volumes of polluted seawater. Therefore, this paper studies a kind of non-experimental method for the efficiency calibration and also carries out the experiment to test the method.

2. Materials and method

2.1. Simulation model

The simulation model of in-situ radioactivity measurement in seawater is mainly focused on the geometric structure of the spectrometer and the description of the marine environment, which is used for calculation of the interaction between gamma ray and atoms in seawater and the absorption of gamma ray in various structures in the spectrometer. A simulation model of NaI (Tl) underwater spectrometer was established by using Monte Carlo software according to the structure, material and size of the instrument developed by ourselves. As shown in figure 1, the 7.5cm x 7.5cm Na (Tl) crystal is connected to the photomultiplier tube. The photomultiplier tube is connected with preamplifier, high voltage and multi-channel processing circuit in turn. The spectrometer is encapsulated in a polyamide shell with a thickness of 0.6 cm.

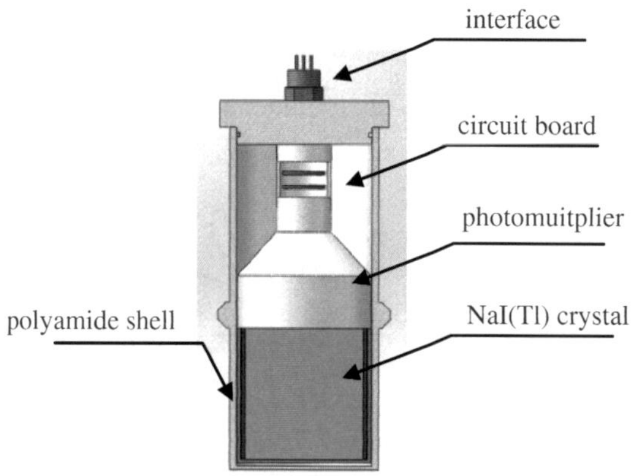

Figure 1. Structure diagram of spectrometer

The in-situ measurement model is shown in figure 2 based on the composition and density of the seawater in the off-shore area of Qingdao. Taking the crystal as the spherical center, the in-situ measurement model is modelled as an infinite source of seawater with natural and artificial radionuclides evenly distributed. Before gamma photons enter the spectrometer, the interaction between gamma photons and various atoms in the seawater is calculated in the model. The gamma photons may change energy due to Compton scattering or electron response, or may disappear due to the photoelectric effect. After gamma photons enter the spectrometer, the model mainly calculates the photoelectric response, Compton scattering and electron pair effect of photons and atoms in the crystal, and also the attenuation of gamma photons in various geometric structures of the spectrometer.

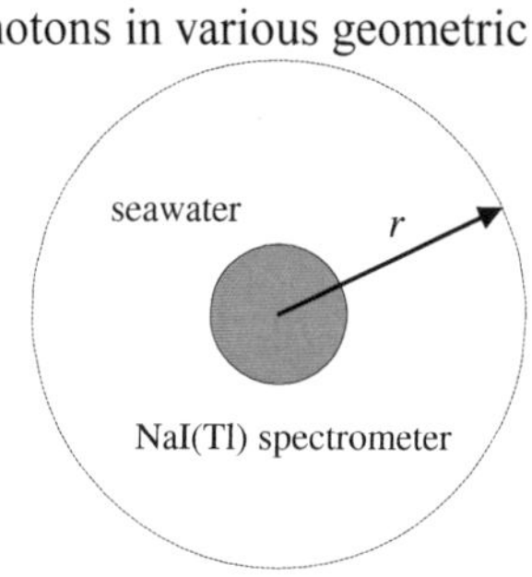

Figure 2. Measurement model illustrated in seawater

2.2. Calculation and analysis method

The gamma photons near the spectrometer contribute more counting to the measurement in the seawater. According to the some research results, the underwater spectrometer using Na(Tl) crystal has different effective detection distances for different radionuclides in the seawater, so corresponding to the different effective detection sources in the model.

The radionuclides follow the different decay laws of gamma rays in the seawater, so the marine detection efficiency of the certain radionuclide is related to the volume of seawater and the energy of the radionuclide. Assuming that the radionuclides of known activity is evenly distributed in the source of seawater, with the radius r = 10, 20, ..., the energy deposition of photons and their secondary electrons in the spectrometer is recorded by using the pulse height counting f8 in the simulation model. The pulse height generated by the energy deposition was then calculated as the detection efficiency of the particle in the specific channel. The radionuclides calculated and gamma-ray characteristics are shown in Table 1. The pulse height generated by the energy deposition is taken as the detection efficiency of the spectrometer for particles. We define that the intensity of gamma rays in the seawater decreases to 0.1% of the initial intensity, and the corresponding detection distance is the marine detection distance. As shown in figure 3, the marine detector distance increases with the energy of the gamma rays [6]. To calculate the marine detection efficiency by simulation method, the in situ measurement in the seawater is simulated only considering the gamma rays equally distributed within the marine detection distance in the seawater. Define V is the marine detection volume. Using $\mathcal{E}_v = \mathcal{E} \cdot V$ as the volume detection efficiency of seawater, the saturated volume detection efficiency of the underwater spectrometer for the different radionuclides in seawater is calculated and then the efficiency calibration curve is fitted. The minimum detectable activity *MDA* of the spectrometer for the radionuclides in seawater (in 95% confidence level) is calculated by the following formula:

$$MDA = \frac{2.71 + 4.65\sqrt{B}}{\varepsilon_v \cdot I_r \cdot t} \tag{1}$$

where t is the measuring time, I_r is the branching ratio of the radionuclides and B is the background count of the photopeak.

Table 1. The radionuclides calculated and gamma-ray characteristics

Radionuclide	Energy (keV)	I_r (%)	Radionuclide	Energy (keV)	I_r (%)
^{109}Cd	88	3.7	^{214}Bi	1764	15.8
^{57}Co	122	85.5	^{137}Cs	661.6	85
^{99m}Tc	140.5	89.1	^{54}Mn	834.8	100
^{214}Pb	295	19	^{65}Zn	1115.5	51
	352	37	^{60}Co	1173.2	100
^{131}I	364.5	81		1332.5	100
^{22}Na	511	180	^{40}K	1461	11
	1274.5	100			

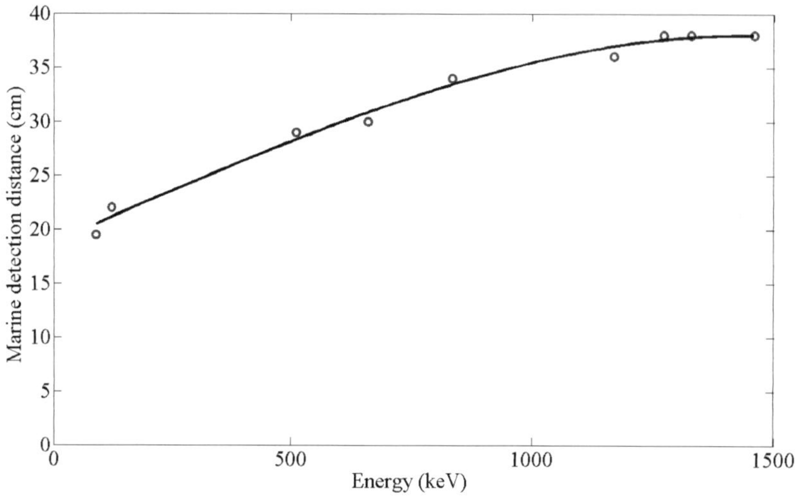

Figure 3. Marine detector distance of the underwater spectrometer developed for the different gamma rays energy

3. Calibration result and experiment

Based on the detection efficiency of the different radionuclide in the seawater simulated, the marine detection efficiency curve for the underwater spectrometer developed using Na(Tl) crystal is acquired and shown in Fig.4. The data points in the figure are the simulated values of the marine detection efficiency for the different gamma ray energies. The simulation errors of little than plus or minus three percent correspond to statistical uncertainty. The solid line represents a fitting curve of the simulated values and can be described in the empirical expression:

$$\varepsilon_V = \frac{a \times E^b}{c + E^d} \tag{2}$$

where *E* is the specific energy of the certain radionuclide in keV, and *a, b, c* and *d* are all fitted parameters. The resulting error (2-3%) was included in the total uncertainty determination of the efficiency by using the fitting curve.

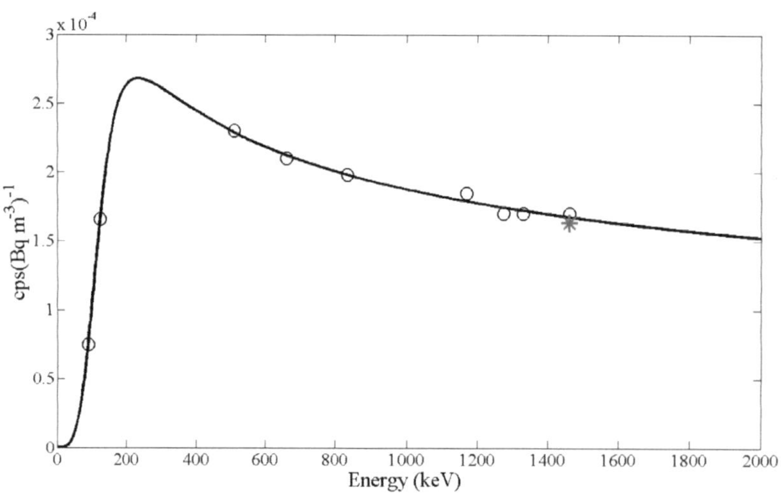

Figure 4. Marine detection efficiency curve of underwater spectrometer developed

To verify the marine detection efficiency calculated by simulation, some field experiments have been carried out by using the underwater spectrometer developed. Before the in situ measurement experiments in the seawater, the underwater spectrometer developed has been energy calibrated by using some radioactive standard sources such as ^{137}Cs, ^{60}Co, ^{40}K and ^{54}Mn in the laboratory. The background spectrum of seawater obtained by continuous cumulative measurement for 24 hours is shown in figure 5. Some seawater was collected and brought back to the laboratory. The activity of ^{40}K in the seawater sample was measured of 1.078×10^4 Bq m^{-3} [7]. The marine detection efficiency of the underwater spectrometer developed for ^{40}K can be calculated by the following formula:

$$\varepsilon_v = \frac{cps}{c\,I} \tag{3}$$

where *cps* is the counting rate of the photopeak, *c* is the activity concentration of the radionuclide in Bq•m^{-3}. The marine detection efficiency value of ^{40}K obtained in field experiment is expressed by a star point in Fig.4. It is clear seen that the marine detection efficiency of ^{40}K from experiment is very near to that from simulation, which showed promising results for the simulation code as well as for the quantitative estimation of radionuclides in seawater.

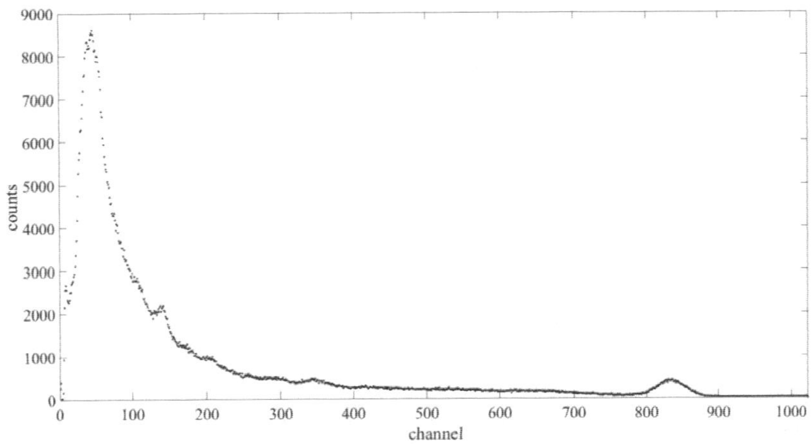

Figure 5. Background spectrum of seawater for 24 hours

4. Conclusion and discussion

A non-experimental calibration method for underwater spectrometer using Na(Tl) crystal is established by simulating the measurement environment and the detection process of gamma photons in seawater using Monte Carlo statistical calculation method in this paper. The experiment result showed promising results for the quantitative estimation of radionuclides in the seawater.

The more quantitative experiments of the radionuclides should be carried out to fully verify the feasibility and accuracy of this kind of non-experimental calibration method. A certain size of sea water tank can be designed and developed, and then the underwater spectrometer is placed in the central position of the tank. However, at present, due to the problem of large-scale contaminated seawater treatment, the specific experiments have not been completed. In the next work, the construction of the standard water tank and some relevant quantitative measurement experiments of radionuclides in the seawater will continue in the next work.

Acknowledgments

This study was financially supported by National Natural Science Foundation of China (No.41606111), National Key Research and Development Program of China (No. 2016YFC1402500), Shandong Provincial National Science Foundation of China (No. ZR2017LD012) and Taishan scholar climbing program.

References

[1] Caffrey J A, Higley K A, Farsoni A T, Smith S and Menn S 2012 *Journal of Environmental Radioactivity* 111 pp 120-125
[2] Dulai H, Kamenik J, Waters C A, Kennedy J, Babinec J, Jolly J and Williamson M 2016 *Journal of Radioanalytical and* Nuclear *Chemistry*. 307 (3) pp 1865–1870.
[3] Hui Z M, Ming P J, Yi Y J and Feng L J *2013 Nuclear Electronics & Detection Technology* 33(2) PP 183-187 (in Chinese)
[4] Zeng Z, Su J, Yi H C, Ma H, Cheng J P and Li J L 2013 *Radiation Protection* 33(1) pp 46-53 (in Chinese)
[5] Ahmadi M, Ahmadpour P M, Rabbani M 2011 *Applied Radiation and Isotopes* 69(7) pp 1079-1083
[6] Zhang Y Y, Li C K, Liu D Y, Zhang Y and Liu Y 2015 *Applied Radiation and Isotopes* 98 pp 44–48
[7] Zhang Y Y, Wu B W, Liu D Y, Zhang Y and Cheng Y 2018 *Applied Radiation and Isotopes* 142 pp 181–186
[8] Tsabaris C and Prospathopoulos A 2011 *Applied Radiation and Isotopes* 69 pp 1546–1553

IOP Conf. Series: Materials Science and Engineering **646** (2019) 012006 doi:10.1088/1757-899X/646/1/012006

A Small Traffic Sign Detection Algorithm Based on Modified SSD

Hongxin Shan and Wenxing Zhu*

Shandong University, Jinan, China

*zhuwenxing@sdu.edu.cn

Abstract. Traffic sign detection is an important part of many systems such as autonomous driving, driver safety and assistance. In this paper, the detection capability of SSD for small targets is analyzed and improved based on ssd_300 model. CSUST Chinese Traffic Sign Detection Benchmark (CCTSDB) dataset is used to train the model for Chinese road traffic conditions. The improved model was compared with ssd_300 model. The experimental results show that the mAP of the improved model on the test dataset achieves 0.85, which is 0.13 higher than ssd_300, and the algorithm can reach real-time detection.The improved model can effectively detect three categories of Chinese traffic signs and has strong robustness against various disturbances.

1. Introduction

To aid safe driving and effective navigation, traffic signs provide essential traffic information that is an important part of autonomous driving, driver safety assistance and traffic monitoring systems[1].Traffic sign detection is a challenging task due to the complexity of the road scene, Light brightness change, occlusion of the object, and the relatively small relative area of the detection object.Therefore, traffic sign detection task has received much attention, and a large number of researchers have conducted in-depth research on it. Liang M et al, present a traffic sign detection model consisting of two modules [2], Extracting regions of interest(ROI) through supervised learning, then detects traffic signs using a combination of Histograms of Oriented Gradients (HOG) and Support Vector Machines (SVM).Zhu YY et al[3], propose two methods of combining neural networks, using fully convolutional network (FCN) to guid traffic sign proposals and deep convolutional neural network (CNN) for object classification, in 2016.

In recent years, many advanced object detection algorithms have achieved huge success, such as R-FCN[4] ,Faster R-CNN[5],SSD[6],YOLO[7],and so on. Among them, the SSD object detection algorithm has a good balance between the detection accuracy and the detection speed.

2. Related work

2.1. Analysis of small target detection by SSD

In SSD, the pixel above each feature map corresponds to several default anchor box. This network trains the anchor to drive the feature training. Select the IOU (Intersection Over Union) of prediction box larger than 0.5 as the positive sample, and the prediction box with the IOU less than 0.5 as the negative sample. Therefore, there are many predicted bounding box with the IOU more than 0.5 on the large target object, the positive and negative samples are balanced. On the small target, there are fewer predicted bounding box with the IOU greater than 0.5, which leads to the imbalance of the number of positive and negative objects in the small target object, which is not conducive to the training of small target objects.

On the other hand, SSD, YOLO and other single-stage multi-scale algorithms, small target detection requires higher resolution, so it is mainly detected at the bottom feature layer, such as conv4_3 in SSD, but the semantic information of the bottom feature layer is not rich enough. For the high-level feature layer, the feature extraction content is more abstract. The more features are extracted, the less corresponding fine-grained information, and less sensitive to the detection of small targets. The size of each feature layer of SSD_300 and the size of the corresponding default anchor box are shown in Table 1.

Table 1. Size of each feature layer and anchor size bounds of SSD_300

Layer	Feature layer		
	Feature shape	Anchor size bound	Anchor size
Conv4_3	38×38	0.07	21×21
Conv7	19×19	0.15	45×45
Conv8_2	10×10	0.33	99×99
Conv9_2	5×5	0.51	153×153
Conv10_2	3×3	0.69	207×207
Conv11_2	1×1	0.87	261×261

It can also be seen from Table 1 that the anchor size (Mapping Region scale) of the bottom feature layer is smaller, while that of the upper layer is larger. In summary, in order to improve the detection ability of small object, it is necessary to set an appropriate anchor size and increase the semantic information of the bottom feature layer[8].

2.2. Dataset

Most of the previous research on traffic sign detection algorithm is based on the German traffic sign detection benchmark (GTSDB)[9],There are many reasons, on the one hand it is widely accepted to compare the traffic sign detection methods in literature, on the other hand a special public challenge has been set for it and a large number of participants have contributed the results. The GTSDB divides traffic signs into three categories: mandatory signs, danger signs and prohibited signs, which contain the natural traffic scenes of various types of roads under different lighting conditions. The data set was composed of 900 images with a size of 1360x800. It was divided into a training set consisting of 600 images (846 symbols) and a test set consisting of 300 images (360 symbols).

In order to conduct research on the detection of Chinese traffic signs, a data set for Chinese traffic conditions was selected. This paper focuses its experimentation on the CSUST Chinese Traffic Sign Detection Benchmark dataset (CCTSDB)[10].Similar to the GTSDB, CCTSDB classifies Chinese traffic signs into three categories: mandatory signs with blue color and circular or rectangular shape, warning signs with yellow color and triangular shape, and prohibitory signs with red color and circular shape. The open source CCTSDB has a total of 15,723 images, including different types of road (highway, street, rural and urban) under the condition of different illumination of natural traffic scene. The picture of the size of 1000 x350 accounted for 77% of the total data set, the size of the rest of the images (1024×768, 1280×720, 912×684, 513×999, 641×936, 457×889, 852×538, 1022×504, 759×497) accounted for 33% of the total data set. Because the open source CCTSDB data set did not divide the training dataset and the test dataset, in this paper, 12,516 images were used as the training dataset, and 1000 images were evenly extracted from all the data sets as the test dataset. It is worth noting in this article, make a small amount of clipping of the width of partly 1000×350 image in the test dataset. The distribution of traffic sign types in the training dataset and the test dataset is shown in Table 2.

Table 2. Distribution of traffic sign types in the training dataset and the test dataset

Types of sign	Training dataset		Test dataset	
	Picture number	Sign number	Picture number	Sign number
warning	3138	3569	239	283
prohibitory	6569	9503	534	734
mandatory	4644	6064	345	454
total	12516	19136	1000	1471

In the natural road scenes, the detection of traffic signs is often the detection of small object. In the object detection algorithm, there are two ways to define small object, one is the relative size, for example, the length and width of the object is less than 0.1 of the original image size, which can be considered as a small object; the other is the definition of absolute size, the object smaller than 32*32 pixels can be considered as a small object. The relative size of the CCTSDB dataset ground truth is shown in Figure 1.

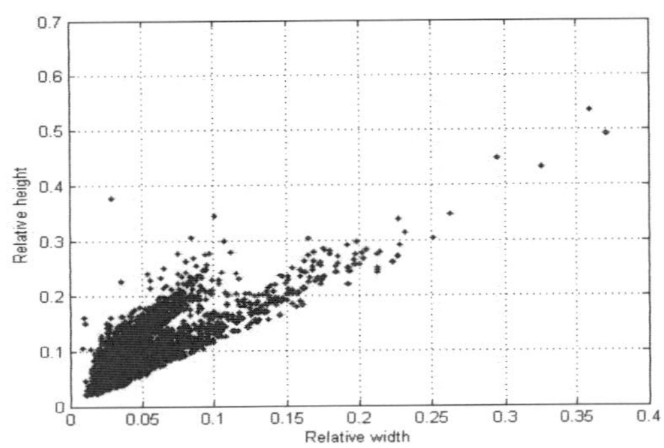

Figure 1. Relative size of the CCTSDB dataset ground truth

It can be seen from Figure 1 that the relative size of most ground truth in CCTSDB dataset is less than 0.1 ,therefore, most ground truth of CCTSDB dataset is a small object. It can also be seen that the aspect ratio of most ground truth is not 1:1.In this paper, the anchor size of the first feature map is set to 10, and the number of default box is set to 6, and the anchor ratios are: 1:1*2,1:2,2:1,1:3,3:1.

3. Traffic sign model

3.1. Architecture

According to the SSD detection of small object and the description of CCTSDB dataset in the second section, in order to improve the detection effect of the model on small object, high-level features map and low-level features map are fused. In our model, the same size as conv4_3 is obtained by deconvolving conv5_3. The following, convolution and Relu activation operations are performed on conv4_3 and the deconvolved conv5_3. Then we merge them and perform a convolution operation to get a new feature map. Similarly, conv7 and block9/conv1×1(size:19×19) perform the same operation for feature fusion. The fusion of high level features and low level features can obtain the characteristics of the receptive field without lacking semantic information. Because the semantic information will affect the detector's determination of whether the detection area is an object or background, feature fusion of the front feature layer can enhance the ability to detect small targets.

It is worth mentioning that there is no large ground truth in the CCTSDB dataset, and it can be seen from Figure 1 that the relative size of the largest object does not exceed 0.6.It can be seen from Table 1 that the anchor size of conv11_2 feature map is much larger than the relative size of the largest object. Therefore, the last feature map(block11) was removed from our model without affecting the detection effect. The improved model architecture is shown in Figure 2.

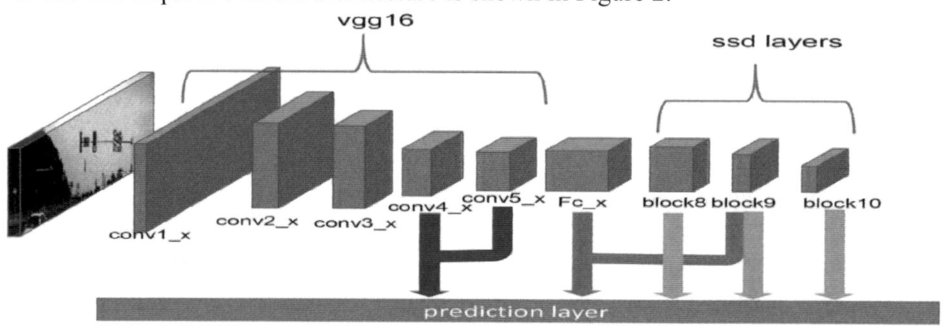

Figure 2. The architecture of our model

3.2. Training

In this paper, two models are trained. One is the original ssd_300 model and the other is our improved model. In order to reduce the time of training model, the transfer learning was adopted to train the model. Through transfer learning, the CCTSDB dataset was used to fine-tune the model to detect and classify the traffic signs based on shape and color: warning(1), prohibitory(2) and mandatory(3). During the training, the public vgg16 weight was used to initialize the weight of our network for feature extraction (conv1_x, conv2_x, conv3_x, conv4_x, conv5_x) to train the weight of the remaining feature map until convergence. Next, the trained model weight is used to initialize the network to train the complete network until convergence. The parameters of the train network is shown in Table 3.

Table 3. Parameter Settings

Parameter	Value
Learning rate	0.001
Optimizer	adam
Batch size	24
Weight decay	0.0005
GPU memory fraction	0.9

4. Results

In this paper, ssd_300 is compared with our improved model. In our experimental evaluation of the model, the training and testing were carried out on a Linux PC, and the machine configuration was described in part B of the section III. The test dataset contains 1000 pictures and 1471 ground truth. The Mean Average Precision (mAP) and the Frame per Second(FPS) of the two models are shown in Table 4.

Table 4.FPS and mAP for each model

Model	mAP	FPS	AP of warning	AP of prohibitory	AP of mandatory
ssd_300	0.715	28	0.694	0.734	0.717
ours	0.85	25	0.859	0.845	0.846

IOP Conf. Series: Materials Science and Engineering **646** (2019) 012006 doi:10.1088/1757-899X/646/1/012006

It can be seen from Table 4 that the detection ability of the improved model for traffic signs is greatly improved compared with the model before the improvement(ssd_300). The mAP achieves 0.85, which is 0.13 higher than the model before the improvement. The detection speed on the experimental platform can achieve real-time detection. The precision-recall curves(PR Curve) of each traffic sign categories are shown in Figure 3.

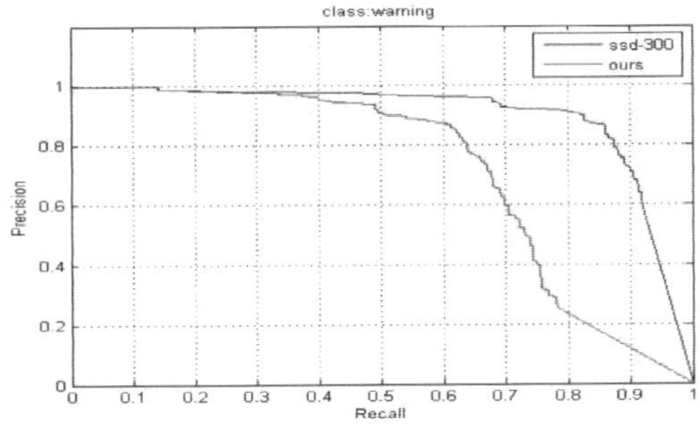

(a). The pr curve of warning

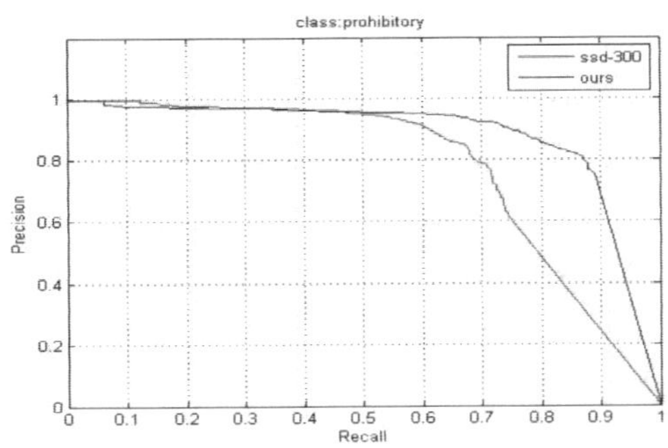

(b). The pr curve of prohibitory

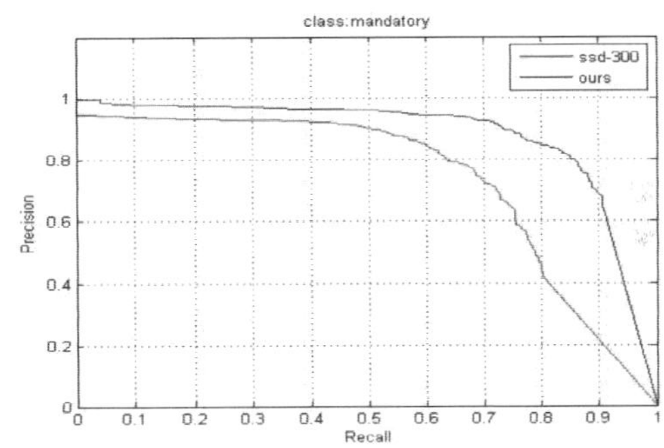

(c). The pr curve of mandatory

Figure 3. PR Curve for each categories

The experimental results under different scenes are shown in Figure 4. It can be seen from Figure 4 that our model can effectively detect three categories of Chinese traffic signs and has strong robustness against various disturbances.

(a)

(b)

(c)

(d)

(e) (f)

Figure 4. The experimental results under different scenarios

5. Conclusion

A traffic sign detection algorithm based on modified SSD is proposed in this paper. This method uses CCTSDB dataset training model for Chinese traffic signs. The improved model was compared with ssd_300 model. The experimental results show that the mAP of the improved model on the test dataset achieves 0.85, which is 0.13 higher than ssd_300, and the detection can reach real-time detection, but the confidence of our model on the small traffic signs is low.

In the future, we will continue to study the detection of small traffic signs to improve the confidence of detection.

References

[1] De L, Moreno L, Salichs M and Armingol J 1997 Road traffic sign detection and classification. *IEEE Transactions on Industrial Electronics* **44(6)** pp 848-859.

[2] Liang M, Yuan M, Hu X, Li J and Liu H 2013 Traffic sign detection by ROI extraction and histogram features-based recognition. *In The 2013 international joint conference on Neural networks* (IJCNN) pp 1-8. *IEEE.*

[3] Zhu Y, Zhang C, Zhou D, Wang X, Bai X and Liu W 2016 Traffic sign detection and recognition using fully convolutional network guided proposals. *Neurocomputing* pp 758-766.

[4] Dai J, Li Y, He K and Sun J 2016 R-fcn: Object detection via region-based fully convolutional networks. *In Advances in neural information processing systems* pp 379-387.

[5] Ren S, He K, Girshick R and Sun J 2015 Faster r-cnn: Towards real-time object detection with region proposal networks. *In Advances in neural information processing systems* pp 91-99.

[6] Liu W, Anguelov D, Erhan D, Szegedy C, Reed S, Fu C Y and Berg A C 2016 Ssd: Single shot multibox detector. *In European conference on computer vision* pp 21-37.

[7] Redmon J and Farhadi A 2017 YOLO9000: better, faster, stronger. *In Proceedings of the IEEE conference on computer vision and pattern recognition* pp 7263-7271.

[8] Jeong J, Park H and Kwak N 2017 Enhancement of SSD by concatenating feature maps for object detection. *In BMVC.*

[9] Houben S, Stallkamp J, Salmen J, Schlipsing M and Igel C 2013 Detection of traffic signs in real-world images: The German Traffic Sign Detection Benchmark. *In The 2013 international joint conference on neural networks* (IJCNN) pp 1-8. *IEEE.*

[10] Zhang J, Huang M, Jin X and Li X 2017 A real-time chinese traffic sign detection algorithm based on modified YOLOv2. *Algorithms* **10(4)** p 127.

The Car-Following Model Based on Fuzzy Inference Controller

Guanzhen Li and Wenxing Zhu*

Shandong University, China

*zhuwenxing@sdu.edu.cn

Abstract. Many existing researches mostly establish a theoretical car-following model from a perspective of statistical physics, but the vehicle driver is the decision-maker of driving behaviors. When modeling the micro-behavior of the vehicle, the driver's driving intention and decision-making basis need to be considered. In order to consider human factors more in microscopic traffic simulation, this paper uses fuzzy inference technology as the basis, selects the headway distance, ideal speed difference, and reasonable spacing difference as the input variables of the control system, considering the free running and car-following state of the vehicle, constructs fuzzy rules and establishes a car-following model.

1. Introduction

This paper mainly studies the free driving behavior and car following driving behavior under single-lane driving conditions. It determines the influencing factors of the following state according to the headway distance and the reasonable distance, and provides support for the decision of the vehicle following behavior.

Since the 1950s, Pipes had explained the car-following behavior from the dynamic point of view, marking the beginning of the car-following theory, and thus building a bridge between micro-traffic behavior and macro-phenomena of traffic flow[1]. In the past 60 years, many experts have made contributions in this field. According to different perspectives, the car-following models can be divided into optimal speed model (Optimal Velocity Model, OV[2]) ,intelligent driving model (Intelligent Driver Model, IDM[3]), artificial intelligence model[4], stimulus-response model[5], etc. Above models based on different factors such as distance and velocity, is in some way to explaining the macro mechanism of all kinds of transport phenomena in traffic flow. However from a microscopic point of view, the actual driving behaviors are the result of a combination of psychological and physiological factors of the driver.

Logic is the basis of human rational decision-making, and fuzzy logic through reasoning of decision-making strategies can effectively describe driving behaviors that cannot be expressed by accurate mathematical models. In this paper, the three variables of the headway distance, the ideal speed difference and the reasonable spacing difference are used as the input of the inference system to construct the decision model of the following behavior.

Kikuchi[6] first proposed the model based on the fuzzy controller. They fuzzed the classic GM model and used relative velocity, relative distance, and leading vehicle's acceleration as model input variables.

If the relative distance is reduced by one level for each fuzzy set, the acceleration is reduced by a fixed value ϕ. In addition, some other scholars have established similar car-following model based on fuzzy logic[7-8].

2. Model Elements

There are three inputs in the vehicle speed decision model: (1) the headway distance T; (2) the ideal speed difference Δv ; (3) the reasonable spacing difference Δd . The output of the model is the acceleration of the vehicle.

The ideal speed difference refers to the difference between the speed of the following vehicle and the ideal speed or the speed of the preceding vehicle. When the headway distance is greater than or equal to 4, the speed difference refers to the difference between the speed of the following vehicle and the ideal speed, and when the headway distance is less than 4, the speed difference is between the speed of the following vehicle and the speed of the preceding vehicle. The ideal speed is based on current road conditions such as weather conditions, traffic congestion, road speed limit and other factors, and is the highest safe driving speed that can be achieved.

The reasonable spacing difference is the difference between the current distance and the reasonable spacing.

$$\Delta d = d_1(t) - d_{ideal}(t) \tag{1}$$

$$d_{ideal}(t) = \max(d_{safe}(t), d_{0.5s}(t), d_{\min}) \tag{2}$$

$$d_{safe}(t) = \frac{v_2^2(t)}{2a} - \frac{v_1^2(t)}{2a} + \delta * v_2(t) \tag{3}$$

$$d_{0.5s}(t) = 0.5 * v_2(t) \tag{4}$$

$$d_{\min} = 3 \tag{5}$$

The linguistic variables and scope can be seen in the table 1.

Table 1. Logic variables and their scopes

Variable	Symbols	Scope
T/s	PS	[0,2]
	M	[1,3]
	PB	[3,8]
	FR	[5,10]
$\Delta v/(m/s)$	NB	[-30,0]
	M	[-3,3]
	PB	[0,30]
$\Delta d/m$	NB	[-25,0]
	M	[-3,3]
	PB	[0,25]
$A/(m/s^2)$	NB	[-4,-2]
	NM	[-3,-1]
	NS	[-2,0]
	M	[-1,1]
	PS	[0,2]
	PM	[1,3]
	PB	[2,4]

IOP Conf. Series: Materials Science and Engineering **646** (2019) 012007 doi:10.1088/1757-899X/646/1/012007

2.1. Fuzzy membership function

With the help of appropriate membership function and correct control rules, the fuzzy controller has good adaptability, stability and fault tolerance, and is suitable for nonlinear, time-varying and incomplete systems. The control strategy of the entire system is based on the setting of the membership function. When setting the membership function, a key point is that adjacent fuzzy sets need to overlap each other, so that the system can smoothly transform when performing fuzzy reasoning, so that the output of the system is stable.

The figure 1 is for membership function.

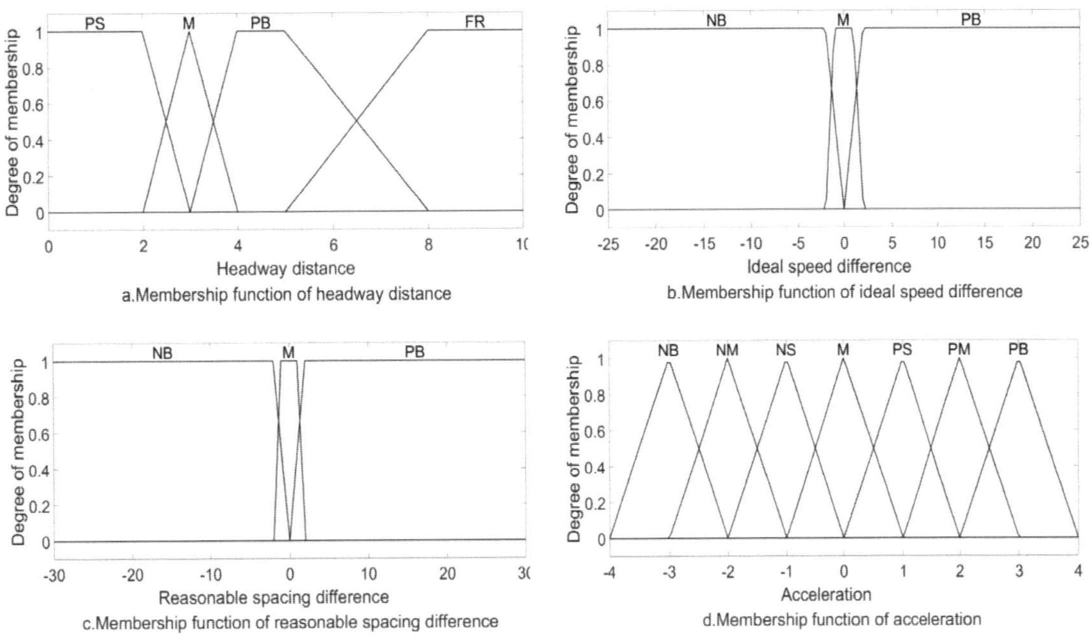

Figure 1. Membership functions for all the variables.

2.2. Fuzzy rules

The fuzzy rule is based on the driving strategy to derive the optimal decision in different driving conditions. These rules maintain the stability of the traffic flow and road capacity at macro view, safety and comfort at micro perspective. The fuzzy rules are shown in table 2.

Table 2. Fuzzy rules based on the driving strategy.

Number	T	Δv	Δd	A
1	FR	NB		PB
2	FR	M		M
3	FR	PB		NS
4	PS		NB	NB
5	PS		M	NB
6	PS		PB	NM
7	M	NB	NB	NS
8	M	M	NB	NM
9	M	PB	NB	NB
10	M	NB	M	PM
11	M	M	M	PS

12	M	PB	M	M
13	M	NB	PB	PS
14	M	M	PB	M
15	M	PB	PB	NM
16	PB	NB		PM
17	PB	M		M
18	PB	PB		NM

The above rules are based on an analysis of the driver's behaviors under different situations. Driving state can be divided into free driving state and car following state without considering lane change. In the free driving state, the driver wants to meet the requirements of ensuring safety and speed and adjusts to the optimal speed as soon as possible. The ideal speed is based on current road conditions such as weather conditions, traffic congestion, road speed limit, and other factors, and is the highest safe driving speed that can be achieved.

In the car following state, the speed of the preceding vehicle and the distance between the cars will have a direct impact on the state of the following vehicle. when the speed of the preceding car is smaller than that of the following car, and the smaller the car spacing, the greater the deceleration of the car. If there is still a distance from the optimal distance at this time, the braking force can be adjusted appropriately considering the comfort of passengers.

When the headway is at PS, due to the difference in the drivers' response at high speed and low speed, we introduce reasonable spacing to regulate the following behavior of the vehicle. When the headway is at PB, the following car is in the transition stage that accelerates to the ideal speed. we need to adjust the acceleration slightly.

In the state of free driving, the driver hopes to adjust the speed to the ideal speed as soon as possible and revise the value of the ideal speed with the change of driving condition. In the car following state, the following vehicle needs to seek the optimal acceleration value according to the speed of the preceding vehicle and the distance between the vehicles. The ultimate goal is to ensure that the speed of the following vehicle is at an optimal value while the distance between the vehicles is safe and stable.

3. Simulation experiment

The following model is built by Matlab simulink app, and simulation experiments are carried out to verify the rationality of the established inference rules and the stability of the car-following model. In this paper, Lilian's[9] experimental scheme was adopted, and the experimental results were used as a control group. The scheme is as follows: Assume that the lane is a single lane straight line of infinite length. The starting point of the leading vehicle is 200 m, the starting point of the following vehicle is the origin, and the initial speed of the rear car is 0 m/s. Speed change of the leading vehicle is shown in equation (6).

$$\mathbf{v}=\left\{\begin{array}{c}\mathbf{15}+\frac{\mathbf{4}}{\mathbf{9}}\mathbf{t},\mathbf{0}\le\mathbf{t}<\mathbf{45}\\ \mathbf{55}-\frac{\mathbf{4}}{\mathbf{9}}\mathbf{t},\mathbf{45}\le\mathbf{t}\le\mathbf{90}\\ \mathbf{15},\mathbf{90}<\mathbf{t}\le\mathbf{120}\end{array}\right\} \tag{6}$$

The leading vehicle running curve is divided into three phases: firstly, it accelerates uniformly from the initial speed of 15 m/s to 35 m/s. Then evenly decelerate to 15 m/s, and finally drive at a constant speed for a while. It can be predicted from the previous analysis that, firstly, the leading vehicle will be in the free driving state and then will accelerate to the ideal speed as soon as possible. If the distance between the vehicles is gradually shortened, the following vehicle is going to step into the car following state. Whether the following vehicle can maintain the car following state and whether the traffic flow can keep stable will be the basis for judging the rationality of membership functions and fuzzy inference rules. The simulation model is shown in figure 2.

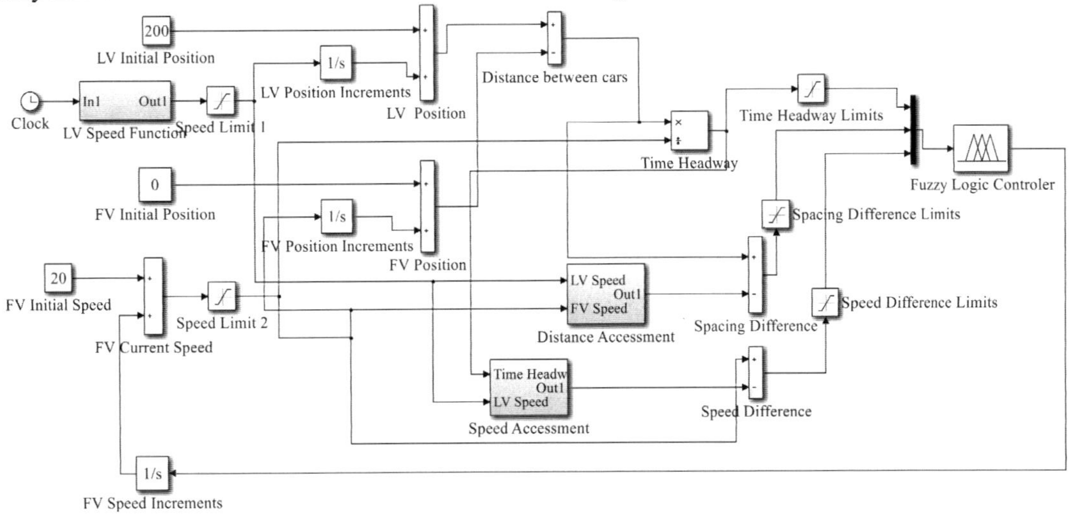

Figure 2. This figure shows the entire experimental model

4. Simulation results

As shown in figure 3, the experimental vehicles are in the free driving state at the beginning of the simulation. The experimental vehicles accelerate from the initial speed to the ideal speed within 5 s and then entered the car following state. In the period of 20-90 s, in all the experimental plans, the speed of the following vehicle closely follows the speed of the leading vehicle.

According to the results, these experiment plans are stable, and the speed curves of the various plans are quite different. The speed curve of the IDM model is almost perfect, and it is almost impeccable in terms of safety, curve roundness and state transition stability. However, it can be seen that the biggest disadvantage of the numerical calculation models that IDM belongs to is unrealistic. In less than 1 s from the start of the experiment, the speed of the following vehicle rapidly increases from 20 m/s to 33 m/s. The acceleration could be hundreds.

Compared with the model proposed by Li Lian, the proposed model in this paper has little difference between the free driving stages. When the leading vehicle accelerates, the FV1 closely fits the speed curve of the preceding vehicle with a certain delay, while the FV2 stabilizes at 30m/s. When the leading vehicle decelerates, the speed curve of FV2 has a steep cliff-like drop at the 70s. While FV1 adopts a slow deceleration strategy to ensure greater safety and comfort while following the speed of the preceding vehicle.

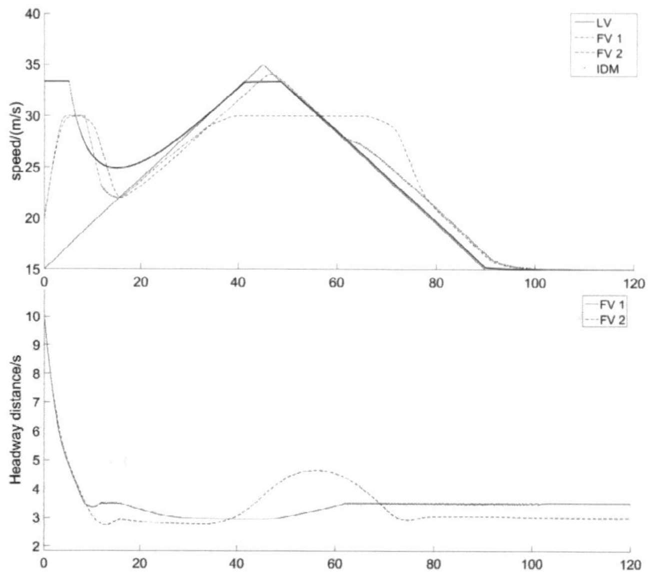

Figure 3. This figure shows the results of the IDM model, Li Lian's model and the model we build.

5. conclusions

The fuzzy following model established in this paper considers the characteristics of free driving behavior and car-following behavior of vehicles, and establishes a complete control system to measure and evaluate the driving environment. The perception layer converts the determined value of the input quantity into the corresponding fuzzy variable value by fuzzification. The inferring layer and the driving strategy layer make the acceleration and deceleration decisions based on current state through fuzzy inference and obtain the accurate value of the acceleration through defuzzification. In the experimental simulation, the system simulates the driving behaviors of the actual vehicle and improves the driving efficiency under the premise of considering safety and comfort.

REFERENCES

[1] Pipes, Louis A 1953 An operational analysis of traffic dynamics J. *Appl. Phys.* **24** p 274.
[2] Bando M, Hasebe K and Nakayama 1995 A dynamical model of traffic congestion and numerical simulation J. *Phys. Rev. E* **51** pp 1035-1042.
[3] Treiber M, Hennecke A and Helbing D 2000 Congested traffic states in empirical observations and microscopic simulations J. *Phys. Rev. E* **62** pp 1805-1824.
[4] Kehtarnavaz N, Groswold N and Miller K 1998 A transportable neural-network approach to autonomous vehicle following J. *IEEE Transactions on Vehicular Technology* **47** pp 694-702.
[5] Wiedemann R 1974 Simulation of Road Traffic in Traffic Flow (Karlsruhe: University of Karlsruhe) pp 1805-1824.
[6] Kikuchi S Chakroborty P 1992 Car-following model based on fuzzy inference system J. *Transportation Research Record Journal of the Transportation Research Board* pp 82-91.
[7] Mar J Lin F J and Lin H T 2003 The car following collision prevention controller based on the fuzzy basis function network J. *Fuzzy Sets and Systems* **139**(1) pp 167-183.
[8] Khodayari A Kazemi R and Ghaffari 2011 *Conf.* A Design of an improved fuzzy logic based model for prediction of car following behavior *IEEE International Conference on Mechatronics*.
[9] Li Lian, Yang Song and Cao Wenjing 2014 Driver's speed decision-making behaviour modeling and aimulation based on fuzzy inference J.*System Simulation Technology* pp 184-190.

IOP Conf. Series: Materials Science and Engineering **646** (2019) 012008 doi:10.1088/1757-899X/646/1/012008

Artificial Intelligence's Turn of Philosophy

Tao Feng

College of Philosophy, Nankai University, Tianjin,300350, China,
China Institute of New generation AI Development Strategy

taofeng@mail.nankai.edu.cn

Abstract. Artificial Intelligence's turn of philosophy emerges because of the new crisis in philosophy. Philosophy should attach importance to technical thinking to supplement the deficiency of philosophical thinking. Philosophy's "language turn" and "AI turn" are based on language. The paradigm shift is also that we can define "intelligence", "consciousness" and "subjectivity" in the way of conventionalism. With the rapid development of AI technology, we also need to guide the development of technology with "Aesthetic reason" and promote the harmony between science and technology and art, human and nature.

1. Introduction

The development and breakthrough of Artificial Intelligence(AI) technology have already set off a new wave in the scientific and technological fields. L. Floridi thinks that this is the "Fourth Revolution" after the industrial revolution. J. Kurzweil, an expert in AI, said in his book *The Singularity Is Near* that the exponential development of AI and other technologies has brought the universe into a new era of human-machine integration [1]. Physicist M. Tegmark believes that human beings are about to enter the era of life 3.0, the software and hardware of life are no longer dependent on evolution but are designed [2]. AI has also attracted great attention in philosophy, and has given birth to many interdisciplinary research directions such as AI Logic, AI Ethics, AI Aesthetics and AI Literary theory [3]. So, is it possible for philosophy to produce a new research paradigm or even an "AI turn" in the face of this emerging technology?

2. the Crisis of Philosophy

I believe that the "AI turn" of philosophy is possible. This turn is not only influenced by technology, but also due to the crisis of philosophy itself. Philosophy has been from all-encompassing wisdom to the current knowledge based on criticism and reflection. It has experienced many crises like ontological crisis, epistemological crisis and expressionist crisis, and results in the "epistemological turn", "linguistic turn" and etc. Today, the research field of philosophy is more and more narrow, and the conclusions drawn by professional philosophers are becoming more and more personalized and decentralized, which makes philosophy lack of universal standards and values. As a result, the existence of philosophy itself has become issue. For another, scientists and technologists such as R. Penrose, A. Turing, M. Minsky and J. Kurzweil all reflect on human consciousness, intention, mind and ethical issues through technology. As they have mastered more professional scientific and technological knowledge, they can better grasp the current new progress in science and technology. They have also held a lot of discussions on basic philosophical issues such as Philosophy of AI,

Quantum philosophy and human consciousness. However, most contemporary philosophers are confined to texts. They used to search for sentences from books of former people as proofs, and finally they can only become literature experts or even become language players.

The functionality of modern philosophy is getting less and less, leaving only criticism and reflection of thinking. There is no denying that criticism and reflection are still the most important thinking ways of human beings. But if there are only these ways without construction, practice, operation, society is difficult to make real progress. In addition, criticism and reflection are also based on in-depth understanding of phenomena. However, most philosophers do not understand new scientific and technological knowledge and even do not want to know either. How can they really criticize them? For example, there are great differences between technologists and philosophers on the issue of the threat of AI. Some philosophers did not understand the development and technical characteristics of AI, so they blindly criticized it. They seem to criticize science fiction literature rather than AI.

Therefore, I believe that the emerging technologies such as AI, network, information, automation, virtual reality and big data, bring new challenges and opportunities to philosophy. These new technologies are changing the relationship between human and nature, raising new issues such as the relationships between artifact with human, intelligence with wisdom, ethics with utility. Philosophers should take the initiative to understand and master new knowledges, or cooperate with scientific and technological experts, in order to grasp the problem of philosophical turn under new technology.

3. Technical Thinking

Since the age of Enlightenment, many philosophers have criticized technology. M. Horkheimer and T. Adorno pointed out in *Dialectic of Enlightenment* that instrumental rationality and computational thinking identify and control everything and action through precise calculations. While human beings repress the others like nature and even regard themselves as tools, human beings are materialized. M. Heidegger believes that technology forces nature, at the meantime, human beings are controlled by technology. Therefore, authentic language has become the medium of signals and the technical language. "Language machines make language work and thus control human beings." [4]

Admittedly, unrestricted technology does bring disaster to mankind. However, technology is also benefiting mankind. Instrumental rationality is still important. Adorno believes that we can correct the suppression and control of the subject to the object in instrumental rationality through mimesis element in art. Many technologists, starting from specific technical examples, tried to find the law of Human mind and understanding. For example, M. Minsky, the father of AI, in his book *The Society of Mind*, tries to explain the real meaning of human's "purpose" and "intention" through the analysis of "difference-engines" [5].

The fundamental difference between technological thinking and philosophical thinking is reflected in the article "Computing Machinery and Intelligence" by technologist Turing and the tit-for-tat article " Can computer think" by philosopher J. Searle. At first glance, the two articles focus on the same question of "can machines think?" But in fact, the two articles are fundamentally different in terms of problem raising, mode of thinking and solutions. Turing's "machine can think" refers to whether machine can think like a human in behaviour and appearance. "Turing Test" has become a superficial imitation of human intelligence in language and even a deception. Searle, on the other hand, believes that machine cannot essentially generate mind intentionality and language intentionality, so it cannot really understand semantics and thus generate intelligence [6]. Turing's technical thinking is an "outside-to-inside" way of thinking. From technical practice to principle, as long as the behaviour conforms to the purpose, it can be regarded as intelligent. Searle, on the other hand, adopted a philosophical approach "inside-to-outside ", which requires some apriori principle such as intentionality. However, this philosophical starting point cannot be verified scientifically.

4. Paradigm Shift

The "linguistic turn" of philosophy wants to solve the problem of how to express the world with language, while technological thinking attempts to express the object and the world with logic, calculation and algorithm. Philosopher W. Leibniz once tried to establish arithmetic relations among concepts and put forward the idea of "constructed language". G. Frege represented sentences as functions, which laid the ideological foundation for later AI. "Turing Machine" can use two numbers 0 and 1 as basic operators to write all calculation instructions. Shannon's conception of logic gate laid the foundation for Turing machine to become an electronic computer. Neural networks, the basis of deep learning, is also based on binary operation. According to Frege's assumption, language can be represented by arithmetic, while electronic computers can represent language by logical operations to some extent. Turing Test indicates that as long as computers can show intelligence in language tests, computers can think. Therefore, language is still the foundation of machines and intelligence, because machines can only show their intelligence in communication with people. This is also why natural language processing (NLP) technology is the core technology and the biggest problem of AI. Thus, I believe that the "linguistic turn" in philosophy to the "AI turn" is not unrelated, and the link between the two turns is "language".

However, philosophers such as H. Dreyfus believe that AI cannot understand human language because it cannot understand the context in which language is used. [7]

How does AI solve this problem? Early AI structuralist experts once imagined whether we can imitate real context by setting grammar and procedures. Later they found it too difficult because the changes of context are endless. The emergence of deep learning and big data enables people to list different contexts through data collection and statistical methods. We can say that structuralism is a top-down approach, the latter is a bottom-up approach of functionalism. In fact, the two factions can be said to be the embodiment of philosophical rationalism and empiricism in the field of AI. AI expert H. Levesque believes that the two approaches can be combined. In other words, to realize true intelligence, machines still need to have innate knowledge and language framework, plus acquired experience and learning.

The paradigm shift also lies in the fact that we can define "intelligence", "consciousness" and "subjectivity" in conventionalism way. The fundamental reason why Turing and Searle hold different opinions is that they have different understandings of "consciousness". Turing insists on behaviourism and believes that the so-called "consciousness" must be combined with behaviour to judge, otherwise, they can only fall into the mire of solipsism. [8] Searle, on the other hand, understood consciousness from the standpoint of "intentionality". They seemed completely unable to talk. However, ethicist J. Kaplan proposed a new possibility. He analyzed the meaning of " Corporation " , whose Latin root "Corpus" means "body", believing that since a corporation can be the subject of limited liability, inanimate AI can also be the subject of moral liability.[9] Just as Kant borrowed "legitimacy" and "laws" from law to build the philosophical foundation, we can also extend this moral and legal subjectivity to the "subjectivity" of the whole epistemology and existentialism, which requires us to jump out of the shackles of "Human-centrism" and "Bio-centrism" and agree and understand from a broader subjectivity.

It is the emergence of AI that makes people have to go beyond their own limitations to reflect on the relationship between life, human beings and intelligence. Long before AI appeared, people were thinking about animals' feelings, morality and Speciesism. P. Singer, author of *Animal Liberation*, believes that as long as life can feel pain, we should face up to its affection and right. As a non-living, AI forces us to consider such problems: Whether a machine without pains or feelings also has moral rights and the right to life? When the right to life of machines conflicts with human rights, do we have the right to sacrifice them to save human beings? These problems pose new challenges to our ethics.

5. Aesthetic Reason

Technical thinking focuses on behavior, results, effectiveness and truth. Artistic thinking is about connection between images, obtaining the highest "beauty" through illusion which seeks truth through falsehood. Philosophical thinking lies in reflection. Dialectics are commonly used. Truth and falsehood are dialectical relations, in the end they also lead to truth.

Only stressing technical thinking and the effectiveness of results, everything becomes calculable. "Reason([G]Vernunft) itself has become an auxiliary tool of the universal economic machine"[10]. Reason has become instrumental rationality, and people have finally been alienated into tools. AI, big-data, gene editing and other technologies may develop excessively while ignoring equality and harmony, such as unemployment, discrimination, privacy and wealth distribution caused by AI technology. Therefore, when we develop AI, we should pay attention to making AI benefit human beings instead of making people become AI's tools. This requires us to attach importance to the regulation and restriction of rationality. We need to combine philosophy, technology and art. The combination of these lies in the realization of "Aesthetic reason". Adorno has presented the concept of "Aesthetic Rationality", [11] which is the rationality in authentic art and the rescue and compensation of instrumental rationality. I developed and extend the meaning of this concept and make it as "Aesthetic Reason".

Aesthetic Reason here contains the meaning of beauty, moderation, equality and harmony, and is the beauty of unity of truth, good and beauty. Aesthetic reason can have multiple meanings. First, aesthetic and sensible things also have elements of rationality, that is, rationality in art and emotion. Can beauty and art be calculated or represented with algorithm? Some philosophers, such as Plato and Kant, attribute art to mysterious inspiration or genius. Wittgenstein believes that art and beauty cannot be expressed and explained with language. On the contrary, AI experts try to analyze and calculate beauty and emotion. For example, AI agent calculates or statistically sums up artistic forms through deep learning and other methods in order to generate AI artworks. "Affective Computing" attempts to imitate human emotions by matching quantifiable forms with emotions. From some examples like Microsoft Xiaobing's poems and paintings by CAN (Creative Adversary Networks), we seem to be able to see the computability of some art elements.

However, Aesthetic reason is not only to calculate and analyse beauty, but also should realize "true" beauty, which is what Adorno called "Truth content" ([G]Wahrheitgehalt). That is to say, what I feel is not the beauty on the surface, but the beauty of unity of truth, good and beauty. Creativity, imagination and emotion are guided by reason. This is what AI cannot achieve and is the real beauty of human beings.

Another important meaning of Aesthetic reason is that "rationality" is beautiful. Adorno and Horkheimer revealed the path of development and degeneration of reason in *Dialectic of Enlightenment*: due to social division of labor, science and technology and art, rationality and imitation, language expression and communication are gradually separated. Reason lost its imitability and became instrumental rationality. The relationship between subject and object, human and nature became dominant and dominated. Therefore, Adorno believes that mimesis preserves the opportunity of equality and reconciliation in art, so rationality should reabsorb the mimesis elements in art so as to return to real rationality. H. Marcuse, another Frankfurt school philosopher, also believes that we should attach importance to life instinct of human and have new sensibility in order to get rid of the bondage of instrumental rationality. Hence, I insist that reason should also be beautiful and equal, free and harmonious. Rationality should be limited when science and technology is highly developing. It is necessary to voluntarily give up some technologies that will threaten the development of mankind and earth like nuclear technology, reproductive Gene Editing and Super AI technology. T. Walsh solicited signatures in 2015 and proposed to abandon the research and development of autonomous weapons [12]. J. Kurzweil said that "Fine-Grained Relinquishment" is possible, that is, abandon those technologies that will cause irreversible and great danger to human beings. [13] For some relatively safe technologies, we also need ethical guidance.

Therefore, the Aesthetic reason not only requires us to explore the true meaning of beauty, but also requires us to develop equal, harmonious and beautiful human reason and realize the harmony between human and nature, science and technology and art, rationality and sensibility. It requires the cooperation of philosophers, technologists and artists. In the process of AI generating artworks and expressing art and beauty with algorithms, it needs the guidance and suggestions of artists. However, we must point out that the purpose of AI painting is to improve the visual technology of AI, and AI literature serves emotional computation and natural language processing. At present, the purpose and foundation of AI art are technology rather than art, and hence it has not realized autonomy. This kind of reflection and criticism on the relationship between technology and art requires the participation of philosophers. Philosophy also needs to rethink the rational problem when facing the problems brought by the new technology of AI, which is why philosophy needs to "turn to AI".

The AI revolution in technology has arrived, and in philosophy a new paradigm shift is coming. We need to reflect on the crisis of philosophical thinking caused by new relationship between science-technology and human being. We might reexamine technological thinking and combine different thinking modes. Kant once made a good example which he reconciled empiricism with rationalism. Can we also unify “from inside out” of philosophy with “from outside in” of technology, structuralism from the top down and functionalism from the bottom up? Technology needs rational guidance, while rationality needs "beauty" to guide. We pursue the harmony and unity of emotion and technology, sensibility and rationality, and human and nature. Only in this way can we know how to make technology benefit mankind, not alienate mankind.

The "AI turn" of philosophy requires the full cooperation of philosophers, technologists and artists. It also requires the crossing and integration of philosophy, science and technology. We need to break down the barriers of disciplines and find the connection between different thinking, so that philosophy can be revitalized in the new technological era.

Acknowledgments
This paper was supported by the National Social Science Fund of China, the name of this project is “The Language Thought in Adorno’s Philosophy”. No. 16BZX118

References
[1] Kurzweil R 2006 *The Singularity Is Near: When Humans Transcend Biology* (Penguin Books)
[2] Tegmark M 2018 *Life 3.0: Being Human in the Age of Artificial Intelligence* (Vintage)
[3] Feng Tao 2018 Why is the Aesthetics of AI Possible? *Literature and Art Forum* vol.5 pp78-85
[4] Heidegger M 1983 Hebel der Hausfreund (1957) *Gesamtausgabe, Band 13, Aus der Erfahrung des Denkens (1910–1976)* (Frankfurt am Main :Vittorio Klostermann) eds.Heidegger H p 149
[5] Minsky M 1988 *The Society of Mind* (Simon & Schuster)
[6] Searle J 2003 *Minds, Brains and Science* (Harvard University Press) 13edition pp28-41
[7] Dreyfus H 1975 *What Computers Can't Do of Artificial Reason* (Harper & Row Publishers) p19
[8] Turing A 1950 Computing Machinery and Intelligence *Mind* Vol.59 pp433-460 p446
[9] Kaplan J 2016 *Artificial Intelligence: What Everyone Needs to Know* (Oxford University Press) pp102-103
[10] Adorno T and Horkheimer M 2003 *Gesammelte Schriften, Band.3: Dialektik der Aufklärung* (Frankfurt am Main: Suhrkamp) ed. Tiedemann p47
[11] Adorno T 1970 *Gesammelte Schriften, Band 7: Ästhetische Theorie*, (Frankfurt am Main: Suhrkamp) ed. Tiedemann p59
[12] Walsh T 2017 *It's Alive Artificial Intelligence from the Logic Piano to Killer Robots* (Australia: Black Inc)
[13] Kurzweil R 2006 *The Singularity Is Near: When Humans Transcend Biology* (Penguin Books) pp 299-300

Biometric Blockchain: A Better Solution for the Security and Trust of Food Logistics

Bing Xu[1], Tobechukwu Agbele[2] and Richard Jiang[2]

[1]Computer & Information Sciences, Northumbria University, Newcastle, UK, NE1 8ST
[2]School of Computing & Communication, Lancaster University, Lancaster, UK, LA1 4WY

Correspondent e-mail: r.jiang2@lancaster.ac.uk

Abstract. Blockchain has been emerging as a promising technology that could totally change the landscape of data security in the coming years, particularly for data access over Internet-of-Things and cloud servers. However, blockchain itself, though secured by its protocol, does not identify who owns the data and who uses the data. Other than simply encrypting data into keys, in this paper, we proposed a protocol called Biometric Blockchain (BBC) that explicitly incorporate the biometric cues of individuals to unambiguously identify the creators and users in a blockchain-based system, particularly to address the increasing needs to secure the food logistics, following the recently widely reported incident on wrongly labelled foods that caused the death of a customer on a flight. The advantage of using BBC in the food logistics is clear: it can not only identify if the data or labels are authentic, but also clearly record who is responsible for the secured data or labels. As a result, such a BBC-based solution can great ease the difficulty to control the risks accompanying the food logistics, such as faked foods or wrong gradient labels.

1. Introduction

Food logistics [1], as the most important industry for everyone's daily life, is always facing great challenges in its security control. This topic has attracted a great attention by industry and researchers following a recent incident reported on BBC and Sky News that a young teenager girl, after eating a sandwich sold at the airport, died of the allergy to the peanuts gradient in the sandwich. Though it has been required by the legislation that food suppliers need to clearly label the sources and gradients in their provided foods, it is hard to monitor the whole food production processes, while multiple gradient providers may be included in one food production, and it is not rare to mislabel one food gradient with others or mistaken one source with another during the food logistics [1, 2].

As it has been widely expected, blockchain technology could provide a solution to the loose control of food logistics, and transform the entire food industry by increasing efficiency, transparency and collaboration throughout the food system [2-10]. With blockchain technology, consumers could be able to trace the source of their lettuce in seconds, shippers could see if a truck is full before they schedule a delivery, and grocery stores could verify if a carton of eggs is actually cage-free. As blockchain gets closer to its marketplace debut in the food system, it is a right time to scrutinize just how the blockchain technology will actually work in food logistics.

Blockchain was initially developed as part of the cryptocurrency, namely Bitcoin [2]. However, the technology in the cryptocurrency context looks different from how it is being developed for the food space. In the paradigm of Bitcoin, blockchain is an immutable digital ledger that works through a consensus of computer systems. Computing on the Bitcoin blockchain are essentially racing to correctly solve a calculation, and when one "wins" the race, it wins a unit of cryptocurrency and a block of data is added to the chain. The huge numbers of computer systems on the Bitcoin blockchain are why there's a huge energy cost associated with Bitcoin, a feature that would be detrimental in the agriculture space, where farmers needs to grow more and use less.

Simply in food logistics, blockchain can be just a digital ledger, a digitized record of whatever data is added by its members, with no ability to verify the accuracy of the underlying data itself. Because the truth of that data isn't actually evaluated, there's no aspect of blockchain technology that can make sure that the cage-free chicken is really cage-free or that the cabbage is actually free from pestcides.

2. Blockchain Technology

Blockchain [2] is a publicly-available and publicly-maintained distributed ledger (somehow equivalent to database) where all verifiable transactions are permanently recorded, and designed in such a way that data in a given block cannot be retroactively altered without altering all sequential blocks.

Originally created as the underlying technology of the Bitcoin protocol to ensure the secure transfer of the cryptocurrency (coins and tokens) without going through a central financial institution, the blockchain technology has recorded a booming success and gained the attention of technology companies doing research on the strengths, weaknesses, opportunities and threats (SWOT) as well as various applications of this technology.

It is a widely recognized that blockchain technology is a cutting-edge technology and an emerging research field with the potential to change our world like what the internet did two decades ago, once there is a strong collaboration among players in the industry. A blockchain is managed by a global peer-to-peer (P2P) network of nodes that validate new blocks using a consensus algorithm. The consensus algorithm ensures that the next block in a blockchain is the one and only version of the genuine one, thus preventing powerful adversaries from successfully branching the chain. As a result, all nodes of the network contain the same replica of data, eliminating the need of a central trusted authority to manage data [3-10].

There has been some evolution of the blockchain technology from version 1.0 to its currently version 4.0 [2-10]. Blockchain 1.0 saw the implementation of cryptocurrencies, while 2.0 was focused more on registration of smart contracts. Blockchain 3.0, on the other hand, brought about decentralized application which uses both a decentralized storage and communication. The current Blockchain 4.0 proposes more alternative approach for implementing the technology in various industries.

New application domains of the blockchain technology [2-10] include telemedicine, data markets, keyless payments, electronic voting system, and banking and financial markets. More recently, blockchain has also been seen to apply to food supply chain management, with great success on solving pending challenges in this industry area with immediate impact on commercialization.

3. Blockchain in Food Logistics

Though blockchain is being touted as the technology that could potentially solve challenges facing food logistics, it is yet not clear why blockchain is better than something like a database or any other form of digital information storage. Companies could simply build a database rather than build a blockchain, particularly as some of the original features of the Bitcoin version, like trustless verification, are not a generic feature.

It may be not entirely clear why blockchain is the best technology for the job of transforming the food industry, and it may or may not be. It may just be that it is the one getting attention right now, particularly as technology experts look for ways to transfer their experience and make their mark in the burgeoning food supply sector.

Where blockchain starts to reach its potential is when it is used with other technologies and systems. At the same time that the blockchain is implemented for food traceability, for example, producers can also put into place systems like enhanced water testing mechanisms or increase buffer zones between leafy green growers and livestock operations.

Blockchain can be used to gather a wealth of data and employ it in the field when used with sensors and precision delivery systems for pesticide and water all connected to a network, as with the Internet of Things. Fig.1 shows a typical framework to use blockchain in the food supply management, while data-oriented blockchain is a match of the food chain as well.

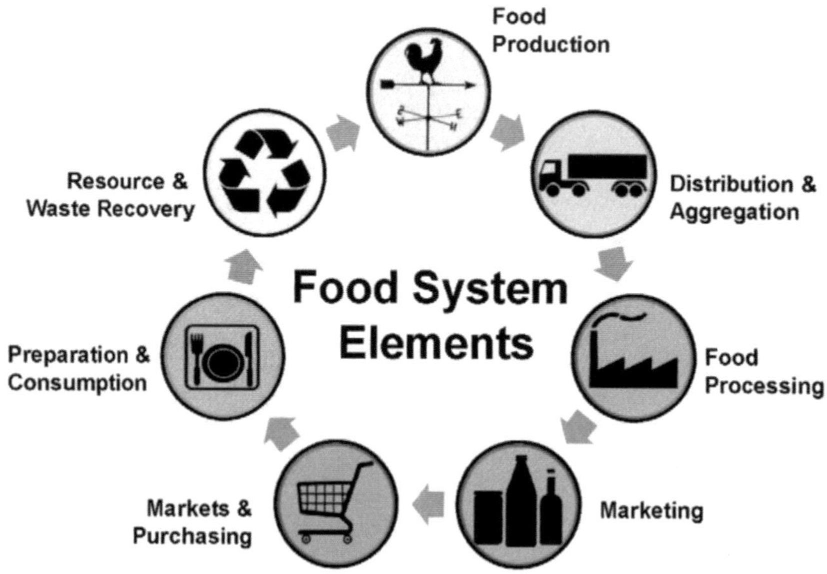

Figure 1. Blockchain can be exploited in many key stages in the food logistics [1].

While blockchain cannot verify that an egg operation is truly cage-free or what that cage-free operation really looks like, it might offer a way for food suppliers to get more verifiable information to consumers. Food suppliers, particularly the ones who do not sell their food to the end users and cannot get an opportunity to interact with consumers, often struggle with how to engage with the customers, looking for ways to explain about details and merits the way they grow or process foods. Blockchain enables food suppliers at all stages in the supply chain to get data to consumers. That's the thing consumers can really be benefited with more information about their purchased foods.

Blockchain could be used to tell consumers that the corn was grown with pesticides, for example. It could even provide a mechanism for explaining why that pesticide is used, or a comparison of that pesticide to other pest prevention methods. The complicated nature of agriculture does not always interpret so well to a smartphone app, which might be a technologic challenge that is yet too big for blockchain to solve anyway.

To help truly understand provenance, the tracking and authentication of food supply chain is critical to finding and helping to address sources of gradients in the food supply chain worldwide. Blockchain provides a permanent record of transactions that are assembled in blocks that cannot be altered. It could serve as an alternative to traditional paper-based tracking and manual inspection systems, which can leave supply chains vulnerable to inaccuracies.

While applied to the food supply chain, digital information on food products such as farm origination details, batch numbers, factory and processing data, expiration dates, storage temperatures and shipping detail are electronically connected to each food item, and the information is entered into the blockchain along every step of the process.

The information captured in each stage in the food supply chain is agreed upon by all members of the food supply network; once there is a consensus, it becomes a permanent record that can't be altered. Each piece of information provides critical data that could potentially associate with food safety issues with the product. The records on the blockchain can also help retailers better organize the shelf-life of products in individual stores, and further strengthen safeguards related to food safety and authenticity.

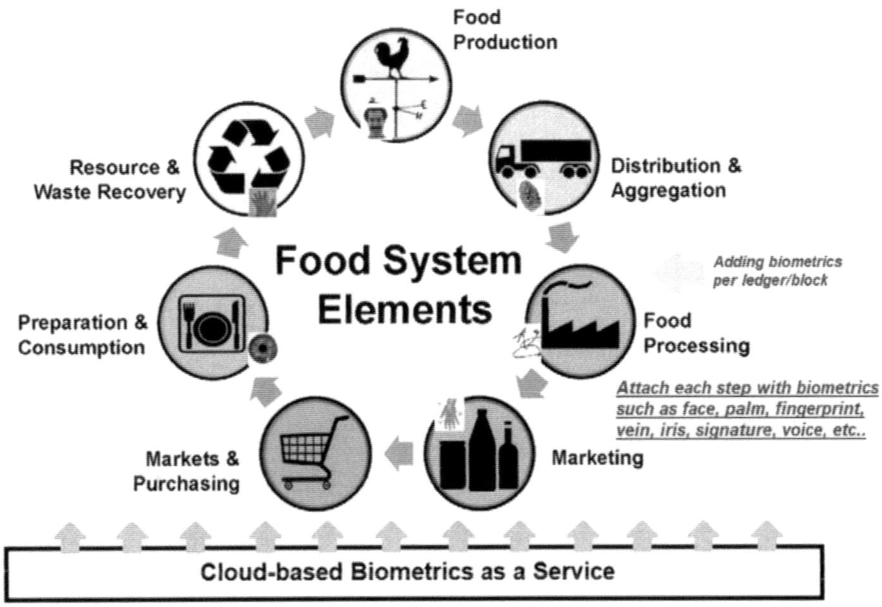

Figure 2. The proposed biometric blockchain framework for food supply chain.

Recent practical use of blockchain in food chain has been witnessed in the Walmart-IBM blockchain project. In October 2018, IBM, Walmart and Tsinghua University signed an agreement to explore food supply chain traceability and authenticity using blockchain technology in Chinese market. This food blockchain project comes as Walmart announced its new Food Safety Collaboration Center in Beijing. It is the first time to fulfil the goal in such a technology way to record how food is tracked, transported and sold to consumers across China harnessing the power of blockchain technology designed to generate transparency and efficiency in monitoring the food supply chain.

The system is yet far more restricted in scale at this initial stage. So far it is only applied to those in Walmart's leafy green food supply chain, which will likely extended into hundreds of users. As a result, the size of data to the data chain is easy to manage, with fewer verification nodes and far less energy costs overall.

With this technology, Walmart can safely tell stakeholders that a particular head of lettuce came from a particular harvest on a particular farm, so if a consumer gets sick from eating it, the police investigation will have a head start on the case. Rather than chasing a paper-based trail for months, they can immediately get to the source of a tainted head of cabbage within seconds, and that should mean a much secured world of our food supply system.

4. Why Biometrics is needed in Blockchain Technology

Despite the seemingly reliable and convenient services offered by the features of blockchain technology for applications, there is a host of security concerns and issues [2-10], and its understanding is relevant to the research community, governments, investors, regulators and other stakeholders.

Given the complexity and the centralised infrastructure of the blockchain, it is not a surprise that it would have some downsides especially in sharing data among different systems or regions, and securing data for a centralised infrastructure is a challenging task since potential attacks and exploits lead to a single-point-of-contact requiring trust for this individual authority. This implies that more research effort needs to be done in order to assure that information are secured in terms of privacy, assuring that only authorized users are able to access the data.

To address these challenges, recent research [11] has started to think about bring biometrics [12, 13] into blockchain with a hope to achieve better security, scalability and privacy. Based on this initiative we proposed a new blockchain framework, namely biometric blockchain (BBC) for food logistics. Fig.2 shows our proposed framework, while each logistic step is associated with a biometric record that can be verifiable over the Biometric-as-a-Service on the cloud.

The benefits of such a BBC-based framework are apparent. First, we can easily pin down the key responsible person in each step, and make the management more transparent, controllable, and safer. Second, the food supply chain could be more robust to any attacks, such as tracing down any unknown inclusion of unwanted gradients in the food logistics. Hence, BBC as proposed in Fig.2 could be a valuable solution for practical applications in food logistics.

5. Privacy-Issues in Biometric Blockchain

Biometrics could secure data, and on the other side it is also sensitive to expose one's privacy. The deployment of BBC based food supply chain framework will inevitably request the biometric information from individuals, such as food makers or delivery drivers. Hence, BBC obvious needs a privacy-protected mechanism when biometric information is collected during food logistics.

Recent research has enlightened the privacy issue with a nice solution by using encrypted biometrics [14, 15, 16]. As shown in Fig.3, a set of biometric features such as faces can be encrypted and combined into a ledger as an encrypted signature. The modern technology [14, 15, 16] has revealed that we do not need to decrypt these biometric information and can directly verify it in its encrypted domain, making the use of biometrics much less sensitive in term of privacy concerns when it is used for blockchain technology.

The implementation of BBC may be associated with the online biometric verification, and may be based on multimodal biometrics [17]. Classically, feature (such as LBP or SIFT) based [18, 19] biometric verification is popular though it could be a bit time-consuming. Recent approaches such as deep neural networks [20, 21, 22] are seen gradually taking over the domain of biometric verification.

There may rise concerns on the computing resources while biometrics-as-a-service is accessed via cloud platforms, as shown in Fig.2. However, in a BBC-based food logistic chain, biometric verification may happen only for necessary check, which means those biometric information could mostly be dormant and hence the requirement on extra computing could be minimized.

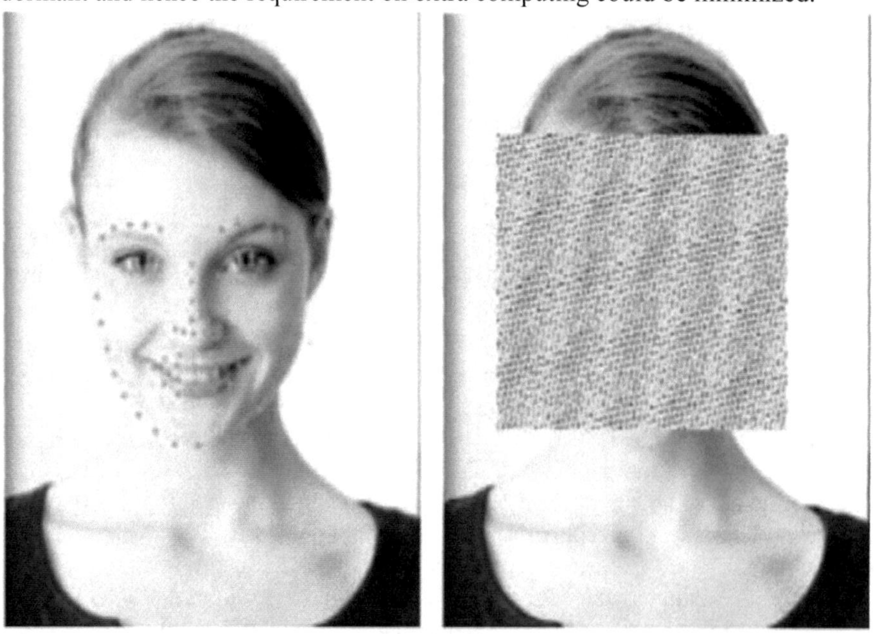

Figure 3. Privacy protection using encrypted biometrics in blockchain [14].

6. Conclusion

Blockchain emerging as a promising technology could totally change the landscape of data security in the coming years, especially for the data access over Internet-of-Things and cloud servers. However, blockchain itself does not contain any information about who owns the data and who uses the data. In this paper, other than simply encrypting data into keys, we proposed a new protocol called Biometric Blockchain (BBC), in which the biometric cues of individuals are explicitly incorporated in the BBC protocol to unambiguously identify creators or users in a blockchain-based system. Such a new protocol can particularly address the increasing needs to secure the food logistics, following the recently widely reported incident on wrongly labelled foods that caused the death of a passenger on airplane. The merits of using BBC in the food logistics are visibly clear: it can not only identify if the data or labels are authentic, but also clearly record who is responsible for the secured data or labels. Hence, such a BBC-based solution can greatly compromise the difficulty to control the critical issues in food logistics, such as food authenticity and gradient mislabelling.

References

[1] Zhong R, *et al*, 2017, Food supply chain management: systems, implementations, and future research, *Industrial Management & Data Systems*, 117(9), p.2085-2114.
[2] Crosby M et al, 2015, BlockChain Technology: Beyond Bitcoin, *Sutardja Center for Entrepreneurship & Technology Technical Report*, University of California, Berkeley.
[3] Jaikaran C, 2018, Blockchain: Background and Policy Issues, *Congressional Research Service*, Washington DC.
[4] Kakavand H, et al, 2016, The Blockchain Revolution: An Analysis of Regulation and Technology Related to Distributed Ledger Technologies, *Luther Systems & DLA Piper*.
[5] Mazonka O, et al, 2016, Blockchain: Simple Explanation, *Journal of Reference*, http://jrxv.net/x/16/chain.pdf.
[6] Tapscott D and Tapscott A, 2016, *Blockchain Revolution: How the Technology Behind Bitcoin Is Changing Money, Business and the World*, London: Portfolio Penguin.
[7] Saito K and Yamada H, 2016, What's So Different about Blockchain? Blockchain is a Probabilistic State Machine, *IEEE 36th International Conference on Distributed Computing Systems Workshops*, Nara, Japan, p.168-75.
[8] Raval S, 2016, *Decentralized Applications: Harnessing Bitcoin's Blockchain Technology*, Oreilly. ISBN 9781491924549.
[9] Bashir I, 2017, *Mastering Blockchain*, Packt Publishing Ltd.
[10] Puthal D, et al, 2018, Everything you Wanted to Know about the Blockchain, IEEE Consumer Electronics Magazine, 7(4), p.06–14.
[11] Garcia P, 2018, Biometrics on the blockchain, Biometric Technology Today, 5, p.5-7
[12] Jiang R *et al* (Editors), 2019, *Deep Biometrics*, Springer-Verlag, in Press.
[13] Jiang R *et al* (Editors), 2016, *Biometric Security and Privacy - Opportunities & Challenges in The Big Data Era*, Springer-Verlag.
[14] Jiang R, *et al*, 2017, Emotion recognition from scrambled facial images via many graph embedding, *Pattern Recognition*, 67, p.245-251.
[15] Jiang R, *et al*, 2016, Face recognition in the scrambled domain via salience-aware ensembles of many kernels, *IEEE Trans. Information Forensics and Security*, 11(8), p.1807-1817.
[16] Jiang R, *et al*, 2016, Privacy-protected facial biometric verification via fuzzy forest learning, *IEEE Trans. Fuzzy Systems*, 24(4), p.779-790.
[17] Jiang R, *et al*, 2010, Multimodal Biometric Human Recognition for Perceptual Human–Computer Interaction, *IEEE Trans. Systems, Man, & Cybernetics Part C*, 40(5), p.676.
[18] Jiang R, *et al*, 2010, Face recognition in global harmonic subspace, *IEEE Trans. Information Forensics and Security*, 5(3), p.416-424.
[19] Jiang R., *et al*, 2010, Live-cell tracking using SIFT features in DIC microscopic videos, *IEEE Tran. Biomed Eng.*, 57, p.2219–2228.

[20] Jiang Z, *et al*, 2019, Social Behavioral Phenotyping of Drosophila with a 2D-3D Hybrid CNN Framework, IEEE Access, p.1.
[21] Storey G, *et al*, 2018, Integrated Deep Model for Face Detection and Landmark Localisation from 'in the wild' Images, IEEE Access, p.2.
[22] Storey G, *et al*, 2017, Role for 2D image generated 3D face models in the rehabilitation of facial palsy, IET Healthcare Technology Letters.

Face Dynamic Modeling Based on Deep Learning and Feature Extraction

Lijing Tong, Jinqiu Yang*, Yuping Lai and Zequn Xiao

School of information, North China University of Technology, No. 5 Jinyuanzhuang Road, Shijingshan District, Beijing, 100144, People's Republic of China

*Corresponding author: 13260050901@163.com

Abstract. Facial modeling is a key step to model visual effects in special movie effects and computer games. In this paper, a method based on the combination of deep learning and feature extraction is proposed for the modeling of 3D face model. Firstly, the face region is located for the captured face image. And then, the facial feature points are extracted by the landmark algorithm and the Convolutional Neural Network (CNN) is used to classify the facial expressions. Next, a special expression 3D face model is created by the deformation of the standard 3D face model based on the facial expressions classification result. Finally, the 3D face model and the extracted facial feature points are combined to perform personalized adjustment of the 3D model to complete a 3D facial expression animation system. The experimental results show that the proposed method can effectively perform the dynamic 3D face modeling which has high reality.

Keywords. Face modeling; Feature extraction; Expression classification; CNN

1. Introduction

Facial expression recognition [1-6] and modeling techniques have broad academic and commercial values. The method of this paper is as follows: Firstly, the images in the video are pre-processed by locating the face and extracting the feature point information. Here, the landmark algorithm is used to extract facial feature points in a face. It can mark 68 feature points in a face area. These feature points can mark the face contour, eyebrow and eyes, bridge of the nose, nostril and mouth outline and other characteristic information of the profile respectively. Then, the convolutional neural network (CNN) algorithm is used to classify the facial expression according to the displacements of these feature points. At the same time, the neutral Candide-3 facial expression model library is established by manually moving the positions of the Candide-3 model vertexes. So that, the corresponding neutral Candide-3 expression model can be called according to the result of the expression classification. After that, the Radial Basis Function (RBF) interpolation algorithm is used to complete the personalized deformation on the neutral Candide-3 expression model. Finally, a facial expression animation system is obtained. The overview of the proposed method is shown in figure 1.

IOP Conf. Series: Materials Science and Engineering **646** (2019) 012010 doi:10.1088/1757-899X/646/1/012010

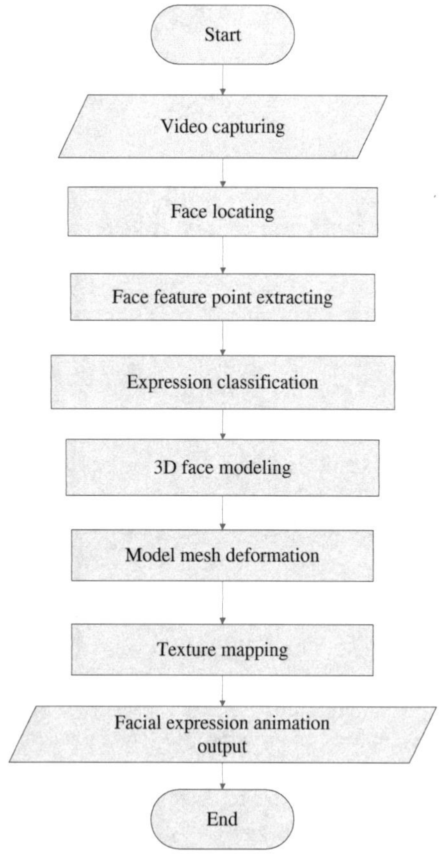

Figure 1. Overview of the proposed method.

2. Extraction of facial feature points and expression classification

2.1. Extraction of facial feature points

The landmark method has the characteristics of fast and stable performance. Therefore, the landmark method is used in this paper to extract facial feature points. In this method, a cascading regression is used to do facial feature alignment by the Ensemble of Regression Trees (ERT) [7]. The prediction of each regression is based on the prediction of the previous image.

In this paper, Let $\boldsymbol{x}_i \in R^2$ be the *x, y*-coordinates of the *ith* facial landmark in an image *I*. The facial shape is denoted by a vector $\boldsymbol{S}$. Furthermore, the vector $\boldsymbol{S} = (\boldsymbol{x}^T{}_1, \boldsymbol{x}^T{}_2, \ldots, \boldsymbol{x}^T{}_p)^T \in R^{2p}$ denotes the coordinates of all the *p* facial landmarks in image *I*. And the current estimate of $\boldsymbol{S}$ is denoted by the $\widehat{\boldsymbol{S}}^{\boldsymbol{t}}$. Each regressor, $\gamma_t(.,.)$, in the cascade predicts an update vector from the image and $\widehat{\boldsymbol{S}}^{\boldsymbol{t}}$ that is added to the current shape estimate $\widehat{\boldsymbol{S}}^{\boldsymbol{t}}$ to improve the estimate.

$$\hat{S}^{(t+1)} = \hat{s}^t + \gamma_t(I, \hat{S}^t) \quad (1)$$

Here, each regressor $\gamma_t(.,.)$, is learned using the gradient boosting tree algorithm. For the training data$(I_1,S_1),\ldots,(I_n,S_n)$, where each I_i is a face image and S_i is its shape vector, a triplets of a face image, an initial shape estimate and the target update step is created from the training data to learn the first regression function r_0 in the cascade.

Assuming the data set $\{(I_{\pi i}, \hat{S}_i^{(t)}, \Delta S_i^t)\}_{i=1}^N$ and the learning rate value *v*, $0 < v \leq 1$, $f_0(I, \hat{S}^t)$ is initialized as:

$$f_0\left(I,\hat{S}^t\right) = arg\ \min_{\gamma\in R^2P} \sum_{i=1}^{N} \left\|\Delta S_i^{(t)} - \gamma\right\|^2 \tag{2}$$

For k = 1,... K, a loop is used to calculated the $r_t\left(I,\hat{S}^t\right)$:

(1) Letting i = 1,..., N, the r_{ik} is calculated as:

$$r_{ik} = \Delta S_i^{(t)} - f_{k-1}\left(I_{\pi i}, \Delta S_i^{(t)}\right) \tag{3}$$

(2) A weak regression function $g_k\left(I,\hat{S}^t\right)$ is given to fit the target regression tree r_{ik}.

(3) Each $f_k\left(I,\hat{S}^t\right)$ result is updated from the last calculation.

$$f_k\left(I,\hat{S}^t\right) = f_{k-1}\left(I,\hat{S}^t\right) + v g_k\left(I,\hat{S}^t\right) \tag{4}$$

At last, the $r_t\left(I,\hat{S}^t\right)$ is learned as the final output result of $f_K\left(I,\hat{S}^t\right)$.

$$r_t\left(I,\hat{S}^t\right) = f_K\left(I,\hat{S}^t\right) \tag{5}$$

After the $r_t\left(I,\hat{S}^t\right)$ is learned, the expressive feature points can be extracted by the iterative calculation from the captured face image. The experimental effect of extracting facial feature points by the landmark method is shown in figure 2(a). The number of facial feature points is indexed in figure 2(b).

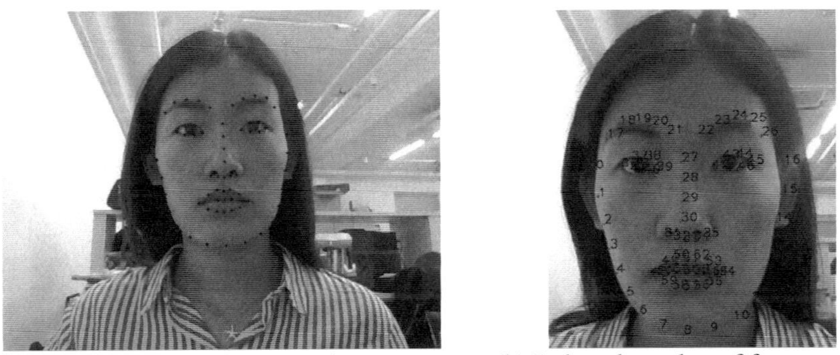

(a) Extracted feature points (b) Indexed number of feature points

Figure 2. Extracting facial feature points.

2.2. Facial expression classification

Facial expression classification is very important in a wide range of applications [8-10]. Convolutional neural network [11] is a kind of deep learning algorithm, which has become a research hotspot in the field of image understanding. In this paper, the second max pooling layer is used to reduce the number of parameters. This reduces the computational complexity of the deep learning network, while the performance reduction is very small. The overview of the deep learning network architecture is shown in figure 3. After designing of the deep learning network, data training is performed based on the network structure. Then the facial expression classification can be fulfilled based on the training result of this deep learning network.

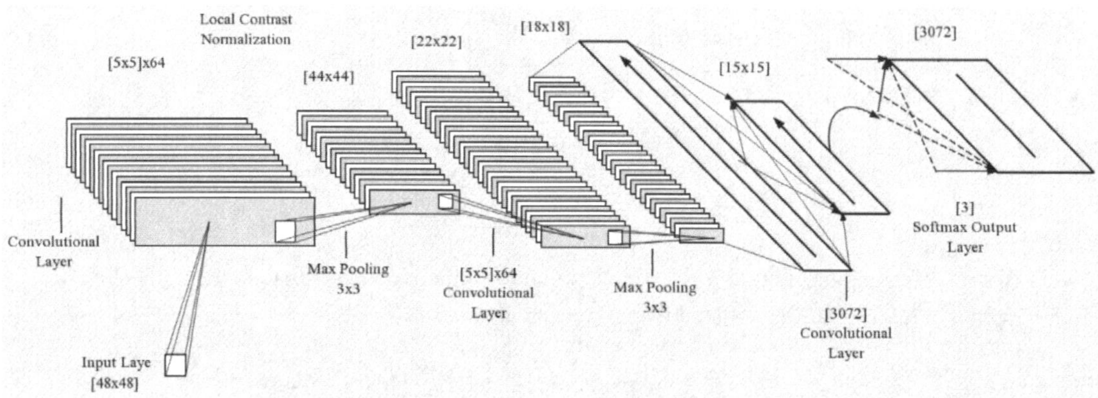

Figure 3. Overview of the deep learning network architecture.

For comparison, experiments were performed using the CNN method and the SVM method respectively. Here, three facial expressions of happy, peace and anger are categorized. Experimental results show that the CNN method can obtain more accurate classification results for expression classification. Sometimes the classification by the SVM method is inaccurate. The following is an example. Among them, figure 4 (a) are the result graphs by the CNN method. Figure 4 (b) are the result graphs by the SVM method. As shown in figure 4 (b), the actual expression is angry, but the classification result is peace. Considering that the CNN method is superior to the SVM algorithm for expression classification, the deep learning based on the CNN is used in the proposed method.

(a) Expression classification using CNN

(b) Expression classification using SVM

Figure 4. Expression classification results.

3. Dynamic modeling of 3D face

3.1. Establishment of 3D expression model library

The Candide-3 model includes 113 feature points and 168 triangular meshes. It is quite detailed to introduce information about the position of the eyes, mouth and nose, but does not include the parameter positioning of the teeth and tongue. Each vertex in the model corresponds to 3 coordinate values x, y, z, $P_i = (x_i, y_i, z_i)$. According to the statistical experiencc, three neutral Candide-3 facial expression models are established by manually moving the positions of the Candide-3 model vertexes. The three neutral Candide-3 facial expression models are shown in figure 5.

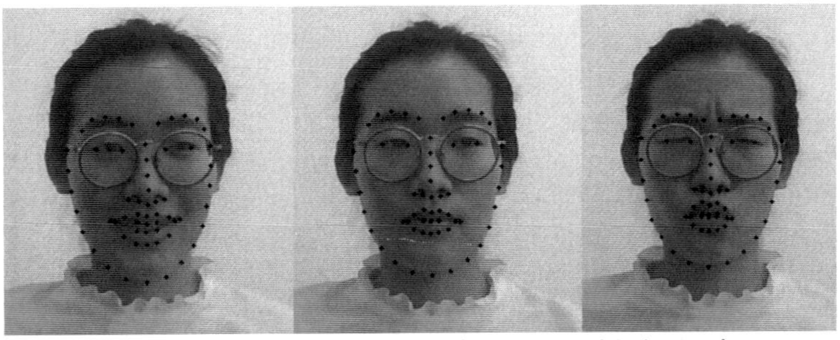

(a) Happy image (b) Peace image (c) Angry image

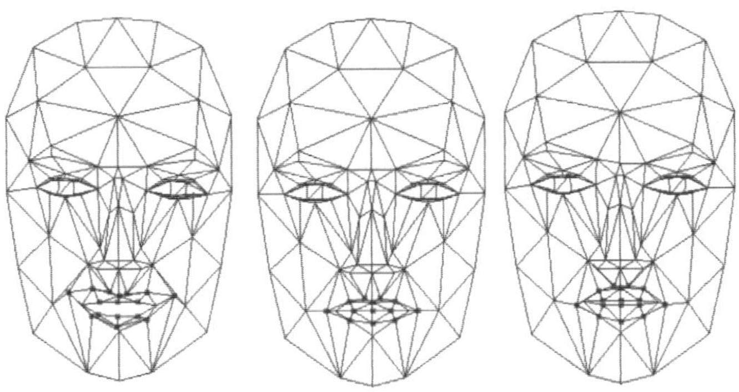

(d) Happy Candide-3 model (e) Peace Candide-3 model (f) Angry Candide-3 model

Figure 5. Candide-3 facial expression models.

3.2. 3D face personalized deformation based on RBF

In order to deform the face mesh model, some feature points should be extracted out from the captured image and their displacements between the positions in the image and the in the neutral expression model should be calculated firstly. Then, the positions of the n feature vertexes in the neutral expression model are moved according to these displacements. Next, the positions of the m non-feature vertexes in the neutral expression mode are moved according to the calculation results of the Radial Basis Function (RBF) based on a spatial interpolation function.

In theory, any function that satisfies the requirements of radial symmetry can be used as an interpolation basis function. The general choices are Gaussian function, Hardy quadratic function, and thin-spline function and so on. In fact, the Gaussian function is used in data interpolation algorithm more widely. And it is reliable for data interpolating in the high-dimensional space. Hence, the Gaussian function in equation (6) are used in the proposed method.

$$g_{(r)} = e^{-(\frac{r}{a})^2} \tag{6}$$

In equation (6), r is the radius of influence. The Gaussian function $g(r)$ will gradually approach 0 with the increasing of r.

The equation (7) is the radial basis function for calculating the displacements of those non-feature vertexes in the neutral expression model.

$$f(\vec{x}) = \sum_{j=1}^{n} c_j g(\|\vec{x} - \vec{x_j}\|) \tag{7}$$

In equation (7), $\vec{x}$ is the original position of a non-feature vertex in the neutral expression model, *g()* is the Gaussian function, c_j is the coefficient of the Gaussian function *g()*, $\vec{x_j}$ is the known original feature vertex position, and n is the total number of the feature vertexes. c_j is calibrated by the known positions of those feature vertexes before and after an expression in the Candide-3 model.

The radial basis function network is a three-layer forward feedback network [10]. The input layer is consisted of an m-dimensional input vector. The intermediate layer is consists of the calculated Gaussian function values. And the output layer is consists of n linear elements.

The calibration process of c_j is as follows.

For the n feature vertexes, assuming the displacement before and after the movement in the neutral expression model is Δp_i, $\Delta p_i = p_i' - p_i = f(p_i)\ \ (1 \le i \le n)$,

$$\Delta p_i = \sum\nolimits_{j=1}^{n} c_j \phi(\|p_i - p_j\|) + Mp_i + t, \quad (1 \le i \le n) \ . \tag{8}$$

And the affine transformation should satisfy these constraints:

$$\begin{cases} \sum_{j=1}^{n} c_j = 0 \\ \sum_{j=1}^{n} (c_j \cdot p_j) = 0 \end{cases} \tag{9}$$

The above two equations can be combined to obtain a linear equation as equation (10).

$$\begin{bmatrix} \phi_{11} & \phi_{12} & \cdots & \phi_{1n} & P_{1x} & P_{1y} & P_{1z} & 1 \\ \phi_{21} & \phi_{22} & \cdots & \phi_{2n} & P_{2x} & P_{2y} & P_{2z} & 1 \\ \cdots & \cdots & \cdots & \cdots & \cdots & \cdots & \cdots & \cdots \\ \phi_{n1} & \phi_{n2} & \cdots & \phi_{nn} & P_{nx} & P_{ny} & P_{nz} & 1 \\ P_{1x} & P_{2x} & \cdots & P_{nx} & 0 & 0 & 0 & 0 \\ P_{1y} & P_{2y} & \cdots & P_{ny} & 0 & 0 & 0 & 0 \\ P_{1z} & P_{2z} & \cdots & P_{nz} & 0 & 0 & 0 & 0 \\ 1 & 1 & \cdots & 1 & 0 & 0 & 0 & 0 \end{bmatrix} \begin{bmatrix} c_1 \\ c_2 \\ \cdots \\ c_n \\ \\ M \\ \\ t \end{bmatrix} = \begin{bmatrix} \Delta p_1 \\ \Delta p_2 \\ \cdots \\ \Delta p_n \\ 0 \\ 0 \\ 0 \\ 0 \end{bmatrix} \tag{10}$$

In equation (8) and (10), $\varphi_{ij} = \Phi(\|p_i - p_j\|) = g(\|p_i - p_j\|)\ \ (1 \le i, j \le n)$ is the Gaussian function. $(P_{ix}, P_{iy}, P_{iz})\ \ (1 \le i \le n)$ are the coordinates of the feature vertex p_i. The coefficients c_j of the Gaussian functions, the affine parts M and t can be solved by the linear system of equation (10).

After the coefficients c_j of the Gaussian functions, the affine parts M and t are solved, the corresponding displacements of all non-feature vertexes can be calculated as equation (8). Binding textures can make the model look more realistic. The OpenGL binding texture functions are used here for the reality of the 3D face personalized graphics.

4. Experimental results and analysis

4.1. Dynamic modeling experiment results

A 3D facial expression animation processing system is implemented using the method proposed in this paper. Figure 6 shows the results of the system. The corresponding 3D face model can be reconstructed with the operator's expression changing. It can be seen that the method proposed in this paper can provide a realistic 3D face model.

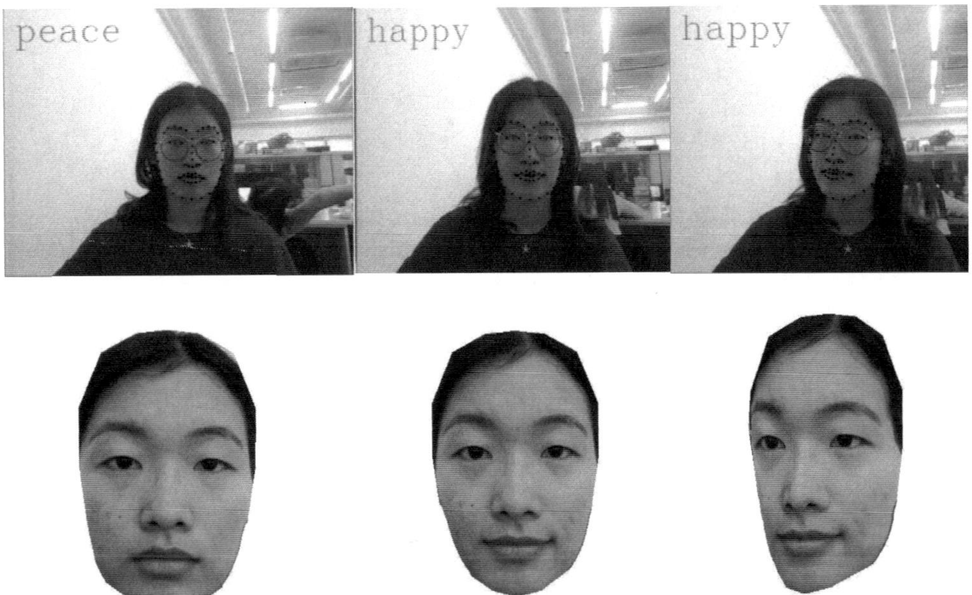

(a) Peace expression (b) Happy expression (c) Happy expression (Rotate 30° to the right)

Figure 6. 3D facial expression animation processing system

4.2. Time performance experiment results

In order to further evaluate the time performance, the run times of different periods are tested during the running of the 3D facial expression animation processing system. The system is implemented in C++ and run on a PC with an Intel(R) Core(TM) i7-4770 CPU @3.40GHz and 4G memory. The graphics card is Intel(R) HD Graphics 4600. The whole processing is divided into 4 periods. They are the face detection + feature extraction, classification + emotion detection, RBF computing, and vertex position updating + model rendering. Table 1 shows the running times of each period in the system processing. The time for RBF computing is a little longer. However, it can be done offline for the facial expression animation generation.

Table 1. Time performance experiment result (sec).

Type	Time taken/proposed method
Face detection + Feature extraction	0.066
Classification + Emotion detection	0.026
RBF Computing	4338
Vertex position updating + Model rendering	0.088

4.3. Comparison of face modeling

In order to verify the effectiveness of the proposed algorithm based on facial expression classification and feature extraction, a comparative experiment on modeling effect is carried out in this paper. In the same experimental environment, for different expressions (happiness, peace, angry respectively), the experimental results of Candide-3 deformation model by the method proposed in this paper is compared with the experimental results of face models invoked from the 3D expression model library.

The experimental results show that the proposed method is more realistic. The movements of the vertexes in the Candide-3 model is more accurate.

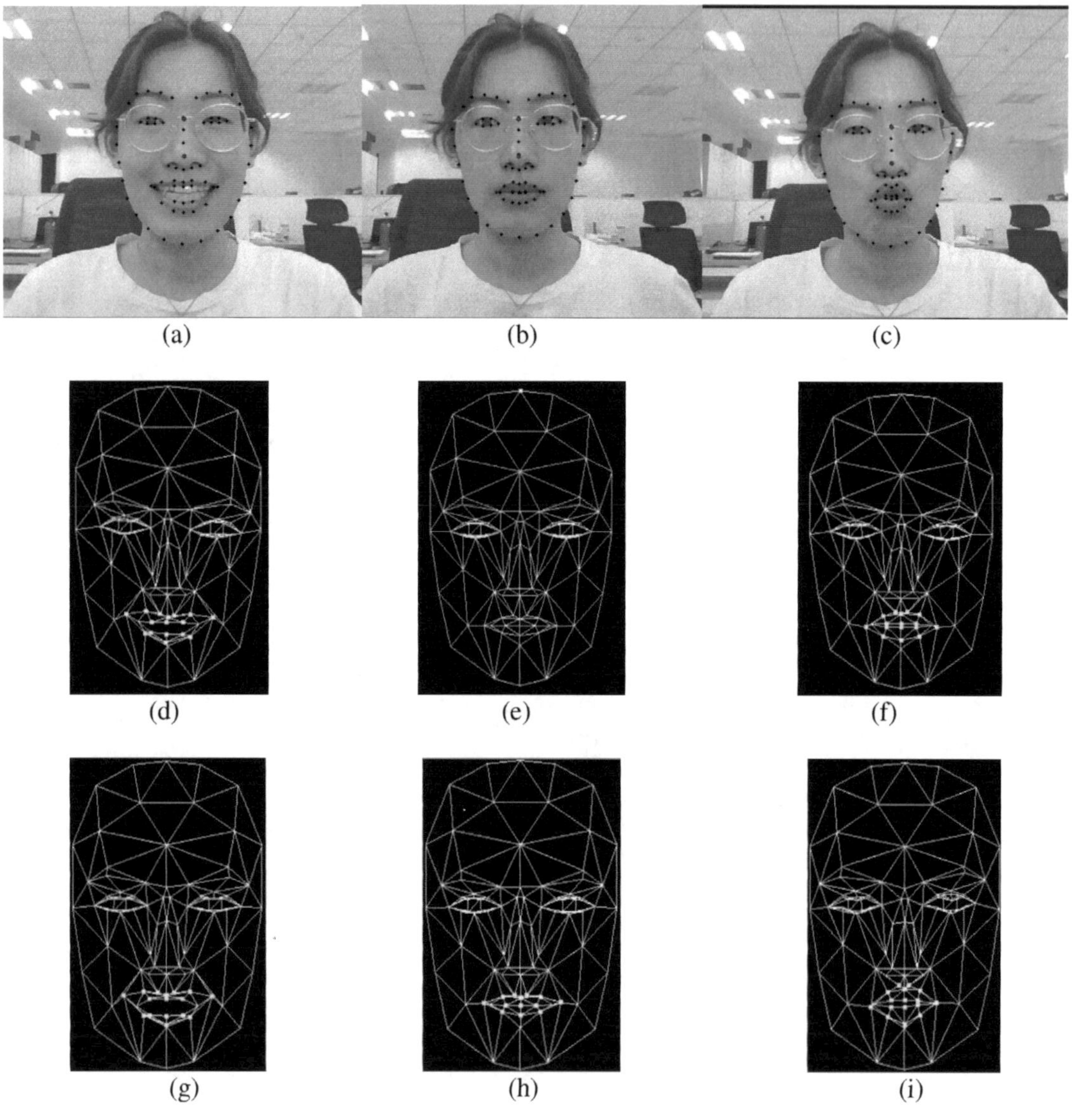

Figure 7. Comparison of modeling effects.

As shown in Figure 7, sub-figures (a) ~ (c) show the expression images of happiness, peace, and angry. Sub-figures (d) ~ (f) represent the corresponding modeling results invoked from the 3D expression model library. Sub-figures (g) ~ (i) present the modeling results by the method proposed in this paper.

5. Conclusion

This paper proposes a method combining deep learning and feature extraction for face dynamic modeling. After the face feature points are extracted and the facial expression is classified by the CNN, the 3D face model is personalized re-constructed based on the 3D neutral face model to complete a 3D facial expression animation system. The experimental results show that the proposed method can more accurately re-construct the face models with the different expressions.

Acknowledgments
The authors would like to acknowledge the support from the National Natural Science Foundation of China (61371142) and the Beijing Natural Science Foundation (4194076, 4182018).

References

[1] Y. Li, Y. Zhao, H. Yao, and Q. Ji, "Learning a discriminative dictionary for facial expression recognition," 2015 International Conference on Affective Computing and Intelligent Interaction (ACII), IEEE, pp. 838-844, 2015.

[2] P. Cui, and T. Yan "A SVM-Based Feature Extraction for Face Recognition," International Conference of Pioneering Computer Scientists, Engineers and Educators, pp. 120-126, 2015.

[3] P. Sarakon, T. Charoenpong, and S. Charoensiriwath, "Face shape classification from 3D human data by using SVM," The 7th 2014 Biomedical Engineering International Conference, pp. 1-5, 2014.

[4] HS. Devi, R. Laishram, and DM. Thounaojam, "Face recognition using R-KDA with non-linear SVM for multi-view database," Procedia Computer Science, vol. 54, pp. 532-541, 2015.

[5] QQ. Tao, S. Zhan, XH. Li, and T Kurihara, "Robust face detection using local CNN and SVM based on kernel combination," Neurocomputing , vol. 211, pp. 98-105, 2016.

[6] M. H. Selamat and H. Md Rais, "Image face recognition using Hybrid Multiclass SVM (HM-SVM)," 2015 International Conference on Computer, Control, Informatics and its Applications (IC3INA), pp. 159-164, 2015.

[7] V. Kazemi, and J. Sullivan, "One millisecond face alignment with an ensemble of regression trees," Proceedings of the IEEE conference on computer vision and pattern recognition, pp. 1867-1874, 2014.

[8] Z. Xie, F. Yang, Y. Zhang, C. Wu, "Face sketch recognition based on edge enhancement via deep learning," LIDAR Imaging Detection and Target Recognition 2017, vol. 10605, pp.106053N, 2017

[9] RK. Yadav, and AK. Sachan, "Application expansion inside optimized RBF kernel of SVM in robust bace recognition system," International Journal of Signal Processing, Image Processing and Pattern Recognition, vol. 8, no. 12, pp. 89-98, 2015.

[10] XH. Wang, C. Xia, M. Hu, and FJ. Ren, "Facial expression recognition under partial occlusion based on fusion of global and local features," Ninth International Conference on Graphic and Image Processing (ICGIP 2017), vol. 10615, 2018.

[11] A. Krizhevsky, I. Sutskever, and GE. Hinton, "ImageNet classification with deep convolutional neural networks," Communications of the ACM, vol. 60, no. 6, pp. 84-90, 2017.

Design and Implementation of Software Simulation System for Dental Orthodontic Robot

Chuanhong Zhou [1,a], Shiyu Pan[1,b] and Tao Zhou[1,c]

[1]School of Mechatronic Engineering and Automation, Shanghai Key Laboratory of Intelligent Manufacturing and Robotics Shanghai University, Shanghai, China

E-mail: [a]13817297415@139.com; [b]15151963013@163.com; [c]1358533088@qq.com

Abstract. With the development of society and the improvement of living standards, people's pursuit of their own beautiful appearance is getting higher and higher, especially the passion for beautiful features. Therefore, the demand for dental orthodontic surgery is growing in daily life. However, due to the limitations of oral physiological structure, the orthodontic brackets designed in traditional orthodontic surgery are often not the optimal dental arch shape for patients. Therefore, a dental orthodontic software system that can interactively simulate a patient's oral model is particularly important. In this paper, a software system of dental orthodontic robot which can simulate dental orthodontics was developed by using Blender secondary development technology combined with orthodontics and tooth extraction algorithm.

Keywords: dental orthodontics, dental robot, Blender, secondary development

1. Introduction

The main research task of orthodontics is to investigate the etiology of malocclusion and then to diagnose and prevent it. The orthodontic process is a process in which the dentist and the orthodontists cooperate to arrange the irregular teeth through the brackets [1]. The correction of dentition malformation mainly relies on the use of an internal or external orthodontic device to exert appropriate "biological force" on the teeth, alveolar bone and jaw bone to generate physiological movement, so as to correct the malformation [2]. The software system of dental orthodontic robot is a combination of orthodontic technology and computer-aided technology, which transforms orthodontic surgery into a simulation operation that can be interactively simulated in a computer.

The software system of the dental orthodontic robot studied in this paper is realized on Blender platform [3], which is further developed by using Python programming language. The simulated operation of orthodontics is performed in the software system studied in this paper, and then the data is transmitted to the robot system. The robot system uses ABB robot, model IRB 140TW. The communication between the software system and the robot system uses the Socket message communication method. The robot system adopts ABB robot, model IRB 140TW.Socket message is adopted to communicate between the software system and the robot system[4][5].

2. The overall structure of the dental orthodontic robot software system

First, we need to import the CT scan models of the patient's oral cavity into the software system, and then rearrange the patient's malformed dentition, which requires the use of the tooth arrangement algorithm. After that, the bracket model is imported from the standard bracket database. Finally, an appliance (braces) suitable for the patient's dental arch is designed according to the arranged dentition.

Therefore, the system structure diagram of the software system is shown in Figure 1. Figure 2 is a software system sidebar function button diagram.

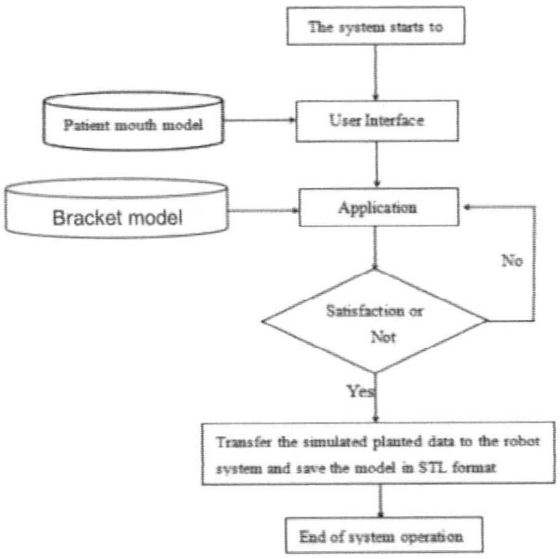

Figure 1. Software system structure diagram

Figure 2. Software system sidebar function button diagram

3. Orthodontic tooth arrangement algorithm

3.1. Acquisition of dental arch curve

Dental arch curve refers to the arched parabola formed by the continuous arrangement of teeth in the upper and lower jaw bone on the alveolar bone [6]. It is a key step to obtain ideal dental arch curve and restore abnormal dental arch. There are many mathematical models of dental arch curve. By comparison, the β- function model is the closest mathematical model to the real tooth arch curve of human body, and the calculation is simple and stable.

The general formula of the β-function model is:

$$Y = 3.0314 \times D \times \left[\frac{X}{W} + \frac{1}{2}\right]^{0.8} \left[\frac{1}{2} - \frac{X}{W}\right]^{0.8} \tag{1}$$

The general form can be simplified to:

$$Y = D \times \left[1 - (\frac{2X}{W})^2\right]^e \tag{2}$$

Among them, W is the width of the third molar arch, that is, the distance between the distal and middle cheek points of the third molar on both sides of the mouth. D is the depth of the third molar arch, that is, the distance between the incisor and the third molar line on both sides of the mouth. The position-length relationship is shown in Figure 3.

It can be seen from the above that when the values of W and D are determined, the point $((X_2 - X_1)/2, (Y_2 - Y_1)/2)$ is substituted into the above equation, and the value of e can be obtained. Figure 4 shows the effect of extracting the arch curve using the β-function model algorithm in the software system.

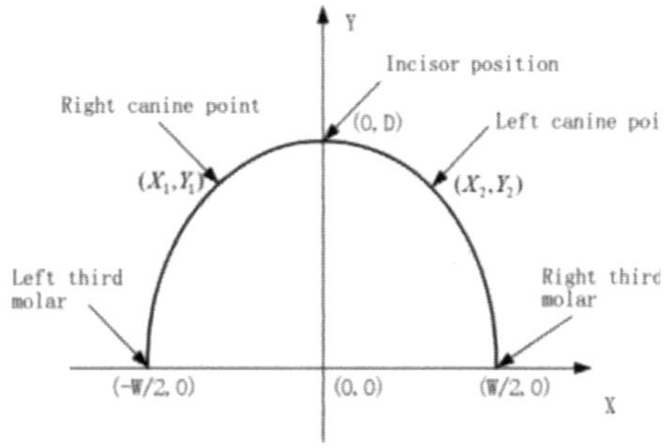

Figure 3. Dental arch curve

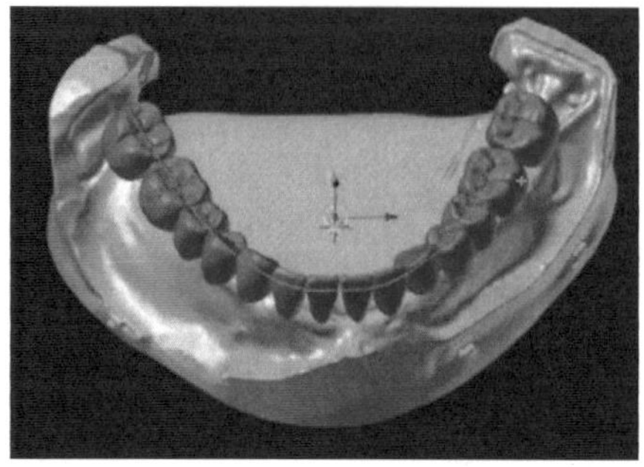

Figure 4. Extraction effect of dental arch curve

3.2. Calculation of tooth position

3.2.1. Calculation of tooth position coordinates. According to the physiological structure of the mouth, the coordinate system is established as shown in figure 5 below. Among them, the Z_0 axis is a straight line perpendicular to the maxillofacial surface, and the upward direction is positive. The intersection point of Z_0 and the maxillofacial is O_0, and the X_0-axis passes through the incisor and is parallel to the line of the third molars on both sides. The right of the observer is positive. The Y_0-axis is perpendicular to the X_0-axis, and the direction is shown in the figure. L is the dental arch curve.

After the reference coordinate system is established, we see the teeth as a rectangular parallelepiped arranged closely in sequence on the arch curve, as shown in Figure 6.

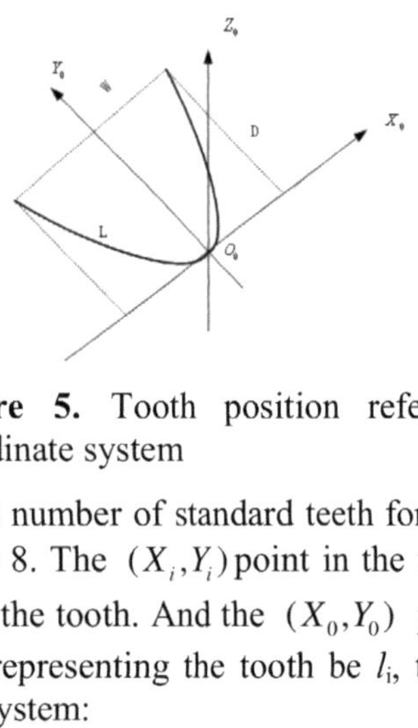

Figure 5. Tooth position reference coordinate system

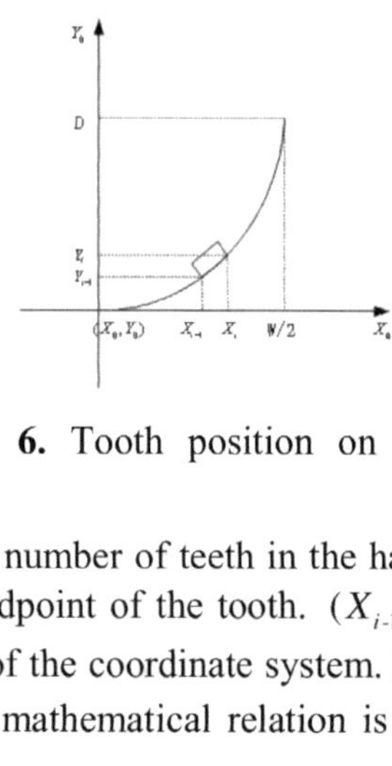

Figure 6. Tooth position on the arch curve

Since the number of standard teeth for humans is 16, the number of teeth in the half arch curve in the figure is 8. The (X_i,Y_i) point in the figure is the far midpoint of the tooth. $(X_{i\text{-}1},Y_{i\text{-}1})$ is the near midpoint of the tooth. And the (X_0,Y_0) point is the origin of the coordinate system. Let the width of the cuboid representing the tooth be l_i, then the following mathematical relation is satisfied in this coordinate system:

$$(Y_i - Y_{i\text{-}1})^2 + (X_i - X_{i\text{-}1})^2 - l_i^2 = 0 \tag{3}$$

Since both points of the cuboid are on the dental arch curve, the following mathematical relation can be obtained:

$$Y_{i\text{-}1} = 3.0314 \times D \times \left[\frac{X_{i\text{-}1}}{W} + \frac{1}{2}\right]^{0.8} \left[\frac{1}{2} - \frac{X_{i\text{-}1}}{W}\right]^{0.8} \tag{4}$$

$$Y_i = 3.0314 \times D \times \left[\frac{X_i}{W} + \frac{1}{2}\right]^{0.8} \left[\frac{1}{2} - \frac{X_i}{W}\right]^{0.8} \tag{5}$$

IOP Conf. Series: Materials Science and Engineering **646** (2019) 012011 doi:10.1088/1757-899X/646/1/012011

Since (X_0,Y_0) is the origin of the coordinate system, X_0=0, Y_0=0.

By using the above conditions and formula and the iterative calculation method, the point (X_i,Y_i) (i=0,1,2,3……8) which is the position coordinate of the tooth on the curve of the dental arch.

3.2.2. Determination of tooth Angle. In order to show the angular attitude of a single tooth on the arch curve, we need to establish the single tooth coordinate system OXYZ as shown in Figure 7. In the figure, the Z axis is located on the plane of the arch curve and is tangent to the arch curve. The X axis is on the curve of the arch curve and perpendicular to the arch curve. The Y axis is perpendicular to the plane of the arch curve, so the angle relationship between the direction of a single tooth and the curve of the dental arch is shown in Figure 7.

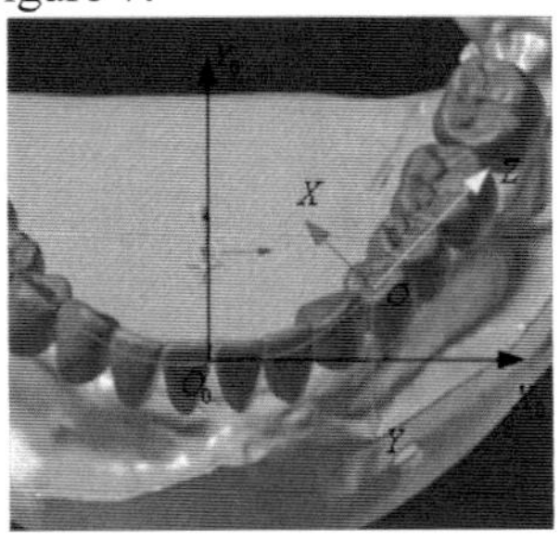

Figure 7. Single tooth angle coordinate system

3.2.3. Application of AABB(axis-aligned bounding box). After determining the position and angle of the teeth, each tooth was need to move to the ideal position on the dental arch curve. However, the teeth will collide with each other in the process of moving the teeth. In order to solve this problem, bounding box was used in this paper[7]. The basic idea of the bounding box technique is to enclose a complex geometric object in a virtual scene with a simple geometric shape (that is, a bounding box), which can be increasingly approached to a real object by constructing a tree structure [8].AABB (Axis-Aligned Bounding Boxes) is the bounding box along the coordinate axis and the geometric object contained is the smallest hexahedron whose sides are parallel to the coordinate axis. During constructing, the maximum and minimum values in the X, Y, and Z directions of the coordinate axes can be obtained according to the shape of the object to determine the highest and lowest boundary points of the bounding box [9], which can be expressed as:

$$R = \{(x,y,x) | a_x \le x \le b_x, a_y \le y \le b_y, a_z \le z \le b_z\} \quad (6)$$

Where, a_x,b_x,a_y,b_y,a_z,b_z represent the minimum and maximum coordinates projected by AABB on X, Y and Z axes respectively.

3.2.4. Tooth movement. After the position and angle of the tooth on the arch curve are obtained, the next step is to move the tooth. In the tooth position coordinate system, when the tooth is moved, the tooth will have a certain distance compensation in the z-axis direction. Let the distance compensation in the Z-axis direction be Z_i, then the translation coordinate of the tooth is (X_i,Y_i,Z_i) , and its transformation matrix is T_1, then:

$$T_1 = \begin{bmatrix} 1 & 0 & 0 & 0 \\ 0 & 1 & 0 & 0 \\ 0 & 0 & 1 & 0 \\ X_i & Y_i & Z_i & 1 \end{bmatrix} \quad (7)$$

According to the teeth arrangement rule, let the rotation angle of the tooth around X, Y and Z axes during its movement be a_i,b_i,c_i respectively, then the corresponding rotation transformation matrix is:

$$T_2=\begin{bmatrix}1 & 0 & 0 & 0\\ 0 & \cos a_i & \sin a_i & 0\\ 0 & -\sin a_i & \cos a_i & 0\\ 0 & 0 & 0 & 1\end{bmatrix} \tag{8}$$

$$T_3=\begin{bmatrix}\cos b_i & 0 & -\sin b_i & 0\\ 0 & 1 & 0 & 0\\ \sin b_i & 0 & \cos b_i & 0\\ 0 & 0 & 0 & 1\end{bmatrix} \tag{9}$$

$$T_4=\begin{bmatrix}\cos c_i & \sin c_i & 0 & 0\\ -\sin c_i & \cos c_i & 0 & 0\\ 0 & 0 & 1 & 0\\ 0 & 0 & 0 & 1\end{bmatrix} \tag{10}$$

In the process of moving the tooth to the ideal position on the arch curve, the final translational rotation of the tooth in the tooth reference coordinate system is:

$$T = T_1T_2T_3T_4 \tag{11}$$

3.3. Implementation of tooth arrangement algorithm

The flow chart of the tooth arrangement algorithm is shown in Figure 8:

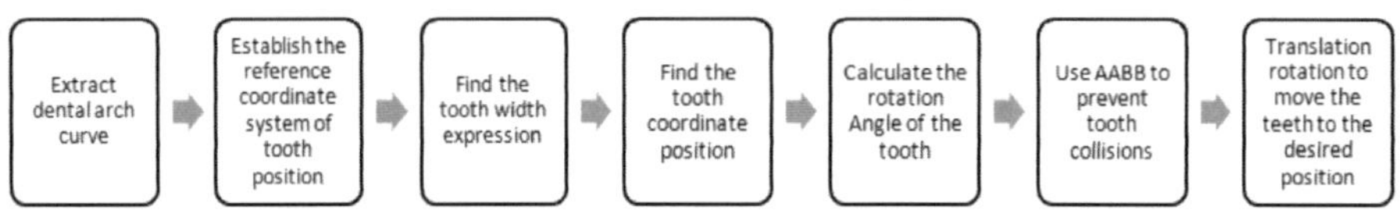

Figure 8. Flow chart of tooth arrangement algorithm

Figure 9 shows the effect of the tooth arrangement algorithm.

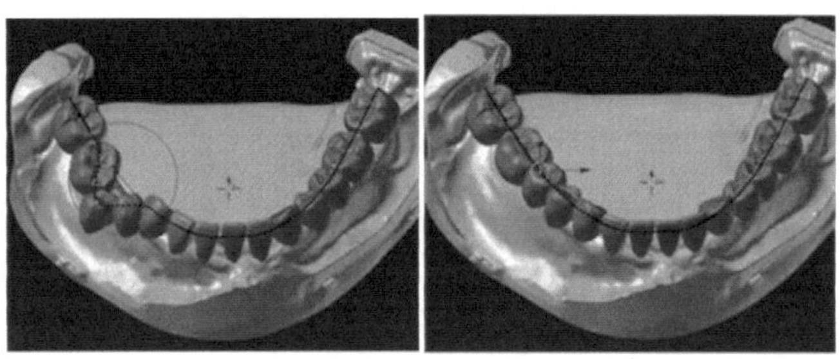

Figure 9. The comparison of the teeth before and after the implementation of the tooth arrangement algorithm

4. Establishment of bracket model library

For the establishment of the standard bracket database, we first need to carry out three-dimensional scanning on the physical model of the standard bracket provided by the hospital to generate point cloud data, and then reconstruct the geometric model to generate a three-dimensional model [10] to establish the bracket model database. Figure 10 shows the model library of orthodontic bracket established in the software system. Figure 11 shows the effect of importing a single bracket into the software system.

Figure 10. Orthodontic bracket model library

Figure 11. A single bracket

5. Conclusions
In this paper, a software system of dental orthodontic robot is developed by using Blender's secondary development technology. This software system can simulate the process of orthodontic process and design orthodontic brackets suitable for patients with dentition malformation. The software system is simple and easy to use, and the user can easily get started. The development and implementation of the software system solves the difficulties of traditional orthodontic surgery in the past and has made certain contributions to the domestic dental repair assistant software system.

Reference
[1] Song Dongqing. On the orthodontics and misunderstanding of the teeth[J].*Journal of Clinical Medicine and Archives*,2017,**4**(88):17418.
[2] Zang Zhuo. The clinical effect of invisible orthodontic treatment in orthodontic treatment[J].*Sichuan Journal of Physiological Sciences*, 2018,**40**(04):283-285.
[3] Hao Zhenhua. Research on Dynamic Simulation Application of Python in Blender Engine[J].*Software Guide*, 2012, **11**(11): 66-68.
[4] Sun Zhiwei. Research on ABB industrial robot network integrated control system[D]. *Yanshan University*, 2017.
[5] Zhao Y. Research and implementation of message queue middleware based on Socket [D]. *Inner Mongolia University*, 2007.
[6] Lu Yuxi. Extraction algorithm of dental arch surface in dental implant surgery [D]. *Shanghai Jiaotong University*, 2007.
[7] Deng Junsheng, Mao Shifeng, Liu Xufeng and Ye Minyou. Collision Detection Algorithm for Fusion Pile Deformation Parts Based on AABB Tree[J]. *Computer Systems*, 2018, **27**(11): 161-167.
[8] Stephen J Adelson, Larry F Hodges Generating exact ray-traced animation on frames by reprojection [J]. *IEEE Computer Graphics and Applications* 1995, **15**(3): 43-52.
[9] Lin Haibo, Wang Wei and Yang Huilin.An improved algorithm based on AABB bounding box method[J].*Tractors & Farming Vehicles*,2008(03):69-70+73.
[10] Song Yali, Li Jia and Huang Tian. Establishment of Standard Crown Database in Denture CAD/CAM System[J]. *Combined Machine Tool and Automatic Machining Technology*, 2003(12): 40-41+44.

Human Motion Sequence Recognition Based on Feature Selection and Support Vector Machine

Yu Yunlei[1], Wang Mei[1,*], Lin Limeng[1] and Zhang Chen[1]

[1]College of Mechatronics Engineering and Automation, Shanghai University, Shanghai, 200444, China

*Corresponding author's e-mail: mwang@shu.edu.cn

Abstract. Aiming at the problem of human motion sequence recognition, algorithm based on feature selection and support vector machine is proposed. Firstly, the feature extraction of human motion sequences is obtained by key frame and human joint angle calculation. Then, based on the Pearson correlation coefficient and CFS evaluation function, the algorithm of relevance feature selection is used to search the optimal feature subset from the original feature set. By reducing the dimension of the feature set, the difficulty of classification recognition is reduced. In the classification process, the support vector machine is used as the classifier to complete the recognition task of the human motion sequence. Through the recognition experiment and the contrast experiment, the effectiveness of the recognition algorithm based on feature selection and support vector machine is proved.

1. Introduction
The main content of motion recognition is to identify human body movements with specific algorithms and these body movements are detected and captured by specific devices and methods. Through the analysis, integration and extraction of the captured information, effective features can be found for efficient classification. The purpose of motion feature extraction is to extract the information related to human motion from the original image, including silhouette, optical flow, gradient, spatial and temporal characteristics, depth characteristics [1-7], etc. Motion recognition algorithms are mainly divided into three categories: probability-based methods, semantic-based methods and template-based methods. Statistical methods include hidden markov model (HMM), dynamic bayesian network (DBN), etc. Template-based methods include template matching [8], dynamic time ordering [9], etc. In this paper, the content will be introduced from the aspects of feature extraction, classification algorithm implementation. And classification experiment will be done for verification.

2. Arm motion feature extraction based on joint Angle calculation
Since the angles of human joint are corresponding to the spatial information of human body, they can reflect human body movements with reduced dimension of feature space. Then the human joint Angle was selected to construct the feature vector.

All joint data trained and tested in this paper are obtained through Kinect sensor and skeleton tracking in OpenNI open source library. According to the requirements of feature extraction, torso (P_t), left shoulder(P_l), right shoulder (P_r), left elbow (P_e) and left wrist (P_w), shown in figure 1, are selected as the key points to analyze human arm movements.

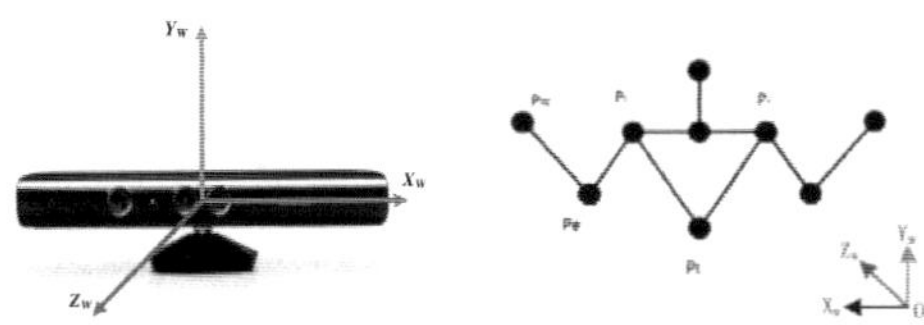

Figure 1. Kinect coordinate system and human body nodes

The coordinate system C_W of Kinect sensor (Figure 1) is used as the reference to collect real-time data of human joint movement. In order to improve the efficiency of angle calculation, the shoulder coordinate system C_S is defined here. For example, the shoulder coordinate system C_S of the left arm is shown in Figure 2. The coordinate of joint points, detected by Kinect, can be transformed into the shoulder coordinate system C_S for joint angle calculation.

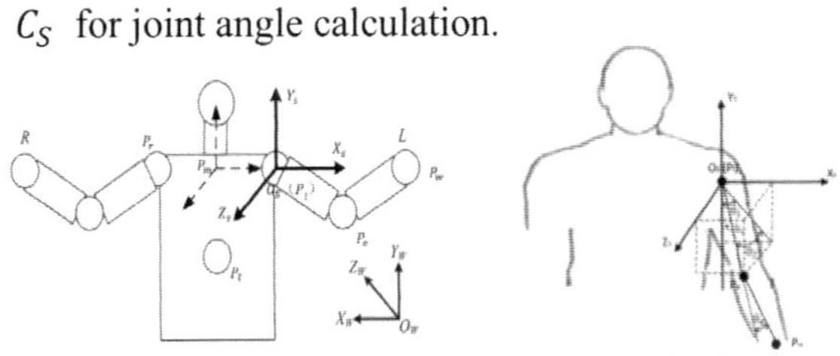

Figure 2. Shoulder coordinate system and joint angle

Several joint angles are defined to reflect the arm movement. θ_1 is defined as horizontally abducted of the shoulder joint. θ_2 is the flexion of the shoulder joint. θ_3 is opposite to the rotation of the arm for the shoulder joint. While θ_4 is the flexion extension of the elbow joint (the angle between the upper arm and the lower arm). As shown in Figure 3, θ_3 is calculating in elbow coordinate system.

The joint angles can be obtained:

$$\theta_1 = arctan\frac{\overrightarrow{p_l p_e}\cdot\overrightarrow{ox}}{\overrightarrow{p_l p_e}\cdot\overrightarrow{oy}}, (\overrightarrow{ox} = \frac{\overrightarrow{p_r p_l}}{|\overrightarrow{p_r p_l}|}, \overrightarrow{oy} = \frac{\overrightarrow{p_t p_m}}{|\overrightarrow{p_t p_m}|}) \quad (1)$$

$$\theta_2 = arctan\frac{\overrightarrow{p_l p_e}\cdot\overrightarrow{oz}}{\overrightarrow{p_l p_e}\cdot\overrightarrow{oy}}, (\overrightarrow{oz} = \overrightarrow{ox}\times\overrightarrow{oy}) \quad (2)$$

$$\theta_4 = \arctan\frac{\overrightarrow{p_l p_e}\cdot\overrightarrow{p_e p_w}}{|\overrightarrow{p_l p_e}|\cdot|\overrightarrow{p_e p_w}|} \quad (3)$$

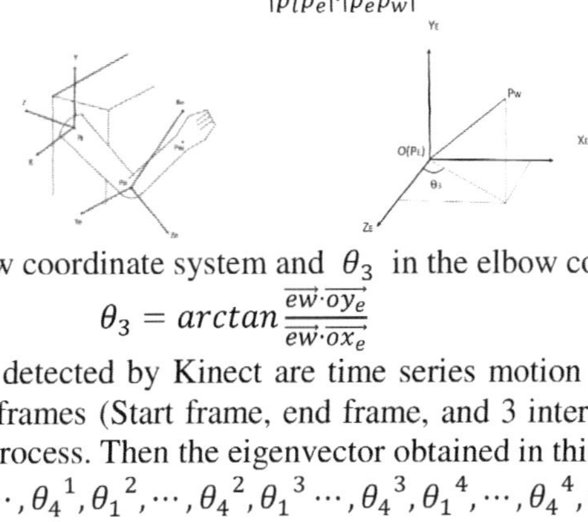

Figure 3. Elbow coordinate system and θ_3 in the elbow coordinate system

$$\theta_3 = arctan\frac{\overrightarrow{ew}\cdot\overrightarrow{oy_e}}{\overrightarrow{ew}\cdot\overrightarrow{ox_e}} \quad (4)$$

Human motion sequence detected by Kinect are time series motion frames. Based on subsequent experimental verification, 5 frames (Start frame, end frame, and 3 intermediate frame) are enough to represent the whole motion process. Then the eigenvector obtained in this paper is:

$$\mathrm{A} = \left(\theta_1{}^1, \cdots, \theta_4{}^1, \theta_1{}^2, \cdots, \theta_4{}^2, \theta_1{}^3 \cdots, \theta_4{}^3, \theta_1{}^4, \cdots, \theta_4{}^4, \theta_1{}^5, \cdots, \theta_4{}^5\right)$$

In which, the Angle of the joint is calculated by means of spatial transformation, whose superscript indicates the number of frames θ_1~θ_4. The number 1 is the starting frame, number 2-4 is the middle frame, and number 5 is the end frame. Figure 4 shows a complete sequence of actions.

Figure 4. A complete sequence of actions

3. Correlation feature analysis and Feature optimization

In order to improve the efficiency of recognition and increase the calculation speed, the feature should be optimized after correlation analyzing.

The feature constructed above is normalized firstly by linear normalization:

$$y = \frac{\theta_i - \theta_{minVal}}{\theta_{maxVal} - \theta_{minVal}} \tag{5}$$

Where θ_{maxVal} and θ_{minVal} represent the maximum and minimum values of one-dimensional variable in original feature vector. θ_i is the value to be normalized. And θ_i is the result, whose value is between 0 and 1. For the convenience of analysis, the characteristic vector, which described the motion sequence, is the normalized feature $(\varphi 1, \varphi 2, \varphi 3, \cdots\cdots, \varphi 19, \varphi 20)$.

To evaluate the correlation of variables in the normalized feature, Pearson correlation coefficient was selected as the basis. For example, in the correlation analysis of the characteristic vectors of n groups of motion sequences, the Pearson correlation coefficient of $\varphi 1$ and $\varphi 2$ is defined below:

$$r_{\varphi 1 \varphi 2} = \frac{\sum_{i=1}^{n}(\varphi 1_i - \overline{\varphi 1})(\varphi 2_i - \overline{\varphi 2})}{\sqrt{\sum_{i=1}^{n}(\varphi 1_i - \overline{\varphi 1})^2}\sqrt{\sum_{i=1}^{n}(\varphi 2_i - \overline{\varphi 2})^2}} \tag{6}$$

Where $\overline{\varphi 1}, \overline{\varphi 2}$ respectively is the average of $\varphi 1$ and $\varphi 2$ of n groups.

Similarly, Pearson correlation coefficients of other variables can be obtained, and the correlation coefficient matrix of characteristic variables are shown in Table 1.

Table 1. Pearson correlation coefficient matrix of each characteristic variable

	φ1	φ2	φ3	φ4	φ5	φ6	φ7	φ8	φ9	φ10	φ11	φ12	φ13	φ14	φ15	φ16	φ17	φ18	φ19	φ20	style
φ1	1	-0.884	-0.398	-0.799	0.878	-0.815	-0.107	-0.265	0.748	-0.661	0.029	0.296	0.819	-0.523	0.038	-0.103	0.416	-0.309	-0.079	-0.211	-0.673
φ2	-0.884	1	0.427	0.863	-0.842	0.962	0.285	0.202	-0.601	0.873	0.045	-0.326	-0.774	0.758	-0.012	0.16	-0.539	0.574	-0.024	0.263	0.48
φ3	-0.398	0.427	1	0.157	-0.286	0.304	0.158	0.255	-0.308	0.396	-0.104	0.068	-0.315	0.32	-0.178	0.344	-0.265	0.257	0.046	0.333	0.128
φ4	-0.799	0.863	0.157	1	-0.815	0.833	0.112	0.224	-0.59	0.678	0.02	-0.367	-0.673	0.56	-0.006	-0.007	-0.407	0.358	-0.019	0.09	0.612
φ5	0.878	-0.842	-0.286	-0.815	1	-0.829	-0.098	-0.213	0.784	-0.626	0.064	0.337	0.872	-0.466	0.026	-0.08	0.397	-0.256	-0.096	-0.186	-0.741
φ6	-0.815	0.962	0.304	0.833	-0.829	1	0.326	0.234	-0.568	0.926	0.086	-0.275	-0.794	0.822	-0.034	0.218	-0.622	0.664	-0.076	0.32	0.413
φ7	-0.107	0.285	0.158	0.112	-0.098	0.326	1	-0.141	0.048	0.397	0.019	-0.03	-0.078	0.434	-0.017	0.089	-0.193	0.433	-0.307	0.112	-0.275
φ8	-0.265	0.202	0.255	0.224	-0.213	0.234	-0.141	1	-0.339	0.186	-0.054	0.663	-0.463	0.133	-0.261	0.694	-0.575	0.151	-0.021	0.662	0.386
φ9	0.748	-0.601	-0.308	-0.59	0.784	-0.568	0.048	-0.339	1	-0.408	0.011	0.142	0.779	-0.23	0.091	-0.107	0.277	-0.022	-0.143	-0.199	-0.702
φ10	-0.661	0.873	0.396	0.678	-0.626	0.926	0.397	0.186	-0.408	1	0.23	-0.187	-0.647	0.934	-0.017	0.281	-0.653	0.818	-0.143	0.372	0.129
φ11	0.029	0.045	-0.104	0.02	0.064	0.086	0.019	-0.054	0.011	0.23	1	-0.006	-0.026	0.202	0.385	0.136	-0.144	0.145	0.082	0.17	-0.252
φ12	0.296	-0.326	0.068	-0.367	0.337	-0.275	-0.03	0.663	0.142	-0.187	-0.006	1	0.034	-0.15	-0.223	0.669	-0.319	0.007	0.02	0.58	-0.12
φ13	0.819	-0.774	-0.315	-0.673	0.872	-0.794	-0.078	-0.463	0.779	-0.647	-0.026	0.034	1	-0.5	0.12	-0.376	0.602	-0.348	-0.09	-0.461	-0.676
φ14	-0.523	0.758	0.32	0.56	-0.466	0.822	0.434	0.133	-0.23	0.934	0.202	-0.15	-0.5	1	-0.062	0.229	-0.676	0.927	-0.257	0.323	-0.03
φ15	0.038	0.012	0.178	0.006	0.026	0.034	0.017	0.261	0.091	0.017	0.385	0.223	0.12	0.062	1	0.068	0.121	0.109	0.138	0.011	0.095
φ16	-0.103	0.16	0.344	-0.007	-0.08	0.218	0.089	0.694	-0.107	0.281	0.136	0.669	-0.376	0.229	-0.068	1	-0.622	0.303	0.115	0.929	0.097
φ17	0.416	-0.539	-0.265	-0.407	0.397	-0.622	-0.193	-0.575	0.277	-0.653	-0.144	-0.319	0.602	-0.676	0.121	-0.622	1	-0.716	0.134	-0.687	-0.258
φ18	0.309	0.574	0.257	0.358	0.256	0.664	0.433	0.151	0.022	0.818	0.145	0.007	0.348	0.927	0.109	0.303	0.716	1	0.303	0.362	0.18
φ19	-0.079	-0.024	0.046	-0.019	-0.096	-0.076	-0.307	-0.021	-0.143	-0.143	0.082	0.02	-0.09	-0.257	0.138	0.115	0.134	-0.303	1	0.084	0.268
φ20	-0.211	0.263	0.333	0.09	-0.186	0.32	0.112	0.662	-0.199	0.372	0.17	0.58	-0.461	0.323	0.011	0.929	-0.687	0.362	0.084	1	0.153

To analyze the feature relevance of each feature subset, Merit is used as the evaluation function in CFS algorithm (correlation-based feature selection), which takes into account of the relevance between features and classes.

$$\mathrm{Merit}_s = \frac{k\overline{r_{cf}}}{\sqrt{k + k(k-1)\overline{rff}}} \tag{7}$$

Where: k is the number of features in the current feature subset, k ∈ (1,20). $\overline{r_{cf}}$ is the average correlation coefficient involving all characteristics and all classes in the current feature subset. $\overline{rff}$ is the average correlation coefficient involving only the characteristics. $\overline{r_{cf}}, \overline{rff}$ can be drew from table 1. Merit values for various subsets of feature vector and the solution are shown in Table 2.

Table 2. Partial Merit and their solution

Feature set	k	$\overline{r_{cf}}$	$\overline{rff}$	Merit
[φ1]	1	-0.673	1.000	$\frac{1\times(-0.673)}{\sqrt{1+1(1-1)1.0}} = -0.673$
[φ2]	1	0.480	1.000	$\frac{1\times0.480}{\sqrt{1+1(1-1)1.0}} = 0.480$
[φ3]	1	0.128	1.000	$\frac{1\times0.128}{\sqrt{1+1(1-1)1.0}} = 0.128$
⋮	1	⋮	⋮	⋮
[φ1 φ2]	2	-0.0965	-0.884	$\frac{1\times(-0.0965)}{\sqrt{2+2(2-1)(-0.884)}} = -0.2$
[φ1 φ3]	2	-0.2725	-0.398	$\frac{1\times(-0.2725)}{\sqrt{2+2(2-1)(-0.398)}} = -0.2483$
⋮		⋮	⋮	⋮
[φ1 φ2 φ3]	3	-0.021	0.24553	$\frac{3\times(-0.021)}{\sqrt{3+3(3-1)(0.24553)}} = -0.029$
⋮		⋮	⋮	⋮

To look for the optimal feature subset, the best-first-search algorithm is adopted to find the feature subset with the largest merit value. Then the optimal subset with only 17 variables (without φ1, φ7 and φ13) is obtained to resemble the human arm movement, and its merit is 0.89.

4. Human motion classification and recognition based on support vector machine

To classify and recognize the human motion, SVM (support vector machine) algorithm is applied here.

The basic principle of SVM is to use hyperplane to conduct sample classification. As shown in Figure5, when faced to the problem of linear indivisibility, it solves by mapping sample data into high-dimensional space by referring to kernel functions. The radial basis kernel function ($K < x_1, x_2 >= exp^{(-\gamma||x_1-x_2||^2)}$) is selected here.

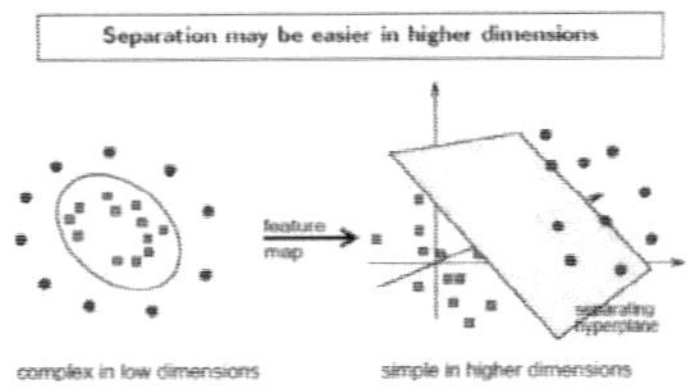

Figure 5. Mapping of low-dimensional data to higher-dimensional data

In the process of classification and recognition, a one-to-one method is selected for SVM to get one label from 6 kinds of motion sequence. Given the training sample set $\{(x_1,y_1),...,(x_i,y_i)|i = 1,2,...,n; y \in \{1,2,3,4,5,6\}\}$, x_i represents the sample vector, which is the optimized feature vector ($\varphi2,\cdots\cdots,\varphi6,\varphi8,\cdots\cdots,\varphi12,\varphi14,\cdots\cdots,\varphi20$)), while y_i is the label result.

The objective function and its constraints are:

$$min\frac{1}{2}||\ w\ ||^2, s.t., y_i(w^T \bullet x_i + b) \geq 1, i = 1,2,...,n \tag{8}$$

The parameters w and b are the normal vector and intercept of the hyperplane respectively. This problem can be solved by Lagrange multiplier method. And according to Karush - Kuhn - Tucker best optimized conditions, w^*, b^* can be calculated below:

$$\begin{cases} w^* = \sum_{i=1}^{n} \alpha_i{}^* x_i y_i \\ b^* = -(\min_{y_i=+1}(w \bullet x_i) + \max_{y_i=-1}(w \bullet x_i))/2 \end{cases} \tag{9}$$

For the unknown test sample input xi, judge the category by the following formula:

$$f(x) = sgn(\textstyle\sum_{i=1}^{n} \alpha_i{}^* y_i\, K < x_i, X > +b^*), X \in \{x_1, x_2, ...x_n\} \tag{10}$$

sgn is the sign function, $< x_i, X >$ is inner product of the vector xi and X.

In SVM, the two important parameters (C, γ) were adjusted to get the best result. Therefore, under the training sample, the best parameters were C = 128 and γ=0.0078125. Thus the prediction accuracy of the theoretical model will reach 96.25%, showing in Figure 6.

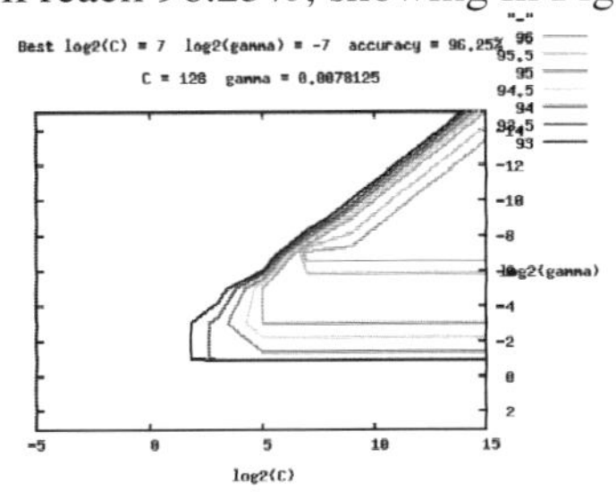

Figure 6. Influence of parameters c and gamma on accuracy

Training to complete the model to the characteristics of the input vector x_i to classify the classification results.

The process of classification recognition is shown figure7 below.

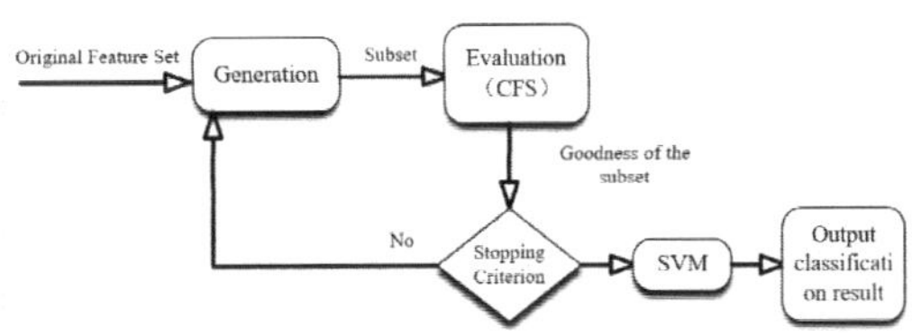

Figure 7. Flow chart of recognition algorithm

5. Experiment

5.1 Verification experiment

In the experiment, six representative movements (including lateral horizontal lift, raising hands, making a phone call, drinking water and eating, shaking hands and shaking hands) are taken as objects and numbered from 1 to 6. In the collection of training samples, the three experimental subjects performed 6 types of movements respectively, and 30 sets of data for each movement were collected, thus 180 samples were obtained to compose the training sample set.

In the verification process, another three experimental subjects performed each type of movements 20 times respectively. Thus 120 human motion sequence was collected to form the test sample set and used as the input of the trained system to judge the accuracy of the algorithm.

The classification results of different action sequences are shown in Table 3:

Table 3. Confusion matrix table of classification results

Original label \ Result	1	2	3	4	5	6	Classification accuracy
1	20						100%
2		16	2	2			80%
3		2	18				90%
4				20			100%
5					20		100%
6						20	100%

It can be seen from the above table that the average recognition rate is as high as 114/120=95%. Among them, the classification error of action 2 and action 3 is 20% and 10%, and the classification of other actions is correct all. Further analysis of the classification result shows that the main reason for misjudgment is the high similarity between action 2 and action 3.

5.2 comparison experiment with KNN and single SVM algorithm

In order to verify the superiority of the algorithm developed in this paper, an experiment compared with KNN (k=6) and SVM is conducted. The training sample set and test set are exactly the same as the verification experiment. The result is shown in Table 4 :

Table 4. Arm motion recognition results of different classification algorithms

Classification algorithm	Recognition rate
SVM	90%
KNN	91.67%
Algorithm of this paper	95%

As can be seen from the table, compared with the KNN classification algorithm and the original SVM classification algorithm, the accuracy of recognition algorithm based on feature selection and SVM was significantly improved to 95%.The verification experiment and comparison experiment prove that the algorithm developed in this paper is effective.

6. Conclusion and Discussion

A motion sequence recognition algorithm, based on the combination of relevance feature selection and SVM for classification, is proposed in this paper. It reduces the dimension of the feature vector, improves the effectiveness of the feature, and guarantees the robustness. The experiment finally proves the validity of the algorithm for human movement classification.

Acknowledgements

The work described in this paper was supported by National Key Research and Development Program Key Project (Project No. 2016YFC0302400).

References

[1] Jiang Zhihao, Li Haoxian, Zhao Chunxia. Pedestrian activity prediction algorithm based on

space-time model video monitoring [J]. Computer applications and software, 2017, 34(1):149-153.
[2] Liu Hechen. Study on accurate recognition and simulation of athletes motion 3d visual image [J], 2016, 33(8):369-372.
[3] Chen Jing. Gesture recognition technology based on Kinect and its application in teaching [D]. Shanghai jiaotong university , 2013.
[4] Yin J Q, Tian G H, Jiang H T, et al. Human action recognition oriented to family service[J]. Journal of Sichuan University, 2011, 42(4):101-107.
[5] Zhu G, Xu C, Huang Q, et al. Player action recognition in broadcast tennis video with applications to semantic analysis of sports game[C]// ACM International Conference on Multimedia. ACM, 2006:431-440.
[6] Johansson G. Visual motion perception [J]. Proceedings of the National Academy of Sciences of the United States of America, 1975, 92(7):2433-2440.
[7] Marr D. Vision: A Computational Investigation into the Human Representation and Processing of Visual Information [M]. Henry Holt and Co. Inc. 1982.
[8] Francomano E, Tortorici A, Calderone V.Regularization of Optical Flow with M-Band Wavelet Transform.Computers&Mathematics with Applications, 2003, 45 (1/2/3) :437-452.
[9] Chen Lifen, Liao H M, Lin Jachen.Wavelet-Based Optical Flow Estimation.IEEE Trans on Circuits and Systems for Video Technology, 2002, 12 (1) :1-12.

IOP Conf. Series: Materials Science and Engineering **646** (2019) 012013 doi:10.1088/1757-899X/646/1/012013

Intelligent Attitude Control of Aircraft Based on LSTM

Bo Li[1,2,*], Peixin Gao[1], Xitong Li[1] and Daqing Chen[3]

[1]School of Electronics and Information, Northwestern Polytechnical University, Shanxi Xi'an, China
[2] CETC Key Laboratory of Data Link Technology, Shanxi Xi'an, China
[3] School of Engineering, London South Bank University, London, UK

[*] libo803@nwpu.edu.cn

Abstract. The flight attitude control is the core part of the maneuvering process in air combats. Traditional flight attitude control methods have high computational complexity, low flexibility and poor ability to learn sequential feature. This paper proposes a flight attitude control model based on long short term memory network, which utilizes its special gates structure to memorize historical information, and acquire the variation law of the attitude control variable from the time sequential data including the battlefield situation and flight parameters automatically. Moreover, the basic framework and training methods of the model are also introduced, and the influence caused by various LSTM network parameters is deeply discussed. The experiment results show that the proposed model has better prediction accuracy and convergence performance than the traditional recurrent neural network.

1. Introduction

It is extremely necessary to choose the appropriate flight attitude control model in air combat. At present, attitude intelligent control methods at home and abroad [1-4] are mostly based on the flight parameters of a single moment. Actually, the attitude change process of the aircraft has strong temporal dependence, so mining sequential feature of the aircraft attitude data is important to control the flight attitude reasonably. Deep Learning provides a new idea for aircraft flight attitude control with its excellent performance in network training, feature extraction and computational accuracy.

The common DL algorithms usually only consider the data characteristics of a single moment, while Recurrent Neural Network (RNN)[5]have good expression of dynamic behavior and sequential feature by introducing neurons with self-feedback[6], but the problem of gradient disappearance is easy to occur during training. The Long Short Term Memory (LSTM) network is one of the most effective RNN improvement models. Therefore, based on the LSTM network, an aircraft flight attitude control model is designed in this paper. By using the air combat data of multiple sequential points as the input signal, mining the sequential variation features, and predicting the value of attitude control variable at the next moment. This model is superior to other traditional algorithms in point of sequential features extraction. Finally, the effectiveness of the proposed model for flight attitude control is verified by experiments.

2. Flight attitude intelligent control framework based on sequential features

Controlling the flight attitude of the aircraft in the air combat needs to mine the flight control laws of both sides from the dynamically changing battlefield situation data, then calculate the estimated value of the attitude control variables at the next moment, and achieve the change of the flight attitude. Meanwhile, the changing of attitude is a dynamic process, which has a strong dependence on the posture of the historical moment and the situation of both sides, so considering the information of the multiple continuous historical points in the dataset is more reliable than considering only a single moment. Therefore, this paper adopts a flight attitude control model based on sequential features using LSTM network. By simulating the pilot's memory and judgment mechanism, using the information of multiple historical moments to predict the attitude control of the aircraft at the next moment, the model can overcome the shortcomings of traditional models in the study of sequential features[7], the framework of LSTM-based attitude control model is given in Figure 1.

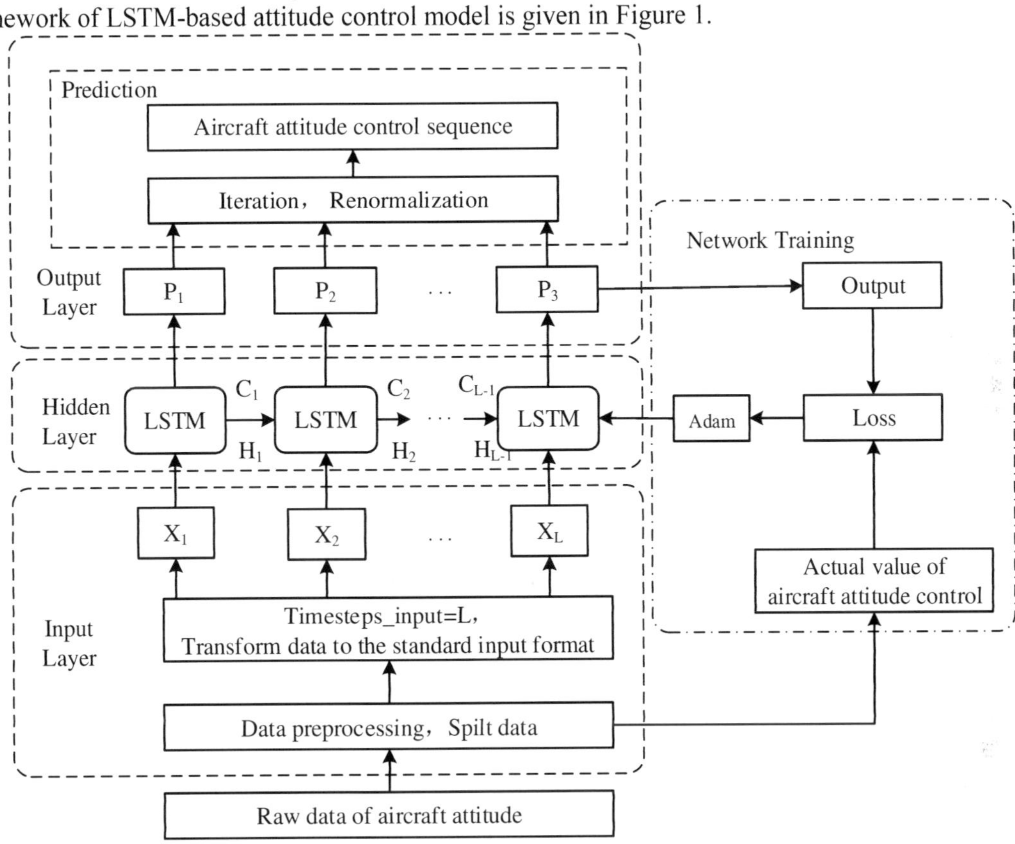

Figure 1 LSTM based fly m line attitude control model framework

3. Establishment of flight attitude intelligent control model based on LSTM

3.1. Related theory

As far as the network structure is concerned, RNN is same as the traditional neural network, but the hidden layer neurons in the RNN are interconnected, so RNN can memorize the previous information and use this information to influence the output of the following nodes. However, in practice, the problem of gradient disappearance or explosion will easily occur during training if the input sequence is too long.

LSTM network was originally proposed by Hochreiter & Schmidhuber as an improved model of RNN, its outstanding advantage is that it can learn the long-term dependence information of sequential data. LSTM replaces hidden layer neurons in RNN with gated memory cells, cells and three control gates ——input gate, output gate, and forget gate can form a memory unit of LSTM, a LSTM network is composed of a group of such memory units. Figure 2 shows the structure of the LSTM memory unit with only a single cell, three gates in the figure let the information pass selectively by the Sigmoid activation function. The output of the function is a value between 0 and 1, which can describe every gate allows how much information to pass[8].

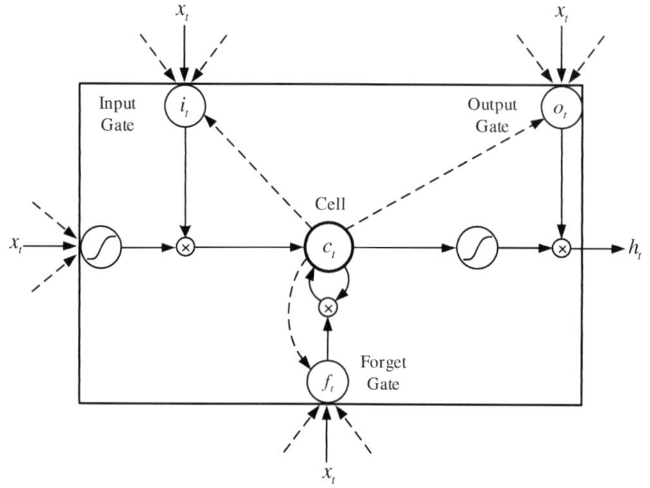

Figure 2. The structure of LSTM memory unit with single cell

3.2. Flight attitude intelligent control based LSTM

3.2.1 Time series data preparation and preprocessing. This paper sets against the backdrop of one-to-one air combat between red side and blue side, and gets data from an air combat simulation system that can simulate the pilot's real maneuver control process. The simulation time is set to 50s, the simulation step is 0.1s, and the simulation confrontation under different initial situation is carried out. Finally, data of 200 confrontations is randomly selected as the dataset of this paper. In order to ensure the comprehensiveness of the collected data, a total of 47 variables representing the state of the aircraft and battlefield situations were selected for collection. Supposing we are red side, these variables include the coordinate component and the velocity components of both sides, the acceleration components of the red aircraft, the heading angle, roll angle, pitch angle, the angle of attack, and the throttle coefficient and brake coefficient of the red aircraft , and 24 Boolean variables indicating the battlefield situation (e.g. whether the radar has target information, whether the pilot sees the target, etc.) are also included in the dataset. Therefore, every time of simulations generates 500 sample, each sample contains values of the above 47 features, 45 of which are characteristic of the state of the enemy aircraft, the other 2 features are attitude control variables of the red aircraft——the angle of attack changing rate and bank angle changing rate, they are the target features to be predicted by the model.

Moreover, to avoid the influence of the data dimension on the model's prediction accuracy, the data needs to be normalized firstly. In this paper, the data is normalized by the max-min normalization method, which can map the data values to [0,1].

3.2.2 Flight attitude control network based on LSTM. The core of the LSTM is the calculation of the input layer, hidden layer and output layer. The number of input layer nodes is equal to the number of features of the input vector, so the input layer in this paper has 46 nodes. The input layer also needs to convert the input time series data into a format of standard supervised-learning data of[Samples, Timesteps, Features], Samples represents the number of input samples; timesteps represents the time step length between the the input and the output of the model, timesteps is the key parameter of the LSTM model, the longer the timesteps, the more the number of loop layers in the model; features represents the dimension of the input sample.

The hidden layer contains several LSTM memory units. This paper selects the tanh function as the activation function for calculating the state value and output value of LSTM cells in the hidden layer. The structure of the hidden layer is the key to affect the accuracy of the model. In this paper, the optimum structure of hidden layer is determined by several experiments with different numbers of hidden nodes and layers.

The number of output layer nodes is equal to the number of the output vector features. In this paper, the LSTM model is used to predict the two attitude control variables respectively, so the output layer has only 1 node. The output layer is also responsible for restoring the model's predictions to the original data format, ie, performing denormalization.

In the training process, this paper takes the Mean Squared Error (MSE) between the output value and the real value as the loss function L, and sets the loss minimization as the optimization target for the training process. At the same time, Xavier is used to initialize the weight of the flight attitude control network[9], and the adaptive moment estimation (Adam) algorithm[10], which can adaptively adjust the learning rate, is used as the optimization method of the loss function. The LSTM based attitude control algorithm flow is given in Table 1.

Table 1. The LSTM based attitude control algorithm flow

BEGIN
Initialize the net parameters $LSTM_{net}$;// Xiaver
Set the max training time: steps;
Convert the format of input data;
Input the training data;
REPEAT
Calculate the output: $Y=LSTM_{foward}(X)$;
Calculate the loss function: $Loss=\frac{1}{T}\sum_{i=1}^{T}(y_i-y_i^{'})^2$;
Back updates the net parameters $LSTM_{net}$;//Adam
UNTIL (Satisfy termination condition) ;
Loss meets the termination condition or training times reach the maximum;
END

4. Simulation and analysis

4.1 Experimental design

4.1.1 Model property indexes. The core task of the flight attitude intelligent control model established in this paper is to predict the value of the attitude control variables at the next moment. This paper selects the MSE and the Mean Absolute Error(MAE) as a performance evaluation index for the model. The smaller the values of MSE and MAE, the better the predictive performance of the model. Meanwhile, this paper suggests that the data in the test dataset satisfying the formula(1) is considered to be predicted successfully, and it can be termed as small error points. Dividing the total number of small error points by the size of the test sample can obtain the small error Point ratio of the model.

$$\frac{|y_t-y_t^{'}|}{|y_t|}<5\% \tag{1}$$

The small error points ratio can directly reflect the ratio of the points predicted accurately. Therefore, the small error points ratio is used as the prediction accuracy of the model in this paper. The higher small error points ratio, the higher prediction accuracy of the model.

IOP Conf. Series: Materials Science and Engineering **646** (2019) 012013 doi:10.1088/1757-899X/646/1/012013

4.1.2 Experimental design. In order to verify the performance of LSTM algorithm in solving aircraft attitude control problem, this paper uses Python to design two sets of simulation experiments based on Tensorflow[11] and Keras.

- Experiment 1: Comparison of RNN and LSTM

Table 2. Initial parameters of RNN and LSTM

Values of paremeters	RNN	LSTM
Hidden layer	1	1
Hidden layer nodes	100	100
Learning rate η	0.01	0.01
Timesteps	80	80
Batch_size	300	300
Training times	10000	10000

Two attitude control models based on RNN and LSTM networks are constructed respectively. The specific network structure and parameters are shown in Table 2.

- Experiment 2: Comparison experiments of timesteps

The optimal network structure obtained from several experiments is used to construct multiple LSTM models with different sizes of input timesteps. The specific network structure settings are shown in Table 3.

Table 3. Parameter settings in the comparison experiments of timesteps

Hidder layer	Hidden layer nodes	Timesteps
1	100	30/50/80/100/150

4.2 Experimental results and analysis

4.2.1 Comparison of RNN and LSTM model results. Table 4 gives the prediction error statistics of two attitude control variables of the RNN and LSTM model. Figure 3 shows the variation of the training error with the iterations of the two models on the different attitude control variables.

Table 4. Statistical table of different models' prediction error

	Alpharate		Bankrate	
Network	RNN	LSTM	RNN	LSTM
MSE	0.05793	0.00923	0.07862	0.02097
MAE	0.09233	0.03513	0.12602	0.06462
Small Error Point Ratio	50.35%	86.94%	41.48%	72.48%

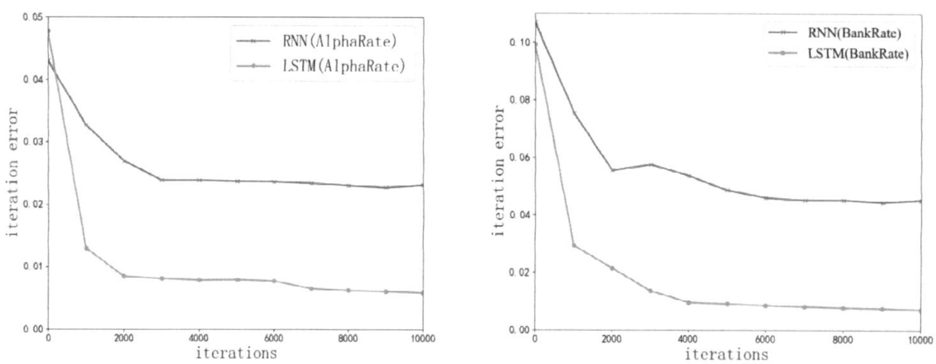

(a)AlphaRate (b)BankRate

Figure 3. Training error variation chart

It can be seen that for two attitude control variables, with the same structure, the convergence speed of the LSTM model is significantly faster than the RNN model. As the training time increases, the error of the RNN model is not greatly improved. Combining the error comparison of the two models on the test dataset in table 4, it is observed that the LSTM attitude control model performs better than the RNN model, which verifies that LSTM can solve the gradient disappearance problem when RNN processes the long sequence data. Besides, it can be concluded that LSTM has higher prediction accuracy for both two attitude control variables from the small error point ratio in Table 4, which proves the feasibility of using LSTM model for attitude control.

4.2.2 The influence of timesteps on the result. Table 5,6 shows the prediction error for the LSTM models with different timesteps. It can be seen that as the input sequence length increases from 30, the prediction accuracy also increases greatly; the accuracy of the two attitude control variables reaches the highest when the timesteps is 80 and 100, respectively. It is visible that timesteps has a significant impact on the performance of the LSTM model. If the timesteps is too small, the data used for training is not enough for the model to learn the essential hidden laws in the sample, so that the accuracy is low. Increasing the timesteps reasonably can improve the prediction performance of the LSTM model, but continuously increasing timesteps can not improve the prediction accuracy sustainably, because the farther the data from the prediction period, the less influence on the prediction data. The cyclic structure of the LSTM network will be more complicated cause by the increase of the timesteps, which affects the training efficiency and accuracy of the model.

Table 5. Prediction error of AlphaRate

Timesteps	Alpharate		
	MSE	MAE	Small Error Point Ratio
30	0.02060	0.06215	68.43%
50	0.01227	0.05142	79.47%
80	0.00923	0.03513	86.94%
100	0.00969	0.03964	84.05%
150	0.00933	0.03525	86.61%

Table 6. Prediction error of BankRate

Timesteps	Bankrate		
	MSE	MAE	Small Error Point Ratio
30	0.04211	0.08154	56.28%
50	0.03451	0.07376	60.65%
80	0.02097	0.06462	72.48%
100	0.02003	0.06314	72.94%
150	0.02106	0.06627	70.10%

5. Conclusion

To the attitude control problem in one-to-one air combat, a LSTM flight attitude control model based on sequential features is established in this paper, the feasibility and accuracy of the model are verified by experiments, the influence of various factors in the LSTM model on the prediction accuracy is analyzed. The LSTM based aircraft attitude control model takes the dynamics and timing dependence of the flight attitude changing process into account, and uses the excellent feature extraction ability of the Deep Neural Network(DNN) to realize the autonomous learning of time series data and the intelligent prediction of the attitude control variables, improving the autonomy and flexibility of the flight attitude control model.

Acknowledgements

This research was supported by The Aeronautical Science Foundation of China (No.2017ZC53021) and The Open Project Fund of CETC Key Laboratory of Data Link Technology (No. CLDL-

20182101).

References

[1] Liu H J, Cai ZX, Liu ZY.: 'Attitude Smoothing Method Based on Least Square Fitting of Orthogonal Function System', Journal of Ordance Equipment Engineering. 2017, 38, (1), pp. 33–35,40

[2] Cui Yue,Zhang Wei.*A two-degree-of-freedom PID control algorithm for aircraft pitch angle[J/OL]. Electronic Technology,*,2020(03):1-6 [2019-05- 17].

[3] Zhou Z J, Wang X S, Wang Y.: '*Spacecraft attitude control based on fuzzy adaptive algorithm*', Electric Machines and Control. 2019, 23, (2), pp. 124–129

[4] Arnab Maity,Leonhard Höcht,Florian Holzapfel. *Time-Varying Parameter Model Reference Adaptive Control and Its Application to Aircraft[J]*. European Journal of Control,2019.

[5] Yang Li, Yuxi Wu,Junli Wang,Yili Liu.*A review of research on cyclic neural networks[J]*.Computer Applications,2018,38(S2):1-6+26.

[6] Palaiahnakote Shivakumara, Dongqi Tang, et al.: '*CNN-RNN based method for license plate recognition',* CAAI Transaction on Intelligence Technology. 2018, 3, (3), pp. 131-132

[7] Ou W, Liu S J, He X Y.: '*Study on the intelligent recognition model of enemy targets tactical intention on Battlefield'*,Computer Simulation. 2017, 34, (9), pp. 10–14

[8] Zhu Junpeng, Zhao Hongli, Du Xing, et al.: '*Application of Long Short-Term Memory Neural Network to Orbit Prediction of Satellite'*, Journal of Ordance Equipment Engineering. 2017, 38, (10), pp.127–132

[9] Greff K, Srivastava R K, Koutnik J,et al.: '*LSTM: A search space odyssey'*. IEEE Transactions on Neural Networks & Learning Systems, 2015, 28,(10), pp.2222-2232

[10] Kingma D P., BA J.: '*Adam: A method for stochastic optimization'.* the 3rd International Conference for Learning Representations, 2015, pp.1-18

[11] MinghaoFan. *Research on the Influence of Different Neural Network Models in TensorFlow on MNIST Datasets[J]*. Modern Information Technology, 2018, 2(11): 75-77.

Research and Implementation of Intelligent ERP Platform for SMEs Based on Cloud Computing

Yong Wen*

Department of finance, GuangDong University of Science & Technology, Dongguan, Guangdong, China

*534235485@qq.com

Abstract. According to the characteristics and business needs of small and medium-sized business management, the paper analyzes and designs the functions and performance of ERP platform, applies Internet of Things and artificial intelligence technology, and builds a dynamic mobile intelligent SME ERP platform through business database, rule database and knowledge database. The platform adopts three-tier B/S architecture, uses MySql as the back-end database, uses Java to encode, implements financial management, supply chain management, manufacturing, human resource management and other functional modules, and finally carries out platform testing.The implementation results show that the intelligent SME ERP greatly reduces the workload of business personnel and improves the efficiency of real-time dynamic management of enterprises.

1. Introduction

The ERP system based on the cloud computing platform is that the system and the database are deployed on the cloud server-side, and the user accesses the cloud server through the peripheral terminal devices such as a smart phone, a notebook computer, a tablet computer, and a desktop computer to access the cloud server to obtain the ERP online service[1].

It has the advantages of on-demand use, low cost, security and stability, easy upgrade and maintenance, etc. It has become the best choice for SMEs to achieve digital management.Emerging technologies such as big data, cloud computing, and artificial intelligence have given life to the traditional ERP system[2].In March 2016, Deloitte, one of the world's four largest firms, took the lead in integrating artificial intelligence into the financial process, launched the Xiaoqin financial robot, and put it into operation at Sinochem International Financial Sharing Center.In May, PricewaterhouseCoopers also developed financial robots, which were developed by Xiaoqin people in the financial sector to other areas such as supply chain and human resources.In June, Ernst & Young and KPMG also launched their own financial robots. In the same year of June, Ernst & Young and KPMG also launched their own financial robots.The above financial robots are designed and developed for large enterprises and financial institutions.There is a lack of SMEs' intelligent ERP platform in the market. SMEs have simple business, clear processes, and flat management , making it easier to build an intelligent ERP platform[3].

2. Platform development process

Firstly, according to the user's needs, build the platform logic model, determine the functions that the platform, and secondly, carry out the overall design, as shown in Figure 1, the functional modules and their interrelationships of the design platform, and then carry out detailed design, define the database model and structure. Again, write the program code, implement the interface and function of the platform, and finally, conduct the platform test and acceptance[4].Due to the limited space,the following only describes the process of platform implementation and platform testing.

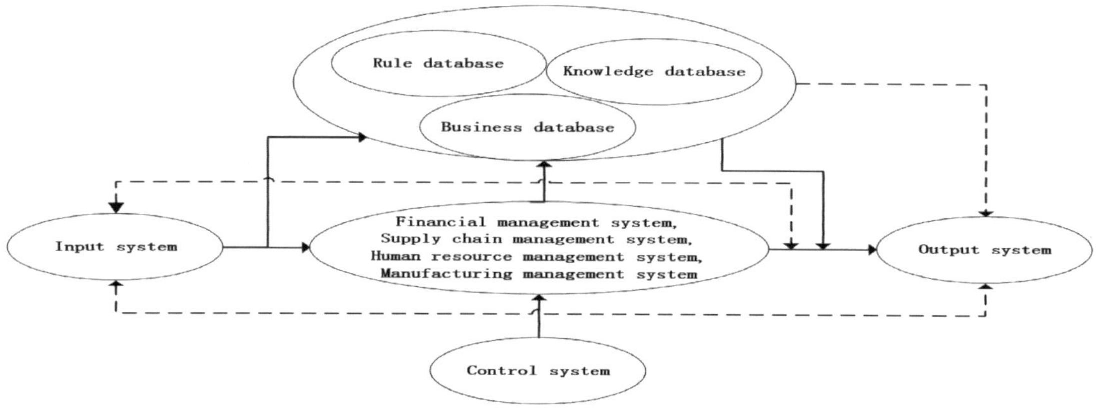

Figure 1. Intelligent cloud ERP platform architecture diagram.

3. Platform implementation

platform implementation is the process of putting the new system into implementation based on the system design[5].The ERP platform mainly includes six modules,such as system management subsystem, financial management subsystem, human resource management subsystem, supply chain management subsystem, production manufacturing subsystem, and daily office subsystem.Because of the limited space,the following is a brief introduction to the implementation process of several sub-modules under the financial management module.And the platform is tested from three aspects: platform function, performance and user interface.

3.1. Platform development environment

*3.1.1.platform development tool selection.*Java with JDK1.6 version or higher is used as the foreground development tool, and the system code is written in accordance with the unified coding mode.UML modeling tools are available: Microsoft Office Visio and Rational Rose Enterprise Edition.

3.1.2.Database selection. At present, the electronic engineering of large and medium-sized enterprises adopts such large and medium-sized relational databases with powerful functions, stable operation and good after-sales service such as Oracle, IBM DB2, Microsoft SQL Server and Mysql.However, considering that MySql is more economical than Oracle, as well as engineering scale and management level, it is determined to use MySql as the back-end database.

3.2. Software architecture

The platform adopts Java development, adopts struts + spring + jstl + jquery + Sql framework and Server2005 database, adopts multi-layer distributed, the architecture is easy to maintain and secondary development, and the system module adopts the idea of plug-in component technology for service-oriented (SOA). Flexible and scalable.

3.3. System management module implementation

System management module is system data management center.After logging in to the system management, you can set up jobs, set up organizational models, user management, enter data dictionary, and download center maintenance.

IOP Conf. Series: Materials Science and Engineering **646** (2019) 012014 doi:10.1088/1757-899X/646/1/012014

3.4.Implementation of sub-module of engineering project

The project sub-module is a sub-module under the general ledger module. Its main function is to complete the entry of cost and expense documents; receivables, payables input or automatic generation; addition, modification and deletion of bank deposits and cash-storage business; shared document maintenance , project cost budget, etc.

3.5. Implementation of report query sub-module

The report query sub-module is a sub-module under the general ledger module. Its main functions include: request summary inquiry, project overview query, current summary inquiry, summary analysis, customer receivable inquiry, business status inquiry, cost expense inquiry and Analysis, detailed account statement inquiry, bank cash journal inquiry, balance sheet inquiry, profit statement inquiry, cash flow statement inquiry, etc.

3.6. Part of the source code display

```
package com.dreamlv.persiste.manager.dao;
import com.dreamlv.exp.RtExp;
import com.dreamlv.iface.sql.IBatch;
import com.dreamlv.persiste.manager.DBMgr;
import com.dreamlv.uti.StrUti;
public class Update {
        private static Update update     = new Update();
        private Update() {
                super();
        }
        public static Update instance() {
                return update;
        }
        public boolean update4SureRows(Connection con, String sql, int updateRows, Object... params)
                        throws Exception {
                if (StrUti.isNull(sql)) {
                        RtExp.alert("Update error: updated related sql statement is empty");
                }
                PreparedStatement ptm = null;
                boolean ok = true;
                try {
                        ptm = con.prepareStatement(sql);
                        int param_num = params.length;
                        for (int i = 0; i < param_num; i++) {
                                ptm.setObject(i + 1, params[i]);
                        }
                        int rows = ptm.executeUpdate();
                        ok = (rows == updateRows ? true : false);
                } catch (Exception e) {
                        throw e;
                } finally {
                        DBMgr.close(null, null, ptm, null);
                }
                return ok;
        }
        public boolean update4SureRows(String sql, String db, int updateRows, Object... params) {
                if (StrUti.isNull(sql)) {
                        RtExp.alert("Update error: updated related sql statement is empty");
                }
                Connection con = null;
                boolean ok = true;
                try {
                        con = DBMgr.getTranConnection(db);
                        ok = update4SureRows(con, sql, updateRows, params);
                        if (ok) {
                                con.commit();
                        } else {
                                con.rollback();
                        }
```

```
        } catch (Exception e) {
            try {
                con.rollback();
            } catch (SQLException e1) {
            }
            DBMgr.logAndAlert(this, "Update data error: ", e, "Update data error:<br/>");
        } finally {
            DBMgr.close(con);
        }
        return ok;
    }
    public boolean update(String sql, String db, Object... params) {
        if (StrUti.isNull(sql)) {
            RtExp.alert("Update error: updated related sql statement is empty");
        }
        Connection con = null;
        try {
            con = DBMgr.getTranConnection(db);
            update(con, sql, params);
            con.commit();
        } catch (Exception e) {
            try {
                con.rollback();
            } catch (SQLException e1) {
            }
            DBMgr.logAndAlert(this, "Update data error: ", e, "Update data error:<br/>");
        } finally {
            DBMgr.close(con);
        }
        return true;
    }
    public boolean batch(Connection con, String sql, int batch_size, IBatch iBatch) throws Exception {
        if (StrUti.isNull(sql)) {
            RtExp.alert("Batch error: batch related sql statement is empty");
        }
        if (batch_size < 1) {
            RtExp.alert("Batch error: batches less than 1");
        }
        PreparedStatement ptm = null;
        try {
            ptm = con.prepareStatement(sql);
            for (int i = 0; i < batch_size; i++) {
                if (iBatch.oneBatch(ptm, i)) {
                    ptm.addBatch();
                }
            }
            ptm.executeBatch();
        } catch (Exception e) {
            throw e;
        } finally {
            DBMgr.close(ptm);
        }
        return true;
    }
    public boolean batch4OneRow(Connection con, String sql, int batch_size, IBatch iBatch) throws Exception {
        if (StrUti.isNull(sql)) {
            RtExp.alert("Batch error: batch related sql statement is empty");
        }
        if (batch_size < 1) {
            RtExp.alert("Batch error: batches less than 1");
        }
        PreparedStatement ptm = null;
        boolean ok = true;
        int updateRows = 0;
        try {
            ptm = con.prepareStatement(sql);
            for (int i = 0; i < batch_size; i++) {
```

```
                        if (iBatch.oneBatch(ptm, i)) {
                                updateRows++;
                                ptm.addBatch();
                        }
                }
                ok = DBMgr.isBatchSuccess(ptm, updateRows);
        } catch (Exception e) {
                throw e;
        } finally {
                DBMgr.close(ptm);
        }
        return ok;
    }
public <T> List<T> listAddParam(Connection con, String sql,List<T> listAddParam,IQuery<T> iQuery, Object... params)
throws Exception {
        if (StrUti.isNull(sql)) {
                RtExp.alert("Error: query related sql statement is empty");
        }
        if(listAddParam == null || listAddParam.isEmpty()){
                return null;
        }
        PreparedStatement ptm = null;
        ResultSet rs = null;
        try {
                ptm = con.prepareStatement(sql);
                int param_num = params.length;
                for (int i = 0; i < param_num; i++) {
                        ptm.setObject(i + 1, params[i]);
                }
                rs = ptm.executeQuery();
                int i = 0;
                T row = null;
                while (rs.next()) {
                        row = iQuery.row(rs, i);
                        if (row != null) {
                                listAddParam.add(row);
                        }
                        i++;
                }
        } catch (Exception e) {
                throw e;
        } finally {
                DBMgr.close(rs, null, ptm, null);
        }
        return listAddParam;
    }
}
```

4. Platform test

4.1. Function test

Introduce the function module test with the general ledger and report module as an example.

The main content of the test: Whether the module can generate accounting vouchers from the original vouchers, register the books, correctly prepare financial statements, etc., whether the module can perform anti-audit and anti-bookkeeping on the voucher, whether each report, detailed account and accounting vouchers can implement an integrated query.Test results: The report module and the general ledger module are seamlessly connected, and each report and detailed account can be linked to the corresponding accounting documents.The functions of the system conform to the business specifications and requirements, and each function runs normally.

4.2. User interface test

The main content of the test: whether the user interface is beautiful, whether the image is clear, whether the text description is appropriate, whether it meets the customer's needs, including: system main menu test, function module test, prompt information test, help window test.Test results: The user interface is user-friendly and can meet customer needs. The interface text and prompt information are correct and the semantics are clear.

4.3. Database running test

The main content of the test: increase the data table to the maximum load to see if it can run correctly.After deleting all the data tables, whether can add new data to the empty table or not. After the database data items are added, deleted, modified, etc., the system whether can run normally or not.

4.4. Safety and reliability test

Security reliability testing includes software fault tolerance testing and security confidentiality testing.Fault tolerance test results: The system can make a correct judgment on the wrong operation and prompt the user to return to the normal state. Security and confidentiality test results: User authentication function is rigorous, and authority allocation meets the requirements of internal control principles.

5. Summary and outlook

According to the current cloud computing technology and the needs of SME business management, the article designs and implements the intelligent SME ERP platform. Due to the limitation of time and business level, the platform still has many shortcomings.For example, the Internet of Things technology is not well integrated into the ERP platform; ERP platform , CRM and e-commerce are not well integrated; the level of intelligent analysis and decision-making of ERP platform is limited. In the future, we will absorb talents from CRM and e-commerce, enhance the research breadth and depth of the team, and improve the above three major deficiencies of intelligent ERP.

Acknowledgement

[Fund Project] Characteristic Innovation Project of “Innovation and Strong School Project” of Higher Education in 2017 in Guangdong Province (Project No.: 2017WTSCX125).

References

[1] Xiqiang M 2018 China. D. *Research on risk evaluation of SaaS ERP service provider selection in small and medium sized enterprises* (Xi'an:Library of Xi'an University of Technology)

[2] Jian W.2013 China. D. *Intelligent three-dimensional dynamic Accounting Information Platform Research* (Qingdao: Ocean University of China Libraries)

[3] Xu H.2013 China. D. *Research and Design of Financial Management Function of ERP System Based on .NET*(Harbin: Jilin University Library)

[4] Annamalai.C and Ramayah.T 2011 *Enterprise resource planning(ERP) benefits survey of Indian manufacturing firms: An empirical analysis of SAP versus Oracle package*.vol 17(3) (London: Business Process Mangemet Jounral)pp 495-509

[5] Huayang M, Mingyun F 2016 *Research on SaaS Model under Management Accounting Informationization* Finance and accounting monthly vol 4 (Wuhan: Hubei) pp 102-104

A New Approach for Short-term Time Series Forecasting

Henghai Fan[1], Shuai Wu[2,*], Ning Chen[1], Bo Gao[1]and Yiming Xu[2]

[1]Xi'an Satellite Control Centre, China

[2]School of Computer Science, Fudan University, Shanghai, China

*wushuaixl@sina.com

Abstract. Short-term series forecasting is one of the essential issues in a variety of tasks, such as traffic flow prediction, stock market tendency analysis, etc. Most current methods based on stable or abundant historical data. In this paper, we proposed a novel model called LA-NN. It takes advantages of both long short-term memory (LSTM) network and autoregressive integrated moving average (ARIMA) by a relation integration of them. So as to deal with the situation of insufficient historical data and sudden abnormal changes in data. A comparison with other representative forecast models validates that the proposed LA-NN network can achieve a better performance.

1. Introduction

A time series is a set of observations x_t, each one being recorded at a specific time t [1].Time series forecasting is a task of collecting past observations, capturing the underlying pattern and then to predict the future. While the objective of short-term time series forecasting is to develop a predictor which could extrapolate the time series into the near future, which ranges from several minutes to dozens of minutes.

Time series data widely exist in various fields, such as transportation, medical, economic, energy, and so on. Time series forecast can promote the development of various fields. At present, many researchers have applied time series forecast technologies to various fields, such as economic forecasting [2], traffic speed forecasting [3], patient flow forecasting [4], electricity-related forecasting [5], and so on.

Although there are many correlational researches, many challenges still exist. At present, methods either need to be improve accuracy or base on massive historical data. But due to some realistic factors, historical data may be insufficient or irregular variations (or unstable pattern), which we simply called mutation period, will emerge in the time series. This characters lead to difficulties in predicting such period of time series. Different from conventional time series forecasting, the accuracy of short-term forecasting in mutation period is extremely important. Therefore, establish a forecasting model that can answer insufficient history data and mutation period is the main task of this work.

To address above issues, we propose a more effective model, called LA-NN. The main contributions of this paper can be summarized as two aspects. Firstly, LA-NN integrate LSTM and ARIMA in a rational way to make use of their advantages. Secondly, we evaluate our model on real traffic data.

The remainder of this paper is organized as follows. Section 2, introduces a general overview of existing literatures on time series forecasting. The methodology is introduced in Section 3. Experiments are shown in Section 4. Section 5, concludes the paper and suggests future directions.

2. Related work

Over the past few decades, a lot of researches and effort has been devoted to develop and improve short-term time series models. ARIMA family of models are quite flexible, they need less data and can accurately predict linear correlation sequence data. But no nonlinear patterns can be captured by the ARIMA models. ARIMAs to complex real-world problem is not always satisfactory. ARIMA and its variants are widely used in time series data. In [6], a new approach called ARIMA-GARCH was used to forecast ridership time series data.

Time series forecast base on deep learning approaches has become a new trend. Compared with conventional learning architectures, deep neural network can model complex non-linear relationship by using distributed and hierarchical feature representation [7]. So far, many NNs variants have been proposed to assist time series forecasting. In recurrent neural network (RNN) has been used in [8] to predict traffic speed time series data. Among all NNs, RNN is widely regarded as a suit model to capture underlying evolution of time series. But many researchers proved that RNN fail to capture long term evolution. To address this problem, long short-term memory (LSTM) are used to predict time series data. In term of time series date, LSTM is excellent. Many LSTM variants or hybrid methods are applied in time series forecast. In [9], a LSTM based deep learning approach is applied in traffic forecast. In [10], a hybrid methods called is proposed to predict traffic speed. The basic idea is to feed the features of surrounding areas extracted by CNN to RNN for learning time-series patterns.

Distinct from other prediction approaches, this paper takes into account the shortage of historical data and sudden changes in data and constructs a new NN model called LA-NN with multi layers based on ARIMA and LSTM. New approach has a better performance on accuracy even in the case of insufficient historical data or sudden changes in data.

3. Methodology

Short-term forecast is a task of modelling underlying patterns of time series, which includes both regular and irregular variations. The major challenge is to predict these irregular variations or unstable patterns. This paper deals with the problem with a combing model, which models ether linear or non-linear relationship in order to make a reliable forecast result. The proposed short-term forecast model is based on the available technologies, which include ARIMA algorithm, RNNs, LSTM network. The detail of the methodology will be explained in this section.

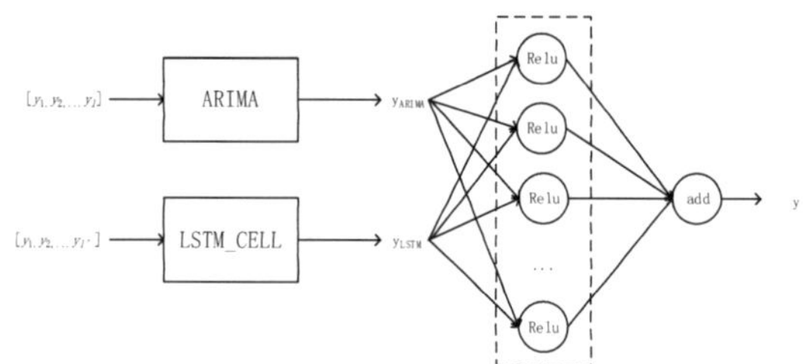

Figure 1. The structure of LA-NN

ARIMA can use very little historical data to predict, but it requires stable data. LSTM can mine time series patterns, but it needs magnanimous historical data. Time series data exist widely, so there are two issues. First, when we need to predict time series, we don't necessarily have enough historical data. Second, when the time series data changes abnormally, the prediction results of the prediction model trained based on historical data are poor. In order to address this situation, a hybrid neural network algorithm called LA-NN based on ARIMA and LSTM is proposed in this paper. The structure is as figure 1.

Figure 1 presents the architecture of LA-NN, which is comprised of one ARIMA layer, one LSTM layer and a fully connected layer, predicting base on small quantities data, modeling long temporal dependency and adjusting weights of y_{ARIMA} and y_{LSTM} respectively. For improve the efficiency and accuracy, the length of the input series can different, because LSTM may need long series to model long temporal dependency, ARIMA may need short series to address sudden changes in data. Finally, a fully connected layer consisting of multiple activation functions is constructed to adjust weights of y_{ARIMA} and y_{LSTM}.

4. Experiment

4.1 Data preparation

The proposed time-series data forecast model was applied to the traffic speed time-series data collected by Shanghai Traffic Information Center as a numerical example. We get the traffic speed of days from 01 May 2017 to 30 August 2017 as original dataset, including 598 viaduct sections with a frequency of 2 min.

In addition, in order to verify the ability of the new model to cope with sudden changes in data, some modifications were made so that some data did not conform to the normal pattern. The selected dataset were divided into four subsets: a small training dataset containing several days of data, a bigger training dataset containing several weeks of data, a test dataset containing several days of data, a test dataset containing modification data.

4.2 Benchmarks

We compare our model with the following baseline methods.

ARIMA: This model [11] is a generalization of autoregressive moving average model with an initial differencing step applied to remove the non-stationarity of the data.

LSTM: LSTM can effectively handle long temporal dependence and reduce gradient vanishing which is appropriate for speed prediction[12].

4.3 Evaluation for forecast result

Three criteria are commonly used to evaluate the performance of traffic forecast model. They are mean absolute error (MAE), root mean square error (RMSE) and mean relative error (MRE).

4.4 Experiment result

The experimental data we selected are periodic. Most of the time, the data has certain regularity, some time period data is larger, some time period data is smaller. Figure 2 show our data intuitively. As mentioned in Section 4.1, part of the test data in this experiment is used to test the prediction accuracy of the algorithm for abnormal mode data, so some of the data are modified. Figure 3 is the modified data of figure 2, and the blue part is the modified data.

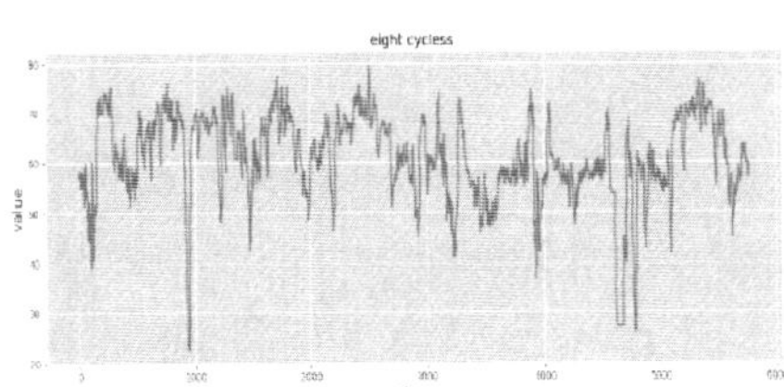

Figure 2. Eight cycles of data

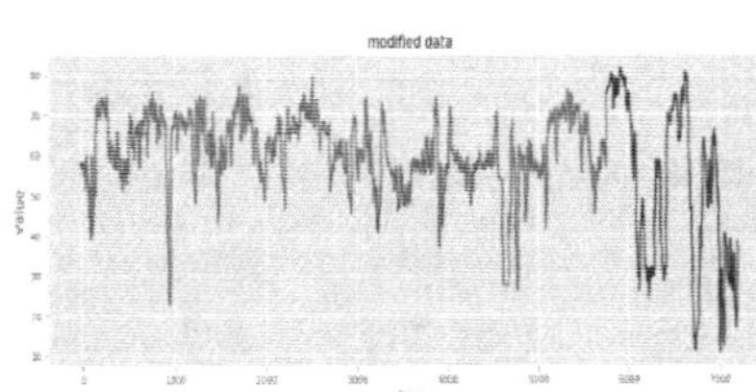

Figure 3. One modified data

First, determine the parameters of ARIMA. The parameters of ARIMA model are determined as follows: p=2, d=1, q=0.Second, Train LA-NN Model and LSTM Model. Each model is trained with small training set. Third, Test ARIMA、LA-NN Model and LSTM Model on normal data and modified data respectively. Evaluate prediction results with MRE、RMSE、MAE. The values given in this paper are the average values of multiple time series data. In order to display the experimental results more intuitively, this paper presents the experimental results of one of the time series data with pictures. The red line is the real value and the blue line is the prediction result.

4.4.1 Experiments on modified data. One of the issues to be solved by the new algorithm is when the law of time series data changes greatly for some reason, the accuracy of time series data is still high. In this stage, the experiment is to predict the blue part data of figure 3. Table 1 show that the accuracy of LAN is higher than that of the other two algorithms. Intuitive prediction results can be seen in figure 4, figure 5, figure 6.

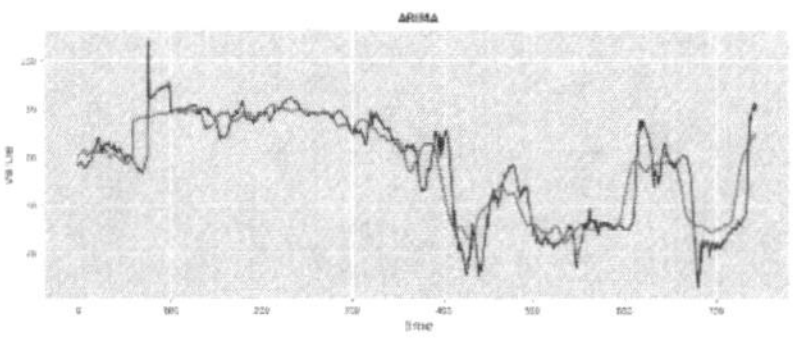

Figure 4. ARIMA experimental results on modified data

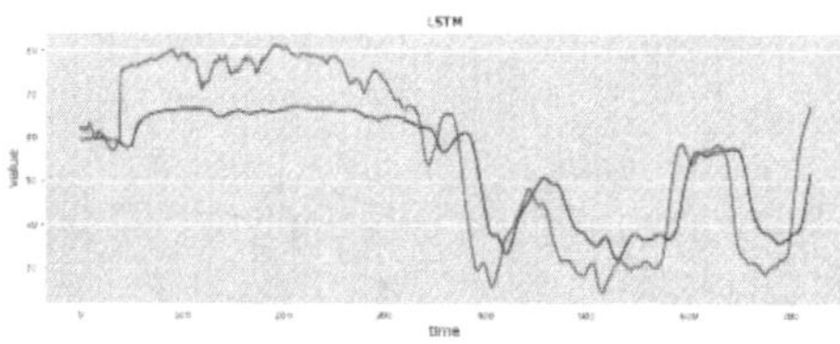

Figure 5. LSTM experimental results on modified data

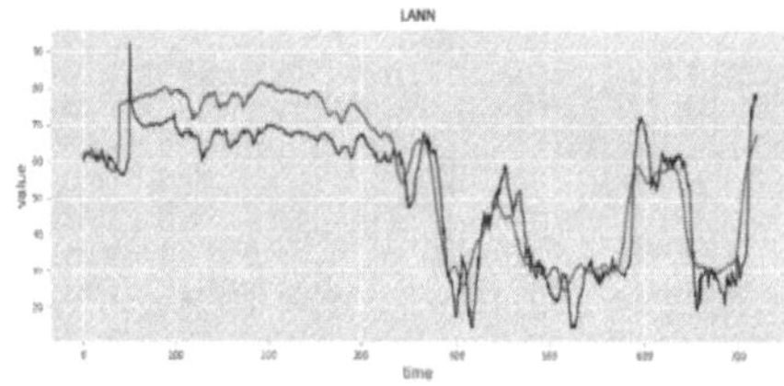

Figure 6. LANN experimental results on modified data

Table 1. Forecast performances of different algorithms for modified data

Algorithm	Evaluation Criteria		
	MAE	*RMSE*	*MRE*
ARIMA	6.70	9.37	0.14
LSTM	8.89	10.43	0.18
LA-NN	7.47	8.92	0.14

4.4.2 Experiments on normal data when training data is insufficient. Another issue we want to address is that models are not well trained when historical data is insufficient. In this stage, the model is trained by using small training sets, and the predicted results are shown in table 2. Because ARIMA model needs less historical data, its accuracy has not been reduced. When training LAN and LSTM models, the same small data sets are used. The results show that the prediction accuracy of LAN is better than that of LSTM, because ARIMA is integrated in LAN, and ARIMA still has higher accuracy when historical data is insufficient. Intuitive prediction results can be seen in figure 7, figure 8.

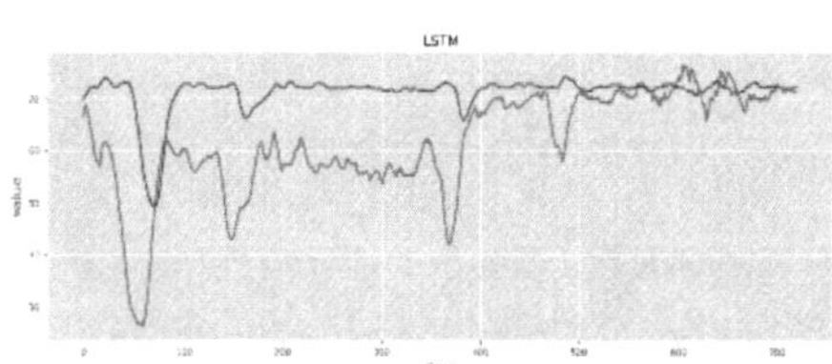

Figure 7. LSTM experimental results on normal data when training data is insufficient

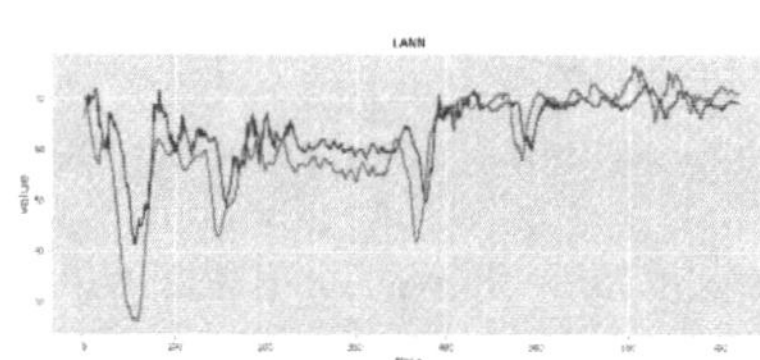

Figure 8. LANN experimental results on normal data when training data is insufficient

Table 2. Forecast performances of different algorithms for normal data when training data is insufficient

Algorithm	Evaluation Criteria		
	MAE	*RMSE*	*MRE*
ARIMA	7.02	9.88	0.13
LSTM	9.77	13.12	0.19
LA-NN	6.53	8.32	0.10

4.4.3 Experiments on modified data when training data is insufficient. Finally, the modified data is predicted using the LAN model and LSTM model trained by ARIMA and small data sets. As shown in table 3, the accuracy of LANN and LSTM decreases in varying degrees, but LANN can still maintain high accuracy. Intuitive prediction results can be seen in figure 9, figure 10.

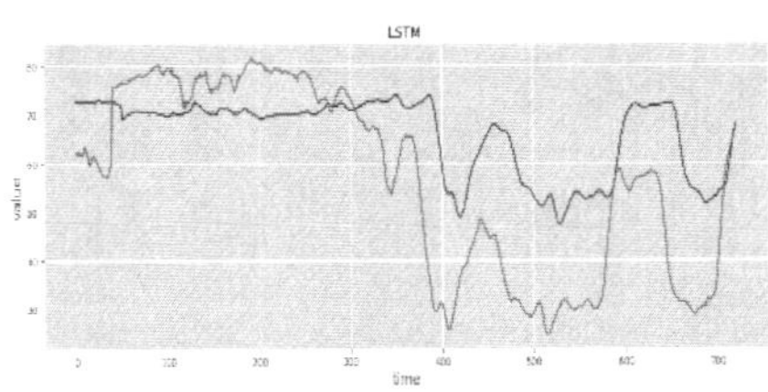

Figure 9. LSTM experimental results on modified data when training data is insufficient

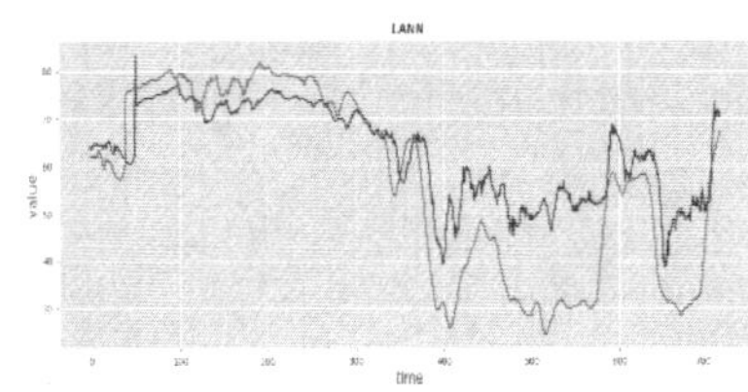

Figure 10. LANN experimental results on modified data when training data is insufficient

5. Conclusion

In order to deal with the problem that the accuracy of prediction model decreases due to insufficient training data and sudden change of time series data, a new neural network model, LANN, is proposed in this paper. Experiments are carried out in four cases. The results show that LANN has higher prediction accuracy and stronger adaptability, and can maintain higher accuracy in various situations.

Table 3. Forecast performances of different algorithms for modified data when training data is insufficient

Algorithm	Evaluation Criteria		
	MAE	*RMSE*	*MRE*
ARIMA	6.70	9.37	0.14
LSTM	13.98	16.98	0.35
LA-NN	9.39	12.07	0.25

Reference

[1] Shanmugam R , Brockwell P J , Davis R A . Introduction to Time Series and Forecasting[J]. Technometrics, 1997, 39(4):426.

[2] Koceska N , Koceski S . Financial-Economic Time Series Modeling and Prediction Techniques – Review[J]. Journal of Applied Economics & Business, 2014.

[3] Lin L , Li J , Chen F , et al. Road Traffic Speed Prediction: A Probabilistic Model Fusing Multi-Source Data[J]. 2018.

[4] Kadri F , Harrou F , Chaabane, Sondès, et al. Time Series Modelling and Forecasting of Emergency Department Overcrowding[J]. Journal of Medical Systems, 2014, 38(9):107-1235.

[5] Hong T , Pinson P , Fan S , et al. Probabilistic energy forecasting: Global Energy Forecasting Competition 2014 and beyond[J]. International Journal of Forecasting, 2016:S0169207016000133.
[6] Ding C , Duan J , Zhang Y , et al. Using an ARIMA-GARCH Modeling Approach to Improve Subway Short-Term Ridership Forecasting Accounting for Dynamic Volatility[J]. IEEE Transactions on Intelligent Transportation Systems, 2017:1-11.
[7] Huang W , Song G , Hong H , et al. Deep Architecture for Traffic Flow Prediction: Deep Belief Networks With Multitask Learning[J]. IEEE Transactions on Intelligent Transportation Systems, 2014, 15(5):2191-2201.
[8] Tang J , Liu F , Zou Y , et al. An Improved Fuzzy Neural Network for Traffic Speed Prediction Considering Periodic Characteristic[J]. IEEE Transactions on Intelligent Transportation Systems, 2017:1-11.
[9] Zhao Z , Chen W , Wu X , et al. LSTM network: a deep learning approach for short-term traffic forecast[J]. IET Intelligent Transport Systems, 2017, 11(2):68-75.
[10] LC-RNN:A Deep Learning Model for Traffic Speed Prediction
[11] Bartholomew D , Box G E P , Jenkins G M . Time Series Analysis: Forecasting and Control[J]. Journal of the Operational Research Society, 1971, 22(2):199-201.
[12] Ma X , Tao Z , Wang Y , et al. Long short-term memory neural network for traffic speed prediction using remote microwave sensor data[J]. Transportation Research Part C: Emerging Technologies, 2015, 54:187-197.

An Improved Location Model for Pedestrian Detection

Tengtao Zou[1,a,*], Chen Cao[1,b] and Shangming Yang[1,c]

[1]School of information and software engineering, University of Electronic Science and Technology of China

E-mail: [a,*]zttjzy@foxmail.com; [b]1471421850@qq.com; [c]minn003@163.com

Abstract. Pedestrian detection in complex scenes has always been a research difficulty in computer vision. The performance of current methods was seriously degraded when the pedestrian is occluded or the size of pedestrians are too small, etc. In this paper, we propose a novel approach based on location model for the detection of multiple pedestrians, which aims to improve the efficiency of algorithms in complex scenarios. In the model, a fully convolutional neural networks for the classification of pedestrian and non-pedestrian are trained to learn pedestrian features first. Then trained model are used to search for pedestrian regions, and the regions where pedestrian may be present will be activated and marked with boxes. Finally, we fine tune the boxes to overlap them with the ground truth more precisely. Compared with the current methods on two pedestrian datasets, experimental results demonstrate the comparable performance of our approach in term of miss rates.

Keywords: Pedestrian Detection, CNN, Location, Global Average Pooling.

1. Introduction

Pedestrian detection as the most popular and challenging problems in computer vision attracts lots of researchers. In the past ten years, a lot of good pedestrian detection algorithms [1,2,3,4,5,6,7] have emerged. However, for CNN-based methods [8,9,10,11,12], such as Faster R-CNN, due to complexity of the scene, the performance of pedestrian detection is not very well. When the pedestrian is occluded or the size of the pedestrian is small, fully connected neural network are difficult to recognize these features so that the detector will neglect them. But in reality, these situations often occur. Thus, fast and accurate pedestrian detection in complex scenes is urgently needed.

In this paper, we propose a location based model for pedestrian detection. In order to obtain more efficient pedestrian detection, we improve the CAM [13] to locate pedestrians. After pedestrian location, the image is separated into some regions, and a threshold can be set up to exclude some regions if their probabilities are lower than a certain threshold. The rest of region will be selected and marked with a box. To locate objects for convolutional neural networks more precisely, by replacing full-connected layers, we use full convolutional layers to be our training model.

The most related work to our algorithm is CAM [13] and Faster-RCNN [11]. CAM proposed by Zhou et.al, is a Channel-Wise Attention Mechanism, which is used to mine high-dimensional feature information of objects. Faster- RCNN as a generic object detector, contains RPN and Fast RCNN, in which RPN network was designed to generate a set of proposed regions instead of Selective Search and made use of Fast-RCNN for object recognition. In this paper, the advantages of both are exploited to construct pedestrian detection model.

The main contributions of this paper are listed as follows:(1) We propose an location based model

IOP Conf. Series: Materials Science and Engineering **646** (2019) 012016 doi:10.1088/1757-899X/646/1/012016

using effective pedestrian region selection and improved pedestrian proposals. (2) The object location capability of CAM is improved by replacing full-connected layers. A full convolutional neural network is trained for object location. (3) Massive experimental results on two datasets demonstrate that the proposed algorithm significantly improves the detection rate pedestrian in complex scene.

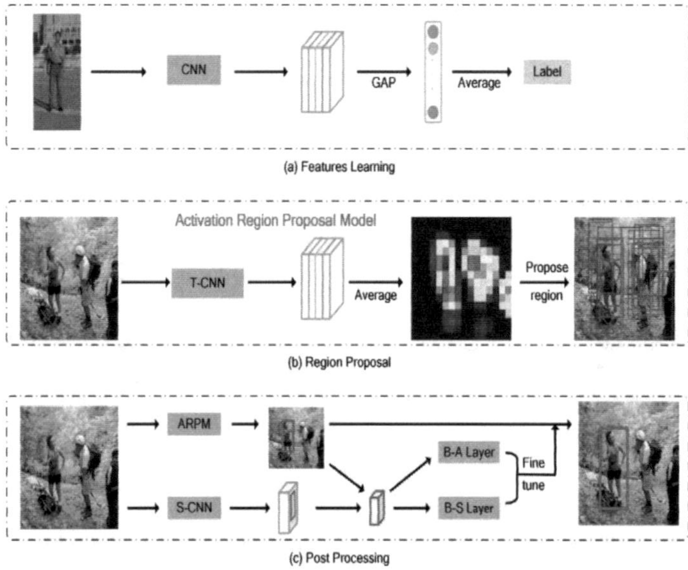

Figure 1: The framework of our model. T-CNN represents the CNN transferred from classification networks. S-CNN represents the shared CNN of detection networks. B-A Layer and B-S Layer indicate box-adjustment layer and box- score layer respectively.

Figure 2: Some selected training samples for the proposed model.

2. The Proposal Model

In this section, we present the detailed implementation of our approach which includes three steps: pedestrian feature learning, region proposal and post processing. Firstly, we train a full convolutional neural network on the datasets that only include positive samples and negative samples (see Figure 1(a)), single pedestrian images and non-pedestrian images (see Figure 2). Secondly, the convolution kernels of trained convolutional neural network are transferred to the detection CNN for locating pedestrians. Then, the model will generate a pedestrian mask which displays the location of people. According to the pedestrian mask, we can use bounding boxes to mark the general position of pedestrians(see Figure 1(b)). At last, we fine tune the size and location of bounding boxes to make them more accurate and employ non- maximum suppression algorithm (NMS) [14] to delete redundant bounding boxes(see Figure 1(c)).

2.1. Pedestrian Feature Learning

To keep convolution nerve units learn pedestrian feature better, we should train a CNN-based network with two classifications. A CNN based on VGG [15] model is selected as our backbone network, and in the last convolutional layer, the network will produce 512 features maps.

After CNN network extracts features, the global average pooling is performed on the feature maps.

Then, the average result from global average pooling is used to indicate whether an input is pedestrian images (see Figure 1(a)). All parameters in network can be learned by minimizing the function:

$$L = \frac{1}{n}\sum_{i=1}^{n}(-y_i \log(F(x_i)) - (1 - y_i)\log(1 - F(x_i))), \tag{1}$$

The datasets for training just include single pedestrian images and background images.

2.2. Region Proposal

In the previous section, we have completed the training of CNN-based network for classification. Then the convolution kernels will be transferred from classification network to detection network for pedestrian locating. An $N * M$ pixels image is input to the detection network to obtain deep convolutional features. the network will obtain 512 feature maps with the size of 1/16 of the input size. An average of all feature maps from the last convolutional layer will be computed to obtain the pedestrian mask (see Figure 1(b)). The pedestrian mask is with $\frac{N}{16} * \frac{M}{16}$ pixels, which respect to $\frac{N}{16} * \frac{M}{16}$ locations. Each element in the pedestrian mask indicates the probabilities of pedestrians appearing.

After the pedestrian mask are obtained, a threshold can be set up to exclude some locations if their probabilities are lower than a certain threshold. The rest of the regions are proposed. Then we map these proposed regions from these pedestrian masks to original image. Due to the proportion of pedestrian's height and width is 2:1 generally, the proposed region is a bounding box whose ratio of height and weight is 2:1 approximately. Then a set of bounding boxes indicating position of pedestrians can be obtained.

2.3. Post Processing

To handle the variations of size and location of bounding boxes, we propose a multi-task regression model referring to RPN network. The multi- task regression model shares the convolutional features of proposed regions with detection model. We slide a small network over these convolutional features, which usually takes a 3*3 spatial window as the input of convolutional feature map. Each sliding window is mapped into lower-dimensional space. The features in low-dimensional space are fed into two sibling fully-connected layers: a box-adjustment layer and a box-score layer. The special model structure is shown in Fig 1(c).

In the box-adjustment layer, we use the parameterizations of four coordinates for bounding box fine-tuning, which are the same as the results in RPN. In the box-score layer, the IOU between predict box and ground truth is used to represent the score of bounding box, which is different from RPN. Therefore, the multi-task loss function is defined as following:

$$L(p_i, t_i) = \frac{1}{n}\sum_{i=1}^{n}(l(p_i, p_i^{\backslash *}) + l(t_i, t_i^{\backslash *})), \tag{2}$$

where Function $l(x, y)$ is Smooth L1 Loss which is described in Fast-RCNN [10].

As training two regression networks, the parameters of the shared convolutional layer is fixed and only the parameters of the multi-task regression model is updated.

3. Experiment

To validate the effectiveness of the proposed method, extensive experiments on the INRIA person dataset [4] and the TUD-pedestrian dataset are conducted. The experimental results show that our method achieves the better performance by comparing with the generic detector (Faster RCNN) in terms of efficiency and accuracy. The MR(miss rate)-FPPI(false positive per image) and precision-recall (PR) are used as the evaluation metric.

Firstly, we compare the performance of the proposed algorithm and CAM on object location. Figure 3 shows a partial class activation maps obtained by CAM and our algorithm respectively. Obviously, our algorithm has better ability of occluded pedestrian location and small instance location than CAM.

From left to right, sub-figures in Figure 4 show the variation of miss rates of different methods [4, 10,11,12,16,17,18,19,20,21,22] on the two datasets. From this figure, we can see that our method obtains relatively good results among all methods in term of miss rate and precision. Figure 5 shows the results of pedestrian detection our method in the case of FPPI = 0.1 on INRIA person dataset and TUD-pedestrian dataset. The blue, green and red bounding boxes represent missing positives, false positives and true positives respectively.

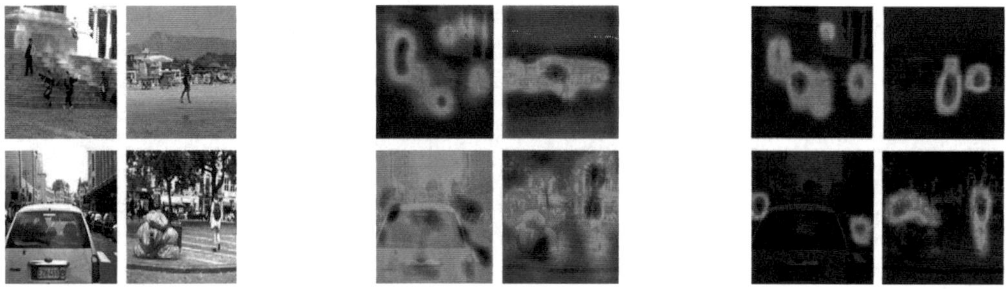

(a) Original images (b) CAM (c) Ours

Figure 3: Some pedestrian activation maps obtained by CAM and our method respectively.

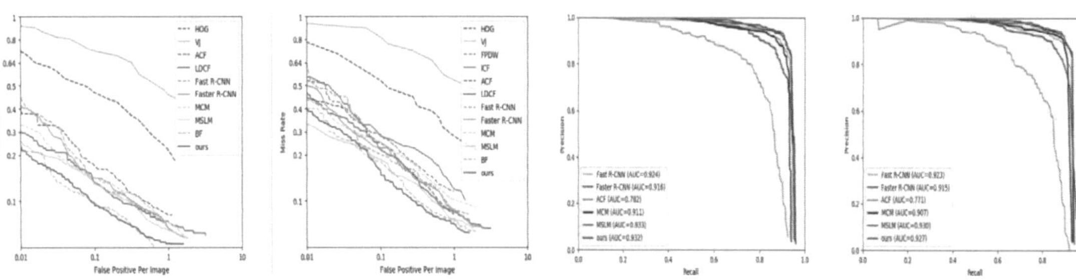

(a) INRIA (b) TUD (c) INRIA (d) TUD

Figure 4: The variations of MR-FPPI of different methods on the INRIA and TUD-pedestrian dataset.

Figure5: Somepedestriandetectionresultsobtainedourmethodonthe INRIA person dataset and TUD-pedestrian dataset.

4. Conclusions

In this paper, we propose an activation based region proposal model for pedestrian detection in complex scene. The model improves the structure of CAM by removing the full connected layer, which significantly improves the accuracy of pedestrian detection in complex scenes. Extensive experiments on two pedestrian datasets demonstrate the efficiency of the proposed model. As a future work, it is necessary to improve the regression model to make the bounding box more accurate.

ACKNOWLEDGMENTS
This research was supported in part by the National Natural Science Foundation of China (NSFC) under grants 61572107, the National Key Research and Development Program of China under grant 2017YFC1703905, and the Sichuan Science and Technology Programs under grant 2018GZ0192, 2018SZ0065.

Reference
[1] Shen X, Sui X, Pan K and Tao Y, 2016, Adaptive pedestrian tracking via patch-based features and spatial-temporal similarity measurement, Pattern Recognit, 163-173.
[2] Andriluka M, Roth S, and Schiele B, 2008, People tracking by detection and people detection by tracking, CVPR, 1-8.
[3] Viola P, Jones M and Snow D, 2003, Detecting pedestrians using patterns of motion and appearance, CVPR, 734-741.
[4] Dalal N and Triggs B, 2005, Histograms of oriented gradients for human detection, CVPR, 886-893.
[5] Gao W, Ai H, and Lao S, 2009, Adaptive contour features in oriented granular space for human detection and segmentation, CVPR, 1786-1793.
[6] Felzenszwalb P, Girshick R, McAllester D and Ramanan D, 2010, Object detection with discriminatively trained partbased models, IEEE Trans. Pattern Anal. Mach. Intell, 1627-1645.
[7] Hinterstoisser S, Lepetit V, Ilic S, Fua P and Navab N, 2010, Dominant orientation templates for real-time detection of texture-less objects, CVPR, 2257-2264.
[8] Yann L, Yoshua B and Geoffrey H, 2006, Deep learning. Nature, 436-436.
[9] Krizhevsky A, Sutskever I and Hinton G, 2012, ImageNet Classification with Deep Convolutional Neural Networks, 1097-1105.
[10] Girshick R, 2015, Fast r-cnn, CVPR, 1440-1448.
[11] Ren S, He K, Girshick R and Sun J, 2015, Faster r-cnn: towards realtime object detection with region proposal networks, NIPS, 91-99.
[12] Li X, Ye M, Liu Y and Zhu C, 2017, Memory-based pedestrian detection through sequence learning, ICME, 1129-1134.
[13] Zhou B, Khosla A, Lapedriza A and Oliva A, 2016, Learning Deep Features for Discriminative Localization, CVPR, 2921-2929.
[14] Neubeck A and Van L, 2006, Efficient non-maximum suppression, ICCV, 850-855.
[15] Simonyan K and Zisserman A, 2015, Very deep convolutional networks for large-scale image recognition, ICLR.
[16] Li X, Ye M, Liu Y, Zhang F, Liu D and Tang S, 2017, Accurate object detection using memory-based models in surveillance scenes, Pattern Recognition, 73-84.
[17] Viola P and Jones M, 2004, Robust Real-time Object Detection, IJCV .
[18] Dollr P, Tu Z, Perona P and Belongie S, 2009, Integral channel features, in: British Machine Vision Conference, 91-98.
[19] Dollr P, Belongie S and Perona P, 2009, The fastest pedestrian detector in the west, BMVC.
[20] Piotr D and Pietro P, 2014, Fast Feature Pyramids for Object Detection, IEEE Trans. Pattern Anal. Mach. Intell, 1532-1545.
[21] Nam W, Dollar P and Han J, 2014, Local decorrelation for improved pedestrian detection, NIPS, 424-432.
[22] Zhang L, Lin L, Liang X and He K, 2016, Is Faster R-CNN Doing Well for Pedestrian Detection, ECCV, 443-457.

IOP Conf. Series: Materials Science and Engineering **646** (2019) 012017 doi:10.1088/1757-899X/646/1/012017

Design and Application of Satellite Automatic Angle Measurement System Based on Robot

Bin Zhang[1,2,3*], Changyu Long[1,2], Wangmin Yi[1,2,3], Xiaoqiang Tang[3], Man Zhang[1,2], Guowei Ruan [1,2] and Wei Wang[1,2]

[1] Beijing Institute of Spacecraft Environment Engineering,Beijing,China, 100094
[2] Assembly Processing Technology Center for Aerospace Product of CASC, Beijing, China, 100094
[3] Tsinghua University, Beijing, China, 100084

*E-mail: zhangbin_bisee@126.com

Abstract: In order to achieve the on-orbit performance of the satellite, the installation positions of some instruments on the satellite are measured when the satellite is integrated on the ground, to ensure the installation position and the accuracy meet the design requirements. With the development of aerospace industry, more and more satellites have the characteristics of difficult development, short development time and heavy position measurement task. In view of the mass produce of satellites, an automatic angle measurement scheme based on robots is designed and applied. Giving full play to the advantages of automation and high measurement efficiency of robots, this system effectively improves the measurement efficiency and reducing the occupancy of personnel.

1. Introduction

Some instruments such as sensors, inertial devices, thrusters , antennas, cameras and etc. have strict requirements of the installation position on spacecrafts. After installation, it is necessary to measure and adjust the installation position to ensure its on-orbit performance. The measurement and adjustment of the installation position accuracy is the alignment of spacecraft, which is an important guarantee for the reliable operation of satellites and on-board instruments[1-3]. In traditional measurement methods, the non-contact large-scale measurement system consisting of several theodolites is used to measure the accuracy of satellite assembly. Theodolite measurement system is a space angle intersection measurement system composed by several high precision electronic theodolites. The hardware equipment of theodolite measurement system is mainly composed of high precision electronic theodolite and reference ruler, interface, on-line cable, computer and so on. It adopts manual operation method.The batch production of satellites have short development cycle and heavy development task. There are a large number of instruments on the satellite which have the requirement of installation position accuracy. It is necessary to replace manual measurement by automatic method to improve measurement efficiency and shorten development cycle[4].

2. Principle and status of satellite precision measurement

Before the measurement, the whole satellite coordinate system is established, that is, the reference cubic prism is installed on the star, and the transformation relationship between the reference cubic prism coordinate system and the coordinate system established by the satellite-rocket interface is obtained by the theodolite station arrangement. When measuring instruments on the satellite, the vector angle and coordinates of the instrument in the whole satellite coordinate system are obtained through the transformation relationship between the cubic prism coordinate system on the instrument and the reference cubic prism coordinate system[5,6].

Theodolite angle measurement is to measure the angle between the mirror normal lines of cubic prisms on different instruments, as shown in Figure 1.

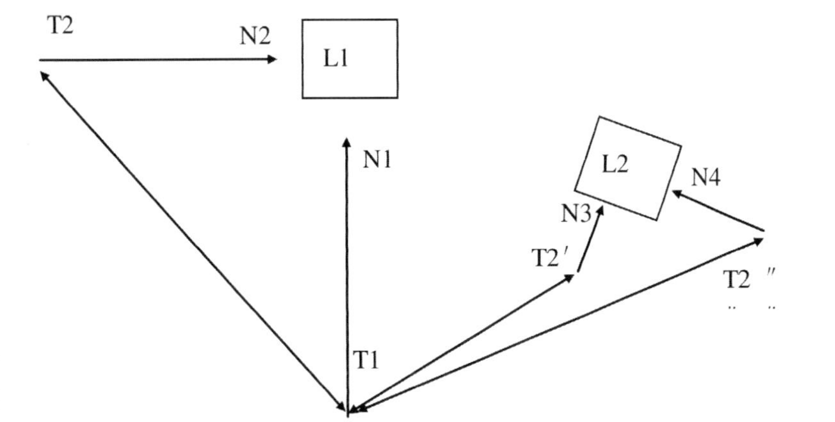

Figure 1. Principle diagram of satellite theodolite angle measurement

L1 and L2 represent two different instruments. It is necessary to measure the angle between mirror normal N3, N4 and L1 of cubic prism L2. In the measurement, firstly, theodolite T1 and T2 are collimated with the mirror of L1 and aiming at each other to obtain the angle vectors of the normal lines N1 and N2 of the cubic prism in the measuring coordinate system, then theodolite T2 and T2', T2', T2 are collimated with the mirror of the cubic prism L2 respectively, and the angle vectors of the normal lines N3 and N4 of the cubic prism in the measuring coordinate system are obtained. Finally, the angle between mirror normal N3, N4 and L1 of cubic prism L2 is obtained by matrix transformation.

The photograph of the angle measurement of the instrument on the satellite by manual manipulation of theodolites is shown in Figure 2. The manual measurement method is inefficient, so it is necessary to find an automatic method.

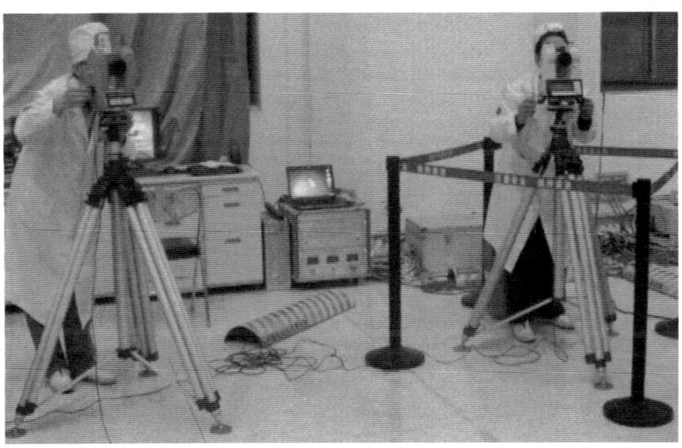

Figure 2. Manual manipulation of theodolite angle measurement

IOP Conf. Series: Materials Science and Engineering **646** (2019) 012017 doi:10.1088/1757-899X/646/1/012017

3. Automatic Angle Measurement Scheme for Satellite

3.1 Measurement System

The composition of the angle measurement system based on industrial robots is shown in Figure 3, which includes industrial robots, laser tracker, theodolite, end-effector, tracking target, vision-guided camera , AGV vehicle, cubic prism of the equipment measured on board.

The system uses robots to carry theodolites to measure the angle of instruments with high position precision on board. The robots can move to all the instruments on board by using AGV vehicle, so as to accomplish the omnidirectional measurement for all the instruments on the satellite. Laser tracker is used as the global guidance and precision control device in the system. Theodolite is the position measurement. The end -effector is used to fix and support the theodolite and the tracking target. The tracking target is used to establish the relationship between laser tracker and the end-effector. Through the end-effector, the relationship of theodolite, robot and laser tracker can be obtained. The vision-guided camera is used to read the collimated light in theodolite instead of human eyes.

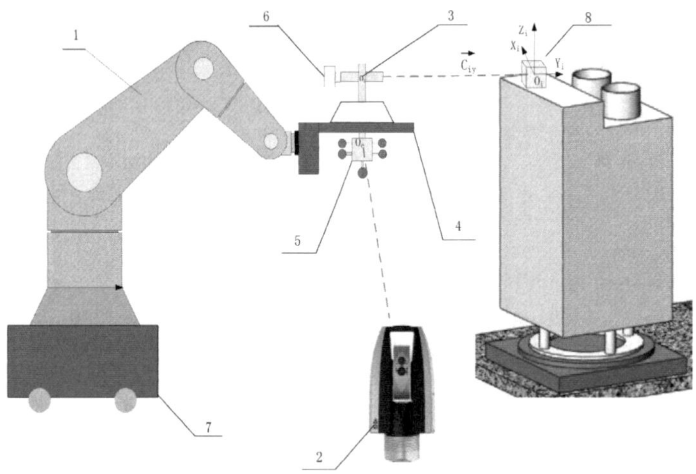

1. robot ;
2. laser tracker ;
3. theodolite; 4.end-effector;
5. tracking target ;
6.vision-guided camera;
7.AGV vehicle;
8.cubic prism。

Figure 3. Automatic angle measurement system based on robot

Firstly, the relationship between theodolite coordinate system, the robot tool coordinate system and the laser tracking target is established by calibration. Secondly, the coordinate system of the tracker and the robot and spacecraft coordinate system need to be calibrated and unified. According to the theoretical position of the instrument to be measured in the satellite coordinate system. With the help of the real-time tracking, positioning and guiding technology of laser tracker, the theodolite can automatically and accurately collimate the measured equipment. Finally, the collimation information of theodolite at various positions is unified into the coordinate system of tracker to realize the automatic measurement of the attitude information of spacecraft equipment.

3.2 Measurement Scheme and Process

According to the measurement principle, the system needs four steps to complete the measurement: acquisition of theoretical position information, system calibration, model driven auto-collimation and coordinate system construction.

1) Acquisition of theoretical position information of equipment to be measured

The theoretical position values of the cubic prism attitude information of all the instruments to be measured are obtained from three-dimensional model, and it is used to drive the robot measurement system to the theoretical position.

2) System calibration

Before measurement, each subsystem needs to be calibrated, which includes three parts: calibration and rapid construction of satellite mechanical coordinate system, calibration and rapid construction of robot body coordinate system, and calibration of theodolite coordinate system at the end of robot. The calibration of satellite mechanical coordinate system is to establish the relative relationship between the satellite mechanical coordinate system and the fixed target on the satellite supporting equipment, so as to facilitate the rapid establishment of the relative orientation relationship between the satellite and the laser tracker in the measurement field. The robot body coordinate system calibration is to establish the relative relationship between the robot body coordinate system and the fixed target on the robot basement. It is convenient to quickly establish the relative orientation relationship between the robot and the laser tracker and calibrate the robot in the field. Calibration of theodolite coordinate system at the end of robot is to determine the relative orientation relationship between theodolite coordinate system and flange coordinate system at the end of robot and laser tracking target respectively.

3) Automated collimation

After completing the overall calibration of the system, it is necessary to import the theoretical position and orientation information of the cubic prism in the satellite mechanical coordinate system into the measurement software. The robot can be driven to carry theodolite to achieve rough alignment of the equipment under test by three-dimensional model transformation operation, and then the automatic and precise alignment of the equipment under test can be achieved by combining the technology of vision-guided alignment.

4) Reconstruction of instrument coordinate system

The normal vector of the cubic prism obtained by collimating is in theodolite's coordinate system. However, for the collimation of different cubic prisms, the orientations of theodolite's coordinate system are different. In order to determine the relative attitude matrix between different cubic prisms, the results of collimation at different positions will be unified in the same coordinate system to complete the reconstruction of instrument coordinate system.

4.Verification

The system is integrated and tested, and the indicators of the system are verified by experiments. Cubic prisms with orthogonal planes whose values were obtained by metrology testing are employed. The two orthogonal planes of Cubic prisms distributed in different positions and attitudes are measured by this automatic measuring system based on robot. The angle between the two planes is calculated according to the measuring datum. The measurement errors are obtained by comparing with the standard values of the cubic prisms. As shown in Table 1, the test results show that the measurement error of the system is better than 15″, and it has high measurement accuracy, which can meet the requirements of the satellite.

Table 1. Testing result by the automatic measuring system based on robot

Planes	**Angles of the normal vector (°)**			**Angles of the two planes by measuring (°)**	**Angles of the two planes by metrology (°)**	**Error (″)**
CP[a1]1-P[a2]1	0.495958	0.8283879	0.26038228	89.99745	89.9992	6.3
CP[a1]1-P[a2]2	-0.85738	0.5146662	-0.0040613			
CP[a1]2-P[a2]1	0.476782	0.8019360	0.35996843	89.99608	89.9993	11.6
CP[a1]2-P[a2]2	-0.85761	0.5142100	-0.0096010			
CP[a1]3-P[a2]1	0.515287	0.8570135	0.00268953	89.99608	89.9995°	12.3
CP[a1]3-P[a2]2	-0.84285	0.5074933	-0.1790332			

[a1] *In this table, CP is short for Cubic Prism.*
[a2] *In this table, P is short for Plane.*

5. Application

The robot-based automatic measurement system is applied to the installation position angle measurement of instruments on the satellite. The application picture is shown in Figure 4. According to the statistics of measuring process time, the measuring time of a single cubic prism on the instrument is less than 15 minutes based on the automatic measuring system of industrial robots. The measuring time is at least 30 minutes by traditional manual method with 3 persons. Therefore, the efficiency is increased[7].

Figure 4. Application of automatic angle measurement system based on robot for satellite

6. Conclusion

In this paper, an automatic angle measurement system based on robots is designed, which is aimed at to be instead of the manual manipulation of theodolites. After design, the system is integrated and applied on the instrument position measurement for satellite. The measurement efficiency is improved and the occupancy of personnel is reduced.

Reference

[1] Wang W, Wan B L, Guo J Y, Tao L, Liu H M, New Method of Spacecraft Integration Precision Measurement Based on Virtual Cubic Prism, *Ordnance Industry Automation*, 2017 36(6):32-36

[2] Liu J X, Discussion on Reliability of Space Position Accuracy Measurement System, *Spacecraft Environment Engineering*, 2008 25(2):183-188

[3] Wang W, Liu , Guo J Y, Sun G, Liu H M, A vector calculation method of irregular prism for spacecraft integration precision measurement, *Spacecraft Environment Engineering*, 2013 30(1) 103: 106

[4] Yi W M, Long C Y, Hu R Q, Design and Application of Spacecraft Assembly and Testing Robot System, Aerospace China, 2019 (2):30-33

[5] Yang Z H, An Analysis of Measurement Technology in Spacecraft Assembly, *Spacecraft Environment Engineering*, 2007 24(6): 390-392

[6] Ye S H, Zhu J G, Zhang Z L, Status And Development of Large-scale Coordinate Measurement Research, *Acta Metrologica Sinica,* 2008, 29(s1): 1-6

[7] Klimehik A, Ambiehl A, Gamier S, et al. Efficiency evaluation of robots in machining applications using industrial performance measure. Robotics&Computer Integrated Manufacturing, 2017, 48: 1229.

IOP Conf. Series: Materials Science and Engineering **646** (2019) 012018 doi:10.1088/1757-899X/646/1/012018

A Glucose Prediction Model based on Variational Mode Decomposition and Least Squares Support Vector Regression

S Wen, H R Li, H H Han and X Yu*

School of Information Science and Engineering, Northeastern University, Shenyang 110819, PR China

*yuxia@ise.neu.edu.cn

Abstract. Online prediction of subcutaneous glucose concentration plays a critical role in glucose management for type 1 diabetes. In this work, a new method combining Variational Mode Decomposition (VMD) and Least Squares Support Vector Regression (LSSVR) is proposed with three main stages to improve the prediction accuracy. Firstly, the time series of blood glucose are decomposed into different frequency series by VMD method. Secondly, the LSSVR model is trained to predict each subsequence. Finally, the predicted sequences are reconstructed to obtain the overall glucose predictions. The experimental results demonstrate the effectiveness and accuracy of the proposed model for short term glucose prediction.

1. Introduction

Type 1 diabetes (T1D) is a disease characterized by the inability of the body to regulate blood glucose concentration[1, 2]. In recent years, with development of Continuous Glucose Monitoring system (CGMs), new opportunities for glycemic management appear by detecting the glucose values of subcutaneous interstitial fluid through continuous glucose sensors and providing blood glucose concentrations with predefined time intervals. In order to design an effective glucose regulations for T1D patients, a reliable and appropriate prediction model becomes necessary to obtain an accurate 30min short-term-ahead blood glucose concentration[3].

Consider of the nonlinear and non-stationary of blood glucose time series, the prediction accuracy and the computational efficiency of the prediction model has attracted more and more attention. In the literatures, various approaches have been proposed to predict blood glucose levels. Compared with physiological model, the data-driven model could neglect the prior knowledge of physiology and only applies time-series signals, from CGM, to predict short-term glucose concentration. Sparacino et al.[4] compared the predictive performance of a first-order polynomial model with a first-order autoregressive (AR) model, which obtained that the AR model with time-varying parameters shown more accurate prediction performance then general physiological model. Pérez et al.[5] used a feed-forward ANN model based on the CGM values, and Hamdi et al.[6] proposed the combination of SVR method and differential evolution algorithm to improve prediction performance. Recently, the decomposition method has been utilized in prediction models to analyze the time-series[7, 8]. The most commonly used signal decomposition methods include wavelet transform, Empirical Mode Decomposition (EMD) and Variational Mode Decomposition (VMD)[9-11]. In this paper, a designed VMD method is used to overcome the mode mixing problem of EMD, and the LSSVR algorithm is applied to learn from different mode component, and obtains 30min short-term-ahead predicted components. Then, superimposing the predicted components to obtain the obtain the overall glucose predictions.

The rest of the paper is organized as follows: Section 2 introduces the general principle of VMD method, and Section 3 describes the LSSVR algorithm. Section 4 proposes the VMD-LSSVR prediction model and the parameter selection regulation. The simulation experiments and the discussion of results are introduced in Section 5 which includes a comparison with previous works to highlight the reliability of the proposed method. Finally, the conclusion is presented in Section 6.

2. Variational Mode Decomposition (VMD) methodology

VMD algorithm was proposed in the basis of EMD algorithm in 2014[12], which could avoid model mixing. It is a decomposition method commonly used for non-stationary signals. The purpose of VMD is to decompose the original signal into K band-limited modes u_k (k=1, ..., K), and it assumes that each subseries is considered to compact around the center pulsation (ω_k). The bandwidth of each subseries is estimated as following steps: The time series of blood glucose f is decomposed by Hilbert transform, and u_k is calculated to have the corresponding unilateral spectrum; The frequency spectrum of each mode $u_k(t)$ is modulated to the corresponding baseband, by mixing $e^{-j\omega_k t}$ tuned to the respective estimated the center frequency (ω_k); The bandwidth is estimated by calculating the squared L^2 norm of demodulated signal gradient. Subsequently, constrained variational problem could be described as follows:

$$\begin{cases} \min_{\{u_k\}\{\omega_k\}} \left\{ \sum_k \left\| \partial_t \left[\left(\delta(t) + \frac{j}{\pi t} \right) * u_k(t) \right] e^{-j\omega_k t} \right\|_2^2 \right\} \\ \text{s.t.} \sum_k u_k = f \end{cases} \tag{1}$$

where $\delta(t)$ is the Fermi-Dirac distribution, $\{u_k\} = \{u_1, \ldots, u_K\}$, $\sum_k = \sum_{k=1}^{K}$ $\{\omega_k\} = \{\omega_1, \ldots, \omega_K\}$, and ω_k are the center frequency of each mode of u_k. In order to obtain the optimal solution, the augmented Lagrange equation is introduced as:

$$L(\{u_k\}, \{\omega_k\}, \lambda) = \alpha \sum_k \left\| \partial_t \left[\left(\delta(t) + \frac{j}{\pi t} \right) * u_k(t) \right] e^{-j\omega_k t} \right\|^2 + \| f(t) - \sum_k u_k(t) \|_2^2 + \langle \lambda(t), f(t) - \sum_k u_k(t) \rangle \tag{2}$$

where α is the quadratic penalty factor and $\lambda(t)$ is the Lagrangian multiplier. Iteratively optimize Lagrange equation of Equation (2) by alternate direction method of multipliers (ADMM). The solutions are expressed as:

$$\hat{u}_k^{n+1}(\omega) = \frac{\hat{f}(\omega) - \sum_{i \neq k} \hat{u}(\omega) + \frac{\hat{\lambda}(\omega)}{2}}{1 + 2\alpha(\omega - \omega_k)^2} \tag{3}$$

$$\omega_k^{n+1} = \frac{\int_0^\infty \omega |\hat{u}_k(\omega)|^2 d\omega}{\int_0^\infty |\hat{u}_k(\omega)|^2 d\omega} \tag{4}$$

where n is the number of iterations, $\hat{f}(\omega)$, $\hat{u}_k^{n+1}$, $\hat{u}(\omega)$, and $\hat{\lambda}(\omega)$ stand for the Fourier transform of $f(t), u_k^{n+1}(t), u(t)$ and $\lambda(t)$.

3. Least Squares Support Vector Regression (LSSVR)

The Support Vector Regression (SVR) is one of the regression algorithms in machine learning based on statistical learning theory. LSSVR is an improved algorithm of SVR, which replaces inequality constraints into equality constraints. Then, the quadratic programming problem in SVR could be converted into linear equations. Therefore, the LSSVR algorithm could reduce computational complexity and increase computational speed.

The training dataset of the LSSVR is $T = \{(x_i, y_i | i = 1,2,3, \ldots. l\}$, where x_i is the input data $x_i \in R^n$, and y_i is the output data $y_i \in R^n$. The optimal decision function can be identified by the high-dimensional feature space. The regression function adopted by LSSVR can be described as follows

$$y = \omega^T\Psi(x) + b \tag{5}$$

where $\Psi(x)$ is the nonlinear mapping function, ω represents the weight coefficient vector, and b stands for bias. The objective function is as follows:

$$\begin{cases} \min Z(\omega, e) = \frac{1}{2}\omega^T\omega + \frac{\gamma}{2}\sum_{i=1}^{l} e_i^2 \\ s.t. y_i = \omega^T\psi(x_i) + b + e_i \end{cases} \tag{6}$$

where e_i represents the error and γ donates the penalty coefficient. The optimization problem in Eq. (6) can be translated into linear equations by means of the Lagrange which solved be based on Karush-Kuhn-Tucher (KKT) conditions. Subsequently, in this study, Gaussian Radial Basis Function (RBF) is chosen as the kernel function of LSSVR, and the expression of the RBF kernel function is as follow:

$$K(x, x_i) = \exp\left\{-\frac{\|x - x_i\|^2}{2\sigma^2}\right\} \tag{7}$$

where σ^2 is kernel parameter. Therefore, the function of LSSVR can be obtained as follows:

$$y = \sum_{i=1}^{l} \lambda_i K(x, x_i) = \sum_{i=1}^{l} \lambda_i \exp\{-\frac{\|x - x_i\|^2}{2\sigma^2}\} \tag{8}$$

For the parameters γ and σ^2, the trial-and-error method is considered and applied to obtain the optimal parameters which has the basic guideline as follows: Firstly, randomly initializes the parameters, and observes the prediction accuracy, Then, fixes one parameter, and changes the other until the prediction accuracy becomes a constant level. Finally, gets the other parameter in the same steps.

4. Overview of the VMD-LSSVR model

The structure of the proposed VMD-LSSVR prediction model is mainly composed of three stages, which could be described as follows:

Stage 1: The data preprocessing is adapted to reset the historical glucose data into 5-minutes sampling points. Then the glucose data are decomposed by the VMD-based decomposition model, and obtain a number of different subseries $\{u_1, u_2, \ldots, u_K\}$, and the training sets and test sets are constructed by subseries.

Stage 2: The forecasting components of subseries are trained by LSSVR algorithm. Then, updating the training sets by current glucose values to achieve online prediction.

Stage 3: The components of all modes are superimposed to obtain the overall glucose prediction.

5. Simulation Experiments and Results

In this section, the prediction method mentioned above were programmed in MATLAB 2019a to verify for online training and short-term glucose prediction. The experiments are based on the blood glucose time series from a silico subject in the FDA-approved UVa/Padova metabolic simulator. The entire dataset has a total of 863 points. The training samples and forecasting samples are made by using a sliding window with a capacity of 500 that moves one step per iteration. In the window, the first 494 blood glucose data are training samples, and the rest are forecasting samples. First of all, the training samples are decomposed into 7 separate sequences by VMD. Next, all subsequences are applied to predict the corresponding prediction values by LSSVR algorithm. Then, the predictions could be obtained from the combined 7 subsequences.

In order to assess the prediction performances of the VMD-LSSVR algorithm, Figure 1 presents a comparison between the actual blood glucose and the prediction results for a T1D patient for 30min prediction horizon (PH=30min). It can be seen that the proposed VMD-LSSVR method is remarkably accurate and could be potentially applied for time series prediction.

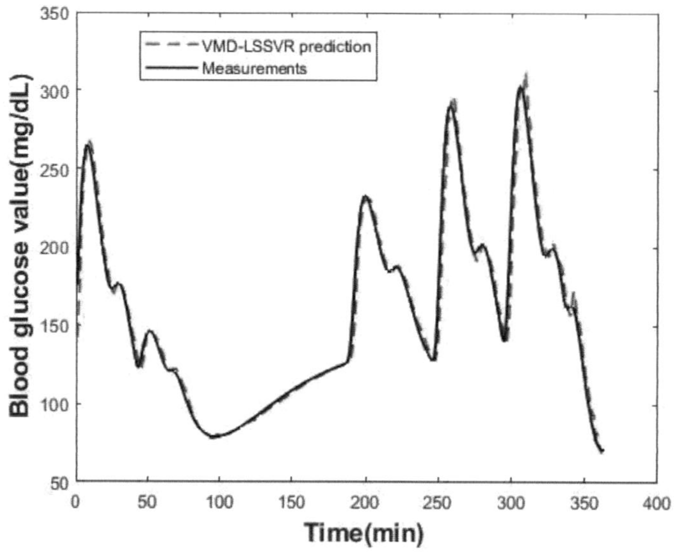

Figure 1. Prediction results of blood glucose for PH=30 min.

The Clarke error grid is shown in Figure 2. As shown, the proposed model has good fitting performance, and most points fall in 'A' area, which indicates that the accuracy of VMD-LSSVR model is high and within the acceptable range.

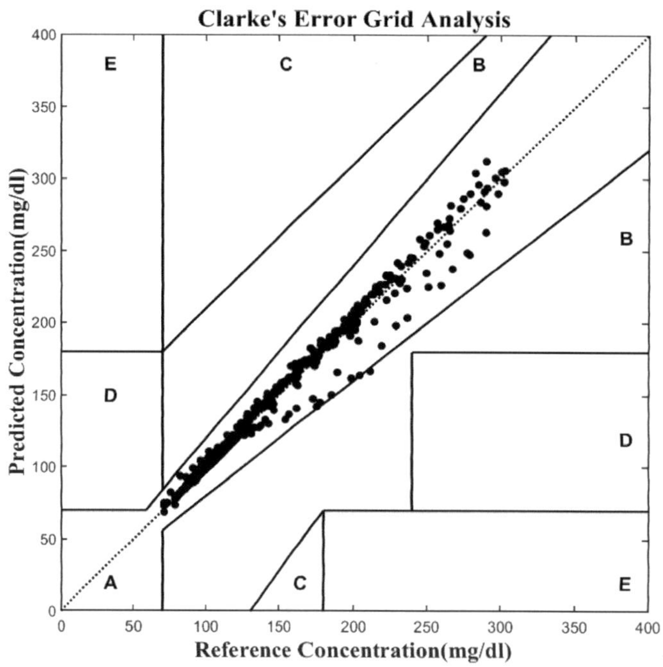

Figure 2. The Clarke error grid of blood glucose prediction for PH=30 min.

To demonstrate that the proposed method can effectively improve the prediction performance compared to another method, a comparable experiment is carried out in this paper. Figure 3 illustrates the comparison between the actual value and the predicted values by LSSVR and VMD-LSSVR. It is obvious that the proposed method VMD-LSSVR has better prediction performance, resulting in higher accuracy.

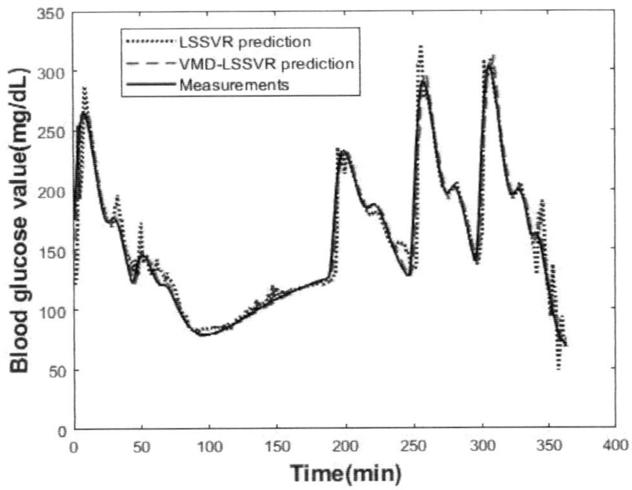

Figure 3. Blood glucose prediction of different algorithms for PH=30 min.

where $\hat{y}_i$ denotes the short-term glucose prediction, y_i donates the test set, i represents the sampling point, N stands for the total number of blood glucose time series and t denotes the number of training sets.

To characterize the prediction accuracy compared with actual CGM values, two metrics, Root Mean Square Error (RMSE) and Mean Absolute Percentage Error (MAPE), are used and calculated as following equations:

$$\mathrm{RMSE} = \sqrt{\frac{1}{N-t}\sum_{i=t+1}^{N}(\hat{y}_i - y_i)^2} \tag{9}$$

$$\mathrm{MAPE} = \frac{1}{N}\sum_{i \in N}\left|\frac{y(i)-\hat{y}(i)}{y(i)}\right| \times 100\% \tag{10}$$

The RMSE and MAPE for LSSVR and VMD-LSSVR are reported in Table 1. It is readily observed that the performance of the proposed prediction algorithm is superior than the conventional LSSVR algorithm.

Table 1. Prediction accuracy metrics of LSSVR and VMD-LSSVR.

Model	RMSE	MAPE (%)
LSSVR	15.9093	5.5652
VMD-LSSVR	8. 9067	3.0537

6. Conclusion

In this paper, a new prediction algorithm, VMD-LSSVR, is developed for online prediction to make the optimum structure and accurate predictions. Through the VMD-based decomposition model, the nonlinear and nonstationary characteristics of blood glucose time-series are considered, and the 30min-ahead glucose concentration is obtained by summing all predictions of each subseries. Simulation results illustrates that the proposed method yielded better online prediction performance in terms of accuracy and stability than the conventional standard LSSVR.

Acknowledgements

This work was supported by the National Key R&D Program of China (2017YFB0304205), the National Natural Science Foundation of China (61533007).

References

[1] A Cinar 2018 *Advances in artificial pancreas systems adaptive and multivariable predictive control*

[2] B W Bequette 2012 *Challenges and recent progress in the development of a closed-loop artificial pancreas* Annual reviews in control, vol. 36, no. 2, pp. 255-266

[3] H Kirchsteiger, J B Jørgensen, E Renard and L Del Re 2015 *Prediction Methods for Blood Glucose Concentration: Design, Use and Evaluation.* Springer

[4] G Sparacino, F Zanderigo, S Corazza, A Maran, A Facchinetti and C. Cobelli 2007 *Glucose Concentration can be Predicted Ahead in Time From Continuous Glucose Monitoring Sensor Time-Series* IEEE Transactions on Biomedical Engineering, vol. 54, pp. 931-7

[5] C. Pérez-Gandía et al 2010 *Artificial neural network algorithm for online glucose prediction from continuous glucose monitoring* Diabetes technology & therapeutics, vol. 12, no. 1, pp. 81-88

[6] T Hamdi, J Ben Ali, V Di Costanzo, F Fnaiech, E Moreau and J M Ginoux 2018 A*ccurate prediction of continuous blood glucose based on support vector regression and differential evolution algorithm* Biocybernetics and Biomedical Engineering, vol. 38, no. 2, pp. 362-372.

[7] A A Abdoos 2016 *A new intelligent method based on combination of VMD and ELM for short term wind power forecasting*, Neurocomputing, vol. 203, pp. 111-120, Aug 26.

[8] G H Li, X Ma and H. Yang 2018 *A Hybrid Model for Forecasting Sunspots Time Series Based on Variational Mode Decomposition and Backpropagation Neural Network Improved by Firefly Algorithm*, Computational Intelligence and Neuroscience

[9] X. D. Wang, Q. B. Yu, and Y. Yang 2018 *Short-term wind speed forecasting using variational mode decomposition and support vector regression*, Journal of Intelligent & Fuzzy Systems, vol. 34, no. 6, pp. 3811-3820

[10] R Bisoi, P K Dash and A K Parida 2019 *Hybrid Variational Mode Decomposition and evolutionary robust kernel extreme learning machine for stock price and movement prediction on daily basis*, Applied Soft Computing, vol. 74, pp. 652-678

[11] P H Ni et al. 2018 *Time-varying system identification using variational mode decomposition*, Structural Control & Health Monitoring, vol. 25, no. 6

[12] C Aneesh, S Kumar, P M Hisham and K P Soman 2015 *Performance comparison of Variational Mode Decomposition over Empirical Wavelet Transform for the classification of power quality disturbances using Support Vector Machine*, Proceedings of the International Conference on Information and Communication Technologies, Icict 2014, vol. 46, pp. 372-380

Design of Dual-Rate PIM-Disturbance Regulators for a DC-DC Switching Converter

Naoya Nishinaga[1], Noriyuki Hori[1,a] and Xinyue Ruan[1,b]

[1]Department of Intelligent Interaction Technologies Systems and Information Engineering, University of Tsukuba, 1-1-1 Tennoudai, Tsukuba, Ibaraki, 305-8573 Japan

[a]hori@iit.tsukuba.ac.jp; [b]genn@digicon-lab.esys.tsukuba.ac.jp

Abstract. Three types of dual-rate digital controllers are designed to suppress the output voltage fluctuations caused by load-current changes for a buck DC-DC switching converter running at 300kHz PWM. The slower-rate outer loop is designed using the IA-PIM method, which stabilizes the outer loop at a slow rate of 25kHz. Since the time lag in sensing the disturbance is critical, a faster loop of 300kHz is added based on three different approaches. The resulting dual-rate control laws are implemented on a single 16-bit fixed-point micro-computer with the clock frequency of 140MHz, via 12-bit A/D and D/A converters. Experiments on the fluctuation of the output voltage caused by load-current transients show that, while the analog regulator suppresses the maximum output fluctuation to be 4.7 % of the output voltage, one of the three digital control methods achieves 4%, without adding extra analog components, such as capacitors as in our previous paper.

1. Introduction

While analog controllers have played a major role in the development of high-performance power converters [1-3], they are being taken over by digital controllers [4-8] due to increased computational power at low costs. As the clock-speed of the processor increases, the frequencies of switching, PWM, and the converter itself follow suits, resulting in never ending quest for algorithms with high accuracy at low or modest sampling rates. Areas of application of such controllers are expanding quickly to higher frequency domains, and demands for control laws that require relatively low sampling rates still exist. A slower sampling rate usually leads to a lower computational load and a reduced power consumption, but often to poorer or less stable performances. This should be overcome in view of looming expectations that digital control methods are practical tools to realize high performances for the next generation of power supplies [9].

A pulse-frequency-modulation (PFM)-based digital controller is designed in [10]-[11] for an LLC resonant DC-DC converter. Since the sudden change in the output voltage can cause shut-down and even break-down of the devices connected to the power supply, the regulator must respond quickly to such changes. As the plant input is a change in the switching frequency, a variable sampling-period is required to achieve soft-switching, for which an adequate digital control theory still lacks in dealing with variable rate technique. In [12], a pulse-width-modulation (PWM)-based digital controller is considered, which can be implemented using a commonly used scheme of a uniform sampling period. The analog controller is modified, taking into account the possible performance degradation in the disturbance rejection properties due to discretization. This is achieved by increasing the output capacitance, and the analog controller is redesigned. While this redesigned analog controller is successfully discretized into an Integral-Action Plant-Input-Mapping (IA-PIM) digital controller, an increase in the number of analog components is not desirable due to higher costs, larger size, and lower reliability. In the present paper, therefore, an internal fast-rate loop is added such that the duty ratio of the PWM is adjusted to counteract the disturbance, thus without adding analog components.

In Section2, the IA-PIM design method is reviewed very briefly. In Section 3, this method is applied to the buck DC-DC converter, to obtain the digital controller running at the slow sampling rate. Since the microcomputer has spare computational capacity for additional calculations, three methods are proposed and implemented. The duty-ratio is adjusted to supress the output-voltage fluctuations against load-current changes. Experimental results are also compared. Conclusions are drawn in Section 4.

2. Low-rate stabilization

2.1. System Models

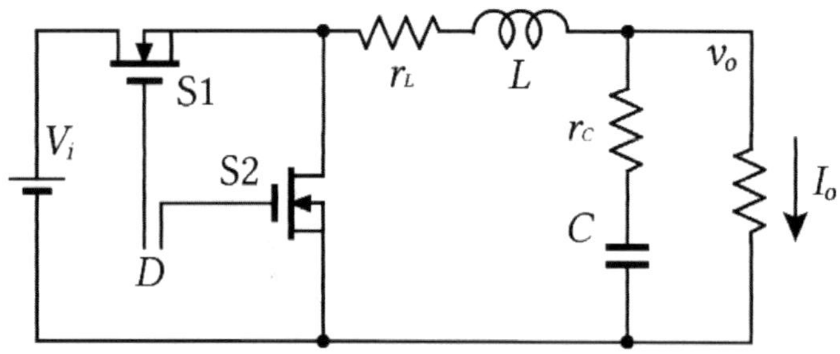

Figure 1. The switching DC-DC buck converter.

Figure. 1 is a schematic of the switching DC-DC buck converter considered in the study, where the control input is the duty ratio D of the PWM signal and the output is the voltage V_O. To design a model-based control system, a plant model is needed, such as a state-space plant model based-on a state averaging method [3], from which the dynamic model can be extracted and the transfer functions can be derived [13]. In the end, the output voltage is determined by ΔD, the change in the duty ratio of the PWM, ΔI_O, the change in the output current, and ΔV_I, the change in the input voltage [14], as

$$\Delta V_O(s) = G_1(s)\Delta D(s) + G_2(s)\Delta I_O(s) + G_3(s)\Delta V_I(s), \quad (1)$$

where

$$G_1(s) = \frac{\Delta V_O(s)}{\Delta D(s)} = \frac{Cr_C s + 1}{P(s)} V_I \quad (2)$$

$$G_2(s) = \frac{\Delta V_O(s)}{-\Delta I_O(s)} = \frac{LCr_C s^2 (L + Cr_L r_C s + r_L)}{P(s)} \quad (3)$$

$$G_3(s) = \frac{\Delta V_O(s)}{\Delta V_I(s)} = \frac{Cr_C s + 1}{P(s)} D, \quad (4)$$

and

$$P(s) = LCs^2 + C(r_L + r_C)s + 1. \quad (5)$$

The device used in the study is TPS40170EVM-758 by Texas Instruments, whose input voltage is 12V, the output 5V, L=8.2μH, C=64μF, and the PWM frequency 300kHz. Transfer function $G_1(s)$ is identified experimentally [14] as

$$G_1(s) = 13.13\frac{5.305\times10^{-7}s+1}{2.585\times10^{-10}s^2+1.961\times10^{-6}s+1}. \quad (6)$$

In the rest of the paper, the deviation Δ terms will be denoted by the lower-case letters.

2.2. Analog Control

The control objective is to maintain the output voltage constant under changes in the load-current. As shown in Figure 2, the analog controller senses the output voltage and compare it with the reference voltage. Their difference is used to adjust the duty ratio to the analog PWM, which in turn activates the gate-driver to switch the transistor S1 and S2 so that the supply voltage of 12V is regulated down to 3V. The analog controller used is TPS40170, for which the frequency response was measured and the transfer function derived [12] as

$$C(s) = 6728\frac{8.331\times10^{-10}s^2+5.793\times10^{-5}s+1}{s(6.608\times10^{-13}s^2+1.626\times10^{-6}s+1)}. \quad (7)$$

The gain margin is 15dB, the phase margin 54 degrees, and the bandwidth 43.5kHz. Experiments have been conducted when the load-current is changed from 2A to 4A. A typical result is shown in Figure. 3, where the maximum amplitude of the deviation is about 140mV from the nominal value of 3V (4.7%).

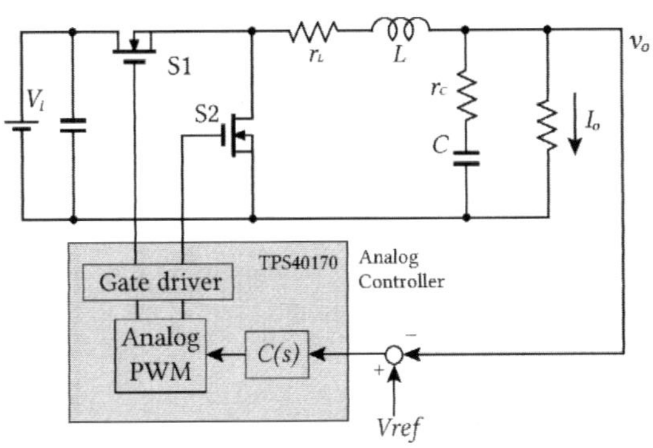

Figure 2. The buck converter under analog control.

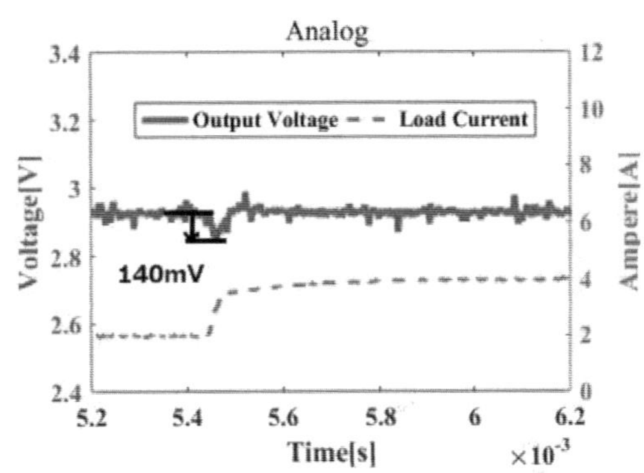

Figure 3. Output-voltage regulation under analog control for load-current change from 2A to 4A

2.3. Tustin Discretizatio

The most popular Tustin's method, where analog controller blocks are individually discretized, by substituting

$$s = \frac{\varepsilon}{\frac{T}{2}\varepsilon+1}, \quad \varepsilon = \frac{z-1}{T} \quad (8)$$

with z the usual zee operator and T the sampling period [15]. These discrete-time blocks can be implemented as in Figure. 4. It is known that its performance approaches that of the analog original as the sampling rate increases. However, unless the sampling rate is sufficiently high, its performance is usually poor and can even be unstable. The device used for digital controllers converts 12V supply to 5V output, while the analog version is from 12V to 3V. Figure. 5 shows the result of Tustin control running at 300kHz. The output voltage is oscillatory and is not regulated.

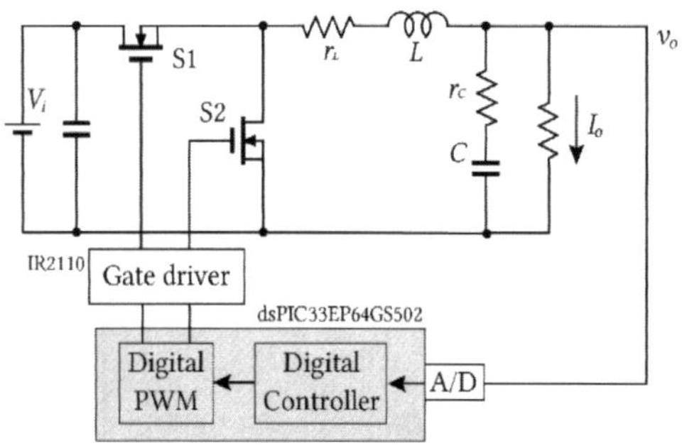

Figure 4. The buck converter under digital control.

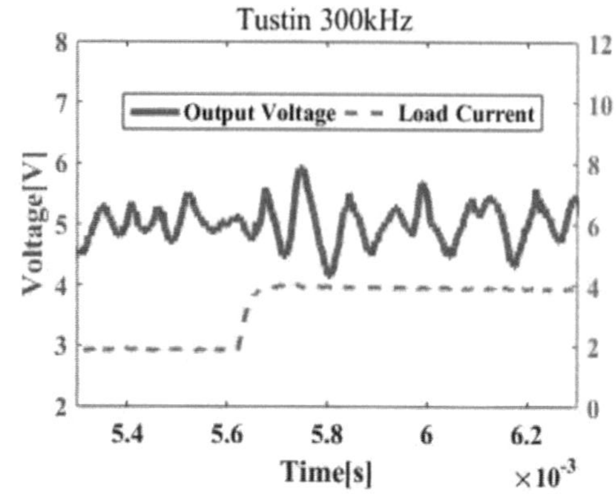

Figure 5. Output-voltage regulation under Tustin control at 300kHz for load-current change from 2A to 4A.

2.4. IA-PIM Discretization

The IA-PIM method [16] is a digital redesign method, where a type-one analog controller is discretized into an equivalent type-one digital controller with guaranteed stability for non-pathological sampling intervals. Thus, the IA-PIM method is used to convert an analog controller in a stable manner at a low sampling frequency. The backbone of this method is the PIM, which is briefly reviewed in the following:

For a plant given by

$$G(s)=\frac{n(s)}{d(s)}, \tag{9}$$

with u(t) the input and y(t) the output, an analog controller (Figure. 6) is designed such that the transfer function from the reference input r(t) to the plant input u(t) can be written as

$$M(s)=\frac{u(s)}{r(s)}=\frac{A(s)C(s)}{1+G(s)C(s)B(s)}. \tag{10}$$

This is called the plan-input transfer-function [17]. In essence, the plant-input-mapping (PIM) method uses the step-invariant-model of the plant, which is given by

$$G(\varepsilon)=SIM\{G(s)\}=\frac{n(\varepsilon)}{d(\varepsilon)}, \tag{11}$$

and the matched-pole-zero model of the plant-input transfer-function, given by

$$M(\varepsilon)=MPZ\{M(s)\}=\frac{\mu(\varepsilon)d(\varepsilon)}{d_M(\varepsilon)}. \tag{12}$$

The PIM digital controller is realized as in Figure. 7 [16], where each control block is determined by

$$A(\varepsilon)=\frac{\mu(\varepsilon)}{\lambda(\varepsilon)}, B(\varepsilon)=\frac{\beta(\varepsilon)}{\lambda(\varepsilon)}, C(\varepsilon)=\frac{\lambda(\varepsilon)}{\alpha(\varepsilon)}. \tag{13}$$

$$d_M(\varepsilon)=\beta(\varepsilon)n(\varepsilon)+\alpha(\varepsilon)d(\varepsilon). \tag{14}$$

These choices achieve eq. (10). The numerator μ of block A is the part of the numerator in Eq. (7) that is not the denominator of the SIM of the plant Eq. (3). This is called the controller zero principle [18]. Eq. (14) is the design equation to be solved for polynomials α and β. λ is an arbitrary stable observer polynomial of appropriate degree [16]. A general PIM method may change the system types before and after discretization. The IA-PIM method [16] solves Eq. (7) such that the resulting block *C* contains the pole at the origin, making the digital controller type-one.

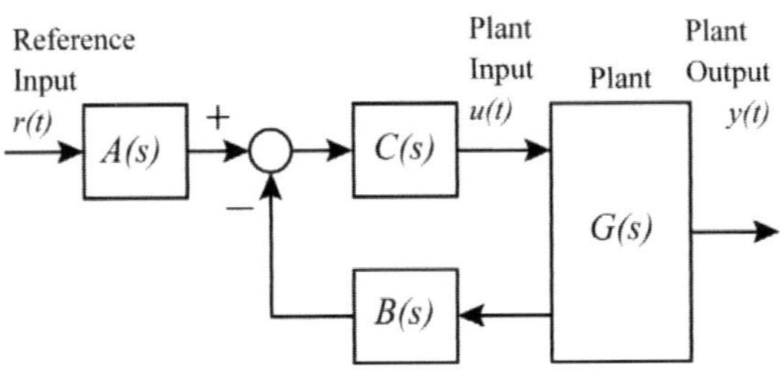

Figure 6. The analog control system.

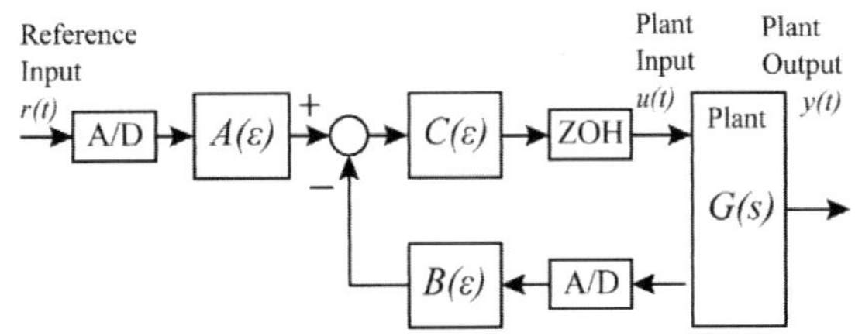

Figure 7. The IA-PIM digital control system.

The IA-PIM controller is designed for the plant, eq. (2), and the controller, eq. (7), at the sampling frequency of 25kHz as

$$B(\varepsilon)=\frac{-1.064\times10^{-9}\varepsilon^{2}+4.520\times10^{-5}\varepsilon+1}{6.400\times10^{-10}\varepsilon^{2}+1.600\times10^{-5}\varepsilon+1} \tag{15}$$

$$C(\varepsilon)=497.9\frac{6.400\times10^{-9}\varepsilon^{2}+1.600\times10^{-4}\varepsilon+1}{\varepsilon(1.461\times10^{-9}\varepsilon^{2}+8.548\times10^{-5}\varepsilon+1)} \tag{16}$$

and implemented as shown in Figure. 8 at the. As shown in Figure. 9, although the resulting output voltage is stabilized at this low sampling frequency, there is a large transient of about 800mV (16%).

The performance of the IA-PIM might improve if the processor were capable of running the control algorithm at a faster rate. However, the CPU used for experiments can handle only up to about 100kHz, at which the result is better but the maximum deflection is still large with the 680mV amplitude (13.6%), as seen in Figure. 10.

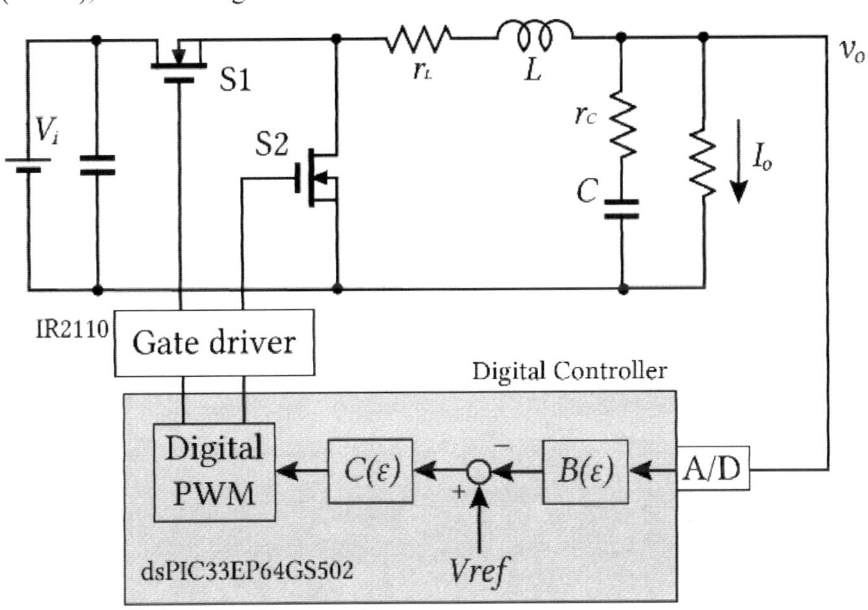

Figure 8. The buck converter under IA-PIM control.

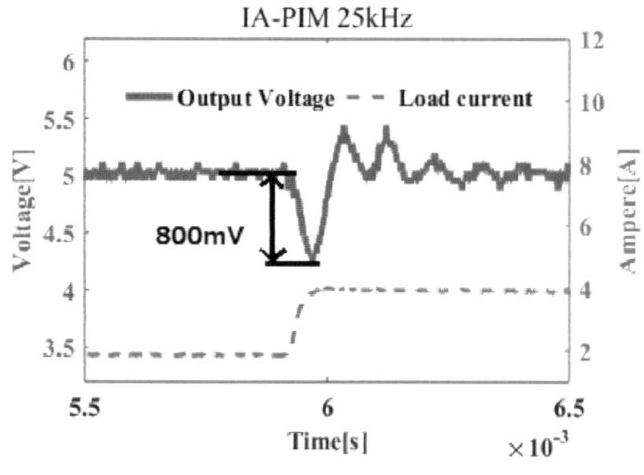

Figure 9. Output-voltage regulation under IA-PIM control at 25kHz.

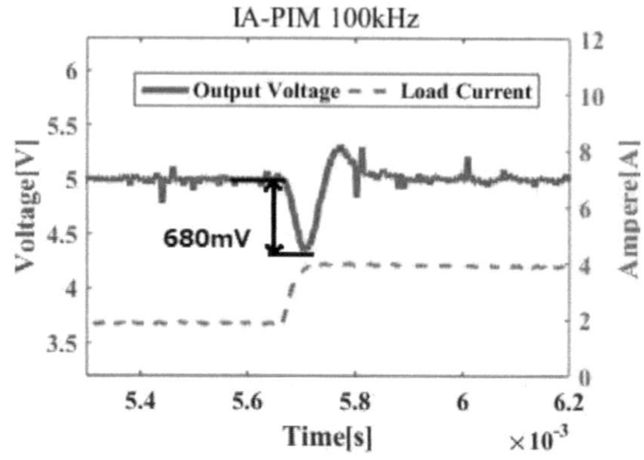

Figure 10. Output-voltage regulation under IA-PIM control at 100kHz (maximum possible).

3. High-rate disturbance alleviation

Since the processor can handle extra computations using the IA-PIM method with 25kHz, a simple algorithm can be added. As it is suspected that a faster sampling rate is required to counteract loading disturbances, three approaches are attempted at a higher sampling rate of 300kHz; two by mimicking current feedback and one by previewing the disturbance using an inverse plant-model. In all cases, the disturbance characteristics of the plant is to be improved and other properties, such as stability and steady-state performances, are not intended to be changed, so that the original controller block $C(s)$ remains to meet requirements other than those for the disturbance. However, $C(s)$ can still be redesigned using the modified plant if desired or necessary. This is not carried out in the present study so that comparisons can be made more easily on the same bases.

3.1. *Addition of Current Feedback*

Methods 1 and 2 are based on feeding back the inductor current in a faster inner-loop, which modifies the plant. By adding the inductor-current feedback as shown in Figure. 11, the characteristic polynomial of the plant, eq. (5), is modified as

$$\hat{P}(s) = LCs^2 + C(r_L + r_C + KV_i)s + 1, \tag{17}$$

which could increase the damping ratio for a positive gain K. When designing IA-PIM controller using the same $C(s)$ for this modified plant model, eq. (17) is used in both Methods 1 and 2 to be explained next.

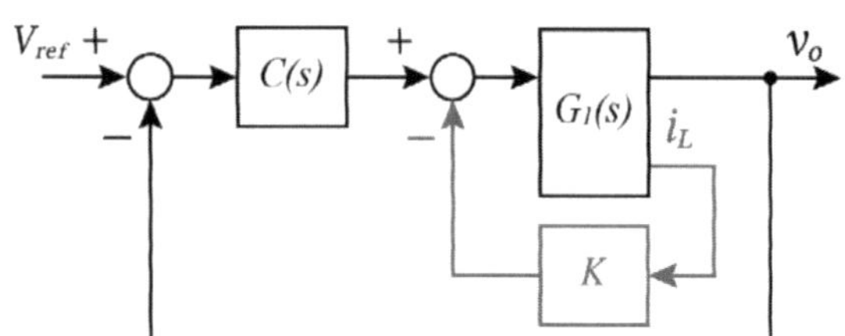

Figure 11. Addition of an inductor-current feedback loop.

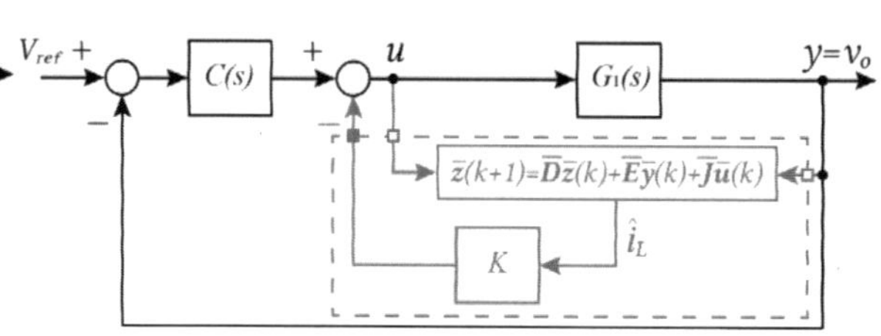

Figure 12. Feedback using the current-observer.

3.2. *Method 1: Current Observer*

Since no extra sensor is available to measure directly the inductor current, it is estimated by using the state-observer or obtained through a filter. In Method 1, the averaged state-space model mentioned in

Section 2.1 is used to design a full-order (second-order) state-observer [15] for estimating and feeding back the inductor current, ignoring the loading-current fluctuation in these equations. It is then discretized as the step-invariant-model in the shift-form at 300kHz sampling frequency, as

$$q\begin{bmatrix} i_L(k) \\ v_C(k) \end{bmatrix} = \begin{bmatrix} 0.8280 & -18.75 \\ 0.03657 & -0.8280 \end{bmatrix}\begin{bmatrix} i_L(k) \\ v_C(k) \end{bmatrix} + \begin{bmatrix} 18.34 \\ 0.1264 \end{bmatrix} y(k) + \begin{bmatrix} 4.838 \\ 0.1264 \end{bmatrix} u(k) \cdot \tag{18}$$

where u is the change in the duty ratio and y the output voltage. This is then implemented as shown in Figure. 12 [14]. This modifies the plant model and eq. (17) is used for both Methods 1 and 2 to redesign the IA-PIM controller at 25kHz, and gives

$$B(\varepsilon) = \frac{3.628\times10^{-9}\varepsilon^2 + 1.306\times10^{-4}\varepsilon + 1}{2.56\times10^{-8}\varepsilon^2 + 3.2\times10^{-4}\varepsilon + 1} \tag{19}$$

$$C(\varepsilon) = 1281\frac{2.56\times10^{-8}\varepsilon^2 + 3.2\times10^{-4}\varepsilon + 1}{\varepsilon(8.151\times10^{-10}\varepsilon^2 + 5.832\times10^{-5}\varepsilon + 1)} \tag{20}$$

A typical experimental result is shown in Figure. 13 with K=0.12. It shows that the maximum amplitude of deviation is reduced to almost one-third (6.4%) during the transient stage of the load-current disturbance compared with one in Figure. 10. However, the output fluctuation continues into the steady-state phases.

It is suspected that a computational delay associated with the second-order observer affect such fluctuation and a lower order approach is attempted next.

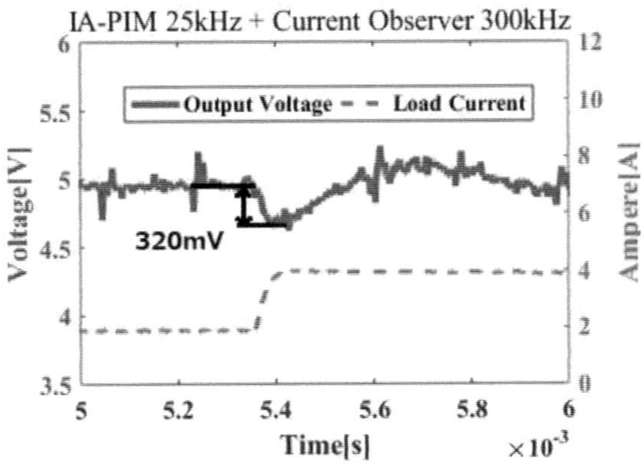

Figure 13. Output-voltage regulation under IA-PIM control at 25kH with current observer at 300kHz.

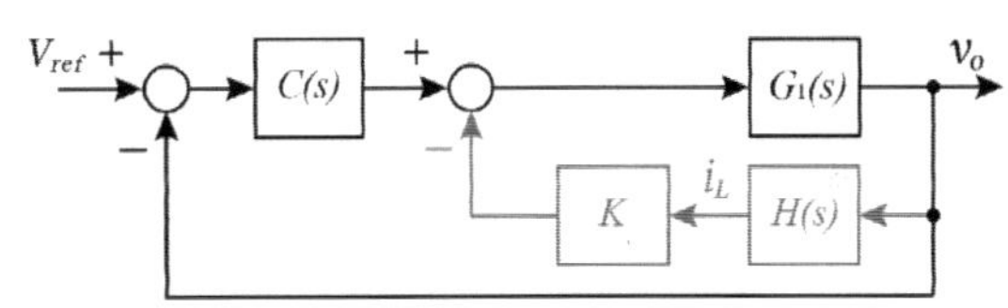

Figure 14. Feedback using the current-filter.

3.3. Method 2: Current Filter

Rather than using the second-order state-observer, the first-order state-equation can be extracted [14] and Laplace-transformed to obtain the current filter (Figure. 14) as

$$i_L(s) = \frac{Cs}{Cr_C s + 1} v_o \cdot \tag{21}$$

Taking noises into account, the low-pass filter with the corner frequency twice the system bandwidth is added, resulting in

$$KH(s) = 0.12\frac{6.4\times10^{-5}s}{3.183\times10^{-7}s + 1} \cdot \tag{22}$$

IOP Conf. Series: Materials Science and Engineering **646** (2019) 012019 doi:10.1088/1757-899X/646/1/012019

This inner-loop current filter is discretized using the Tustin's method at 300kHz and implemented. As for the outer-loop, the same controller $C(s)$ is used with the modified plant of eq. (15) for IA-PIM at 25kHz, resulting in the same block as given by eq. (19) and (20).

A typical experimental result is shown in Figure. 15. It is observed that the maximum deviation during the transient is 200mV (4%) and the output voltage regulation is satisfactory. A better performance may be achieved by re-designing $C(s)$, which was not attempted this time for comparison with other two methods and left for further studies.

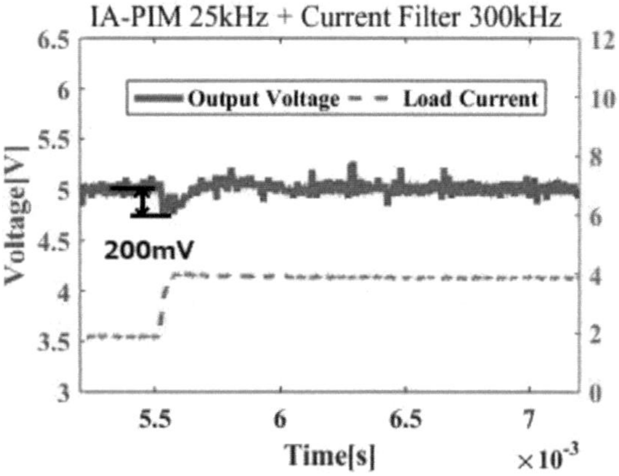

Figure 15. Output-voltage regulation under IA-PIM control at 25kHz with current filter at 300kHz.

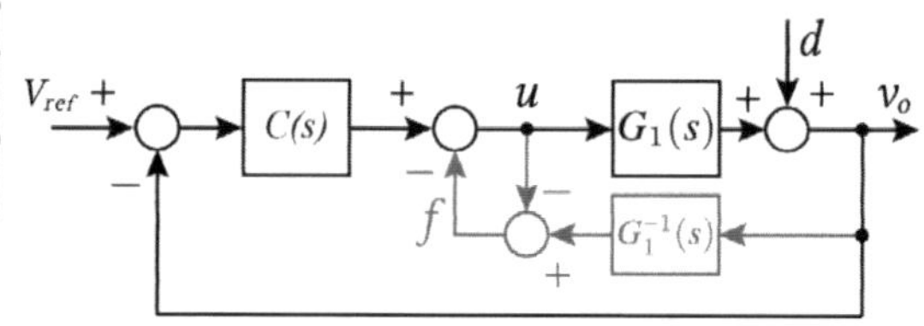

Figure 16. Feedback using the inverse-filter.

3.4. *Method 3: Inverse Filter*

To improve the disturbance rejection property, the inverse model of the plant is used in the third method, rather than using current-based feedback. An estimate $\hat{d}$ of the actual disturbance d may be calculated from the measured output v_O and the available plant input u, as

$$\hat{d}(s) = v_O - \hat{G}_1^{-1} u \,, \tag{23}$$

where $\hat{G}_1^{-1}$ is the inverse of the plant model. For most plant, this inverse is non-proper and needs to be filtered to make this to be bi-proper, as explained shortly. Now, let f be the fictitious portion of the plant input that is capable of causing the plant output equivalent to the disturbance d. Subtracting f from the plant input a priori, the effect of the actual disturbance d on the plant output can be expected to be reduced. Figure. 16 shows such a control system, which includes an algebraic loop. However, when implemented in a discrete-time form, the computation of f introduces one sample delay, which makes the loop dynamic.

Since the inverse filter $G_1^{-1}(s)$ is improper and cannot be implemented on-line, a low-pass filter $F(s)$ given by

$$F(s) = \frac{1}{3.183 \times 10^{-6} s + 1} \tag{24}$$

is multiplied to make it bi-proper and then discretized at 300kHz using the Matched-Pole-Zero method [17], so that

$$G_1^{-1}(\varepsilon) = 0.0833 \frac{5.273 \times 10^{-10} \varepsilon^2 + 4.263 \times 10^{-6} \varepsilon + 1}{1.715 \times 10^{-11} \varepsilon^2 + 8.475 \times 10^{-6} \varepsilon + 1} \,. \tag{25}$$

(For some reason, Tustin's method does not work well.) The IA-PIM controllers for the plant G_1 at the sampling frequency of 25kHz are the same as those given by eqs. (15) and (16). A typical experimental result is shown in Figure. 17, which illustrates that after the load current changes, the output voltage fluctuates and the transient dip has the magnitude of 230mV (4.6%).

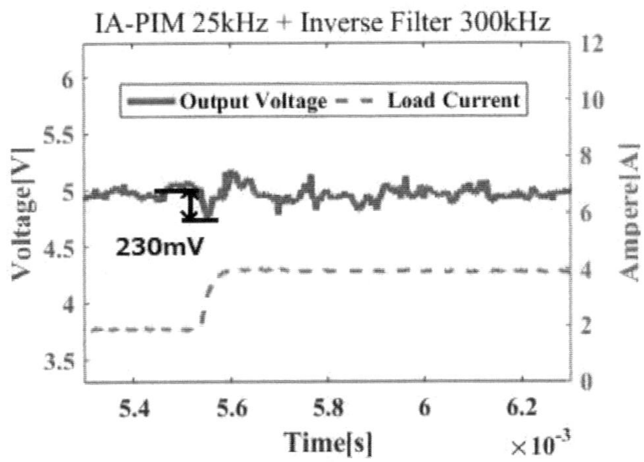

Figure 17. Output-voltage regulation under IA-PIM control at 25kHz with inverse filter at 300kHz.

4. Conclusion

Through dual-rate digital control, improvements on disturbance rejection properties have been attempted for a switching DC-DC buck converter. As in the previous study, a slower outer-loop controller was designed by discretizing an existing analog controller, which satisfies all the design requirements, using the so-called IA-PIM method, which guarantees the stability for any non-pathological sampling interval. In the present study, to suppress the amplitude of a sudden dip in the output voltage for a change in the load-current, a faster inner-loop controller was added. For this purpose, three types of digital controllers have been designed and tested experimentally; those using the current-observer, the current-filter, and the inverse-filter. The results obtained using the current-filter was superior to the current-observer and the inverse-filter, at this stage. However, further tuning of parameters in the digital controllers should be attempted. Besides, since the analog performance is achieved by fine-tuning various component values, a better digital performance may be possible by re-tuning the parameters and taking some measures to counter such issues as computational and conversion delays and quantization errors. Larger transients with the digital controllers are more eminent in high-frequency applications as in the present study. Nevertheless, since the conventional Tustin method running at 300kHz did not provide stable results, the stable performance of the proposed methods at 25kHz is a good start. In terms of percentage maximum fluctuation from the output voltage, better performances of the digital regulator (4% and 4.6%) than those of the analog regulator (4.7%) have been experimentally confirmed. However, the digital performance has small ripples compared with the analog performance. The approaches presented in the present study are based on pole-relocations using feedback. Those based on zero-relocations using feedforward are under way, to further improve the digital performance.

References

[1] Erickson R W and Maksimovic D, *Fundamentals of Power Electronics*, Kluwer Academic Publishers, Norwell, MA, 2004.
[2] Basso C P, Designing control loops for linear and switching power supplies, Artech House, Boston MA, 2012.
[3] Basso C P, Switch-Mode Power Supplies. McGraw Hill, New York, NY, 2014.
[4] Prodic A, Maksimovic D and Erickson R W, Design and implementation of a digital PWM

controller for a high-frequency switching DC-DC power converter, in IECON, IEEE Industrial Electronics Society, Denver, CO, USA, IEEE, p 893-8, 2001.

[5] Choudhury S, Designing a TMS320F280x based digitally controlled DC-DC switching power supply. Application Report, Texas Instruments, July 2005.

[6] Wang K, Rahman N, Lukic Z and Prodic A, All-digital DPWM/DPFM controller for low-power DC-DC converters. in IEEE Applied Power Electronics Conference and Exposition, Dallas, TX, USA, IEEE, p 5, 2006.

[7] Guo S, Lin-Shi X, Allard B, Gao Y and Ruan Y Digital sliding-mode controller for high-frequency DC/DC SMPS, IEEE Trans. *Power Electronics*, **25**(5), 1120-3, 2010

[8] Corradini L, Maksimovic D, Mattavelli P and Zane R Digital control of high-frequency switched-mode power converters. IEEE Press. John Wiley & Sons, Hoboken, NJ, 2015.

[9] Buso S and Mattavelli P, Digital Control in Power Electronics, Morgan & Claypool Publishers, 2015.

[10] Isobe Y, Hori N and Someya K, Digital regulation of an LLC resonant converter undergoing dynamic load transients – Model-based PIM design and simulation, in SICE Annual Conference, (Tsukuba, Japan, 2016), IEEE, 270-5.

[11] Higashiya R and Hori N, Model orders and numerical issues in LLC resonant DC-DC converters, in IASTED Conference on Modeling, Identification and Control, (Innsbruck, Austria, 2017), 166-171.

[12] Nishinaga N and Hori N, "PIM-Discretization of Analog PWM Control for DC-DC Switching Converters," Proc. 2017 The 5th International Conference on Control, Mechatronics and Automation, Edmonton, Canada, pp. 90-4, October, 2017.

[13] Sain M K and Schrader C B, The role of zeros in the performance of multiinput, multioutput feedback systems, IEEE Trans. Educ, **33** no. 3, pp. 244–257, 1990.

[14] Abe S and Zaitsu T, The Core of Control in Switching Power Supplies, *Nikkei Electronics*, pp. 80-94, Oct, 2014. (in Japanese).

[15] Middleton R H and Goodwin G C, *Digital Control and Estimation*, Prentice-Hall, Englewood Cliffs, NJ, 1990.

[16] Nishinaga N, A Multi-Rate PIM-Disturbance Regulator for Suppression of Voltage Fluctuations in a DC-DC Converter, Master's Thesis, Department of Intelligent Interaction Technologies, University of Tsukuba, Japan, March 2019.

[17] Kashimura T and Hori N, Digital redesign based on Diophantine equation with assured integral control and stability, IASTED Conference on Control and Applications, (Quebec City, Canada, 2008), 234-239.

[18] Markazi A H D and Hori N, A new method with guaranteed stability for discretization of continuous time control systems. in American Control Conference, (Chicago, IL, USA, 1992), IEEE, 1397-1402.

IOP Conf. Series: Materials Science and Engineering **646** (2019) 012020 doi:10.1088/1757-899X/646/1/012020

Fault Diagnosis Method of Photovoltaic Array Based on BP Neural Network

Junjie Wang[1], Shan Wang[1], Haixiong Liu[1],Jianbo Hong[2] and Dedong Gao[1,2,*]

[1]School of Mechanical Engineering, Qinghai University, Xining 810016, China
[2]Haixi Huahui Chemical and Machinery Co.,LTD

[*]Corresponding author: gaodd@zju.edu.cn

Abstract. Photovoltaic arrays are prone to various failures due to long-term work. In order to quickly and accurately diagnose the type of failure of the PV array and implement online monitoring of the PV array, this paper proposes the BP neural network for PV array fault diagnosis, and proposes a network search method when training BP neural network. And the K-cross-validation method is used to select the number of hidden layer nodes. The BP neural network fault diagnosis model designed and trained by this method is proved to have high precision.

1.Introduction

Solar energy is a kind of carbon-free renewable energy, which has been used more and more widely in the field of photovoltaics. However, since photovoltaic cells are installed in outdoor environments and they are exposed to sunlight and wind and rain throughout life of PV array, so the PV array is prone to various failures. If a method can be found to monitor the working status of the PV system in real time and quickly find various faults of the PV system then it can reduce the cost of maintenance of the PV power plant and reduce the burden on workers[1-3].

At present, methods for fault diagnosis of photovoltaic systems mainly include traditional diagnostic methods, model algorithms and intelligent algorithms[4].The traditional fault diagnosis methods mainly include two kinds: online diagnosis and offline diagnosis. Traditional fault diagnosis methods usually require other additional hardware equipment to complete the accurate diagnosis for PV system, and the cost is high, but the traditional method can accurately locate the fault of the PV system[5-8].Online diagnosis methods mainly include infrared thermography method and multi-sensor data fusion method. Infrared thermography method detects the fault by using the temperature difference on the surface of the photovoltaic panel. And it is necessary to set up the infrared imager to judge the photovoltaic panel by processing the image returned by the infrared imager. The existing fault can accurately be detected, but the cost is high. The infrared thermography method requires the imager to have high temperature resolution accuracy, and it is only suitable for the detection of hot spot fault of the photovoltaic panel[9-11].The multi-sensor data fusion method achieves the purpose of fault detection by installing multiple voltage and current sensors at each node of the photovoltaic system. By collecting the data returned by multiple sensors and integrating the data, the fault can be monitored in real time in the photovoltaic system. The multi-sensor data fusion method can accurately diagnose the location of the fault in the PV system, but the cost is too high due to the need to install the sensor at each node of the PV system. The offline fault diagnosis method mainly includes the ground capacitance method and the time domain reflection method[12].The ground capacitance method locates the fault by measuring the capacitance of each node of the series battery to ground. This method can accurately locate the fault, but the photovoltaic system need to be offline when the fault is detected. The time domain reflection method need to input an electric pulse to the photovoltaic system, determine the position of the fault by the waveform and delay time of the returned signal, use the time domain reflection method can accurately determine the fault location, but because it is analysed by the signal waveform, so the requirements for the instrument are higher, the cost is higher, and online fault diagnosis cannot be achieved[13-15].

The intelligent algorithm and the model algorithm do not need to install other additional equipment, and only need to obtain the photovoltaic system output electrical parameters and real-time environmental parameters. Model algorithm fault diagnosis must be based on accurate modelling[16].Model algorithm method requires accurate establishment of the PV system model, the diagnostic accuracy is high, but the modelling is very difficult. Intelligent algorithm fault diagnosis methods mainly include cluster analysis method and BP neural network method. Clustering analysis is an unsupervised learning algorithm. In the fault diagnosis, the unsupervised learning of the sample data is performed first, then the data is divided into K categories, and the classification results are interpreted based on the corresponding prior knowledge. To discover the failure of the photovoltaic system, this method requires a certain prior knowledge, but BP neural network is a supervised learning machine[17, 18].There is no need for prior knowledge of PV array fault diagnosis using BP neural network. It only needs to collect the daily operation data of PV array, and BP neural network can solve any complex nonlinear problem[19-21].

The BP neural network is proposed to diagnosis the PV array faults in this paper. The BP neural network is a deep learning algorithm. It is widely used as pattern recognition algorithm. It has been used in photovoltaic system fault diagnosis. When performing PV system fault diagnosis, the data is usually not properly processed, so the accuracy of algorithm classification has not been improved reliably. Therefore, this paper proposes data protocol for training data. At the same time, when the BP neural network is established, the network search method and the K-cross-validation method are proposed to optimize the number of hidden layer nodes, which solves the problem that the hidden layer cannot be selected when using BP neural network for fault diagnosis[22-24].

2.PV array failure analysis and feature parameter selection

2.1. Failure analysis

Photovoltaic arrays work in outdoor environments and are exposed to strong ultraviolet and sand environments for a long time. Due to the special working environment of photovoltaic panels, various faults occur in photovoltaic arrays. Photovoltaic array failures mainly include hot spots, shadows, aging, short circuits and open circuits.

The hot spot fault is mainly caused by long-term shadow of a part of the photovoltaic panel. When a certain area of the panel is blocked, the battery cell in this area will be converted into a load and exist in the circuit. The battery unit will output a negative voltage. At the beginning, it consumes a certain amount of power and generates heat. The battery cells that have been in such a shade for a long time cause the battery cells to burn out. When the shading disappears, it cannot work normally again.

Aging is caused by the long-term operation of the panel and the erosion of wind and rain. The aging of the panel usually shows an increase in the series resistance of the panel. The output characteristic curve is usually similar to the shadow. The principle of shadow formation of the photovoltaic panel is shown in Figure 1. Since aging and hot spots are consistent on the output characteristic curve, aging, hot spots, and shadows are collectively classified as insufficient irradiation failure. Therefore, photovoltaic array failures include insufficient irradiation, open circuits, and short circuits.

The PV array failure analysis is shown in Table 1.

Table 1. Failure analysis.

Fault type	Cause of formation	Harm
Open circuit	Installation wiring error	Reduce power generation efficiency
Short circuit	Cable aging	Accelerate aging and cause fire
Insufficient irradiation	Aging, shadows and cracks, etc.	reduce power generation efficiency

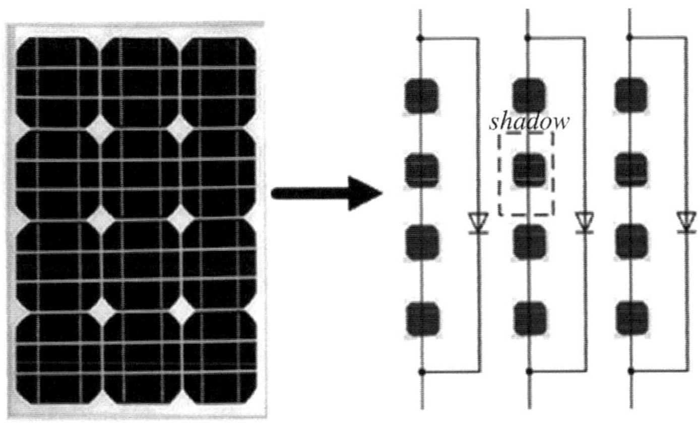

Figure 1. Photovoltaic panel shadow failure diagram

2.2. Feature parameter selection

The photovoltaic array is composed of a plurality of photovoltaic panels. The output characteristic curve of the photovoltaic array directly reflects the working state of the photovoltaic array. The output characteristic curve of the photovoltaic array can be drawn by the short-circuit current, open circuit voltage, maximum power current and maximum power voltage of the photovoltaic cell, such as Figure 2 shows the output characteristics of the PV array. As can be seen from the figure, the PV array output characteristic curve is basically determined by the PV array short-circuit current, open circuit voltage, maximum power current and maximum power voltage. Therefore, this paper determines the most characteristic PV array fault characteristics using Isc, Uoc, Imp, and Ump parameters.

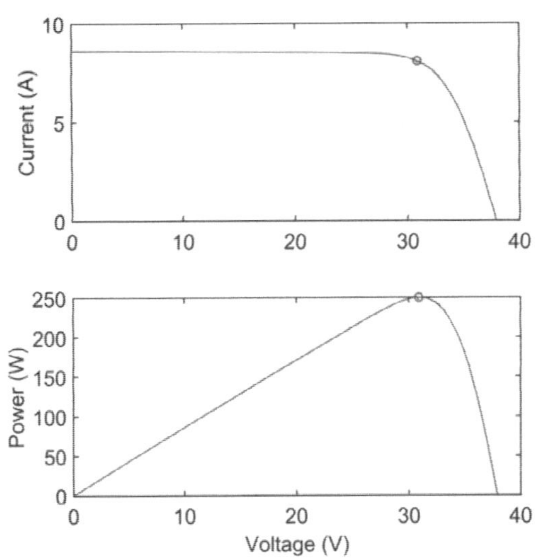

Figure 2. PV array output characteristic curve

3. Modelling BP neural network

3.1. BP neural network

BP neural network is one of the intelligent algorithms widely used in the field of fault diagnosis, which can solve the problem of highly nonlinear fault pattern recognition. BP neural network has efficient self-learning, self-adaptive ability, and can complete contact memory. Since the input and output layer nodes of the BP neural network model can be determined according to the input parameters and the number of fault states, the BP neural network is suitable for solving complex fault diagnosis problems with multiple faults and multiple modes.

Since the PV array fault diagnosis problem is a highly nonlinear and complex fault diagnosis problem, there is a high degree of nonlinearity between the PV array output characteristic curve and various fault modes, and there is a high degree of coupling between various faults and output characteristic parameters. Sexuality, while BP neural network can complete the nonlinear mapping problem of input to output with arbitrary complexity, and has good generalization ability. Therefore, this paper proposes BP neural network for PV array fault diagnosis.

BP neural network is a multilayer feedforward error back propagation neural network system. BP neural network algorithm structure diagram shown in Figure 3. Its training learning process mainly includes two processes: the positive propagation process and the reverse propagation process. The positive propagation process is that the input data is directly transmitted to the output layer through the hidden layer. When the forward propagation is performed, the weights of the neurons in the neural network remain fixed. When there is an error between the output and the actual fault output, the network enters. In the anti-propagation process, the weight of the network during the anti-propagation process is adjusted according to the error so that the output is close to the desired output.

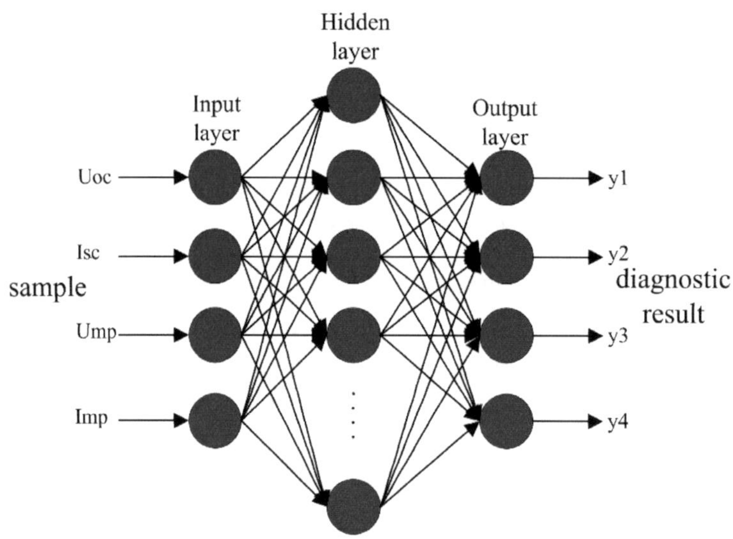

Figure 3. BP neural network structure

As can be seen from Figure 3, the BP neural network includes an input layer, an implicit layer, and an output layer. In the BP neural network model training, four problems need to be determined: the number of nodes of the input layer, the output layer and the hidden layer, and the hidden layer basis function. In actual use, the number of input layer nodes is determined according to the number of fault data characteristic parameters, and the number of fault data characteristic parameters is the number of input nodes. The number of output layer nodes is determined by the number of faults, and the number of faults is the number of output layer nodes.

There are three methods for determining the number of hidden layers, which are shown in Table 2 below. Since the method of determining the number of hidden layers is still not fully studied, there is no most suitable method for determining the number of hidden layers. Therefore, this paper uses the network search method to determine the number of hidden layers. This method can find the optimal number of hidden layers. Before performing network search and optimization, we must determine the scope of optimization. Use the three methods in the table to calculate the number of hidden layer nodes, and take the maximum and minimum values as the scope of network. Searching for the range value, as can be seen from the above analysis, the number of input layer nodes is 4, and the number of output layer nodes is 4.

Table 2. Hidden layer node calculation function

Serial number	Implicit layer calculation formula	Implicit layer node value
1	$m=\sqrt{n+l}+\alpha$	m=3.8~12.8
2	$m=\log_2 n$	m=2
3	$m=\sqrt{nl}$	m=4

The hidden layer basis function uses a Gaussian function, which is a nonlinear function, which has the advantages of good convergence and fast calculation speed. The Gaussian function is shown in equation (1).

$$F(\sigma)=\exp(\frac{\|x-c\|^2}{-2\sigma^2}) \quad (1)$$

Where c is the initial training value and σ is the width of the Gaussian function.

By using a Gaussian function to transform the input into the hidden layer, it is known that the hidden layer value is as shown in equation (2).

$$s_i=\exp(\frac{\|x_i-c_i\|^2}{-2\sigma_i^2}) \quad (2)$$

Where s is the calculated value of the hidden layer.

Since there is no connection between the output and the input, and there is a linear relationship between the output and the hidden layer, the output calculation formula is as shown in equation (3).

$$y_i=\sum_{i=1}^{4}\omega_i s_i \quad (3)$$

3.2. Hidden layer node number optimization

The number of hidden layer nodes is optimized by network search and K-cross-validation. The minimum range of network search is 2, the maximum value is 13, and the search step is 1, the number of nodes is increased each time. The number of nodes obtained by the search is verified using the K-cross validation method.

The K-cross-validation process is to mix the collected fault sample data into the same sample training set, and then divide the sample set into K-classes, and use one set of data each time the number of searched nodes is verified. As a test set, the K-1 group is left as a training set.

The fault diagnosis accuracy of BP neural network under the number of hidden layer nodes is shown in Figure 4. The figure shows that the model has the best effect when the number of hidden layer nodes is m=5, and the accuracy rate is 97 %.

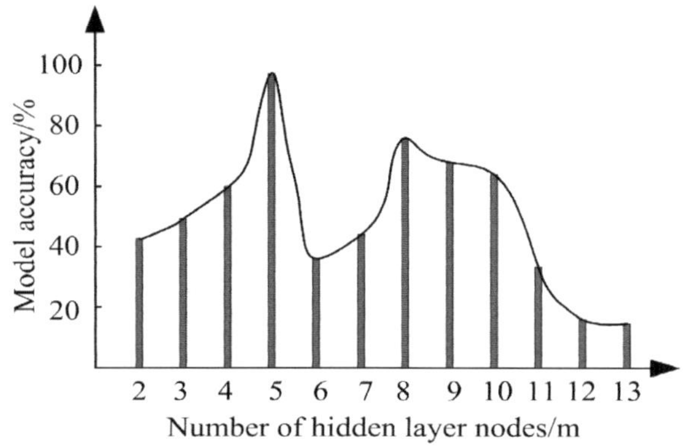

Figure 4. Implicit layer node optimization

4.Experiment analysis

4.1.Data collection

Collecting data is the first step in big data analysis. The use of BP neural network based PV array fault diagnosis requires a large amount of high quality training data. Therefore, a 3 × 4 photovoltaic array was built, as shown in Figure 5. The 3×4 photovoltaic array includes a total of 12 battery boards, and the type of the battery board is STS-156P-255W. Each battery board is composed of 60 battery cells. The basic parameters of the battery board are: Isc=8.95A, Uoc= 37.8V, Imp=8.32A, Ump=30.65V, Pm=255W.

Figure 5. 3×4 photovoltaic array

There are three kinds of fault data that need to be acquired in this experiment: insufficient irradiation, open circuit and short circuit. A total of four data are collected for data including the normal state of the PV array. In order to ensure the reliability of the experiment and collect high-quality data, this experiment must ensure that the irradiance is 1000W/m2 at the time of collecting the normal working data of the PV array. This experiment uses the PV array IV curve tester produced by Kewell Company for data acquisition. The Cowell IV curve tester model is IVT-30-1000, and the maximum current that can be tested is 30A. The maximum voltage that can be tested. It is 1000V.

In order to collect the PV array state data under the condition of insufficient irradiation, it is necessary to ensure that the irradiance is less than 1000 W/m2 at the time of collecting data, and at the same time, in order to ensure the uniformity between the experimental collection data and the working state of the official photovoltaic array, the experiment is being carried out. The PV array operating status data under different temperature and humidity conditions were collected.

In the open circuit data acquisition, since the state data obtained when the circuit is open at any different position is similar, the data acquisition of the PV array operating state is performed in three open modes, as shown in Figure 6 below.

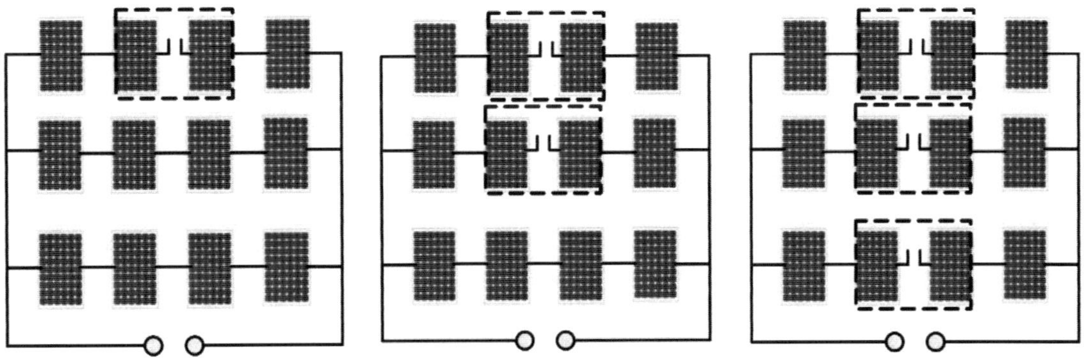

(a) First open circuit (b) First and second open circuit (c) First, second and third open circuit

Figure 6. Open circuit experiment

In order to ensure that the battery board is not burned out during the short-circuit data acquisition, in the process of simulating the short circuit, at least one string of batteries will always be guaranteed to be free of short-circuit conditions. At the same time, it is necessary to ensure that a whole series of battery boards cannot be completely ensured. The true and random nature must ensure that the illumination is not less than 1000W/m2, and the temperature and humidity are in various different situations. The process of short-circuit test is shown in Figure 7.

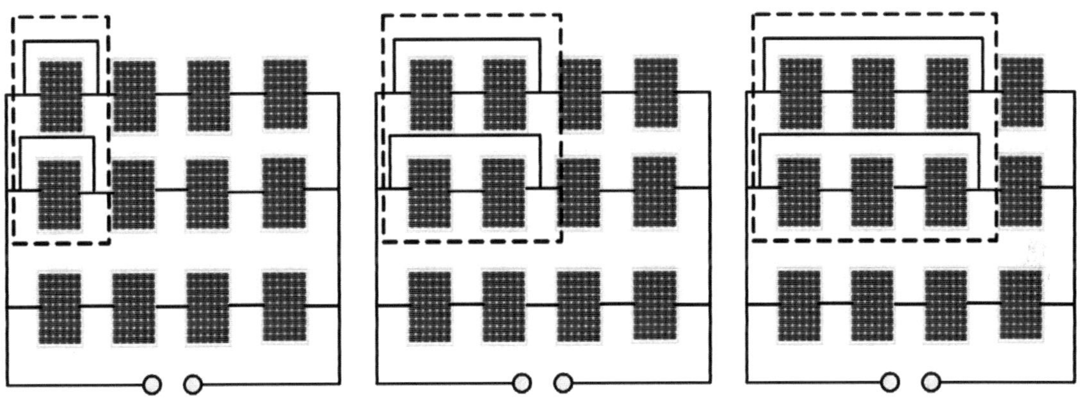

(a) The first and second strings each have a short circuit (b) The first and second strings each have two short circuits (c) The first and second blocks each have a single short circuit

Figure 7. Short circuit experiment

4.2.Data normalization

Due to the limitations of the BP neural network model, the input data must be a unit value between [0, 1]. Therefore, the fault sample data obtained by the above experimental methods cannot be directly input into the algorithm for training, and must be normalized by data. The main calculation method of data normalization is shown in equation (4).

$$p = \frac{x - x_{\min}}{x_{\max} - x_{\min}} \quad (4)$$

Where P is the normalized value of the to-be-stated value X.

IOP Conf. Series: Materials Science and Engineering **646** (2019) 012020 doi:10.1088/1757-899X/646/1/012020

4.3. Training and learning

In this paper, based on SPSS data analysis software, the neural network model is trained, and the preprocessed data is input into the BP neural network model. From the above, the number of hidden nodes in the neural network is 5, and the implicitness of BP neural network is determined. The number of layers is 1, and the number of input and output nodes is 4.

A total of 400 data were collected in this experiment, including 100 normal, insufficient, open and short circuit data. The data set was divided into two groups before the model training, 70% of which was used as the training set, 30%. Proportional data as a test set. The BP layer neural network activation function is a hyperbolic tangent function, and the output layer activation function is a softmax function. The maximum number of training times is set to 1000 times, the training step size is set to 0.05, and the training error is not more than 0.001. Figure 8 shows the error curve of the training data during the training process. It can be seen from the figure that the training error of the BP neural network will decrease with the increase of the training times. When the 420 trainings, the model error reaches 10-3. The magnitude, at which point the desired target error is reached.

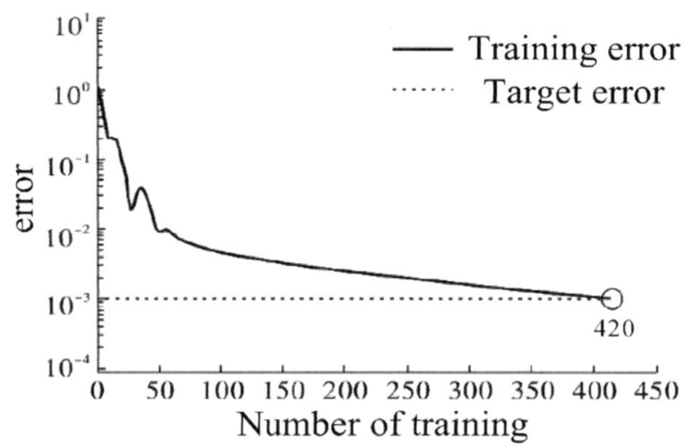

Figure 8. Training error

4.4. Test verification

In order to test the accuracy of the BP neural network fault diagnosis model after training, the remaining 120 data for the non-participating model training will be tested as a test set. The test results are shown in Table 3. It can be seen from the table that the normal data test accuracy is 80%, the insufficient irradiation data test accuracy is 91%, the open circuit data test accuracy is 96%, and the short circuit data test accuracy is 87%, where in the open circuit data test accuracy is the highest, and the model has the resolution capability. It is best, and the short-circuit data is more difficult to distinguish, and it is easy to blur the data with insufficient irradiation, but the test effect is still good.

Table 3. Test Results

label	0	1	2	3	Precision
0	27	0	3	0	80%
1	1	28	0	1	91%
2	0	0	29	1	96%
3	0	3	1	26	87%

5. Conclusions

In this paper, a fault diagnosis method for PV array based on BP neural network is proposed. After analysis, the open circuit voltage, short circuit current, maximum power voltage and maximum power current of the PV array are determined as input parameters of BP neural network. These four parameters can be very good to reflect the working state of the PV array. The network search method and the K-cross-validation method are proposed to optimize the number of BP hidden nodes in the network. It is found that the model accuracy can reach 97% when the number of nodes in the hidden layer is 5. The network search method and the K-cross-validation method are combined, and the generalization of the BP neural network can be improved by using K-cross-validation method . However, due to the complexity of the working environment of the PV array, the complex and variability of the PV array output characteristic curve information seriously affects the accuracy of the PV array intelligent diagnosis algorithm. Therefore, the intelligent fault diagnosis algorithm based on BP neural network needs to be improved.

Acknowledgments
This research work is sponsored by the National Natural Science Foundation of China(No. 51566015), Qinghai Science and Technology Projects(No. 2019-ZJ-7002 and 2019-HZ-822).

References
[1] Huang Zhiqiang and Guo Li. 2009 Research and implementation of microcomputer online fault detection of solar array *Computer Science & Education, 2009. ICCSE '09. 4th International Conference on*
[2] Huang Z and Li G. 2009 Research and implementaion of microcomputer online fault detection of solar array *2009 4th International Conference on Computer Science &Education(ICCSE'09)* 1052-55
[3] Cheng Z, Zhong D and Li B. 2011 Research on fault detection of PV array based on data fusion and fuzzy mathematics *2011 Asia-Pacific Power and Energy Engineering Conference (APPEEC). IEEE* 1-4
[4] Cheng Z, Zhong D and Li B. 2011 Research on fault detection of PV array based on data fusion and fuzzy mathematics *2011 Asia-Pacific Power and Energy Engineering Conference(APPEEC)* 2011 1-4
[5] Xiaoli Xu, Huan Wang and Yunbo Zuo. 2011 Method for Diagnosing Photovoltaic Array Fault in Solar Photovoltaic System *Power and Energy Engineering Conference (APPEEC), 2011 Asia-Pacific*
[6] Guinjoan-Gispert F Velasco-Quesada G, Pique-Lopez R,et al. 2009 Electrical PV array reconfiguration strategy for energy extraction improvement in grid-connected PV systems *IEEE Transactions on Industrial Electronics* 2009,**56** 4319-31
[7] Takumi Takashima, Junji Yamaguchi, Kenji Otani, Takashi Oozeki, Kazuhiko Kato and Masayoshi Ishida. 2008 Experimental studies of fault location in PV module strings *Solar Energy Materials and Solar Cells* **93**
[8] Chowdhury S R and Saha H. 2010 Maximum power point tracking of partially shaded solar phtotvoltaic arrays *Solar Energy Materials&Solar Cells* **94** 1441-47
[9] Agarwal V Patel H. 2008 MATLAB-based modeling to study the effects of partial shading on PV array characteristics *IEEE Transactions on Industrial Electronics* **23** 302-310
[10] Ducange P, Fazzolari M and Lazzerini B. 2011 An intelligent system for detecting faults in photovoltaic fields *IEEE11th International Conference on Intelligent Systems Design andApplications. Cordoba, Argentina: IEEE* 1341-46
[11] Togashi S Noguchi T. 2002 Short-current pulse-based maximum-power-point tracking method for multiple phtotovoltaic-and-converter module system *IEEE Transactions on Industrial Electronics* **49** 217-23
[12] Noguchi, T., Togashi, S., Nakamoto and R. 2000 Short-current pulse based adaptive maximum-power-point tracking for photovoltaic power generation system *Industrial Electronics, 2000. ISIE 2000. Proceedings of the 2000 IEEE International Symposium on*
[13] Lehman B Nguyen D. 2008 An adaptive solar photovoltaic array using model-based reconfiguration algorithm *IEEE Transactions on Industrial Electronics* **55** 2644-54

[14] Inada T Miyatake M, Hiratsuka I,et al. 2004 Control characteristics of a fibonacci-search-based maximum power point tracker when a photovoltaic array is partially shaded *The 4th International Power Electronics and Motion Control Conference 2004(IPEM2004)* 2004 816-21

[15] Yanli Liu, Bingfeng Li and Dan Zhong. 2010 Research on domestic PV module structure based on fault detection *Intelligent Control and Automation (WCICA), 2010 8th World Congress on*

[16] Stellbogen D. 1993 Use of PV circuit simulation for fault detection in PV array fields *Conference Record of the Twenty Third IEEE on Photovoltaic Specialists Conference 1993* 1302-07

[17] Lin C, Jie G and Wu Z. 2011 Design of networked monitoring system of PV grid-connected power plant *2011 International Conference on Electronic and Mechanical Engineering and Information Technology. IEEE*

[18] Nian B, Fu Z and Wang L. 2010 Automatic detection of defects in solar modules: image processing in detecting *6th International Conference on Wireless Communications Networking and Mobile Computing. Chengdu, China: IEEE* 1-4

[19] King D L, Kratochvil J A and Quintana M A. 2000 Applications for infrared imaging equipment in photovoltaic cell, module, and system testing *Conference Record of the Twenty-Eighth Photovoltaic Specialistis Conference* **456** 1487-90

[20] Patel H and Agarwal V. 2008 MATLAB－based modeling to study the effects of partial shading on PV array characteristics *IEEE Transactions on Energy Conversion* **23** 302-10

[21] Chao K H, Ho S H and Wang M H. 2008 Modeling and fault diagnosis of a photovoltaic system *Electric Power Systems Research* **78** 97-105

[22] Boztepe M Karatepe E, Colak M. 2007 Development of a suitable model for characterizing phtotovoltaic arrays with shaded solar cells *Solar Energy* **81** 977-992

[23] Nalin K, Gautam N D and Kaushika. 2002 Reliability evaluation of solar photovoltaic arrays *Solar Energy* **72**

[24] Shigeo H, Toshiali Y and Tatsuo T. 2010 A New Stand-alone Hybrid Power System with wind Generator and photovoltaic Modules for Radio base station photovoltaic Energy Conversion *The IEEE International Conference* **2** 155-58

Modelling Soft Tissue Deformation Based on Spatial Kriging

Mengxiao Zhao[1], Dedong Gao[1,*], Shijian Zhao[1], Linze Wang[1] and Lirong Li[1]

[1]School of Mechanical Engineering, Qinghai University, Xining, China

[*]Corresponding author: gaodd@qhu.edu.cn

Abstract. People can only obtain incomplete information when needle get into soft tissue. The actual conditions limit monitoring results. The kriging, as a statistical interpolation method, can effectively transform discrete point information into facial continuous information. On the basis of displacement data of pork soft tissue, the correlation parameter setting and optimal semi-variance model selection of kriging were studied to determine the kriging interpolation results. The results show that the accuracy of the kriging model can reach millimetre level, the maximum error is about 0.8mm, the average standard error is 0.35mm, and the calculation is simple. It can help doctors provide references when the soft tissue cannot be directly observed, and predict the deformation within the soft tissue.

1. Introduction

Percutaneous puncture technique is a common treatment method in modern medicine. Needle get into the soft tissue, reaching the target to achieve diagnosis, treatment, sampling, stimulation and other purposes. Most of target are concentrated in the internal organs of the body, the clinical application of tissue biopsy, local anaesthesia and blood RT, etc [1-2]. In these applications, the main causes of error are Imaging equipment resolution limit, image coordinate deviation, target measurement error, artificial error , as well as the position error caused by tissue deformation and needle deflection [3-5]. It is necessary to learn the biological characteristics of soft tissue to improve the precision and reduce the system error. The deformation prediction model of soft tissue, which connects needles and soft tissue, is of great significance for improving accuracy and reducing error [6]. At present, the common soft tissue models are mass-spring model, finite element model, meshless model, boundary element model and so on [7]. The mass-spring model and finite element model are the most widely used [8-9].

The mass-spring model is a method to construct the soft tissue model based on several sets of particle network structures with springs, and its deformation is mainly depending on the elastic coefficient of the spring [10]. Modelling is simple and classic, so it can support large-scale and high-real-time deformation calculations, many scholars has been improved it [11-17]. However, the accuracy of the calculation is low precision, which is limited by the precondition. Llinger et al [18] using genetic algorithm and others to optimize the mesh parameters of the mass-spring model to reduce the deformation error, but the computation is complex. Qiao et al [19] the mass-spring model of grid topology based on regular hexagon has been improved in stability and real-time, but the number of springs and particles is more. Choi et al [20] use parametric anisotropic spring-particle model to simulate the mechanical reaction of tissue. The simulation is very good, When the external force is stable. However, Once the external force is unstable, the deformation simulation will be different from the actual situation. Natsupakpong et al [21] study the elastic parameters of the spring-particle model, using tetrahedron and hexahedron to construct two-dimensional and three-dimensional grids. the ideal experimental results are obtained, but the calculation was more complicated. Basafa et al [22] proposed to simulate the deformation of nonlinear and viscoelastic soft tissue by combining damper and spring-particle model, but the algorithm cannot solve the simulation of large deformation, it needs further optimization. Liu et al [23] put forward that the spring-particle model and nonlinear anisotropic elasticity theory are combined to reflect the nonlinearity, anisotropy, viscoelasticity and quasi-compressibility of soft tissue, and the ideal results are obtained by experiment, but the calculation is more.

The finite element model is a kind of model which is popular with the development of computer technology, and this common simulation method is to solve the simulation results by analysing the numerical solution of partial differential equations, and the finite element method provides a very high precision soft tissue model [10]. However, the accuracy of the finite element model is directly related to the number of mesh generation, high-precision need to divide a large number of grids, resulting in too large a computation, so that real-time is very poor. Tang Xiangen [24] uses finite element model to simulate soft tissue deformation based on OpenHaptic environment, which ensures the authenticity and real-time requirements of deformation simulation, and provides important reference value for virtual surgery. Yin Lulu [25] builds the soft tissue of human face with finite element model, forming a multilayer finite element model containing skin, muscle and fat, provides a strong basis for further deformation after facial surgery, but there is no study on the bonding between skin and subcutaneous two layers. Moita et al [26]improve the damping optimization of viscoelastic tissue by finite element simulation, and the experimental results show that the algorithm can improve the structure of model. Woraratsoontorn et al [27] use finite element model to simulate the deformation of the real skull and soft tissue of the human body, and still needs to further improve the boundary conditions to improve the deformation effect and consider the influence of the surrounding tissues in the future. Sangpradit et al [28] use the finite element model to simulate the movement and indentation of soft tissue under external forces. It can not only obtain high accuracy, but also accurately determine the depth and location of the tumor, and the model algorithm can conform to soft tissue and irregular organization in the future. Xie et al [29] use the finite element model to simulate the three-dimensional friction contact analysis of Cosserat material, and the experiment proves the feasibility of the algorithm, but the calculation amount is very large.

In this paper, we propose a soft tissue prediction method based on Kriging to balance the contradiction between precision and real-time. Kriging is an excellent interpolation method. Only a small amount of information is required, the whole working face can be estimated by the semi-variance function, and is the optimal linear unbiased estimation. It is possible to simplify the calculation and maintain high precision in the modelling by Kriging interpolation method. Therefore, this paper proposes to build a soft tissue model by Kriging, which greatly simplifies the calculation and improves the accuracy of the model. It reduces the puncture error to provide a reference for percutaneous puncture technique.

2. Kriging

2.1. Kriging model

Kriging model is one of the best interpolation models at present, and has been widely used in spatial interpolation in GIS in recent years. As one of the most widely used methods in kriging, the ordinary kriging interpolation mainly uses the raw data and the variation function of regional variables to estimate the linear unbiased optimal estimation of the regionalization variable of the unused sample point. On the premise that the expected data is normal distribution, it is considered that the expectation of the regionalization variable is unknown, and the approximate value of the interpolation point can be determined by the weight of the surrounding sampling point, and the regionalization variables with both randomness and structural, or spatial correlation in the spatial distribution are studied. The interpolation estimation formula is:

$$\hat{z}_0 = \sum_{i=0}^{n} \lambda_i z_i \tag{1}$$

As long as you give an expression of $\boldsymbol{\lambda}=[\lambda_1\ \lambda_2\ \lambda_3 \text{L}\ \lambda_i]^T$, any region variable in the interpolation space can be known. In order to minimize the difference between the estimate of $\hat{z}_0$ and the true value of z_0, the weighted coefficient λ must meet two conditions, namely: the optimal coefficient $\min_{\lambda_i} Var(\hat{z}_0 - z_0)$ and the unbiased estimate $E(\hat{z}_0 - z_0) = 0$.

2.2. Semi-variance function

Semi-variance function, as show in Figure 1, is the most important step in constructing kriging model, and its function selection directly affects the interpolation accuracy of the model. The selection of the theoretical model of semi-variance function is mainly based on the relationship between distance and semi-variance function calculation value, as well as the professional theory or experience to determine the appropriate theoretical model, but also can use scatter plot to speculate the appropriate theoretical model [30]. The common theoretical models are: exponential model, Gaussian model, spherical model and so on. The regional variables that are two points apart in space are recorded as $z(x_i + h)$ and $z(x_i)$ respectively, and the semi-variance function of $N(h)$ pairs of zone variables with the same spacing of h in space is:

$$r(h) = \frac{1}{2N(h)} \sum_{i=1}^{N(h)} [z(x_i + h) - z(x_i)]^2 \tag{2}$$

In the formula, $r(h)$ is the value of the semi-variant function corresponding to all points of space distance h, $N(h)$ is the point logarithm of space distance h, $z(x_i + h)$ is the measured value of space $x_i + h$, and $z(x_i)$ is the measured value at space x_i.

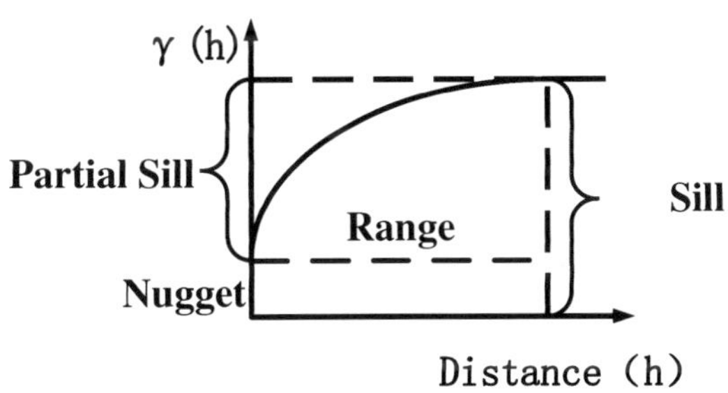

Figure 1. Semi-variance function diagram

The parameters such as nugget, partial sill, sill and range of the semi-variance function indicate the degree of spatial variation and correlation of regionalization variables. According to the theory of semi-variance function, the values of the two points in the same position should be equal, and with the increase of the distance (h), the numerical difference between the two points increases, until the initial base value is tended, and the interval distance between the sampling points is range. Because the sampling error and the existence of spatial variation make the two points very close, so that there is a nugget with a semi-variance function value of not zero. The Nugget/ Sill is called the substrate effect and is used to represent the variation characteristics between samples, reflecting spatial correlation, and if the value is greater, it means that there is more variation between samples caused by random factors. Then, we will use experimental data to build an accurate model.

3. Material

3.1. Test process

The choice of biological soft tissue material is about 25cm long, about 10cm wide, about 5cm tick of a shape of the regular pork leg(as show in figure 2). The meat in pork leg is evenly thick and easy to process into a shape-regular piece of meat to facilitate the insertion of identifiers. Pork leg meat has typical soft tissue mechanical characteristics of anisotropy, and its mechanical properties in the fiber direction of identical and vertical are different. Its mechanical properties can be obtained by SHPB experiment [31], in the case of strain rate of 0.02/s, the modulus of fiber direction of identical the is 121.00±28.76KPA, the ultimate strength is 63.73±18.53kpa, the failure strain is 0.934± 0.189; the modulus of fiber direction of vertical is 47.60±19.30kpa, the ultimate strength is 22.94±3.63kpa, failure strain is 1.077±0.111[32]. The markers (as show in figure 3) use an iron nail with a diameter of about 2.3mm and is arranged at a 1.5cm distance interval.

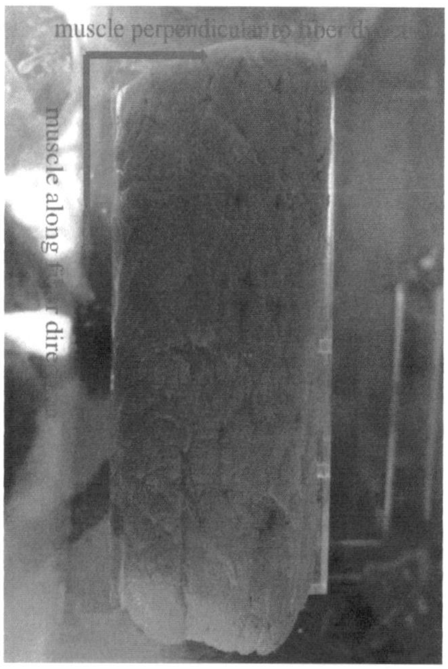

Figure 2. The soft tissue used

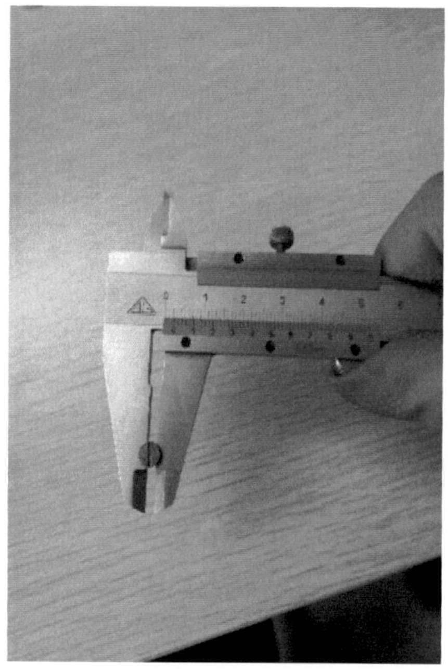

Figure 3. Marker

The meat of the inserted markers is placed on the special fixture, and the puncture experiment is carried out by using the three-coordinate puncture system, and the B-ultrasound image is collected by the ultrasonic probe placed on the meat. Because of the limitation of the scanning width of the ultrasonic probe, the internal situation of the whole organization cannot be observed with one scan, The B-ultrasound images collected sequentially must be stitched together. Stitched image is generally jpg format, not convenient for direct image processing. PS is generally used to convert its format to BMP format and then image processing. Using MATLAB to binarization, improve the contrast of the image to reduce the difficulty of subsequent processing, corrosion removal of small, discrete, large area of noise in the image, expansion is amplifying and recover the identifier image which remove the noise after, noise reduction to remove the large noise spots which corrosion still exists after. The marker marks the centroid of the remaining image and sends its coordinates to the specified location, and finally converts the coordinate units on the image from pixels to millimetres. The position of the identifier in the soft tissue can eventually be obtained. The displacement of the identifier in the puncture process can be obtained by comparing the coordinate position of the image with different puncture depth. Figure 4 is a B-ultrasound image when the needle puncture is completed, and the identifier is not completely identified in the image due to equipment limitations and other reasons (as shown in Figure 5). The displacement of 37 markers can be obtained by comparing it with the B-ultrasound image when the puncture is not beginning.

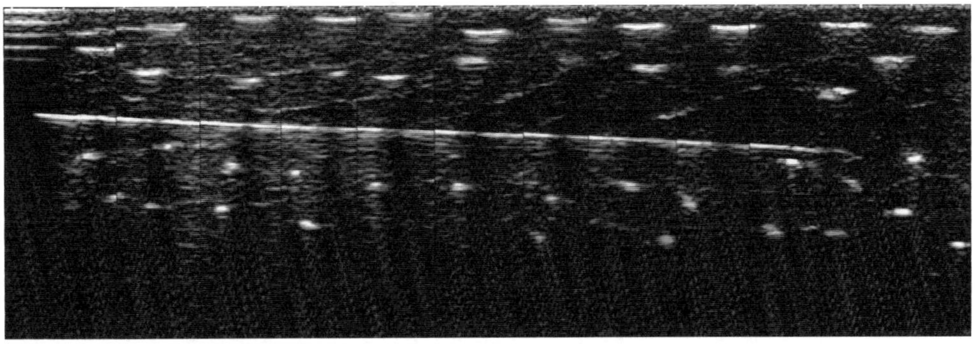

Figure 4. needle puncture soft tissue B-ultrasound image

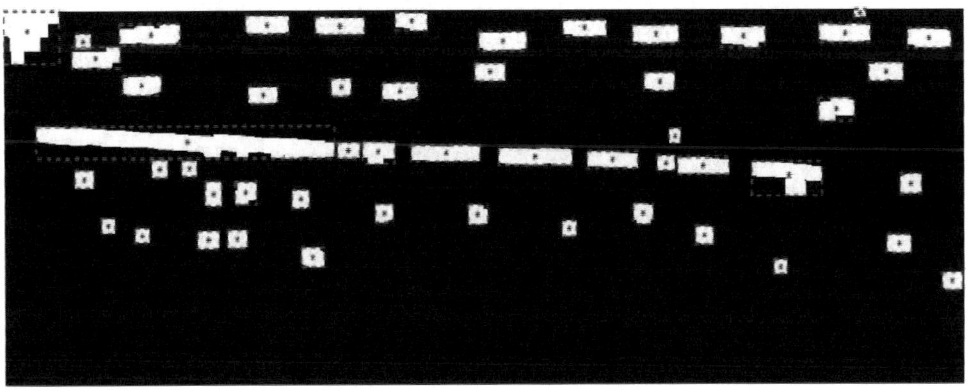

Figure 5. Marker coordinates

3.2. Data processing

Data normal distribution is the premise of applying spatial interpolation of kriging method, using histogram to verify the normal distribution of the data. In this paper, logarithmic transformations are adopted for each identifier displacement data, and the skewness and kurtosis of the data before and after the transformation are compared (as shown in table 1). The closer the skewness and kurtosis are to 0, the more the distribution of data tends to be normal. Therefore, in this paper, the displacement data is used as the basic data of the spatial interpolation semi-variation function of kriging method by logarithmic transformation.

Table 1. Comparison of displacement data before and after logarithmic transformation

	Sample points (PCs)	Average (mm)	Minimum value (mm)	Maximum value (mm)	skewness		kurtosis	
					Before conversion	After conversion	Before conversion	After conversion
Displacement	38	0.7353	0.09881	3.00501	2.783	-0.455	13.473	4.769

3.3. Kriging values

By substituting the experimental data into different semi-variation function models, the performance of different semi-variation function models under 5 indexes has its advantages and disadvantages, and the accuracy evaluation results are shown in table 2.

Table 2. interpolation accuracy analysis of different semi-variation functions

Semi-Variance Model	Mean	Root-Mean-Square	Mean Standardized	Root-Mean-Square Standardized	Average Standard Error
Stable	-0.0289	0.4497	-0.0756	1.3086	0.3482
J-Bessel	-0.0318	0.4572	-0.1486	1.5345	0.3403
K-Bessel	-0.0296	0.4496	-0.0771	1.3109	0.3470
Hole Effect	-0.0120	0.4549	-0.1075	1.4546	0.3645
Rational Quadratic	-0.0354	0.4492	-0.0892	1.2990	0.3387
Gaussian	-0.0144	0.4601	-0.0938	1.4115	0.3603
Exponential	-0.0256	0.4512	-0.0624	1.2732	0.3578
Pentaspherial	-0.0270	0.4478	-0.1018	1.4102	0.3387
Tetrasphenrical	-0.0206	0.4482	-0.0837	1.3732	0.3460
Sperical	-0.0116	0.4466	-0.0691	1.3439	0.3529
Circular	-0.0198	0.4501	-0.0571	1.3120	0.3591

Accuracy standards includes 5 evaluation indexes of Mean, Root-Mean-Square, Mean Standardized, Root-Mean-Square Standardized, Average Standard Error, in which the closer the Mean and Root-Mean-Square is 0, the better, the smaller the Average Standard Error and the Mean Standardized, the better the Root-Mean-Square Standardized is closer to 1. Because the spatial interpolation scale range of this paper is very small, the average standard error and the root-mean-square of each point are the main precision evaluation indexes, combined with the mean, the mean standardized, the root-mean-square standardized and so on. The semi-variation function model of the identification point displacement is the highest precision of the sperical model.

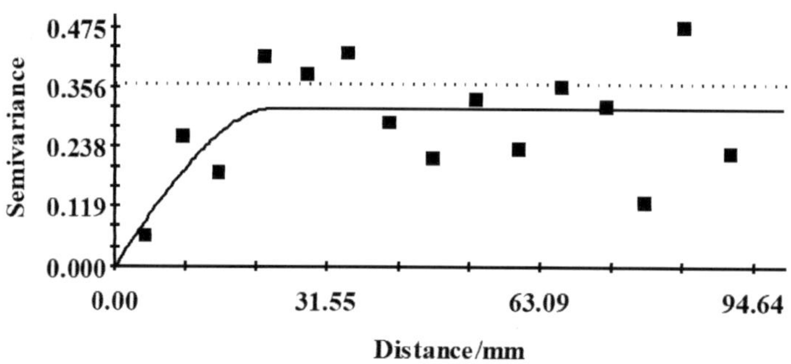

Figure 6. Spherical model semi-variance function model

Figure 6 is a spherical model semi-variation function diagram, in which the scatter represents the semi-variance of the data, the x-axis is the distance between the data, the curve is based on the scatter plot of the spherical model, the dotted line represents the variance of the data. The specific expressions of the spherical model are:

$$\gamma(h) = 0.0001 + 0.3121(\frac{3}{2}\frac{h}{23.7} - \frac{1}{2}\frac{h^3}{23.7^3}) \quad (3)$$

The specific parameter nugget 0.0001, the partial still 0.3121, the still 0.3122, the range 23.7. When the ratio of nugget to sill is ≤25%, the spatial autocorrelation degree of variable space is strong, and if the ratio is between 25% and 75%, it is medium space autocorrelation level, and if the ratio is

≥75%, it is weak space autocorrelation level. The nugget/still is 0.0003. Therefore, the experimental data has a strong spatial correlation, and it is very suitable to use the Kriging model.

4. Results and analysis

In order to better reflect the change of the displacement of the marker, ArcGIS uses the non-isometric grading method to classify the displacement of the identifier, which is divided into 10 levels, of which the first 7 levels are small and detailed, reflecting the displacement change, and the latter level 3 is relatively large to reflect the deformation of the needle tip site. From Figure 7, it can be seen that the displacement takes the needle tip part as the core and gradually weakens to both sides. The concrete performance is that the displacement of the needle tip part changes the most, followed by the needle shaft. The active needle axis is centred and the displacement decreases to both sides. and presents a strong anisotropic characteristic, in the direction of the CIS displacement is significantly greater thanthe vertical fiber direction, and the results of block gold effect analysis is consistent.

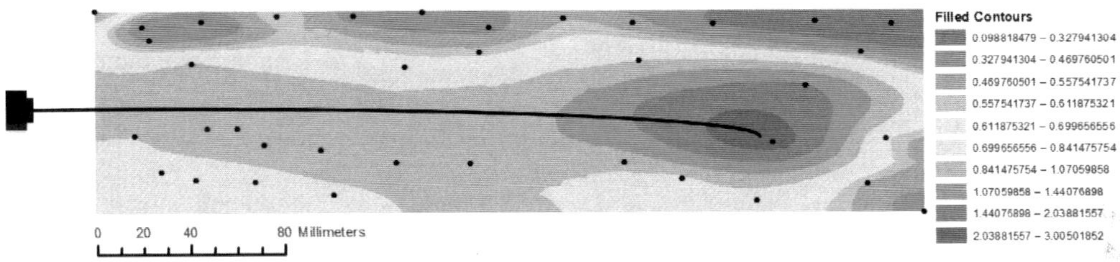

Figure 7. Soft tissue kriging model and schematic diagram

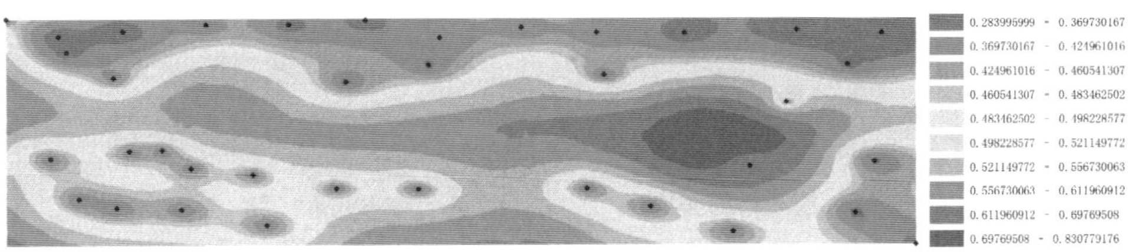

Figure 8. error distribution

From the error distribution(as shown in Figure 8), it can be seen that the model precision constructed by Kriging interpolation method fully meets the requirements. The maximum error is about 0.8mm, which fully meets the requirements of millimetre grade accuracy for needle puncture surgery. Most of the errors are relatively small, about 0.5mm and distributed on both sides of the needle body. This part of the sampling point is denser, so the interpolation accuracy is higher. In the needle part, in order to prevent the needle tip from colliding with the marker, the identifier is set less. There are few sampling points around these sampling points and there is a lack of spatial information for interpolation, so the error is larger than elsewhere. Especially the needle tip site spatial information is the most lacking, the error is the greatest.

5. Conclusions and future work

The displacement of soft tissue when the needle get into it has a strong correlation in space. The suitable semi-variation function can improve the precision of kriging model. In this paper, displacement data of soft tissue are obtained by image processing. The data are transformed by logarithmic transformation to make the data obeys normal distribution. Comparing the accuracy of each semi-variation function model to select the most suitable one. It is most appropriate semi-variation function is spherical model by comparing. Finally, the Kriging interpolation results and error distribution are obtained. The results are consistent with the actual situation, and the error also meets the requirements.

Compared with the traditional soft tissue model, such as the mass-spring model and the finite element model, the advantage of the Kriging is that it can simplify the calculation with high precision, so its practicality is greatly improved. In the clinic process or training system, the deformation is required to be feedback to the physical. In future work, kriging model will add time factors to the to improve prediction and provide the real time information.

Acknowledgments
This research work is sponsored by the National Natural Science Foundation of China (No. 51665049), Qinghai Provincial Natural Science Foundation of China (Grant No.2015-ZJ-906).

Reference

[1] Zivanovic, A., & Davies, B. L. (2000). A robotic system for blood sampling. *IEEE Transactions on Information Technology in Biomedicine A Publication of the IEEE Engineering in Medicine & Biology Society,4*(1), 8.
[2] Rizun P R, Mcbeth P B, & Louw D F,. (2004). Robot- assisted neu- rosurgery. *Surg Innov, 11*(2): 99-106.
[3] Jeffrey Carr, J. , Hemler, P. F. , Freimanis, R. I. , Choplin, R. H. , & Halford, P. W. . (1997). Stereotactic localization of breast lesions: how it works and methods for improving accuracy. *Academic Radiology, 4*(12), 862-863.
[4] Roberson, P. L. , Narayana, V. , Mcshan, D. L. , Winfield, R. J. , & Mclaughlin, P. W. . (1997). Source placement error for permanent implant of the prostate. *Medical Physics, 24*(2), 251.
[5] Hussain, H. K. , Kingston, J. E. , Domizio, P. , Norton, A. J. , & Reznek, R. H. . (2001). Imaging-guided core biopsy for the diagnosis of malignant tumors in pediatric patients. *Ajr Am J Roentgenol, 176*(1), 43-47.
[6] Gaodedong. (2017). *Study on deformation mechanism and dynamic trajectory planning method of soft needle puncture.* (Doctoral dissertation).
[7] Chen. (0). *Study on the modeling method of soft tissue in virtual surgery.* (Doctoral dissertation).
[8] Batteau, L. M. , Liu, A. , Maintz, J. B. A. , Bhasin, Y. , & Bowyer, M. W. . (2004). *A Study on the Perception of Haptics in Surgical Simulation. Medical Simulation.* Springer Berlin Heidelberg.
[9] Cotin, S. , Hervé Delingette, & Ayache, N. . (1999). Real-time elastic deformations of soft tissues for surgery simulation. *IEEE Transactions on Visualization and Computer Graphics, 5*(1), 62-73.
[10] Sun Yanxia. (2005). *Modification of elastic model in human soft tissue modeling.* (Doctoral dissertation, Southeast University).
[11] Zhong, Y. , Shirinzadeh, B. , Smith, J. , & Gu, C. . (2012). Soft tissue deformation with reaction-diffusion process for surgery simulation. *Journal of Visual Languages and Computing, 23*(1), 1-12.
[12] Liu Xuemei, Mao Lei, Li Yunhua, & Amini. (2013). *Coupling non-mesh gad gold and particle springs to achieve soft tissue deformation simulation. Journal of Computer Aided Design and graphics* (1), 1-6.
[13] Hammer, P. E. , Sacks, M. S. , Nido, P. J. D. , & Howe, R. D. . (2011). Mass-spring model for simulation of heart valve tissue mechanical behavior. *Annals of Biomedical Engineering, 39*(6), 1668-1679.
[14] Morin, J. B. , Tomazin, K. , Edouard, P. , & Millet, G. Y. . (2011). Changes in running mechanics and spring–mass behavior induced by a mountain ultra-marathon race. *Journal of Biomechanics, 44*(6), 1104-1107.
[15] Mi, S. , Hou, Z. , Yang, F. , Xie, X. , & Bian, G. . (2013). *A Multi-body Mass-spring Model for Virtual Reality Training Simulators Based on a Robotic Guide Wire Operating System. A multi-body mass-spring model for virtual reality training simulators based on a robotic guide wire operating system.*

[16] Zhongmin, Huangfu, & Luoheng Yan, (2013) An Improved Mass-spring Model for Simulation of SoftTissue Deformation. *Journal of Information & Computational Science, 10*(17): 5551–5558.

[17] Ouyang, H. , & Zhang, J. . (2015). Passive modifications for partial assignment of natural frequencies of mass–spring systems. *Mechanical Systems and Signal Processing, 50-51*, 214-226.

[18] V?Llinger, U. , Stier, H. , Priesnitz, J. , & Krause, F. L. . (2009). Evolutionary optimization of mass-spring models. *CIRP Journal of Manufacturing Science and Technology, 1*(3), 137-141.

[19] Qiao, B. , Chen, G. , & Ye, X. . (2009). The research of soft tissue deformation based on mass-spring model. *International Conference on Mechatronics & Automation.* IEEE.

[20] Choi, K. S. . (2010). Toward realistic virtual surgical simulation: using heuristically parameterized anisotropic mass-spring model to simulate tissue mechanical responses. *International Conference on Education Technology & Computer.* IEEE.

[21] Suriya Natsupakpong, M. & Cenk Çavus og˘lu.(2010) Determination of elasticity parameters in lumped element (mass-spring) models of deformable objects[J]. *Graphical Models, 72*: 61–73.

[22] Basafa, E. , & Farahmand, F. . (2011). Real-time simulation of the nonlinear visco-elastic deformations of soft tissues. *International Journal of Computer Assisted Radiology & Surgery, 6*(3), 297-307.

[23] Liu, X. P. , Xu, S. , Zhang, H. , & Hu, L. . (2011). A new hybrid soft tissue model for visio-haptic simulation. *IEEE Transactions on Instrumentation & Measurement, 60*(11), 3570-3581.

[24] Tang Xiangen, Shanyua, Tan Ke, & Pan Wan. (2012). *A finite element deformation model of soft tissue based on openhaptic. Journal of System Simulation* (7), 1434-1437.

[25] Yin Lu-lu, Riochime, & Lu Hong Ice. (2010). Establishment of three-dimensional finite element model of facial soft tissue. *Science and technology and engineering, 10* (12).

[26] Moita, J. S. , Araújo, A.L., Mota Soares, C. M. , & Mota Soares, C. A. . (2013). Finite element model for damping optimization of viscoelastic sandwich structures. *Advances in Engineering Software, 66*(Complete), 34-39.

[27] Woraratsoontorn, P. , & Sooraksa, P. . (2012). Development of finite element model for analysis in the realistic human skull impact. *Biomedical Engineering International Conference.* IEEE.

[28] Sangpradit, K. , Liu, H. , Dasgupta, P. , Althoefer, K. , & Seneviratne, L. D. . (2011). Finite-element modeling of soft tissue rolling indentation. *IEEE Transactions on Biomedical Engineering, 58*(12), 3319-3327.

[29] Xie, Z. Q. , Zhang, H. W. , & Chen, B. S. . (2012). A finite element model for 3d frictional contact analysis of cosserat materials. *Finite Elements in Analysis and Design, 57*(none), 92-102.

[30] Clark, D. S. . (1981). Adjudication to administration: a statistical analysis of federal district courts in the twentieth century. *S.cal.l.rev.*

[31] Wang Baozhen, Zheng Yuxuan, & Hu time win. (2010). Dynamic tensile properties of porcine hind leg muscles. *Explosion with shock, 30* (5).

[32] Wang Baozhen. (2010). *Study on dynamic mechanical properties of muscle soft tissue.* (Doctoral dissertation, China University of Science and Technology).

Deep Neural Networks in the View of Ecological Holism

YUE Qiang[1,b] and YUE Xun[2,a]

[1] College of Water Conservancy and Civil Engineering, Shandong Agricultural University, Tai'an, Shandong 271018, China
[2] College of Information Sciences and Engineering, Shandong Agricultural University, Tai'an, Shandong 271018, China

[a] yuexun@sdau.edu.cn, [b]yueqiang406@163.com

Abstract. In the view point of ecological holism, deep neural networks can be view as a machine cognitive agent with overall multi-level nested structure, the representation learning and problem-solving ability of machine cognitive agent with multi-modal perception function is also the results from the perception and judgment of ambient situation environment. In the paper, a machine cognitive method based on ecological holism was deconstructed. Then, two core questions to characterize learnability of AlphaGo were answered, one is how to explain the characteristics of the hierarchical gradient and update parameters between different layers of networks? Two is how to understand the effective ability to choose the drop sampling at the strongest level of game of Go. These core controlling structural and functional roles proposed in this study may be further improved to a new type of artificial intelligence law of machine cognitive system.

1. Introduction

Decision making system for cognitive machines is becoming the basic service system of digital ecosystem, deep neural network is considered to be the important advancement of machine cognitive system, which rely on its ability to extract complex patterns of underlying dataset [1][2]. For example, the mathematical model of single hidden layer fully-connected neural network is as follows [3] [4]:

$$F(x;\Theta) = \sum_{m=1}^{M} \beta_m \tanh(w_m \bullet x + b_m) \quad (1)$$

X is the input representing the neural network, $\Theta = \{\beta_m, w_m, b_m\}_{m=1}^{M}$ represent the parameter set of the neural network, M is the number of hidden layer neurons. x_i, y_i represent the training sample and the label assigned to the sample manually, an optimized parameter set Θ^* can been gotten by the training sample set $T = \{x_i, y_i\}_{i=1}^{N}$. How to understand the extracted hierarchical gradient and update parameters of the sample metadata features between different layers of networks, and how to explain the physical meaning of the distributed representation represented by the function parameter set Θ^* ?

In the view point of ecological holism, deep neural networks can be view as a machine cognitive agent with overall multi-level nested structure, the representation learning and problem-solving ability of machine cognitive agent with multi-modal perception function is also regarded as the perception and judgment of ambient situation environment. Can ecological holism explain machine cognitive system of AlphaGo correctly? In the paper, a machine cognitive method in the view point of ecological holism was deconstructed, which is different from reductionism and expand and upgrade the basic law of machine cognitive system. Then, two core questions of AlphaGo were answered, one is how to explain the characteristics of the hierarchical gradient and update parameters between different layers of networks? Two is how to understand the effective ability to choose the drop sampling at the strongest level of game of Go.

2. Machine cognitive method in the view point of ecological holism

Machine cognitive agent with multi-modal perception function can "understand" the unstructured data, such as language, images, video et al, the problem-solving ability of machine cognitive agent is regarded as the perception and judgment of ambient situation environment. The machine cognitive agent has the following cognitive characteristics:

- The external environment of machine cognitive agent is the set of external connected objects (things). The internal environment refers to the internal control elements, information transfer and the dynamic evolution process of spatial organization. The bounded time and space is called the system boundary.
- The relationship between machine cognitive agent and the external environment determines a dialectical relationship of interdependence. On the one hand, the changes of the external environment will stimulate and regulate the overall dynamic evolution of the internal environment; on the other hand, the dynamic evolution of information transfer and spatial organization will feedback the surrounding situation through information exchange with the external environment. Therefore, the representation learning and problem-solving ability of machine cognitive agent in specific application areas are in the process of dynamic exploration to unify the changes of the external environment.
- The spatial organization of internal control elements have the overall vertical multi-level nested structure, which follow the characteristics of stratification, classification, and subsection, each level of control elements has its own unique concepts, characteristics, and rules.
- Due to the vertical dimension hierarchy superposition non-linear structure, the internal information transfer of machine cognitive agent is not the single nature and the same direction, the gradient difference of the time space forms the difference of the information flow path or the movement mode, which include "top-down" and "bottom-up" in a single vertical dimension. "Top-down" reflects the system's downward causal relationship. "bottom-up" reflects the system's upward causality. The representation learning and problem-solving ability of machine cognitive agent lies in the hierarchical gradient difference of the time space and update parameters of the information flow path.
- Machine cognitive agent is often limited to a specific vertical dimension, however, decision making system for cognitive machines of digital ecosystem have the horizontal multi-dimensional dimensional synergy. The competition and cooperation of information transfer paths in different vertical dimensions will present two orderly rules of machine cognitive agent, one is the order of the information flow path; another one is the order of spatial organization.
- From the view point of dynamics and histology, the order of machine cognitive agent must be the dynamic evolution process of the information flow path and spatial organization. Orderly process of dynamic evolution depend on the direction of information transfer, which include "top-down" and "bottom-up" in a single vertical dimension , and the horizontal multi-dimensional dimensional synergy. The core technology of deep neural networks using deep reinforcement learning for machine cognitive system lies in the control parameters of information transfer paths and re-integration model of spatial organization.
- To cope with the external environment, machine cognitive agent at certain times of orderly process of dynamic evolution, which through a multistep of information transfer pathway in a single vertical dimension and the horizontal multi-dimensional dimensional synergy, can use

sophisticated and complex mechanisms to modulate the homeostasis between "top-down" and "bottom-up".

- Machine cognitive agent with multi-modal perception function has the effective ability to deal with specific field problems through interaction with the external environment. In fact, the problem-solving ability of machine cognitive agent should be regarded as the perception and judgment of ambient situation environment, which depend on the dynamic evolution of information transfer paths and spatial organization.

3. The representation learning and problem-solving ability of AlphaGo under the view of ecological holism

AlphaGo can be view as a machine cognitive agent, play at the strongest level of game of Go. Using multi-modal perception function, each features of the input and outcome of game of Go was pre-processed into a set of 19 × 19 feature planes. Every board position or state determines the integer feature value, which is split into multiple 19 × 19 planes of binary values (one-hot encoding). Neural network architecture of AlphaGo (the policy and value networks) is the deep convolutional neural networks, which use convolutional layers to construct a representation of the position. The effective ability of AlphaGo, which choose the drop sampling at the strongest level of game of Go, selects actions to combine the policy and value networks.

3.1. Policy network sample actions

The effective ability to choose the drop sampling comes from policy gradient reinforcement learning of policy network, which depends on how to train the weight-sharing structure of the policy network [4] [5, 6]. As shown in Fig. 1: (a), 13 hidden layer SL policy network use stochastic gradient ascent to maximize the likelihood of the human move a selected in state. The parameter update of SL policy network training was based on mean squared error between the predicted values and the observed rewards.

For the first stage of the training pipeline, AlphaGo build on prior work on predicting expert moves in the game of Go using supervised learning, and applied an asynchronous stochastic gradient descent update to maximize the log likelihood of the action. The SL policy network alternates between convolutional layers with weights, and rectifier nonlinearities. A final soft-max layer outputs a probability distribution over all legal moves. Owing to its enormous search space (the number of legal moves per position and game length) and the difficulty of evaluating board positions and moves, the second stage of AlphaGo aims at improving the policy network by policy gradient reinforcement learning. The training pipeline is to receive high rewards between the current strategy network and the strategy network that randomly selects a previous iteration [5]. From the view point of dynamics and histology, The results using deep reinforcement learning for AlphaGo lies in the weights parameters (the characteristics of the hierarchical gradient and update parameters between different layers of nodes) of 13 hidden layer policy network, which will be "mapped" to the "top-down and "bottom-up" of vertical information transfer paths and the horizontal multi-dimensional dimensional synergy.

The representation learning of AlphaGo is not only to accumulate knowledge, when AlphaGo "read and understand" the change of board position or state, how to select the next-hand strategy? Problem-solving ability is more important. However, AlphaGo learn how to drop the sub-sampling for a specific situation and predict expert moves on a held out test set with an accuracy of 57.0% using the policy network.

3.2. Value network evaluate positions

As shown in Fig. 1: (b), the value network has a similar architecture to the policy network, the hidden layers 2 to 11 of the value network are identical to the policy network, hidden layer 12 is an additional convolution layer, hidden layer 13 convolves 1 filter of kernel size 1 × 1 with stride 1, and hidden layer 14 is a fully connected linear layer with 256 rectifier units. The output layer is a fully connected linear layer with a single tanh unit, which outputs a single prediction instead of a probability distribution.

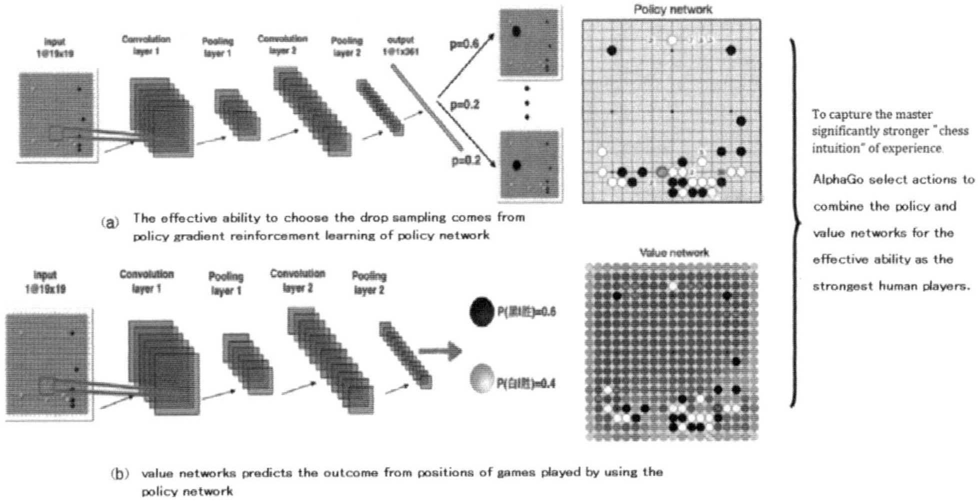

Figure 1. AlphaGo select actions to combine the policy and value networks to capture the master significantly stronger “chess intuition” [4] [5]

3.3. AlphaGo select actions to combine the policy and value networks for the effective ability as the strongest human players

As alternating Markov games, the highest state of the game of Go is the “balance and harmony”, which is not just kill the other party. The change of the game in the process of chess is unfolded step by step, the homeostasis of the chess process depend on the whole process of the game, which from balance to imbalance and then from balance to balance. In order to play at the level of the strongest human players, a way was found to capture the master significantly stronger “chess intuition” of experience. As shown in figure 1, using the regression, reinforcement learning of value networks, which is the final stage of the training pipeline, focuses on position evaluation, estimating a value function that predicts the outcome from positions of games played by using the policy network. AlphaGo select actions by evaluating positions using a value network, and sampling actions using a policy network. Systematic homeostasis of AlphaGo master the significantly stronger “chess intuition” of experience, AlphaGo has finally reached a professional level in Go.

4. Transcendence of reductionism: the symbolic framework s of machine cognitive agent from the view point of ecological holism

As the deep neural networks (DNNs), the main challenge feature powerful representation capability of AlphaGo is that the position evaluation functions, which means the hierarchical gradient and update parameters of the sample metadata features between different layers of networks, are typically highly non-linear and non-convex, non-linear model free reinforcement learning algorithms may not have enough representation capacity in modeling large-scale or high-dimension systems with nonlinear dynamics, which causes great difficulty for following decision making. It is really linear and non-linear algebra fashion for the mathematical model of deep neural networks? In the view point of ecological holism,the weights parameters of hidden layer for the policy and value networks can be evaluated by changes system of the whole concept of Chinese traditional cultures. The symbolic system of “Zhouyi: The book of Changes” can be further improved to a new type of artificial intelligence program. As mentioned mathematical model of single hidden layer fully-connected neural network, the parameter set of the neural network $\Theta = \{\beta_m, w_m, b_m\}_{m=1}^{M}$ can be explained by following parameters:

M is the number of vertical multi-level nested structure

$\{w_m, b_m\}$ represent the hierarchical gradient and update parameters between different layers of the information flow path.

$\{\beta_m\}$ represent the horizontal multi-dimensional dimensional synergy.

The parameter set of the neural network $\Theta = \{\beta_m, w_m, b_m\}_{m=1}^{M}$, which represent the hierarchical gradient and update parameters between different layers of the information flow path, will "mapped" to the "top-down and "bottom-up" of vertical information transfer paths and the horizontal multi-dimensional dimensional synergy, The dynamic evolution process of the information flow path and spatial organization is

$$Number(M; \beta_m) = 2^M \times \beta_m \tag{2}$$

Instead of linear and non-linear algebra fashion for the mathematical model of deep neural networks, the outcome of the game can been determined by the unified systematic model based on the assignment of numbers, binary or decimal between different layers of the information flow path of the policy and value networks. Let $\{M = 3, \beta_m = 1\}$, the number of the orderly process of dynamic evolution of machine cognitive system is $2^3 \times 1 = 8$. $M = 3$ represent the three vertical dimension hierarchy superposition non-linear structure, "1、0" represent the hierarchical gradient and update parameters between different layers of the information flow path of the policy and value networks,which represent the direction of information transfer of "top-down" and "bottom-up" in a single vertical dimension. (☰ 111), (☴ 110), (☲101), (☶ 100), (☱ 011), (☵ 010), (☳ 001), (☷ 000) are used to describe the eight orderly process of machine cognitive system. This systematic construction framework seek out dynamic functional activity rather than to look for the fixed structures, which constitute an alternate way of understanding the machine cognitive agent from the view point of ecological holism. The representation learning and problem-solving ability of machine cognitive agent lies at the orderly process of (☰ 111), (☴ 110), (☲101), (☶ 100), (☱ 011), (☵ 010), (☳ 001), (☷ 000), which depend on the direction of the information flow path and spatial organization, which include "top-down" and "bottom-up" in a single vertical dimension , and the horizontal multi-dimensional dimensional synergy.

As shown in figure 2, the nonlinear coupling structure between the orders of the hexagrams among eight orderly process is acceptance of the inevitability of change, the internal harmony through a system of mutual checks and balances cycles, and the dynamic evolution constitute the self-organized dynamic evolution global model of machine cognitive system. $\{M = 13, \beta_m = 2\}$ is the policy and value networks of AlphaGo, the number of search control strategy in AlphaGo is $2^{13} \times 2 = 16384$, which are from (0000000000000) to (1111111111111). Instead of linear and non- linear algebra fashion to evaluate the characteristics of the hierarchical gradient and update parameters between different layers of nodes, the symbolic system of the orders of the hexagrams among 16384 orderly proces can be further improved to a new type of artificial intelligence program.

IOP Conf. Series: Materials Science and Engineering **646** (2019) 012022 doi:10.1088/1757-899X/646/1/012022

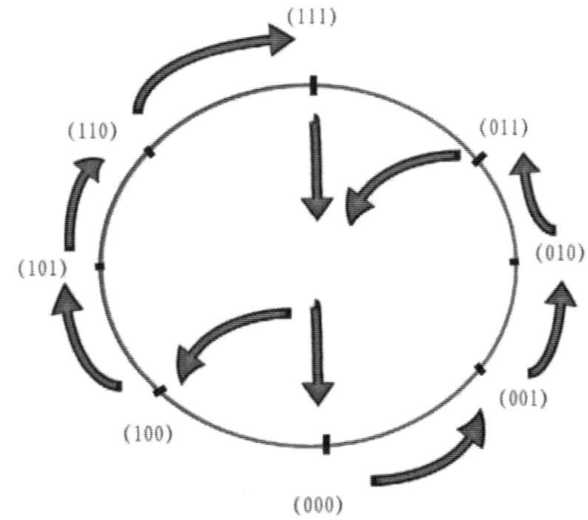

Figure 2. the nonlinear coupling structure between the orderly process of the hexagrams among (☰ 111), (☱ 110), (☲ 101), (☳ 100), (☴ 011), (☵ 010), (☶ 001), (☷ 000), constitute the self-organized dynamic evolution global model of machine cognitive system

5. Conclusions

With the development of the application (Internet of Things, 5G, machine cognition, big data, cloud computing technology and artificial intelligence) in the 21st century,, machine cognitive system will become the basic service system of the digital ecosystem. In particular, machine cognitive system consisting of a certain number of agents will effectively improve the ability to solve large-scale complex problems. Recently, deep neural networks have achieved unprecedented performance in visual domains: for example, image classification, face recognition. AlphaGo and Alpha Zero make a better chess playing algorithm, but, it is not the right approach for most problems. However, machine cognitive system would discern and learn tasks without human intervention. Based on ecological holism, the core controlling structural and functional roles proposed in this study may be further improved to a new type of artificial intelligence law of machine cognitive system.

Acknowledgments

This work was financially supported by the State Key Laboratory of Crop Science, China (Grant No. 2013KF15), the National Key Technology Research and Development Program of the Ministry of Science and Technology of China (2015BAB07B05) and the Significant Application of Agriculture Technology innovation program of Shandong Province (SDNYCX1531963).

References

[1] CHEN Wei-Hong, AN Ji-Yao, LI Ren-Fa 2017 *ACTA AUTOMATICA SINICA* Review on Deep -learning-based Cognitive Computing (in Chinese). 43(11): 1886–1897
[2] David Silver, Thomas Hubert, Julian Schrittwieser 2017 Mastering Chess and Shogi by Self-Play with a General Reinforcement Learning Algorithm. arXiv:1712.01815v1,
[3] Luo T, Liu S L, Li L 2017 A neural network supercomputer. *IEEE Transactions on Computers* 66(1) 73–8
[4] Isma Hadji and Richard P. Wildes 2018 What Do We Understand About Convolutional Networks? https://arxiv.org/abs/1803.08834
[5] Manchun Tan and Desheng Xu 2016 Synchronization of Coupled Neural Networks with Nodes of Different Dimensions *Advances in Neural Networks* 9719 135–142
[6] David Silver, Aja Huang, Chris J Maddison 2016 Mastering the game of Go with deep neural networks and tree search. *Nature.*529 484 - 489

Knowledge-oriented Hierarchical Neural Network for Sentiment Classification

Yanliu Wang[1] and Pengfei Li[2,*]

[1] School of Smart City, Beijing Union University, No.97 Beisihuan East Road, Chao Yang District, Beijing 100101, China
[2] School of Electrical and Electronic Engineering, Nanyang Technological University, 50 Nanyang Avenue, 639798, Singapore

[*]E-mail: pli006@e.ntu.edu.sg

Abstract. Sentiment classification aims to classify the sentimental polarities of given texts. Lexicon-based approaches utilize lexical resources to explore the opinions according to some specific rules, whose effectiveness strongly depends on the goodness of the lexical resources and the rules. Traditional machine-learning methods tightly rely on feature engineering and external NLP toolkits with unavoidable errors. Deep learning models strongly rely on a large amount of labelled data to train their numerous parameters, which often suffer from overfitting issue since it is difficult to obtain sufficient training data. To address the issues, we design a model that combines Knowledge-oriented Convolutional Neural Network (K-CNN) and bidirectional Gated Recurrent Neural Network (biGRU) in a hierarchical way for sentiment classification. Firstly K-CNN is used to capture the n-gram features in sentences. Sentiment word filters are constructed in the knowledge-oriented channel of K-CNN based on the linguistic knowledge from SentiWv ordNet, which can capture the sentiment lexicons and alleviate overfitting effectively. Then biGRU with attention mechanism is utilized to model the sequential relations between sentences and obtain the document-level representation based on the relevance of each sentence to the final sentiment classification. Experiments on two datasets show that our model outperforms other classical deep neural network models.

1. Introduction

The overwhelming of text data on the Internet leads to the revolution of automatically exploiting and extracting valuable information from them. Sentiment classification is a subfield of natural language processing (NLP) which is designed for automatically determining the overall sentiment orientation of the given document based on the opinions in it. The task is to classify the given document into different sentimental polarities (e.g., positive and negative), depending on the contents. With the rapid growth of available subjective texts on the Internet in the form of product reviews, blog posts and comments in discussion forums, business analysts are turning their eyes on the Internet in order to obtain factual as well as more subtle and subjective information (opinions) on companies and products [1]. Therefore, sentiment classification is becoming progressively crucial along with the data blooming.

Current approaches for sentiment classification include lexicon-based and machine-learning-based methods. Lexicon-based techniques generally use the opinion words in the dictionary combined with complex rules and work on the assumption that the collective polarity of a sentence or a document is the sum of polarities of the individual phrases or words [2]. However, with the difficulties of constructing and maintaining the expert rules, the precision and recall obtained by lexicon-based models are not satisfied. In machine-learning-based approaches, sentiment polarities are automatically inferred from a large amount of labelled data. In early years (before 2013), people construct lexical, syntactic and semantic features by complex feature engineering, where the handcrafted features are specifically designed and adjusted for the specific domain. The performance is limited by extensive feature engineering, domain dependency, and unavoidable errors of external NLP toolkits. Recently, researchers started to focus on developing deep compositional models such as Convolutional Neural Network (CNN) and Recurrent Neural Network (RNN) for sentiment classification [3-6]. Deep learning models have demonstrated excellent results because the features are automatically extracted by learning from a large amount of training data, and they are more adept at understanding complicated semantics of a document. However, deep learning models typically have a large number of free parameters which are required to train with a lot of labelled data. It is tough to obtain sufficient training data that covers all the variants of sentiment expressions due to the complexity of natural language. With the limited training data, deep learning models frequently suffer from overfitting issue which hinders their performances.

To tackle the problems mentioned above, we take the advantages of both lexicon-based and machine-learning-based methods and propose a Knowledge-oriented Hierarchical Neural Network (KHNN) that combines Knowledge-oriented CNN (K-CNN) [7] and bidirectional GRU (biGRU) [8] hierarchically for sentiment classification. The idea of hierarchical modelling of the document comes from the theory that the meaning of a long expression is determined by its compositions [9]. KHNN first extracts n-gram features in sentence-level using K-CNN, which reflect the sentimental polarities of sentences; then a bidirectional GRU is used to model the sequential and semantic relations between sentences in the document; finally, a document-level representation is obtained by combining the biGRU outputs using attention mechanism based on the relevance of each sentence to the final sentiment classification. The contributions of the paper are summarized as follows:

- Sentiment word filters in K-CNN are built based on SentiWordNet [10], which can adequately capture the words with strong sentimental polarities appeared in the sentences. Meanwhile, with the fixed weights of sentiment word filters, the number of free parameters is significantly reduced, which effectively minimizes the reliance on training data to alleviate overfitting.
- The proposed model is constructed hierarchically by combining K-CNN which focuses on local features, and biGRU with attention mechanism which captures the global features by sequentially combing the local features. To be specific, K-CNN with sentiment word filters is designed to capture the linguistic clues about the sentiment in each sentence. Then a biGRU with attention mechanism is used to capture the sequential relation and semantic relatedness between sentences to obtain the overall sentiment representation of the document.
- Real-world datasets including IMDB movie reviews and Amazon apparel reviews are used to evaluate our model. Results show that our model can adequately capture the sentimental polarities of documents and outperforms other classical deep learning models for sentiment classification.

2. Related work

The approaches for sentiment analysis are divided into two categories: lexicon-based and machine-learning-based methods. The lexicon-based approaches use the words in the dictionaries and base on some complicated rules. The machine-learning methods include traditional machine-learning and deep learning methods. The difference is that complicated feature engineering is necessary for traditional machine-learning methods. However, it is automatically fulfilled in deep learning methods.

For the lexicon-based method, some works presented approaches which detect the sentiment polarity of a text depending on the semantic orientation of words or phrases appearing in the document [11, 12]. Also, Hanen and Salma [13] presented a lexicon-based approach using lexicon words to determine the sentimental orientation of Facebook comments. The accuracy of the lexicon-based method is not satisfied because it is tough to construct and maintain complicated rules. The machine-

IOP Conf. Series: Materials Science and Engineering **646** (2019) 012023 doi:10.1088/1757-899X/646/1/012023

learning-based approach performs better since sentiment polarities can be automatically inferred from a large amount of labelled data. Some early works focused on creating effective features and used Naive Bayes and support vector machine (SVM) for sentiment classification [14, 15, 16]. However, sophisticated feature engineering is indispensable, and the performance of such system strongly depends on the quality of designed features. Recently, deep learning is becoming more and more popular because the features are automatically extracted from input text by learning from a large amount of data. Convolution Neural Network (CNN) and Recurrent Neural Network (RNN) are footstones for the sentiment classification tasks. Dos Santos et al. [17] designed a model, which combines two CNN layers and Wang et al. aggregate CNN and RNN for sentiment classification [18]. Some works focused on hierarchical neural network models for sentiment classification [19, 20], which perform state-of-the-art accuracy. Even though the deep learning models can achieve high performance in sentiment classification, they rely on quite a large amount of labelled data, which are hard to obtain. Hence, the overfitting problem often perplex researchers.

In order to addressee the issues of lexicon and machine-learning-based approaches, some researchers took a benefit of both methods and proposed some hybrid approaches for sentiment analysis. Zhang et al. [21] firstly use the lexicon-based method for entity-level analysis and then they use additional opinionated indicators to analyze the result of the previous lexicon-based method. Shin et al. [22] suggested a sentiment analysis method that integrates lexicon embeddings and an attention mechanism into Convolutional Neural Networks. In addition, similar hybrid models were proposed and achieved high accuracy for sentiment classification [23, 24]. Therefore, our concentration is on the hybrid method with a novel hierarchical model structure.

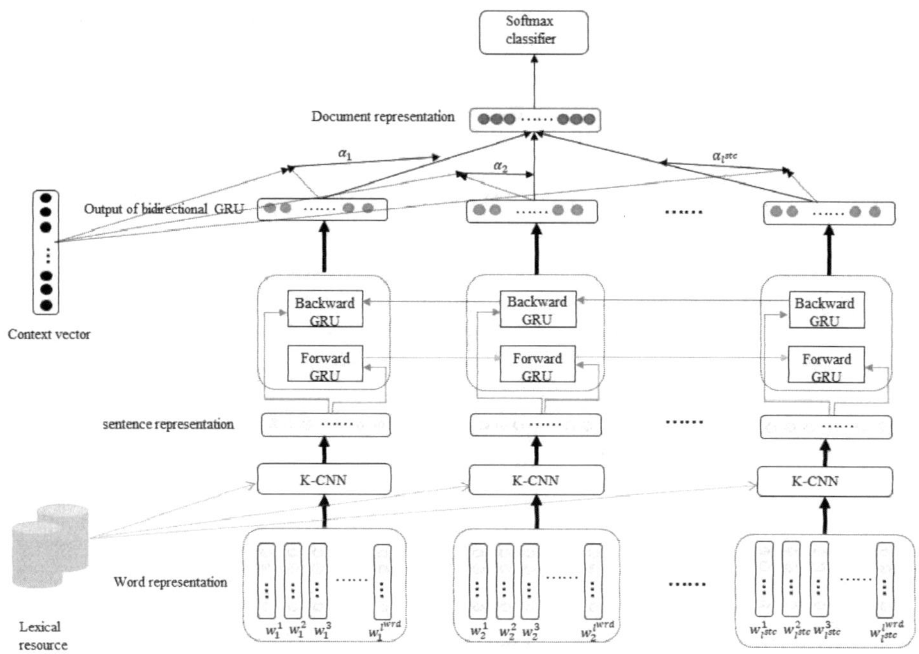

figure 1. The overall structure of the Knowledge-oriented Hierarchical Neural Network (KHNN). w_i^j is the vector representation of the j-th word in the i-th sentence, l^{stc} is the number of sentences in the document, l^{wrd} is the number of words in each sentence and α_i is the weight for the i-th sentence representation vector in attention mechanism.

3. Methodology

The proposed Knowledge-oriented Hierarchical Neural Network (KHNN) has two main components, which are the Knowledge-oriented Convolution Neural network (K-CNN) and bidirectional Gated Recurrent Neural Network (biGRU) with an attention mechanism. K-CNN extracts local features such as significant sentiment lexicons or linguistic clues about sentimental polarity from sentences and obtains the sentence representations. biGRU captures the sequential semantic relations between sentences and obtains the final document representation through an attention mechanism. Finally, the model detects the sentimental polarity based on global features. The overall model architecture is shown in Figure 1.

3.1. Local feature extraction using K-CNN

The main difference between K-CNN and conventional CNN is that K-CNN contains additional knowledge-oriented channel, whose convolutional filters (word filters) are constructed from prior human knowledge instead of learning from data. The benefit of K-CNN is that linguistic knowledge is incorporated into the model which allows it to capture the linguistic clues of sentimental polarity more effectively. Also, K-CNN alleviates overfitting issue since the number of free parameters is significantly reduced. The detailed structure of K-CNN is shown in Figure 2.

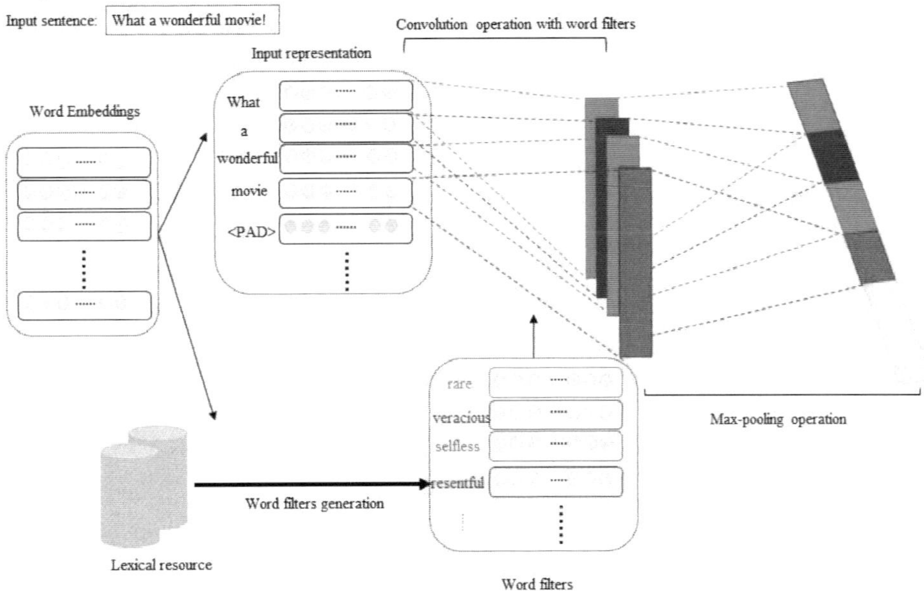

figure 2. The structure of knowledge-oriented channel of K-CNN

3.1.1. Sentence representation. The input document D is composed of a sequence of sentences $D = \{s_1, s_2, s_3, \ldots\}$, and each sentence consists of a sequence of words. We first tokenize each sentence s_i into word tokens with its lower case $s_i = \{x_1, x_2, x_3, \ldots\}$. Next, we map each word into a continues vector space using pre-trained word embeddings to capture the semantic and syntactic information of words. To be more specific, each word x_i is represented as a word vector $w_i \in R^d$ according to word embedding matrix $W^{word} \in R^{d*|s|}$, where $|s|$ is the size of the vocabulary and d is the dimensionality of word embeddings. Since our model only receives the fixed length vectors as input, the lengths of words in each sentence and sentences in the document are padded as l^{wrd} and l^{stc} using special padding character <PAD> with zero word embedding vector, where l^{wrd} is the maximum number of words in a sentence and l^{stc} is the maximum number of sentences in a document. Finally, a document is represented as $D = \{e_1, e_2, \ldots, e_{l^{stc}}\}$, where e_i is the vector representation of sentence s_i: $e_i = \{w_1, w_2, \cdots, w_{l^{wrd}}\}$.

3.1.2. Sentiment word filters generation. SentiWordNet [10] is one of the most frequently utilized and advantageous lexical resources for sentiment analysis. It is a publicly available dictionary based on WordNet [25] which assigns to each WordNet synset three sentiment scores (positivity, negativity, objectivity) ranging from 0 to 1. SentiWordNet includes a relatively large number of terms compared with other popular lexicons [26]. In addition, SentiWordNet provides polarity scores to each sense of a term, whereas the other lexicons only cover a single polarity score. Based on SentiWordNet, we extract significant sentiment lexicon whose sentimental score (either positivity or negativity) is greater than or equal to 0.8. Table 1 shows examples of positive and negative lexicons extracted from SentiWordNet.

Table 1. Examples of lexicons extracted from SentiWordNet

positive	selfless attractive pretty great gracious admirable fabulous honorable superb excellent reputable intellectual good veracious
negative	wrong malevolent unreliable troublesome low depressed foul incompatible unfortunate improper negative miserable fouled

After the extraction of sentiment lexicons, each sentiment lexicon is transformed into a sentiment word filter $\boldsymbol{f} \in \boldsymbol{R}^{d}$ by looking up the same embedding matrix $\boldsymbol{W}^{word}$ as used for input words. The sentiment word filters are used as the convolutional filters in the knowledge-oriented channel of K-CNN to capture the significant sentiment lexicons appeared in the sentence.

3.1.3. Convolution operation. Given the input sentence matrix is $\mathrm{e_i} = \{\mathrm{w_1}, \mathrm{w_2}, \cdots, \mathrm{w_{l^{wrd}}}\} \in \mathrm{R}^{\mathrm{d}*\mathrm{l^{wrd}}}$, convolution operation is performed on the input sentence using convolutional filters $\mathrm{f_i} \in \mathrm{R}^{\mathrm{d}*\mathrm{k}}$, where k is the convolution window size which also corresponds to the k-grams in input texts. In the knowledge-oriented channel, the convolutional filters $\mathrm{f_i}$ are the sentiment word filters whose weights are pre-defined and $\mathrm{k} = 1$ to capture the words with high sentimental polarities. In the data-oriented channel, the weights of $\mathrm{f_i}$ are random initialized and adjusted by learning from data through back-propagation. We set convolution window $\mathrm{k} = 3$ to capture the long-term dependencies between words and other sentiment lexicons which could be ignored in the knowledge-oriented channel.

Mathematically, a feature map $\boldsymbol{m} \in \{\boldsymbol{m_1}, \boldsymbol{m_2}, \cdots, \boldsymbol{m_{l^{wrd}-k+1}}\}$ is obtained for each convolutional filter $\boldsymbol{f}$, where $\boldsymbol{m_i}$ is calculated from input words of window size $\boldsymbol{k}$ $\boldsymbol{e_i}[*, \boldsymbol{i}: \boldsymbol{i}+\boldsymbol{k}]$ following equation (1).

$$m_i = \sigma(\sum(e_i[*, i: i+k] \odot f) + b) \quad (1)$$

where $\boldsymbol{\sigma}$ is a non-linear function, such as tanh and relu, $\boldsymbol{b}$ is the bias and $\odot$ is the Hadamard product between two matrices.

After the convolution operation, max-pooling is performed on each feature map to capture the most significant feature. We use global max pooling as shown in equation (2).

$$p = \max\{m\} = \max\{m_1, m_2, \cdots, m_{l^{wrd}-k+1}\} \quad (2)$$

With the complementary effects of the knowledge-oriented channel and data-oriented channel, K-CNN has outstanding performance in extracting local n-gram features from each sentence in the document.

3.2. Global feature extraction using biGRU

Apart from the sentiment feature vector of each sentence which represents the sentiment information of a sentence obtained via K-CNN, the sequential information between a sequence of sentiment feature vectors also affects the sentimental polarity of a document. For example, "I hated fiction movies because they are ridiculous and boring. However, one day I reverse my opinion after watching Avengers: Infinity War". If we do not consider such sequential information between sentences, the result of final sentiment classification may be wrong. Hence biGRU with attention mechanism is utilized to capture the sequential information and obtain the global feature vector in document-level.

3.2.1. Bi-directional GRU to model document. GRU [27] is a type of RNN which is designed to model sequential data, where the hidden state $\mathrm{h_t}$ at time step t depends on the current input $\mathrm{x_t}$ as well as a hidden state at time step $\mathrm{h_{t-1}}$. To overcome the gradient vanishing and exploding [28] problem of

RNN, two special gates including update gate and reset gate are introduced. The detailed process of GRU is shown as following:

$$r_t = \sigma(V_r * x_t + W_r * h_{t-1} + b_r) \quad (3)$$

$$z_t = \sigma(V_z * x_t + W_z * h_{t-1} + b_z) \quad (4)$$

$$\widehat{h_t} = tanh(V * x_t + r_t(W \odot h_{t-1}) + b_h) \quad (5)$$

$$h_t = (1 - z_t) * h_{t-1} + z_t * \hat{h}_t \quad (6)$$

where $\boldsymbol{\sigma}$ is a sigmoid function, $\odot$ represents the element-wise multiplication, $\boldsymbol{r_t}$ represents the reset gate, $\boldsymbol{z_t}$ represents the update gate and $\widehat{\boldsymbol{h_t}}$ represents the candidate hidden layer.

In our model, we use a bi-directional GRU [8] which stacks two GRUs: forward GRU and backword GRU. The main idea is that the output of each time step may not only depend on the previous elements of the sequence but also depends on the subsequent elements. The forward GRU processes the input sentence from $\boldsymbol{e_1}$ to $\boldsymbol{e_{l^{stc}}}$ obtaining the forward hidden states $\overrightarrow{\boldsymbol{h_t}} = \overrightarrow{\boldsymbol{GRU}(\boldsymbol{e_t})}$ where $\boldsymbol{t} \in [\boldsymbol{1}, \boldsymbol{l^{stc}}]$. The backward GRU processes the input from $\boldsymbol{e_{l^{stc}}}$ to $\boldsymbol{e_1}$ calculating the backward hidden states $\overleftarrow{\boldsymbol{h_t}} = \overleftarrow{\boldsymbol{GRU}(\boldsymbol{e_t})}$ where $\boldsymbol{t} \in [\boldsymbol{l^{stc}}, \boldsymbol{1}]$. Then, we concatenate the hidden states at each time step to obtain the sentence representation $\boldsymbol{h_t} = [\overrightarrow{\boldsymbol{h_t}}, \overleftarrow{\boldsymbol{h_t}}]$.

3.2.2. Combine biGRU outputs using the attention mechanism. Yang et al. [29] proposed a model with an attention mechanism for constructing the document representation according to the semantic importance of contents. The motivation is that not all the sentences play the same role to the overall sentiment polarity and the same sentence may play a different role in a distinct context. With the attention mechanism, we combine the bi-direction GRU outputs h_i at each time step to obtain a global document-level feature vector. The calculation process is as follows:

$$u_i = tanh(W_s * h_i + b_s) \quad (7)$$

$$\alpha_i = \frac{e^{u_i^T * u_s}}{\sum_i e^{u_i^T * u_s}} \quad (8)$$

$$d = \sum_i \alpha_i * h_i \quad (9)$$

$\boldsymbol{u_i} \in \boldsymbol{R^\alpha}$ is the hidden representation of $\boldsymbol{h_i}$ by passing $\boldsymbol{h_i}$ to a fully connected neural network. $\boldsymbol{u_s} \in \boldsymbol{R^\alpha}$ is a context vector which is used to calculate the importance of each sentence ($\boldsymbol{\alpha_i}$) for the overall sentiment orientation, and $\boldsymbol{d}$ is the final document representation. The context vector is randomly initialized and learned during the training process.

3.3. Regularization and sentiment classification

Firstly, we apply dropout regularization [30] to our model for alleviating it from overfitting. The main idea is to randomly drop the connection between units from the neural network during training. After the dropout operation, the obtained final feature vector is fed into a softmax classifier to get the probability distribution over sentiment categories $\boldsymbol{p(D)}$.

We use categorical cross-entropy loss function to calculate the loss between real sentiment distribution $\boldsymbol{p^r(D)}$ and output distribution $\boldsymbol{p(D)}$, as shown in equation (12):

$$\text{loss} = -\sum_{D \in T} \sum_{i=1}^{c} p_i^r(D) \log p_i(D) \quad (10)$$

where T is the training dataset, $\boldsymbol{c}$ is the number of sentiment classes. $\boldsymbol{p^r(D)}$ is a $\boldsymbol{c}$ dimensional one-hot vector whose element corresponding to the real document sentiment category is 1 and other elements are set as 0. Adam [31] optimizer is used to minimize the loss function.

4. Experiment

4.1. Dataset description

We use two real-world sentiment classification datasets to evaluate our model. One dataset IMDB[3] which contains 50,000 movie reviews with high sentimental polarities [32]. Training set and test set are separated equally with balanced positive and negative movie reviews. Another dataset is apparel reviews from Amazon[4] which consists of 1000 positive and 1000 negative reviews [33]. For IMDB dataset, the average number of words in a sentence and number of sentences in a review is 90 and 14 respectively. Whereas for Amazon apparel review dataset, the average number of words in a sentence and number of sentences in a review is 60 and 5 respectively. Hence, Amazon review dataset is relatively smaller compared with IMDB dataset.

4.2. Experiment settings
To investigate the effectiveness of our proposed KHNN model, we compare KHNN with several baseline models including non-hierarchical models and hierarchical models. For non-hierarchical models, we compare the performances of CNN, LSTM, GRU, and bi-directional GRU (biGRU) with an attention mechanism. For hierarchical models, we study biGRU(atten)+CNN which is biGRU with attention for sentence modeling and CNN for document modeling, CNN+biGRU(atten) which is CNN for sentence modeling and biGRU with attention for document modeling, as well as our proposed KHNN which is K-CNN for sentence modeling and biGRU with attention for document modeling.

For CNN, the number of convolutional filters is set to 100, and the window size of the filter is 3. For LSTM, GRU, and biGRU, the dimension of hidden units is 128. In the data-oriented channel of K-CNN, the number of convolutional filters is set to 50 for IMDB and 75 for Amazon reviews since less trainable filters are needed for K-CNN. In the attention mechanism, the dimension of the context vector is set to 100 for IMDB dataset and 50 for Amazon review dataset. Batch size is 100 for IMDB and 50 for Amazon reviews. Word2Vec [34] word embeddings pre-trained on Google News dataset with about 100 billion words are used for all the models. The dropout rate is set to 0.3, and we train each model for 15 epochs.

The evaluation metrics we used include accuracy, macro-average precision, recall, and F1 score. For IMDB, we evaluated our model on test set; for Amazon reviews dataset, we used 5-fold cross-validation to evaluate the models since no train-test split is provided. We report the average results of the 5 runs.

4.3. Experiment results and analyses
The experiment results of all the models described in Section 4.2 on IMDB dataset and apparel reviews dataset are shown in Table 2.

Table 2. Results of various models for sentiment classification on IMDB and Amazon review dataset. "atten" means attention mechanism which is associated with biGRU.

Model	IMDB				Amazon			
	Accuracy (%)	Precision (%)	Recall (%)	F1 (%)	Accuracy (%)	Precision (%)	Recall (%)	F1 (%)
CNN	88.02	88.38	88.02	87.99	74.60	75.05	74.60	74.49
LSTM	85.45	85.56	85.45	85.44	55.25	55.67	55.25	54.32
GRU	87.65	87.75	87.65	87.64	57.20	57.62	57.20	56.62
biGRU(atten)	89.10	89.46	89.10	89.07	68.85	69.45	68.85	68.63
biGRU(atten)+ CNN	90.08	90.18	90.08	90.08	84.95	85.16	84.95	84.93
CNN+biGRU(atten)	90.23	90.24	90.23	90.23	87.90	87.97	87.90	87.90
Our KHNN	**90.90**	**90.92**	**90.90**	**90.90**	**88.75**	**88.88**	**88.75**	**88.74**

[3] The dataset can be downloaded from http://ai.stanford.edu/ amaas/data/sentiment/

[4] The dataset can be downloaded from http://www.cs.jhu.edu/ mdredze/datasets/sentiment/

4.3.1. Comparison of non-hierarchical models. We first study the performances of non-hierarchical models according to the experiment results shown in Table 2. CNN has demonstrated better performance than LSTM and GRU. One reason is that CNN is good at capturing n-gram features such as words and phrases expressing high sentimental polarities, which are particularity important for sentiment classification. Another reason is due to the limitation of LSTM and GRU in dealing with long sequences: even through gate operations are introduced to alleviate gradient vanishing problem, the models still cannot cope with long sequence modelling, resulting in information loss for document-level sentiment classification. However, GRU outperforms LSTM due to the fewer model parameters, which is advantageous in training with small datasets.

To address the limitation of GRU, bidirectional modelling and attention mechanism are added to GRU, and significant improvement is observed. biGRU considers not only forward sequences but also backward sequences, which considers the whole context information. Attention mechanism combines biGRU outputs based on the contribution of each hidden state to the final sentiment classification. This solves the gradient vanishing problem since every time step is considered for the final document representation. Therefore, in hierarchical models, we only consider biGRU with attention mechanism for RNN-based models.

4.3.2. Comparison of hierarchical models. Based on the results in Table 2, hierarchical models yield better performance than the non-hierarchical models, which proves that hierarchical structure is indeed beneficial in dealing with document-level sentiment analysis since hierarchical models process texts from sentence-level to document-level and preserve their semantic compositionality. Results show that CNN+biGRU(atten) performs better than biGRU(atten)+CNN. The reasons are that CNN is good at capturing local n-gram features instead of global features, and biGRU with attention mechanism does well in extracting global features but may not be able to capture n-gram features that are important for sentiment classification effectively. Hence, it is beneficial to firstly using CNN to capture local n-gram features in each sentence and then utilize biGRU with an attention mechanism to extract a global feature from the sequential sentence representations.

Comparing our KHNN model with CNN+biGRU(atten), the performance improvement shows that K-CNN with sentiment word filters is more effective than conventional CNN in capturing local features from sentences. Since the sentiment word filters are generated from SentiWordNet lexicons with high sentimental polarities, they can adequately capture such lexicons appeared in the sentence and alleviate overfitting issue.

5. Conclusion

In this paper, we propose a Knowledge-oriented Hierarchical Neural Network (KHNN) for sentiment classification, where a document is modelled from sentence-level to document-level. A Knowledge-oriented CNN (K-CNN) is used to capture n-gram local features from sentences. Sentiment word filters are constructed in the knowledge-oriented channel based on SentiWordNet, which can effectively capture lexicons with high sentimental polarities and alleviate overfitting issue. A bidirectional GRU with attention mechanism is used to model the sequential semantics between sentences and obtain a document-level global representation for sentiment classification. Experiments on two real-world datasets show that our model yields better results than other classical deep neural networks.

In future work, more external knowledge will be explored and incorporated into KHNN to model the document. Furthermore, we plan to apply KHNN to more challenging tasks such as aspect-based sentiment analysis.

References

[1] Alaa Hamouda, Mohamed Rohaim, 2011. Reviews Classification Using SentiWordNet Lexicon. *The Online Journal On Computer Science and Information Technology (OJCSIT)* **Vol. (2)** - No. (1).

[2] Devika M D, Sunitha C, Amal Ganesh, 2016. Sentiment Analysis: A Comparative Study on Different Approaches. In: *Fourth International Conference on Recent Trends in Computer Science & Engineering. Procedia Computer Science* 87 (2016) 44 – 49.

[3] Huimin Chen, Maosong Sun, Cunchao Tu, Yankai Lin, and Zhiyuan Liu, 2016. Neural sentiment classification with user and product attention. In: *Proceedings of the Conference on Empirical Methods in Natural Language Processing (EMNLP 2016).*

[4] Xu J, Chen D, Qiu X, and Huang X, 2016. Cached long short-term memory neural networks for document-level sentiment classification. In: *Proceedings of the Conference on Empirical Methods in Natural Language Processing (EMNLP 2016).*

[5] Monalisa Ghosh and Goutam Sanyal, 2018. Document Modeling with Hierarchical Deep Learning Approach for Sentiment Classification. In: *Proceedings of the 2nd International Conference on Digital Signal Processing*. Pages 181-185

[6] Guozheng Rao, Weihang Huang, Zhiyong Feng, and Qiong Cong, 2018. LSTM with sentence representations for document-level sentiment classification. In: *Neurocomputing*, **Volume 308**, Pages 49-57.

[7] Pengfei Li, Kezhi Mao, 2018. Knowledge-oriented convolutional neural network for causal relation extraction from natural language texts. *Expert Systems with Applications.* **Volume 115** (2019).Pages 512–523.

[8] Dzmitry Bahdanau, Kyunghyun Cho, and Yoshua Bengio, 2014. Neural machine translation by jointly learning to align and translate. *arXiv preprint arXiv:1409.0473.*

[9] Gottlob Frege. 1892. *On sense and reference*. Ludlow (1997), pages 563–584.

[10] Stefano Baccianella, Andrea Esuli, and Fabrizio Sebastiani, 2010. SentiWordNet 3.0: An enhanced lexical resource for sentiment analysis and opinion mining. In: *Proceedings of the Seventh International Conference on Language Resources and Evaluation, LREC '10*, Valletta, Malta, pages 2200–2204.

[11] Peter D. Turney, 2002. Thumbs up or Thumbs down? Semantic orientation applied to unsupervised classification of reviews. In: *Proceedings of 40th Annual Meeting of the Association of Computational Linguistics*, Philadelphia, PA, pages 417–424.

[12] Maite Taboada, Caroline Anthony, and Kimberly Voll, 2006. Method for Creating semantic orientation dictionaries. In: *Proceedings of 5th International Conference on Language Resources and Evaluation (LREC)*, Genoa, Italy, pages 427–432.

[13] Hanen Ameur, Salma Jamoussi, 2013. Dynamic construction of dictionaries for sentiment classification. In: *13th IEEE International Conference on Data Mining Workshops. ICDM workshops,* TX, USA (2013), pages 896-903.

[14] Bo Pang, Lillian Lee, and Shivakumar Vaithyanathan, 2002. Thumbs up? sentiment classification using machine learning techniques. In: *Proceedings of the Conference on Empirical Methods in Natural Language Processing (EMNLP)*, Philadelphia, July 2002, pages. 79-86.

[15] Bingwei Liu, E. Blasch, Yu Chen, Dan Shen, and Genshe Chen, 2013. Scalable sentiment classification for Big Data analysis using Naïve Bayes Classifier. In: *Big Data, 2013 IEEE International Conference,* Silicon Valley, CA, 2013, pages. 99-104.

[16] Joseph D. Prusa, Taghi M. Khoshgoftaar, and David J. Dittman, 2015. Impact of Feature Selection Techniques for Tweet Sentiment Classification. In: *Proceedings of the Twenty-Eighth International Florida Artificial Intelligence Research Society Conference*, pages 299-304.

[17] Cicero Nogueira dos Santos and Maira Gatti, 2014. Deep Convolutional Neural Networks for Sentiment Analysis of Short Texts. In: *Proceedings of COLING 2014, the 25th International Conference on Computational Linguistics*: Technical Papers, Dublin, Ireland, pages 69–78.

[18] Xingyou Wang, Weijie Jiang, and Zhiyong Luo, 2016. Combination of convolutional and recurrent neural network for sentiment analysis of short texts. In: *Proceedings of COLING 2016, the 26th International Conference on Computational Linguistics*: Technical Papers, Osaka, Japan, pages 2428–2437.

[19] Duyu Tang, Bing Qin, and Ting Liu, 2015. Document modeling with gated recurrent neural network for sentiment classification. In: *Proceedings of the Conference on Empirical Methods in Natural Language Processing,* Lisbon, Portugal, pages 1422-1432.

[20] Sebastian Ruder, Parsa Ghaffari, and John G. Breslin, 2016. A Hierarchical Model of Reviews for Aspect-based Sentiment Analysis. *arXiv preprint arXiv:1609.02745.*

[21] Lei Zhang, Riddhiman Ghosh, Mohamed Dekhil, Meichun Hsu, and Bing Liu, 2011. Combining Lexicon-based and Learning-based Methods for Twitter Sentiment Analysis. *Technical Report HPL*-2011-89, HP, 21/06/2011.

[22] Bonggun Shin, Timothy Lee, and Jinho D. Choi. 2016. Lexicon Integrated CNN Models with Attention for Sentiment Analysis. *arXiv preprint arXiv:1610.06272.*

[23] Vikash Nandi, Suyash Agrawal. 2016. Political Sentiment Analysis using Hybrid Approach. In: *International Research Journal of Engineering and Technology*, **Vol.3**, issue. 5, pages 1621-1627.

[24] Amira Shoukry, Ahmed Rafea. A Hybrid Approach for Sentiment Classification of Egyptian Dialect Tweets *2015. In: First International Conference on Arabic Computational Linguistics (ACLing)*, Cairo, Egypt, pages 78-85.

[25] G. A. Miller, 1995. Wordnet: A lexical database for English. In: *Communications of the ACM,* **(11)**:39–41. http://wordnet.princeton.edu/.

[26] Liviu Adrian Cotfas, Camelia Delcea, Irina Raicu, Ioana Alexandra Bradea, and Emil Scarlat, 2017. Grey sentiment analysis using SentiWordNet. In: *2017 IEEE International Conference on Grey Systems and Intelligent Services, GSIS 2017*, pages 284-288.

[27] Kyunghyun Cho Bart van Merrienboer Caglar Gulcehre, Dzmitry Bahdanau, Fethi Bougares Holger Schwenk, and Yoshua Bengio, 2014. Learning Phrase Representations using RNN Encoder-Decoder for Statistical Machine Translation. In: *Proceedings of the 2014 Conference on Empirical Methods in Natural Language Processing (EMNLP)*, pages 1724-1734.

[28] Yoshua Bengio, Patrice Simard, and Paolo Frasconi. 1994. Learning Long-Term Dependencies with Gradient Descent is Difficult. In: *IEEE Transaction on Neural Networks*, **Vol**. 5, NO.2., pages 157-166.

[29] Zichao Yang, Diyi Yang, Chris Dyer, Xiaodong He, Alex Smola, and Eduard Hovy, 2016. Hierarchical Attention Networks for Document Classification. In: *Proceedings of the 2016 Conference of the North American Chapter of the Association for Computational Linguistics*: Human Language Technologies, pages 1480–1489.

[30] Nitish Srivastava, Geoffrey Hinton, Alex Krizhevsky, Ilya Sutskever, and Ruslan Salakhutdinov, 2014. Dropout: A Simple Way to Prevent Neural Networks from Overfitting. In: T*he Journal of Machine Learning Research 15 (2014)*, pages 1929-1958

[31] Diederik P. Kingma, Jimmy Ba, 2014. Adam: A method for stochastic optimization. *arXiv preprint arXiv:1412.6980*

[32] Maas, A. L.; Daly, R. E.; Pham, P. T.; Huang, D.; Ng, A. Y.; and Potts, C, 2011. Learning word vectors for sentiment analysis. In: *The 49th Annual Meeting of the Association for Computational Linguistics: Human Language Technologies, Proceedings of the Conference*, 142–150.

[33] Blitzer, J.; Dredze, M.; and Pereira, F, 2007. Biographies, Bollywood, Boom-boxes and Blenders: Domain adaptation for sentiment classification. In: *ACL 2007, Proceedings of the 45th Annual Meeting of the Association for Computational Linguistics. June 23-30, 2007, Prague, Czech Republic.*

[34] Tomas Mikolov, Kai Chen, Greg Corrado, and Jeffrey Dean, 2013. Efficient estimation of word representations in vector space. In: P*roceedings of International Conference on Learning Representations (ICLR)*, pages 1-12.

IOP Conf. Series: Materials Science and Engineering **646** (2019) 012024 doi:10.1088/1757-899X/646/1/012024

A Hybrid Chaotic Oscillatory Neural Network (HCONN) Based Financial Time Series Prediction System

Yifu Qiu[1] and Raymond S. T. Lee[1,*]

[1] Beijing Normal University-Hong Kong Baptist University United International College, 2000 Jintong Road, Tangjiawan, Zhuhai, Guangdong Province, China

[*]Corresponding author email: raymondshtlee@uic.edu.hk

Abstract. Financial time series prediction is one of the most complex and challenging problems in both AI and finance engineering. In our research, we proposed a Hybrid Chaotic Oscillatory Neural Network (HCONN) model by replacing the traditional sigmoid-based activation function with chaotic oscillatory activation function, which provides significant performance in the global minimum convergence through the application of Adaptive Moment Estimation optimizer. In addition, by integrating the latest R&D on Quantum Finance Theory (QFT) and its Quantum Price Level (QPL) as the deep features' extraction, we add the daily 8 nearest QPLs together with the time series price variables as the input of our HCONN. In terms of system implementation, 12 different forex products including the AUDCHF, AUDUSD, CADCHF, EURAUD, EURCHF, EURGBP, EURUSD, GBPAUD, GBPCAD, GBPUSD, USDCAD and USDCHF are used. System performance results reveal that HCONN outperforms other financial models including: Feedforward Backpropagation Neural Network (FFBPN) and Chaotic Oscillatory Neural Network (CONN) in terms of training performance and forecast accuracy.

1. Introduction

Nowadays, one of the major areas for artificial intelligence (AI) applied in finance is the analysis and prediction of the financial time series, which is also the trend and the hottest topic for many decades in both financial industry and academia. In the past decades, AI tools such as machine learning and deep learning, are popularly utilized in financial time series analysis and forecasting, which have also been proved with high degree of validity and correctness [1].

In general, prediction of financial instrument can be categorised into two related directions based on the scope we observe the market: long-term market trend prediction and short-term price prediction. In long-term trend prediction, we try to predict the future trend of the market movement according to the selected financial variables such as technical indicators in a certain previous trading day timeframe. In the short-term prediction, we try to predict the price in the future day based on the previous several days price data. These two kinds of prediction are tightly related with each other in finance, the long-term market trend prediction provides the direction for the short-term prediction. On the other hand, short-term prediction can be considered as a kind of microscopic analysis of the long-term prediction.

For both kinds of prediction in financial time series data, the essence is that we construct a learning algorithm model which could learn and extract some key patterns from the large volume of the past financial time series. Then based on these patterns and current financial time series, we apply machine learning algorithms to do the pattern matching or so-called prediction for future market movement.

In machine learning perspective, Artificial Neural Networks (ANN) and Deep Neural Networks (DNN) are commonly used on financial time series prediction. In summary, ANN is a kind of simulation of the human brain's operation by simulating the connection among various neurons in the human brain. The neural dynamics are represented by the regulating of the weights between neurons and transfer of information by means of activation function. Through network training, ANNs can learn complex patterns and perform various classification and even prediction operations [2]. In short, DNN can be considered as a kind of extension of ANN. In terms of network structure, DNN possesses more complex hidden layers and neural dynamics, which believes can be used to tackle with more complex problem, financial prediction for example [3].

With the development of neural networks, the architecture of contemporary neural networks is no longer limited in the fully connected neural network structure. Other neural networks structure such as Convolutional Neural Networks (CNN), Recurrent Neural Networks (RNN) and Generative Adversarial Networks (GAN) are gradually demonstrating their powerful capabilities in various fields of application. Specially, in finance, recurrent neural networks which intuitively accept the time series as input, have gradually become the most popular neural networks in the financial industry and academia.

In this paper, we proposed an integrated financial prediction system called Hybrid Chaotic Oscillatory Neural Network (HCONN), which integrates the Chaotic Oscillatory Neural Network (CONN) and Quantum Price Levels (QPL) from the latest research on Quantum Finance Theory (QFT) [4], which are considered as the intrinsic properties of the financial markets. From the implementation perspective, HCONN takes the nearest 8 QPLs and the past 9 trading-days timeseries of open, high, low and closing prices as inputs to forecast the next-day high and low prices. For system testing, 2048 trading-day timeseries from 5 Oct 2012 to 12 May 2019 are used, together with 12 major forex products include: AUDCHF, AUDUSD, CADCHF, EURAUD, EURCHF, EURGBP, EURUSD, GBPAUD, GBPCAD, GBPUSD, USDCAD and USDCHF for system implementation. For performance analysis, HCONN is compared with Feedforward Backpropagation Neural Network (FFBPN) and Chaotic Neural Network (without QPL) as benchmarks for training performance and forecast accuracy performance testing.

2. Literature Review

Since using neural network to do the forecasting for the financial time series is a popular and developing topic in the application of neural network, there are substantial number of scholars and studies working in this field. In this literature review, the authors will attempt to summarize those significant works on the financial time series data in the past few years.

2.1. Quantum Finance Theory (QFT)

In order to extract the deep features of each financial time series data, we apply the author's latest research on Quantum Price Levels (QPL) of the Quantum Finance Theory (QFT)[4]. By modelling the market movements in the secondary financial market using quantum anharmonic oscillator (QAHO), and through the solving of Quantum Finance Schrödinger Equation (QFSE), we calculate the energy levels for the Quantum Financial Particle (QFP) in each financial market directly – the QPL. According to the QFT, QPL could be used in both financial forecasting and trading strategy design as analogue to the resistance and support lines in technical analysis. As compared with traditional features we used in the prediction problem, QPL reflects the nature of the particle's dynamic, which sheds new lights in the exploration of the intrinsic characteristics of the QFPs in financial markets.

2.2. Chaotic Oscillatory Neural Network (CONN)

Due to the highly degree of chaotic and complex fluctuation of financial markets, they are difficult to be modelled successfully by traditional neural networks. According to our experiment, by using the traditional activation function in neural network such as ReLU, tanh and sigmoid functions, the fitting results for the financial time series are not so promising, so we introduce the Lee-oscillator model [5].

Lee-oscillator is a chaotic neural oscillator with remarkable transient-chaotic property which provides a perfect solution to serve as Bifurcation Transfer Unit (BTU) to model highly chaotic and complex problems such as financial timeseries prediction [6].

3. Methodology

From the system framework perspective, HCONN is the integration of: 1) 8 closest Quantum Price Levels (QPL) evaluated from quantum finance model; 2) 9-day time series of the financial products; 3) Chaotic Oscillatory Neural Network (CONN) model with the adoption of Lee-oscillators as model framework.

3.1. Quantum Price Levels (QPL)

With the continuous development of the econ-physics in the recent years, Quantum Finance Theory (QFT) based on Quantum Mechanics and Quantum Field Theory to extract the deep features of the motion of the quantum financial particles (QFP) become one of the hottest topics in Fintech and quantum computing technology. Based on the author's latest research on QFT, the Quantum Finance Schrödinger Equation (QFSE) is given by [5]:

$$\left[\frac{-h}{2m}\frac{d^2}{dr^2} + \left(\frac{\gamma\eta\delta}{2}r^2 - \frac{\gamma\eta v}{4}r^4\right)\right]\psi(r) = E\psi(r) \tag{1}$$

In QFSE, the first term denotes the kinetic energy term, and the second term represents the potential energy term of the quantum finance market, ψ (r) is the wavefunction and E represents the corresponding quantum energy levels of the QFPs. By solving QFSE, we could calculate the discrete QPLs which directly analogue to the energy level of the QFPs. From the perspective of the QFT, if there is no outside stimulus, the financial particle will remain at its own energy level and perform chaotic oscillation. Once there is outside stimulus, the QFPs will discretely absorb/release the quantized energy and jump to other QPLs.

In our financial forecasting problem, we take both time series price variables and the QPLs as the inputs since the QPL is a direct reflection corresponding to the result of price movement, which symbolizes the possibility and stages of the movement of financial product prices within the trading day. In this way, we extract the deep characteristics of price movements by calculating the QPLs and input it into the chaotic neural network together with the time series price variables for system training and next-day prediction.

3.2. Time series to time sequence

In order to tackle the gradient descent deadlock problem during the training period and reflect the intrinsic relationship among different prices and QPLs, many methods have been proposed. In our research, at the pre-processing phase, we do the normalization for the time series data based on the Lasso normalization method and the equation for normalization factor is given as follow:

$$\mathrm{x}' = \frac{x}{\sum_{i=1}^{n}|x_i|} \tag{2}$$

After applying this normalization method to the timeseries, we reflect our two groups of feature values into the range between 0 and 1. In this way, we could maximize the usage of our chaotic activation function and its chaotic region to jump out from the local minimum which we will show in the following section.

In order to forecasting the future price movement, we train the HCONN by capturing the patterns of price movements in the past 2048 trading-days. Therefore, in the process of the data pre-processing, we spilt the time series continuously through a sliding window method shown as followed, which enables us to get 2038 fixed scale training samples. We take the past 9-day daily OHLC (Open, High, Low and Close) financial time series variables as input and predict the next day highest price and lowest price because in the actual trading, daily high/low prices are commonly for reversal and breakout trading strategy. The working mechanism of the sliding window is given in Figure 1.

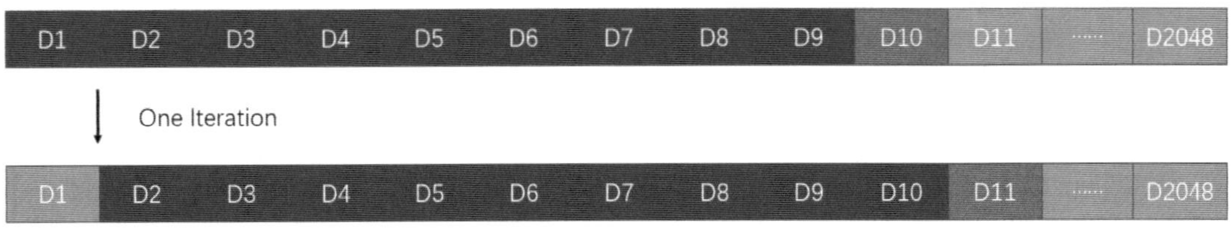

Figure 1. Spilt time series into sequence based on 10-day fixed size sliding window.

In order to prevent the market trend influence of our model to learning the past price movement pattern, for example, if the market is in the bullish or bearish, the divided 10-day sequence's pattern might be influence by the current market trend, and ensure that our model has sufficient generalization predictive ability in different market condition, we randomly choose the sequence and append it into the training set and testing set. In here, we follow the traditional 75% and 25% rule to construct the training set and testing set.

3.3. Hybrid Chaotic Oscillatory Neural Network (HCONN)

The structure of our chaotic neural network follows the traditional fully-connected proposed by Rumelhart and McClelland in 1986, which is the fundamental structure for all the neural networks but contains power fitting ability to the complex and non-linear data. In our Hybrid Chaotic Oscillatory Neural Network (HCONN), by considering with the volume of the training data, we construct an eight-layer based fully connected neural network containing three chaotic oscillatory layers, three batch normalization layers and two dropout layers, for each layer it contains 20 neurons.

In our HCONN, the input layer contains 102 neurons which includes two major parts: 1) 9-day time series data, which include Open, High, Low and Closing prices; 2) each day's 8 nearest Quantum Price Levels (QPLs). The 8 hidden layers including three different types: 1) Chaotic oscillatory layer, 2) Batch normalization layer and 3) Dropout layer. The output layer contains 2 neurons corresponding to the next day High/Low prices.

In order to solve the problem in fitting the highly random and stochastic time series data, we adopt the Lee-Oscillator [5] as the activation function in our HCONN. Its neural dynamics is given by:

$$E(t+1) = sigmoid(e1 \cdot \mathrm{E(t)} - e2 \cdot I(t) + S(t) - \zeta_E) \quad (3)$$

$$I(t+1) = sigmoid(i1 \cdot \mathrm{E(t)} - i2 \cdot I(t) - \zeta_i) \quad (4)$$

$$\Omega(t+1) = sigmoid\big(S(t)\big) \quad (5)$$

$$L(t) = [E(t) - I(t)] \cdot e^{-kS^2(t)} + \Omega(t) \quad (6)$$

Where E, I, Ω and L represent the Excitatory, Inhibitory, Input and Output neurons, e_1, e_2, i_1 and i_2 are the weights, ζ_E and ζ_i are the threshold values and S(t) is the external input stimulus. By simulation in the Matlab program, the bifurcation diagram is shown in Figure 2.

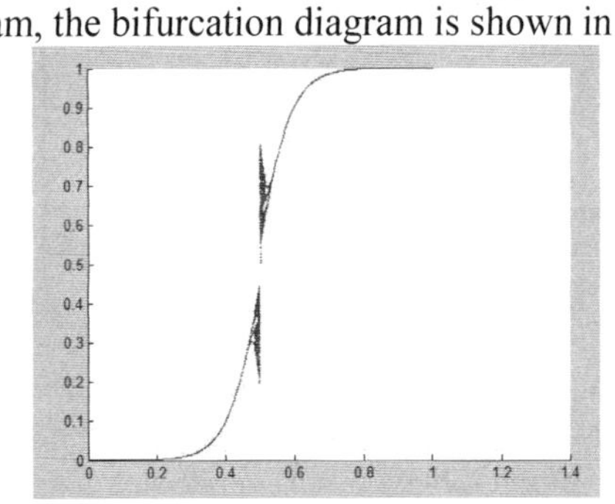

Figure 2. Bifurcation diagram of the Lee-oscillator

By applying the oscillator model as the activation function, we effectively convert our traditional neural network into a chaotic neural network. In this kind of transformation, during the training of our chaotic oscillatory neural network, the chaotic region provides different degree of "hysteresis" during network training, which provides the ability for the network to jump out of the local minimum and to resolve the "deadlock" and "over-training" problems which are commonly found in traditional neural networks.

Inside the hidden layer structure, in order to fully use the chaotic region and prevent the gradient problem to stuck our chaotic activation function, we need to limit the input distribution of the chaotic neurons within the range between 0.3 to 0.7 as much as possible, by combing with the training acceleration technique [7], we add the batch normalization layer between each chaotic oscillatory layer. In addition, considering our input contains lots of feature of the time series, which are 9-day time series price variables and 9-day eight nearest daily quantum price levels, in order to prevent the overfitting problem [8], we construct the dropout layer with 0.8 as the keep probability inside the hidden layers. The hidden layers architecture is given in Figure 3.

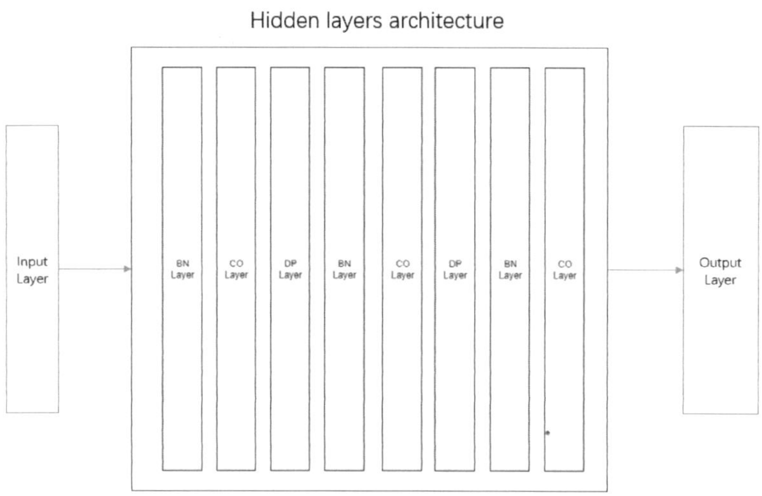

Figure 3. Architecture of the hidden layers in the HCONN.

Where the BN, CO, DP Layers represent the batch normalization layer, chaotic oscillatory layer and dropout layer respectively.

During system training, we use the ADAM optimizer with default learning rate 0.01 to minimize the cost function, which has the state-of-art performance in the deep neural network training [9]. The main reasons to adopt ADAM optimizer include: 1) ADAM optimizer could adjust its learning rate based on different parameters, which can be applied in several different models; 2) In our chaotic activation function, the calculation is much more complex than the traditional activation function. So, in order to ensure the effective usage of the neural network in the actual trading and forecasting, the time consuming of the training of the network should be minimize; 3) For the chaotic activation function, it could jump out of the local minimum because of its chaotic region and the stochastic gradient calculation. However, in order to prevent it from jumping out of the global minimum and ensure it could converge in the global minimum, we need to choose the optimizer which could adjust the learning rate with the process of convergence.

4. Experimental Tests

In this section, HCONN is evaluated in two major perspectives following the traditional deep learning model evaluation principle, which are 1) the accuracy of the prediction test which quantified by the mean square error. 2) the speed of convergence test quantified by the training epochs. In the model comparison, the HCONN will be compared with FFPBN and CONN (without QPLs).

4.1. Experimental data

All the experimental data are fetched from the MetaTrader 4 forex trading platform historical data centre. In total, 12 forex products are used: AUDCHF (2051), AUDCHF (2051), AUDUSD (2051), CADCHF (2051), EURAUD (2051), EURCHF (2051), EURGBP (2051), EURUSD (2047), GBPAUD (1531), GBPCAD (1531), GBPUSD (2047), USDCAD (2047), USDCHF (2051), where numbers in () shows the number of trading-day of the timeseries being used for system training and testing.

4.2. Experimental environment

For the QPL calculation, we conduct this process in the MetaTrader 4 platform by using the MQL4 programming language. For the data pre-processing, we conduct it in the Python 3.6, relying on the packages numpy and pandas. In the model construction, we develop the HCONN and its training comparison model CONN and FFBPN with Google tensorflow with version 1.31 which is a powerful framework for the machine learning research and development. In addition, for the accuracy comparison model including support vector machine regressor and random forest regressor, we construct it by using the sci-kit learn package. All the models including the HCONN and its comparison model are running in CPU cluster with Windows 10 operating system and Intel(R) Core(TM) i5-6300HQ CPU @ 2.30GHz, 2304 Mhz processor.

4.3. Comparison model

By following the two measurement methods which are 1) the accuracy of the prediction test which quantified by the mean square error. 2) the speed of convergence test quantified by the training epochs. In the model comparison, we evaluate and compare the three model: HCONN, CONN (without QPL) and FFBPN in both training set and testing set.

4.3.1 Convergence comparison. During the convergence comparison, we plot the mean square error versus the training iteration for three different models to compare the speed of convergence. Due to the huge numerical difference during the convergence process between the 3 models, we draw a point representing each 5 epochs, in this way we have 300 points in the training process graph. Then we separate the 1500 training epochs into 3 phases define as: 1) 1 epoch to 20 epoch; 2) 21 epoch to 100 epoch; 3) 101 epoch to 300 epoch, these three-period graphs are given in Figure 4.

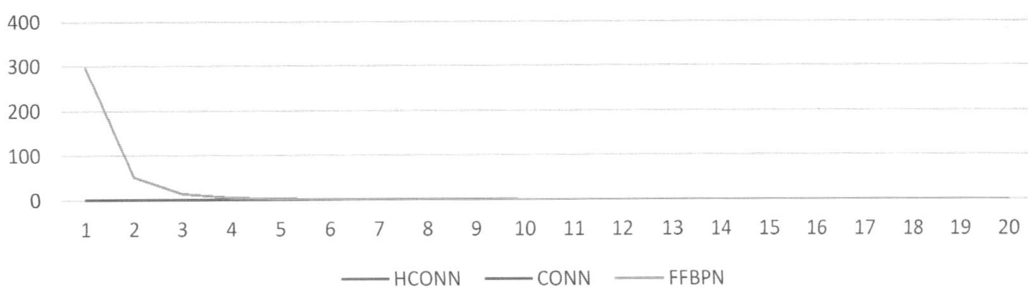

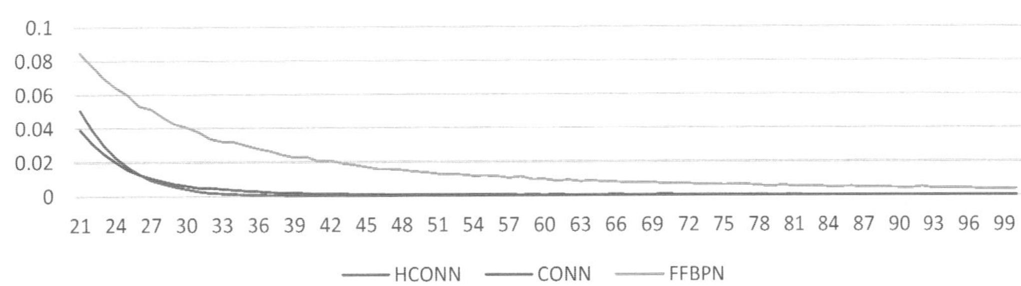

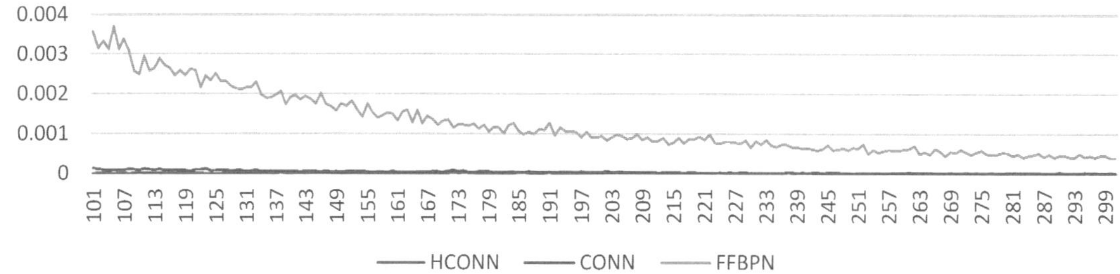

Figure 4. Average mean square error among 12 products versus the training iteration graph in three period: Top) 1 epoch to 20 epoch; Middle) 21 epoch to 100 epoch; Bottom) 101 epoch to 300 epoch.

From these three charts, it reveals that in the early phase, FFBPN almost has the highest mean square error. Also, both HCONN and CONN have fast convergence speed and better global minimum to convergence, one of the main reasons is that for the HCONN and CONN, the chaotic oscillatory activation function could enable the neural network to jump out of the local minimum during the training process. By combing with ADAM optimizer, its learning rate will be limited in an appropriate range during the training process. In conclusion, the combination of the oscillatory activation function with ADAM optimizer is one of the best training techniques combination in the financial prediction question case, the oscillator enable the model to have the ability to escape from local minimum without too much sacrifice in the accuracy by ADAM's mechanism.

In order to compare HCONN with CONN in detail, we plot the average mean of Mean-Square-Error (MSE) for the network training of the 12 forex products versus training epoch in Figure 5, in which we could see that the quantum price level will accelerate the speed to convergence to the global minimum. But for the local minimum result, the HCONN and CONN have no significant difference for the training set. The detail comparison we will discuss in the accuracy comparison and analysis.

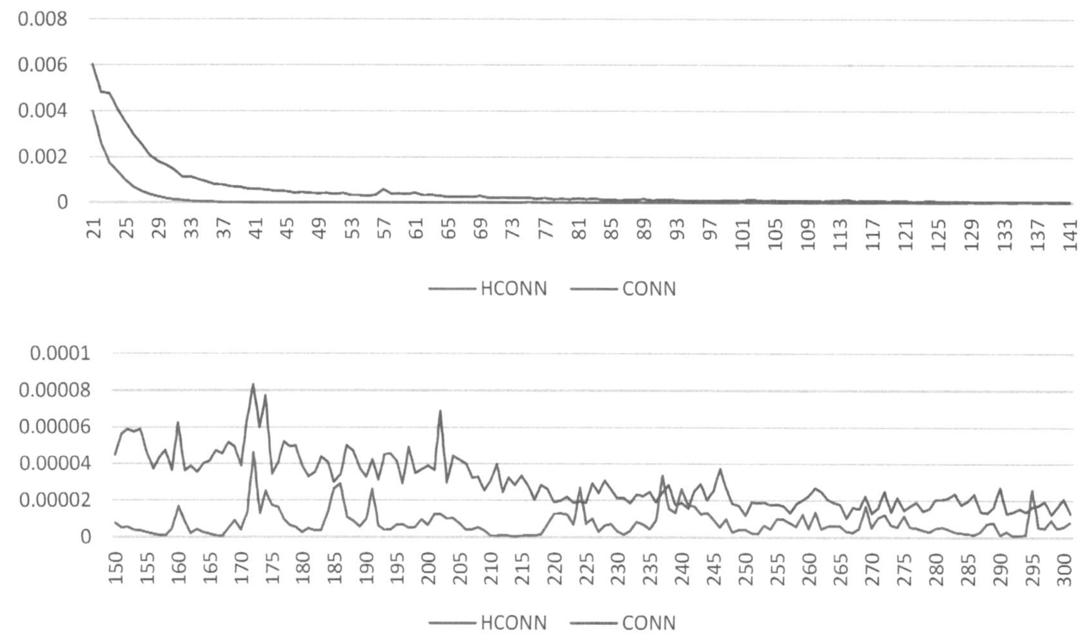

Figure 5. HCONN and CONN comparison: Average mean square error among 12 products versus the training iteration graph in two period: Top) 20 epoch to 150 epoch; Bottom) 150 epoch to 300 epoch.

4.3.2 Accuracy comparison. For each forex product, we show the mean square error of the prediction in the corresponding testing set for the three models (Table 1).

Table 1. Mean square error in each product for three models

	AUDCHF	AUDUSD	CADCHF	EURAUD	EURCHF	EURGBP
HCONN	2.36E-06	1.81E-06	5.58E-06	1.72E-06	2.17E-05	1.72E-06
CONN	0.00032897	0.00045184	1.87E-06	1.73E-06	1.76E-06	1.84E-06
FFBPN	3.22E-06	0.14337366	0.01719210	4.33E-03	1.19E-03	1.18E-05
	EURUSD	GBPAUD	GBPCAD	GBPUSD	USDCAD	USDCHF
HCONN	2.69E-06	7.83E-06	1.02E-05	1.75E-06	1.79E-06	3.65E-05
CONN	1.73E-06	2.53E-05	3.48E-05	4.200E-05	7.09E-05	0.00012943
FFBPN	4.52E-03	4.77E-03	0.09754947	0.01744503	0.00107840	1.67E-06

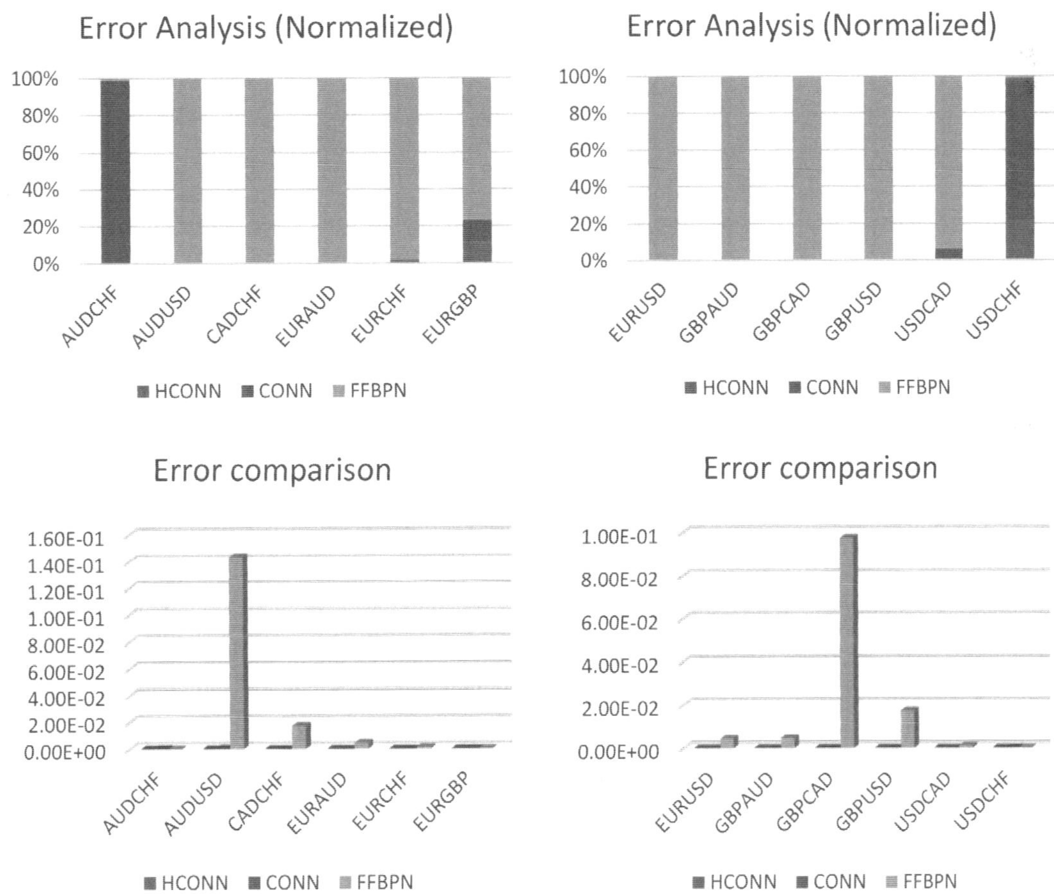

Figure 6. Normalized mean square comparison, each column means the proportion of three model's error in a certain product, the bottom two graphs are the absolute error comparison for three models.

In order to illustrate the error analysis among the three forecasting models, we plot the proportion of three model's error for each certain product in Figure 6 and plot the absolute error value comparison between HCONN and CONN in Figure 7. From the result of the normalized error analysis, we could see that for each product, most of the error is contributed by the FFBPN model and CONN model. By comparing the HCONN with two benchmarks model, we could see that for the HCONN, it contains the most powerful and stable ability in prediction. The accuracy difference between CONN and FFPBN shows that by applying the chaotic oscillatory activation, we could efficiently enable the neural network to fit the complex, chaotic and highly random time series patterns. In addition, the stability in prediction among different products show that by applying the chaotic oscillatory activation, the generalization ability of the neural will increase as well, which means the model could adapt the different product well in the market, which is important for reduction of consuming in the actual trading.

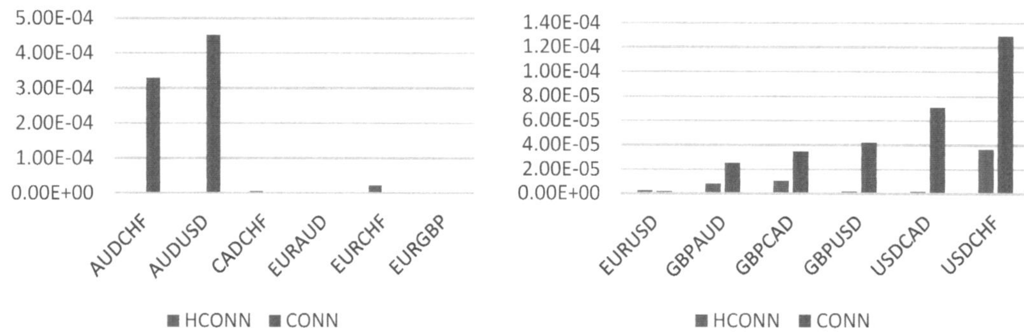

Figure 7. Absolute mean square error comparison between HCONN and CONN.

By comparing the MSE difference between CONN with HCONN as shown in Figure 7, it reveals that by applying the quantum price level as the feature extraction method the accuracy of the prediction can be improved significantly.

5. Conclusion

In this paper, we apply the Hybrid Chaotic Oscillatory Network (HCONN) which integrate chaotic neural network with the quantum finance theory to extract the intrinsic financial property of each financial product to tackle short term daily prices prediction problem. We take the past 9-day time series prices and quantum price levels as the input, to predict the next day highest and lowest price.

Specially, by integrating with Lee-oscillator as the chaotic activation function with ADAM optimizer, the model has the more powerful ability in the convergence to the global minimum and training acceleration. By comparing with the original chaotic oscillatory neural network, the quantum price level has the significant positive contribution to minimize the loss value by extracting the deep feature of the financial product.

Overall, based on the experiment result, we could see that the hybrid chaotic oscillatory neural network (HCONN) model could successfully extract the price movement pattern from the past chaotic, stochastic and complex data. Compared to other models, it has the significant ability in both the short term prediction accuracy but also the generalized prediction ability, which produces the reliable result in terms of trading analysis.

Acknowledgments

The author wants to thank for the MetaTrader 4 platform for the historical data providing. The authors also wish to thank Quantum Finance Forecast Center of UIC for the R&D supports and the provision of the channel and platform qffc.org for system testing and evaluation. This paper is supported by

Research Grant R201948 of Beijing Normal University-Hong Kong Baptist University United International College (UIC).

References

[1] Bao W, Yue J and Rao Y 2017 A deep learning framework for financial time series using stacked autoencoders and long-short term memory PloS One 12(7) pp. e0180944
[2] Zhou Z 2016 Machine Learning, Tsinghua University Press, Beijing
[3] Cucu GS, Ciotec AD and Neagoe VE 2018 Deep Convolutional Neural Networks versus Multilayer Perceptron for Financial Prediction *2018 International Conference on Communications (COMM)* pp 201-206
[4] Lee RST 2019 Quantum Finance - Intelligent Financial Forecast and Program Trading Systems, Springer NATURE, Singapore (in printing).
[5] Lee RST 2004 A Transient-chaotic Auto-associative Network (TCAN) based on Lee-oscillators *IEEE Trans. Neural Networks* **15**(5) 1228-1243
[6] Lee RST 2019 Chaotic Type-2 Transient-Fuzzy Deep Neuro-Oscillatory Network for Worldwide Financial Prediction *IEEE Transactions on Fuzzy Systems* DOI:10.1109/TFUZZ.2019.2914642
[7] Sergey I Christian S 2015 Batch Normalization: Accelerating Deep Network Training by Reducing Internal Covariate Shift I*nternational Conference on International Conference on Machine Learning* 448 – 456
[8] Srivatava N, Hinton G, Krizhevsky A, et al. 2014 Dropout: A Simple Way to Prevent Neural Networks from Overfitting. *Journal of Machine Learning Research, 15*(1), 1929-1958
[9] Kingma D and Ba J 2015 Adam: A Method for Stochastic Optimization. *The 3rd International Conference for Learning Representations* San Diego. 1–13

A Deep Fully Convolutional Network for Distal Radius and Ulna Semantic Segmentation

Shuqiang Wang[1,*] , Wei Liang[2] , Hongfei Wang[1], ZhuoChen[1] and Yiqian Lu[1]

[1]Shenzhen Institutes of Advanced Technology, Chinese Academy of Sciences, China
[2]Liaocheng Institute of Water Resource Survey and Design, Shandong, China

[*]sq.wang@siat.ac.cn

Abstract. Semantic segmentation is an essential step to do further image analysis and scene understanding tasks. In medical imaging analysis applications, it is even more challenging to do automatic segmentation due to tissues' complicated boundaries. In this paper, a fully convolutional network (FCN) based model is constructed to segment distal radius and ulna (DRU) areas from hand X-ray images. We evaluated the proposed network on a clinical DRU dataset with different network configurations. The proposed network can achieve 98% accuracy and 96% mean Intersection over Union (IoU).

1. Introduction

The purpose of image segmentation is trying to cluster pixels into distinct regions according to interest features. Conventional segmentation methods including region-based algorithms, boundary-based algorithms and graph-based image segmentation methods. To acquire accurate segmentation results is still a challenging problem. In recent years, deep learning techniques have dramatically improved the state-of-the-art in visual object detection and recognition [1]. In medical image analysis areas, research groups have tried to apply CNNs to a broad range of applications and got promising results[2]. For example, Pereira *et al.* proposed an automatic brain tumors segmentation method based on convolutional neural networks. Their method achieved top results on a 2013 public challenge dataset [3]. However, CNN-based segmentation algorithms always suffer from low efficiency and low accuracy. In 2015, Long *et al.* proposed a full convolutional network (FCN) which can classify each pixel of an image [4]. From then on, FCNs became more popular to do semantic segmentation tasks than conventional CNN models.

To evaluate bone age in patients will guide physicians to make clinical decisions. Luk*et al.* have proposed a novel skeletal maturity assessment method based on distal radius and ulna (DRU) radiographs [5]. Extracting accurate DRU regions from hand X-ray images is the key step to apply this new DRU classification scheme to assess bone age. In this paper, we propose a fully convolutional network (FCN) to do distal radius and ulna semantic segmentation. The main contributions of this paper include: (1). To our best knowledge, this is the first work to apply FCN to DRU segmentation and get satisfactory results. (2). We deployed numerous experiments to investigate the proposed network performance with different network configurations and different training samples.

2. Related work

Convolutional neural networks have been widely used in medical image segmentation tasks. Kamnitsas*et al.* proposed a 3D fully connected Conditional Random Field to post-process the network's soft segmentation [6]. In 2016, Moeskops*et al.* presented a CNN based system to automatically segment MR brain images into multiple tissue classes.[7]. Also, Roth *et al.* proposed a multi-level deep CNNs to do pancreas segmentation in abdominal CT scans [8]. From the above-mentioned works, it can be concluded that more CNN related methods have been employed to medical image segmentation and achieved promising results in the recent years.

In 2015, fully convolutional network (FCN) was presented to do semantic segmentation. For example, Ronneberger*et al.* presented a fully convolutional network for segmentation of neuronal structures in electron microscopic stacks. Their proposed network was the winner of ISBI cell tracking challenge 2015 [9]. In 2017, Andrearczyk*et al.* developed a fully convolutional network for texture segmentation. [10]. Moreover, FCN can be applied to video segmentation problems. Valipour*et al.* built a fully convolutional network and used a recurrent unit to work on temporal data [11]. These cited works have proved FCN has strong convincing performance on semantic segmentation. In this paper, we will try to apply FCN to distal radius and ulna segmentation from hand X-ray images. The proposed method and corresponding experiments will be presented in detail in the following sections.

3. Method

3.1 Fully Convolutional Neural Network and Semantic Segmentation

In convolutional neural networks, fully conventional networks (FCNs) were proposed to replace all the fully connected layers by convolutional layers to prevent spatial information damage. Both the input and output data of FCNs are two-dimensional images. The output results contain spatial structure information according to the input images. FCNs use deconvolution operations to reconstruct the output results from the final feature map which is a heatmap. Then, softmax will be applied to each pixel of generated images to do pixelwise classification. The final semantic segmentation is built on pixelwise classification results of the reconstructed image. We fuse feature maps generated from convolutional layers and feature maps from previous pooling layers together. Then the mixed feature maps will be upsampled back to the image. The purpose of these operations is to utilize the shallow and deep features for semantic segmentation.

Compared to traditional convolution neural networks, fully convolutional networks have two remarkable advantages: (1). arbitrary input size accepted: FCNs do not require training data have identical size; (2). high training efficiency: FCNs avoid repetitive computation and storage overhead of small surrounding regions. Based on the above description, the integrated working flowchart of FCNs semantic segmentation is illustrated in the Figure 1.

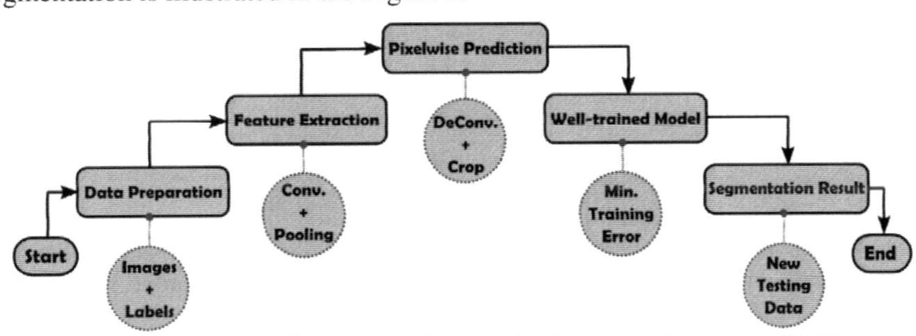

Figure 1. The flowchart of Semantic Segmentation using FCNs

3.2 Network Configurations

To achieve reasonable segmentation results, we choose VGG16 [12] as the basic network configuration. We replace the original fully-connected layer with convolutional layers and also add several new operations such as deconvolution, fusion and cropping. Input images and corresponding label information are processed as network input and the corresponding output is two-dimensional images which contain each single pixel classification label.

In such network configurations, feature maps after several convolutional and pooling operations become coarse and the sizes are much smaller. To get back to the original image resolutions, deconvolution (upsampling) is applied to the generated feature maps. We will fuse feature maps after pooling. Then a upsampling operation is used on the fusion results to get the segmentation results. Even we can upsampling the previous results of different times and then sum this larger feature maps with previous pooling layer output to generate new feature maps. Then, this new feature maps will be upsmapled to produce more smooth and detailed output. These different network configurations are called FCN-32s, FCN-16s and FCN-8s respectively.

4. Experiment

4.1 Data Preprocessing

In the experiment section, all the hand radiographs collected from Li Ka Shing Faculty of Medicine, The University of Hong Kong. We have got 1189 pieces images and saved them as .jpg files. When using DRU classification scheme to evaluate skeletal maturity, physicians only focus on the distal radius and ulna regions. However, the raw data images contain many other regions information such as fingers and black background. These regions will slow down segmentation processing speed and will also decrease the segmentation accuracy. Therefore, a preprocessing step is necessary to extract only the regions of interest (ROI) i.e. the distal radius and ulna regions. After that, all the extracted hand regions will be resized to be an identical resolution. Some extracted DRU region samples are displayed in Figure 2.

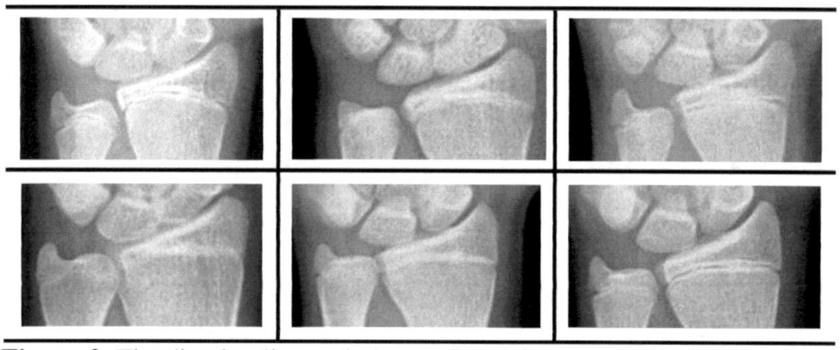

Figure 2. The distal radius and ulna regions after preprocessing the raw data

4.2 Data Labeling

When training the FCNs and later testing the network performance, we need to label these data. In this experiment, we choose to use LabelMe [13] which is an open source annotation tool. We label the distal radius and ulnaregions under physicians' guidance. In Figure 3, a labeled sample is presented. The ROI is covered by green dot lines and other regions will be considered as background.

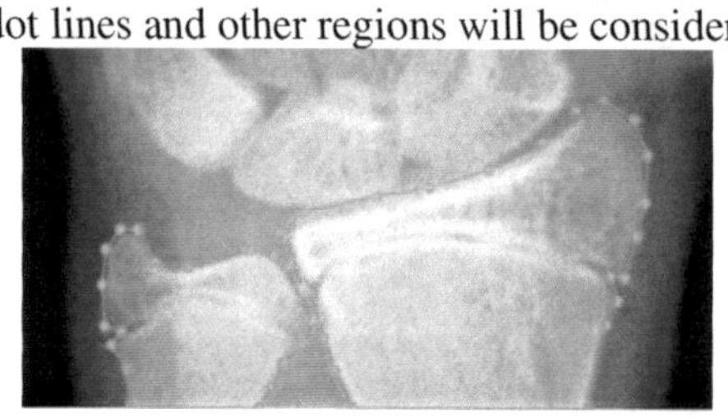

Figure 3. A labeled sample of distal radius and ulna regions (ROI)

4.3 Segmentation Evaluation Metrics

In our experiment, we use two common metrics for semantic segmentation evaluation. The first one is Precision and Recall. Precision reflects the percentage of classified positive samples of all true positive samples and Recall denotes the fraction of interested samples which have been retrieved over all the data samples. Definitions of Precision and Recall can be written as equation (1) and equation (2)

$$Precesion=\frac{TP}{TP+FP} \tag{1}$$

$$Recall=\frac{TP}{TP+FN} \tag{2}$$

Where TP is True Positive, FP is False Positive, and FN is False Negative. The second evaluation metric is called Intersection over Union (IoU). It represents the overlapping rate of segmentation results and ground truth. The definition of IoU can be formulated as equation (3)

$$IoU=\frac{A\cap B}{A\cup B} \tag{3}$$

Where A and B represent the segmentation result and ground truth value respectively. Ideally, we want IoU to approach to 1 as much as possible. $\mathbf{IoU = 1}$ refer to the segmentation results match the ground truth regions perfectly.

5. Result and Analysis

The DRU dataset contains 1189 hand radiographs in total. During the experiment section, 951 images (80%) are randomly selected as training set, and the left 238 images (20%) are used to test trained network performances. All the images are normalized before fed into the proposed network. The normalization operations include mean value subtraction and variation division. To accelerate the network training convergence speed, we choose to use the VGG-16 model which were welled trained on ImageNet dataset [14] and finetune it with our DRU dataset. All the models are implemented in Caffe[15] on NVIDIA Quadro M4000. When training deep neural networks, it is easy to occur gradient exploding problems. It is important to initialize learning rate to be a moderate value such as 1.0×10^{-8} in this paper. To overcome loss functions falling into local minima, the momentum value is set to be 0.99, weight decay is 5×10^{-4}, the maximal iteration is determined to be 20000 and batch size is 1 during our experiment.

5.1 Network Performance of Different Training Iterations

We use DRU training dataset to train FCN-8s model in various iterations and use our testing dataset to test FCN-8s model. Figure 4 and figure 5 show the radius and ulna pixel-wise accuracy, recall and IoU of the testing dataset.

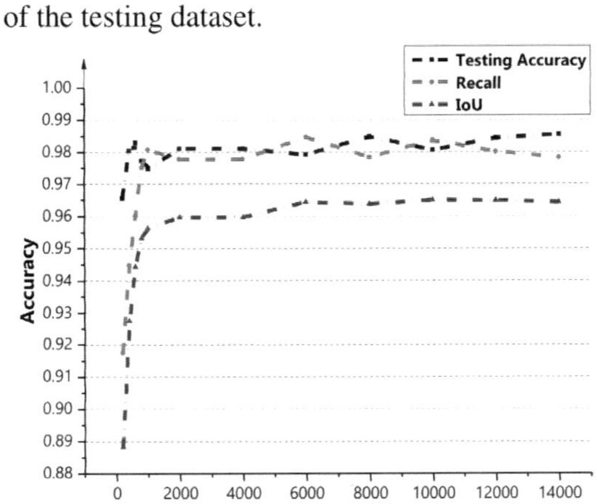

Figure 4. Radius Testing Accuracy, Recall and IoU of different iterations

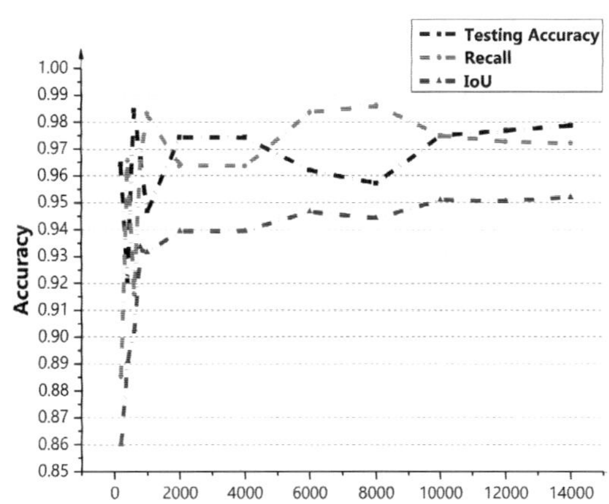

Figure 5. Ulna Testing Accuracy, Recall and IoU of different iterations

From the above several figures, the total testing accuracy and recall achieve around 98%, IoU is about 97%. The Radius testing accuracy and recall is 97% and radius IoU is around 95%; Ulna only testing accuracy, recall and IoU are 98.5%, 98% and 96.6% respectively.

5.2 Network Performance of Different Configurations

We want to know the influences of DRU semantic segmentation of different network configurations. Therefore, we will test FCN-2s, FCN-4s, FCN-8s, FCN-16s and FCN-32s. In Figure 6 and figure 7, the testing accuracy, recall and IoU are recorded according to different network configurations.

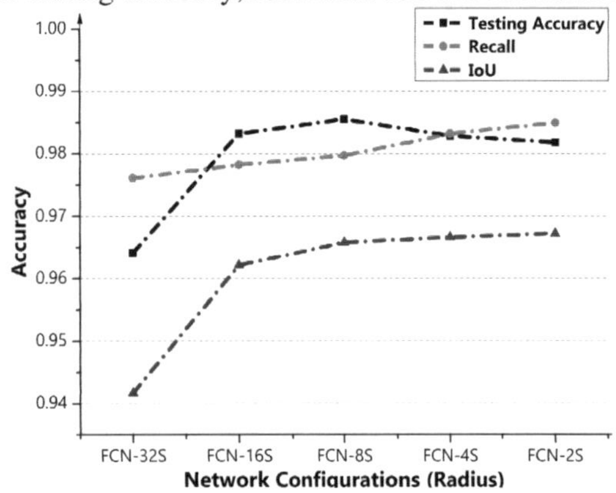

Figure 6. The Radius Accuracy, Recall and IoU of different network configurations

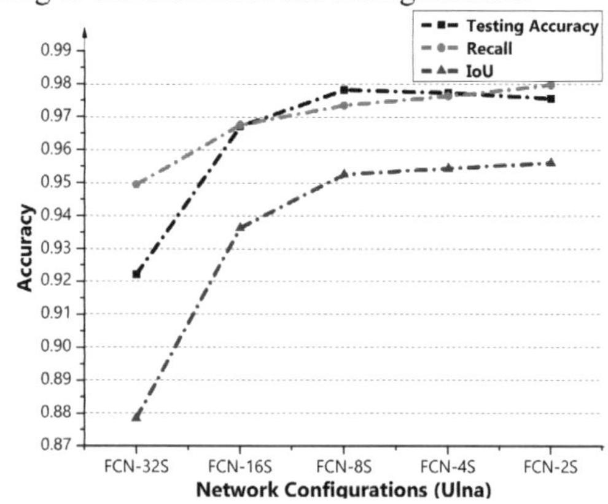

Figure 7. The Ulna Accuracy, Recall and IoU of different network configurations

From the above figures, we know that FCN-16s and FCN-32s models utilize shallow layer features. They combine the deep features with shallow features and then produce the segmentation results. In the same way, FCN-8s uses more shallow features to assist segmentation performance. Generally speaking, according to the above experiment results, the testing accuracy of FCN-8s is higher than FCN-16s and FCN-32s models. On the other hand, although FCN-4s and FCN-2s involve more shallow features, the network performance has no significant improvement. These two kinds of models have more network parameters and more complicated network structures. After comprehensive considerations, FCN-8s is the best performed network model among all the others for this task.

Finally, the FCN-8s model is used to do DRU semantic segmentation. The results are shown in the Figure 8. The gray region denotes for ulna, while region represents radius, and the black region stands for the image background. The proposed network segment all the 238 testing images in 174 second. Each image only costs 0.73 second to be processed in the experiment.

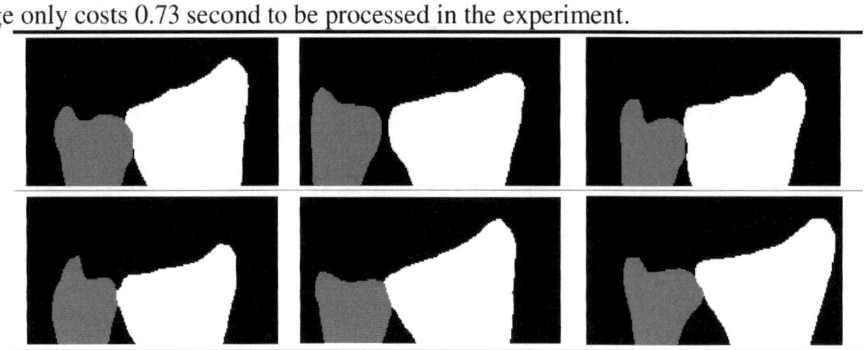

Figure 8. Radius and ulna segmentation results

6. Conclusion

In this paper, we trained a fully convolutional neural network to extract distal of radius and ulna regions of hand X-ray images. The well-trained network can segment bones boundaries subtly in the experiment. Also, we have compared different network configurations and chosen the one with the best performance i.e. FCN-8s. Additionally, we have found that the proposed network achieved high testing accuracy with a small number of training samples.

Acknowledgments
This work was supported partly by National Natural Science Foundation of China (No. 61872351), Shenzhen Peacock Plan (No.KQJSCX20170331162115349), and Shenzhen Key Basic Research Project (No.JCYJ20180507182506416).

References

[1] LeCun, Y., Bengio, Y., & Hinton, G. (2015). Deep learning.*Nature*,521 (7553), 436-444.
[2] Greenspan, H., van Ginneken, B., & Summers, R. M. (2016). Guest editorial deep learning in medical imaging: Overview and future promise of an exciting new technique. *IEEE Transactions on Medical Imaging* 35(5), 1153-1159.
[3] Pereira, S., Pinto, A., Alves, V., & Silva, C. A. (2016). Brain tumor segmentation using convolutional neural networks in MRI images. *IEEE transactions on medical imaging*, 35(5), 1240-1251.
[4] Long, J., Shelhamer, E., & Darrell, T. (2015). Fully convolutional networks for semantic segmentation. *In Proceedings of the IEEE Conference on Computer Vision and Pattern Recognition* (pp. 3431-3440).
[5] Luk, K. D., Saw, L. B., Grozman, S., Cheung, K. M., &Samartzis, D. (2014). Assessment of skeletal maturity in scoliosis patients to determine clinical management: a new classification scheme using distalradius and ulna radiographs. *The Spine Journal*, 14(2), 315-325.
[6] Kamnitsas, K., Ledig, C., Newcombe, V. F., Simpson, J. P., Kane, A. D., Menon, D. K., ...&Glo-cker, B. (2017). Efficient multi-scale 3D CNN with fully connected CRF for accurate brain lesion segmentation. *Medical image analysis*, 36, 61-78.
[7] Moeskops, P., Viergever, M. A., Mendrik, A. M., de Vries, L. S., Benders, M. J., &Išgum, I. (2016). Automatic segmentation of MR brain images with a convolutional neural network. *IEEE transactions on medical imaging*, 35(5), 1252-1261.
[8] Roth, H. R., Lu, L., Farag, A., Shin, H. C., Liu, J., Turkbey, E. B., &Summers, R. M. (2015, October). Deeporgan: Multi-level deep convolutional networks for automated pancreas segmentation. *In Internnational on Medical Image Computing and Computer-Assisted Intervention* (pp. 556-564). Springer, Cham.
[9] Ronneberger, O., Fischer, P., &Brox, T. (2015, October). U-net: Convolutional networks for biomedical image segmentation. *In International Conference on Medical Image Computing and Computer-Assisted Intervention* (pp. 234-241). Springer, Cham.
[10] Andrearczyk, V., & Whelan, P. F. (2017). Texture segmentation with fully convolutional networks. arXiv preprint arXiv:1703.05230.
[11] Valipour, S., Siam, M., Jagersand, M., & Ray, N. (2017, March). Recurrent Fully Convolutional Networks for Video Segmentation. *In Applications of Computer Vision (WACV)*, 2017 IEEE Winter Conference on (pp. 29-36). IEEE.
[12] Simonyan, K., & Zisserman, A. (2014). Very deep convolutional networks for large-scale image recognition. arXiv preprint arXiv:1409.15-56.
[13] Torralba, A., Russell, B. C., & Yuen, J. (2010). Labelme: Online image annotation and applications. Proceedings of the IEEE, 98(8), 1467-1484.
[14] Deng, J., Dong, W., Socher, R., Li, L. J., Li, K., & Fei-Fei, L. (2009, June). Imagenet: A largescale hierarchical image database. In Computer Vision and Pattern Recognition, 2009. CVPR 2009. IEEE Conference on (pp. 248-255). IEEE.
[15] Jia, Y., Shelhamer, E., Donahue, J., Karayev, S., Long, J., Girshick,R., ... & Darrell, T. (2014, November). Caffe: Convolutional architecture for fast feature embedding. In Proceedings of the 22nd ACM international conference on Multimedia (pp. 675-678). ACM.

Research on Tourism Destination Attraction Based on Deep Learning

Shanshan Li[1], Canhua Qiu[1] and Mingming Jiang[2,*]

[1]School of Economics and Management, Tongji university, Shanghai, China
[2] Bilibili Inc.

[*]stupidhod@gmail.com

Abstract. Mining and analyzing online travel reviews and travel information is playing an increasingly important role in the tourism industry. Accurately capturing the uniqueness and attractiveness of the tourist destinations recorded in the travel notes is the key to tourism analysis and application. The current way to obtain the attraction of tourism is easy to cause bias due to the use of simple statistical methods. This paper proposes a model based on deep learning, which uses Bert pre-training method, based on Transformer, and mines travel notes through Attention to find the attraction point. The model can understand the chapter-level semantics of travel notes based on the context, so much so that the extracted features are closer to the meaning of the text. It also exhibits good performance in generating unique labels of tourist destinations and similar tourism clusters. The experimental results are consistent with the facts, the validity of the model is also proved.

1. Introduction

The motivation of tourists to visit a city is often the uniqueness and attraction of the tourist destination. The scientific positioning and management of the attractiveness to the tourism image by managing teams can reduce the substitutability of the tourism destination and form a stable competitive advantage of the brand. With the development of Internet technology, the availability of extensive travel online reviews provides an unprecedented opportunity to analyse the emotions, preferences, feelings and opinions expressed by visitors. Therefore, it is important to determine the attractiveness and uniqueness of the image of a tourist destination by obtaining the text data of online travel platforms. It can help the tourists to choose the right place to travel and optimize the image of the tourist destination.

Currently, a large number of studies have been conducted through online analysis and application of online travel, and Serna [1] uses sentiment analysis techniques to analyse online hotel reviews and understand user preferences or requirements. Li Ping et al. used the website review data to extract the high-frequency feature words of the researched area and conduct semantic network analysis to obtain the image perception of five tourism communities in Beijing [2]. The measurement of the image of tourist destination is the subjective perception of the tourist towards the image of the tourist destination. It is also the basis for evaluating whether the tourist destination is attractive. The traditional method for measuring the image of a tourist destination is the surveying method by means of questionnaires or to carry out word frequency statistical analysis of the dimensional words to which the image of a destination belongs, etc. These methods all use simple statistical methods, which do not have a good understanding of the user's comments or travel notes and are likely to cause deviations in

IOP Conf. Series: Materials Science and Engineering **646** (2019) 012026 doi:10.1088/1757-899X/646/1/012026

results. From n-grams [3] to recurrent neural networks, technologies applied in the field of natural language processing are constantly evolving. Language modeling is the task of predicting the next word in a text given the preceding words, with practical applications in the NLP field. Transformer is a feature extractor that can train long text semantics and produce better results [4]. Google's BERT pre-training structure [5] solves the problem of long text dependence based on Transformer model. In view of this, this paper uses Bert pre-training model, based on Transformer structure to let the model learn the method of attraction mining with Mask's corpus. Then we use the trained model to extract the attraction of the travel notes. Finally, we show the attraction effect of the tourist attraction we extracted and the clustering results of similar tourist destinations.

2. Related work

2.1. Related technology application

Currently, word2vec[6] is one of the most widely used word embedding models for various applications related to text processing. But it still has some limitations. For example, it can't solve the problem of polysemous words. The learned word vector can't represent the global meaning. Moody proposed a model called lda2vec by mixing Dirichlet theme models and word embedding, which greatly improved the representation ability of standard word vectors [7]. The attention mechanism is used to learn to attribute the different weights of words to the overall meaning of the text, and is an effective technique for implementing long-term memory (LSTM). Li et al. proposed a two-way gated recursive unit neural network model (BiGRULA) by combining the topic model (lda2vec) and attention mechanism for analyzing hotel review data [8]. Dieng proposed TopicRNN, which integrates the advantages of RNN and potential topic models to achieve remote semantic dependence [9]. Zhou et al. [10] combined a convolutional neural network (CNN) with a long-term and short-term memory network (LSTM) to propose a C-LSTM model, which achieved good results on both the emotional and problematic classification tasks. The addition of LSTM, GRU models, and the Encoder-Decoder framework extends the application of RNN, but the sequence structure of RNN itself is difficult to perform large-scale parallel computing. CNN has strong parallel computing capabilities to solve text classification tasks, but CNN is more difficult to obtain long-distance features of text due to its structure. In the Transformer model [4], all calculations can be performed in parallel, thereby increasing the speed of training. Because of the existence of self-attention, there is direct interaction between any two frames, which can establish direct long-distance dependence.

2.2. Model structure

The model in this paper consists of four layers, which is shown in the figure1.The first layer is the input layer. Each [CLS]special symbol is added to the front of the vector representation, and the [SEP] special symbol is added at the end of the word embedding vector to mark and split the data. The vector is then position-coded with Position Embeddings and multiplied by the word vector of the mask corpus. The obtained characterization information enters the second layer of Transfromer structure to obtain attractive points. These attractive points constitute a vector of 1x512 dimensions, multiplied by the vector of 512 cities, with 500 travel notes per city, Then softmax decide with which city's attraction it is best matched. The softmax parameter can represent the city's vector.

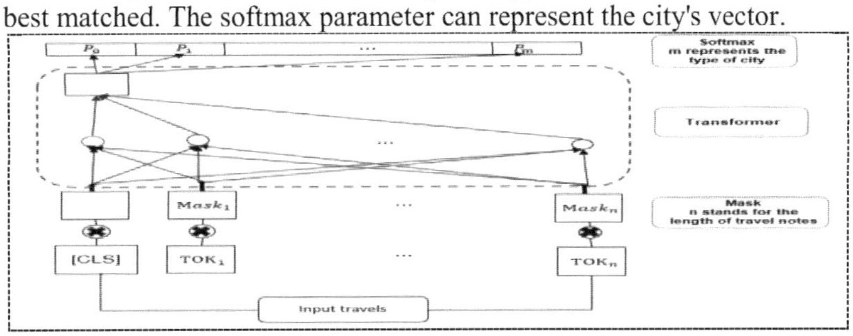

Figure 1. The model structure .

The bert pre-training method used in this paper only requires the Transformer model's Encoders architecture. This section will focus on the encoder architecture of the Transformer model. The encoder architecture is shown in the figure2.

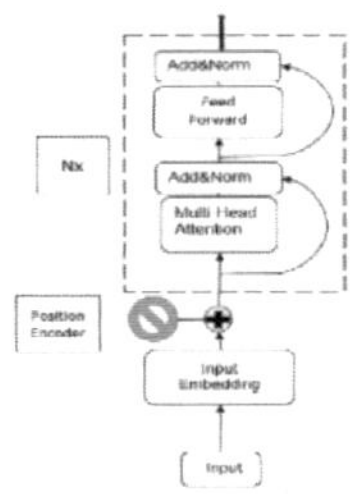

Figure 2. Encoder Architecture.

Each layer of the Encoder consists of two sublayers, a self-attention layer and a feed-forward neural network. At the bottom of the encoder input, the input word is first converted into a word embedding vector. The length of the input sequence is usually the length of the longest sentence of the training set, and each element is a 512-dimensional word vector. The bottom-level encoder takes the received sequence as input, then passes it through the self-attention processing, sends it to the forward network, and finally passes the input to the next encoder. After the input sequence is embedded in the word, the feature vector Z is first obtained through the "Self-attention" module, and then passed to the fully connected layer of the feedforward neural network, and the output is passed to the next encoder. The formula for Z is expressed as:

$$\text{Attention(Q,K,V)=softmax}\left(\frac{QK^T}{\sqrt{d_k}}\right)\text{v} \quad (\text{Q,K} \in R^{d_n \times d_k}\ ;\text{V} \in R^{d_n \times d_v}) \tag{1}$$

Multi-Head Attention is equivalent to the integration of h different self-attentions and stitching together the final results. The multi-attention mechanism extends the ability of the model to focus on different locations, and multiple sets of randomly initialized query/key/value-matrix trained vectors can be mapped to different sub-expression spaces, giving multiple way of expression to attention mechanism. The formula is expressed as:

$$[Multihead(h) = Concat[head_1, \ldots\ldots, head_m]W^o \tag{2}$$

$$\text{where } head_i = Attention(Q, K, V) \qquad \text{where Q, K, V} = \text{h}W_i^Q, \text{h}W_i^k, \text{h}W_i^v$$

In addition, the location information of the word is added in the process of encoding the word vector, that is, the position code is introduced, so that the distance between two words is better expressed. The coding formula is as follows:

$$\text{PE(pos, 2i)} = \sin\left(\frac{pos}{10000^{\frac{2i}{d_{model}}}}\right) \quad \text{PE(pos, 2i + 1)} = \sin\left(\frac{pos}{10000^{\frac{2i}{d_{model}}}}\right) \tag{3}$$

pos: Location of words I: dimensions of words

The add&norm layer is a combination of layer normalization [11] plus residual network in each sublayer. Add represents Residual Connection, which is to solve the problem of difficult training in multi-layer neural networks. By passing the information of the previous layer to the next layer, it can effectively focus only on the difference. Norm represents Layer Normalization. Normalization of the layer's activation values speeds up the model's training process and allows it to converge faster. The related formula of feed-forward neural network is expressed as:

$$\text{FFN(h)} = \text{ReLU}(\text{h}W_1 + b_1)W_2 + b_2 \tag{4}$$

$$\text{where } W_1 \in R^{d_{model} \times d_{ff}}\ , W_2 \in R^{d_{ff} \times d_{model}}\ , \text{h} \in R^{d_n \times d_{model}}$$

2.3. Pre-training model bert

In October 2018, Google released a large-scale pre-training language model, BERT [11], the coding model structure of the two-way Transformer. In order to obtain the unique label of the tourist destination, the more accurate and efficient extraction of the information in the travel notes is the focus of this paper. This paper will use the BERT pre-training model. The BERT adopts a two-stage dual-language model, combining the dual advantages of the Transformer depth model and two-way information.The purpose of this pre-training is to learn the intrinsic logic of the text through two unsupervised tasks: Masked Language Modeling - predicting missing words by giving left and right contexts, and Next Sentence Prediction - Predict whether a sentence follows another sentence.

Before the data are put into the model, each word and special symbol of the input sentence need to be converted into a word embedding vector. The characterization information of the model input is obtained by three layers of Embedding. The first layer is Token Emdeddings, and the Token Emdeddings is a word vector. That is, the special character [CLS] and [SEP] are added the sentence word embedding vector, the special character [CLS] is used for the classification task, and the special character [SEP] is the sentence segmentation. symbol. The final position coding Position Embeddings adds positional information to each word, allowing Transformer to perceive the positional relationship between different words.

In our model structure, softmax is the final output layer, and with the cross-entropy (CE) loss function. In this paper, we want to maximize the value of the p(y|x) probability in the model and the value of the cross entropy where y represents a tourist destination and x represents a travel note. By maximizing the value of p(y|x) probability, the difference between the travel information of this tourist destination and other travel destinations is found. In the Bert pre-training phase of our model, the input sequence of the fixed dimension is denoted as $C \in R^{H}$. During the fine-tuning, the only new parameter added is regarded as a classification layer $W \in R^{K \times H}$, where K is the number of classification labels, and the probability of the label $P \in R^{K}$ is calculated using the standard softmax, ie $P = \text{softmax}(CW^{T})$. Fine tune all parameters in BERT and W to maximize the log probability of the correct tag.

$$p\left(y^{(i)} = j | x^{(i)}; \theta\right) = \frac{\exp\left(\theta_j^T x^{(i)}\right)}{\sum_{l=1}^{k} exp\left(\theta_l^T x^{(i)}\right)} \quad (5)$$

$$l_{CE} = -\sum_{i=1}^{C} t_i \log(y_i) \quad (6)$$

$p\left(y^{(i)} = j | x^{(i)}; \theta\right)$is the probability of categorizing x into category j

3. Data sources and processing methods

3.1. Data sources

Mafengwo is one of the relatively well-known travel websites in China and is the source of the data in this article. This article gathered the travel notes on the website of Mafengwo and obtained travels notes of 512 cities, each one with 500 to them, which means the total amount of the travel notes 256,000 travel notes. In addition to travel notes, this article also collects Chinese and English names for all tourist destinations in 512 cities.

3.2. Data processing

Because the neural network can only perform numerical calculations, this article will cut the numbers, English and each Chinese character in each travel note that we collected, and perform the word2vec operation to obtain a dense vector representation of the travel. In this paper, the BERT pre-training model is used. The length of the input sequence in the BERT pre-training model is 512 or less, including the [CLS] that the model needs to add before the input characterization information and [SEP] special symbols. Since the length of each travel note is long, in order to obtain the information of the data in the travel note as much as possible, this paper defines the vector length representation of the travel note as 512. After obtaining the word embedding vector representation of each travel note, if the length of a travel vector exceeds 510, the following information is discarded. If the length of a travel vector is less than 510, the remaining length is replaced by the number "0". After that, the word embedding vector of each travel is multiplied by the [mask] vector with Chinese and English tourist destinations, and the information with Chinese and English tourist destinations in the travel notes is masked to obtain the travel words embedded vector after removing the names of Chinese and English tourist destinations.

4. Model results

4.1. Visualization of tourism destination attractions

In order to verify that our model can explore the uniqueness and attraction of the tourist destination in the travel notes, we randomly selected a travel note introducing Sichuan. We visualized the attention probability after each word in the travel note obtained by the model. And used the color scale in excel from red to gray to indicate probability of the last layer of the 12 headers from the minimum to the maximum. The red box of the 12 boxes behind the word means that the word is the attraction that the model notices. The deeper the red color, the more the word represents the uniqueness of the tourist destination. For reasons of space, we excerpted the two segments in this travel note to illustrate the attraction points of the model to Chengdu. Because the word "Chengdu" has been processed by [mask], the probability of attention after "Chengdu" shown in the figure is zero, which is consistent with the purpose of our operation. Figure 3 shows that our model can choose words with emotions as attractiveness, such as “easy”, “gentle”, “leisure”, “trend”, “food”, “life”, and these words are also integral with our impression of the city. When it comes to Chengdu, people generally think of "panda" and "hot pot". These two words are also noticed by the model, as shown in the figure. Our model can learn the attraction of the tourist destination through a certain travel note, it has also verified the possibility that the model can learn the full attraction of a tourist destination by training all the travel notes of a tourist destination.

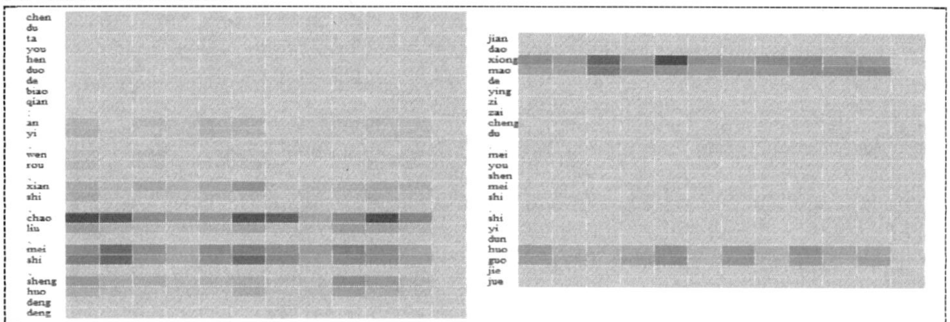

Figure 3. Chengdu attraction points noticed by the model in a random travel note.

4.2. Generation of all tourism destination attractions

Instead of a single character, the attraction of a tourist destination often consists of a word formed by more than one character. we get the travel notes of the Jieba word segmentation, and the Jieba participle belongs to the probabilistic language model segmentation, that is, to find the maximum probability in all the results obtained from the full segmentation. We add the attention information of the last layer of the multi-head to get the attention information on each character, and then re-add the attention information of each character in the dimension of words. Then we add attention information to the words of each of the 500 travel notes to get the probability of attraction of each word in this city. Then we sort them according to this weighted value W, the top word represents the attraction point of the tourist destination.

$$C_{j-\,attention_k} = \sum_{i=0}^{11} P_{L-i-j} \tag{7}$$

$$W_{attention} = sum(C_{i-attention}, C_{j-attention}) \tag{8}$$

IOP Conf. Series: Materials Science and Engineering **646** (2019) 012026 doi:10.1088/1757-899X/646/1/012026

P_{L-i-j}:attention information of the jth word on the i head
$C_{j-\,attention_k}$:attention information for the jth word in k tokens i—j:index of the word in travel notes
$W_{attention}$:attention information that makes up the word in all travel notes of a tourist destination

We selected two cities, namely Chengdu, Shanghai,and ranked the attraction points according to the attention scores of each city's attraction, and get the 10 most attractive points with local characteristics. These attractive points are provided in Figure 4 and Figure 5. "Taikoo Li", "Panda", "hot pot", "Delicious food", " the land of abundance" and "Dujiang Dam" are the top six words that the model noticed in travel notes about Changdu. "Taikoo Li" is known as the most popular business street in Chengdu. Watching "Panda" and eating "hot pot" are two things that tourists almost always do when they come to Chengdu. Chengdu is also known as the "land of abundance". The attractions of Shanghai include "Pearl of the Orient", "Disney", "Magic Capital" and "City God Temple". These words also represent the characteristics and culture of Shanghai. "Oriental Pearl" is Shanghai's iconic cultural building. Shanghai "Disneyland" is the first Disney theme park in China. "Magic Capital" is synonymous with Shanghai, which is well known. Therefore, it is proved that the model excavates the attractive points of tourism destinations, including the subjective impressions of tourist destinations, special snacks, tourist attractions and humanistic scenes.

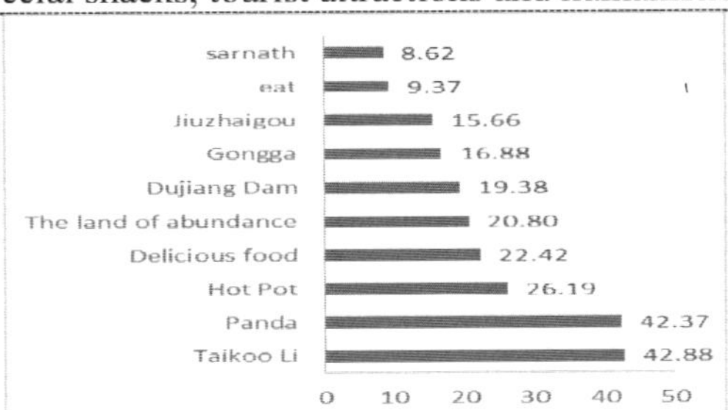

Figure 4. Chengdu Attraction Point in Top 10

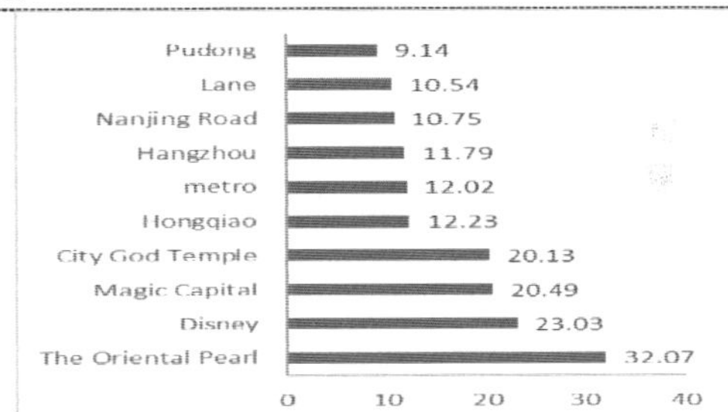

Figure 5. shanghai Attraction Point in Top 10

5. Conclusion

In this paper, we propose a model based on deep learning that can explore the attraction and uniqueness of tourism. The model is characterized by the use of Google's latest BERT pre-training model, combined with the Transformer and Attention. The application of our model in online travel analysis shows that it can extract rich information from the text dataset, making it closer to the meaning of the article. The Attention scores the extracted tourism information to obtain the unique attraction point of the tourist destination. Using the parameters of the last layer of softmax as the embedding information of the urban tourism destination, the results obtained by the model are consistent with the facts and remarkably improves the traditional practice of conducting statistical analysis based on word frequency, making the attraction of tourist destinations more convincing and more in line with the essential characteristics of tourism destinations.

References

[1] He W, Tian X, Tao R, Zhang W, Yan G, and Akula V 2017 Application of social media analytics: a case of analyzing online hotel reviews. Online Information Review, 41(7), 921-935
[2] Li P, Chen T, Wang F and Wang Xinge 2017 Text Mining-based Perception of Urban Tourism Community Image: A Case Study of Beijing. Geographical Research 36 (6), 1106-1122.
[3] Kneser R and Ney H 1995. Improved backing-off for m-gram language modeling. In 1995 International Conference on Acoustics, Speech, and Signal Processing Vol. 1 pp 181-184
[4] Vaswani A, Shazeer N, Parmar N, Uszkoreit J, Jones L, Gomez A. N and Polosukhin I 2017 Attention is all you need. In Advances in Neural Information Processing Systems pp 5998-6008
[5] Devlin J, Chang M W, Lee K and Toutanova K 2018 Bert: Pre-training of deep bidirectional transformers for language understanding. arXiv preprint arXiv:1810.04805

[6] Mikolov T, Chen K, Corrado G and Dean J 2013 Efficient estimation of word representations in vector space. arXiv preprint arXiv:1301.3781
[7] Moody, C. E 2016 Mixing dirichlet topic models and word embeddings to make lda2vec. arXiv preprint arXiv:1605.02019
[8] Li Q, Li S, Hu J, Zhang S and Hu J 2018 Tourism Review Sentiment Classification Using a Bidirectional Recurrent Neural Network with an Attention Mechanism and Topic-Enriched Word Vectors. Sustainability, 10(9), 3313
[9] Dieng A B, Wang C, Gao J and Paisley J 2016 Topicrnn: A recurrent neural network with long-range semantic dependency. arXiv preprint arXiv:1611.01702
[10] Zhou C, Sun C, Liu Z and Lau F 2015 A C-LSTM neural network for text classification. arXiv preprint arXiv:1511.08630
[11] Ba J L, Kiros J R and Hinton G E 2016 Layer normalization. arXiv preprint arXiv:1607.06450

IOP Conf. Series: Materials Science and Engineering **646** (2019) 012027 doi:10.1088/1757-899X/646/1/012027

Short-term Time Series Data Prediction of Power Consumption Based on Deep Neural Network

Kang Xu[1,*], Ruichun Hou[1], Xiangqian Ding[1], Ye Tao[2] and Zhifang Xu[3]

[1]School of Information Science and Engineering,Ocean University of China, Qingdao,China

[2]School of Information Science and Technology, Qingdao University of Science and Technology, Qingdao,China

[3]Qingdao Haier Technology Co., Ltd. , Qingdao, China

[*]xukang@stu.ouc.edu.cn

Abstract. Under the background of the rapid development of Internet technology and the popularity of smart grids, the analysis and prediction of short-term time series data of users' power consumption has important guiding significance for grid planning, management decision of economic sector and optimization and allocation of power resources. Considering that the traditional statistical-based time series analysis method is weak in generality and can not handle the complex linear problem in prediction, the long-term dependence of the ordinary cyclic neural network model is insufficient, and the time series data has multidimensional problems, a deep neural network is proposed. The PCA-LSTM model is used for time series data prediction. The model firstly uses the PCA (principal component analysis) method to reduce the dimensionality of the electricity consumption time series data, optimizes the number of input variables, and inputs the data into the long- and short-term memory network LSTM for training prediction. The experimental results show that the LSTM network prediction based on PCA improves the accuracy of short-term time series data prediction, and also improves the convergence speed of LSTM network. It proves that the method has better prediction performance and versatility.

1. Introduction

In recent years, with the rapid development and wide application of monitoring, information communication management and Internet technology, the integrated energy system [1] with smart grid as the core has achieved relatively rapid development. The concept of "energy internet" with the goal of achieving multi-energy open interconnection, energy free transmission, energy greening, marketization and energy efficiency has emerged as the new international academic and industrial circles. And the development of emerging industries such as the Internet of Things, big data, and cloud computing [2] has also been greatly pushed to the automation process of power companies such as power grids. In the process of power supply automation, the power grid will generate corresponding big data according to the electricity consumption information collected by the sensors collected by the home appliance (air conditioner) Internet of Things. These data have typical time series data characteristics, which are generally defined as the time according to the occurrence time. In the order of the order, the ordered sets formed by the values of the same statistical indicators are arranged. Time series data also has important application value and is one of the more popular research topics in recent years [3].It is extremely important to predict the short-term time series data of the above-mentioned power consumption, not only as one of the key technologies of the energy Internet, but also to predict that we can know the user's power demand in a certain period of time in the future. To meet the needs of users, while improving energy utilization and optimizing and distributing power resources, it plays an important role in the energy Internet.

At present, the research methods for time series data are mainly divided into two categories. One is a statistical-based approach [4], including traditional autoregressive models, Autoregressive Moving Average (ARMA) models [5], and Autoregressive Integrated Moving Average (ARIMA) model [6]. The ARMA model is obtained by combining a moving average process and a linear variance equation for predicting data with linear stationary properties, and the ARIMA model is used to predict linear non-stationary properties. Although this type of method is simple, it cannot handle complex nonlinear problems in prediction, so it will produce higher prediction errors when the variables change greatly. The other is based on artificial intelligence, mainly based on support vector regression algorithm [7], BP neural network [8] and artificial depth neural network algorithm [9] as a representative of machine learning methods and suitable for complex nonlinear time sequence. Among them, the support vector regression algorithm adopts the idea of minimizing structural risk. The model structure is easy to determine and the performance is stable. BP neural network also has this advantage, but the prediction accuracy is limited. With the advent of the era of big data, the improvement of computing power and the substantial increase of training data have provided support for deep learning [10], and the deep neural network represented by Recurrent Neural Network (RNN) [11] The advantages of strong versatility and high prediction accuracy have gradually become the hot research direction of time series prediction. The purpose of the Recurrent Neural Network (RNN) is to process sequence data. In the traditional neural network model, from the input layer to the hidden layer and then to the output layer, the layers are fully connected, and the nodes between each layer are disconnected, so this common neural network There is nothing that can be done about many complicated problems. RNN is called a cyclic neural network, that is, the current output of a sequence is also associated with the previous output, that is, the nodes between the hidden layers are no longer connectionless but connected. In theory, RNN can process sequences of any length, but when the number of layers in the network increases, RNN will appear as gradient disappearance and gradient explosion. To solve the above problem, long-term memory neural network (Long Short-Term Memory Neural Network) is introduced. Network, LSTM) [12]. The emergence of long-and short-term memory neural network (LSTM) is actually a new deep learning network based on RNN, which can be used to solve the problem of gradient disappearance and gradient explosion that RNN can't solve, thus effectively making up for the defects of ordinary RNN. And controls the cumulative speed of information and demonstrates superior capabilities in predicting long-range dependent timing data.

Long and short term memory neural network (LSTM), which has certain information mining capabilities for long-distance time series data, is also widely used in speech recognition, machine translation, fault detection, and time series data prediction. For LSTM, the longer the input, the more information it contains [13].However, for the selected raw data, having multiple measurable influencing factors, that is, having multiple dimensions, greatly increases the computational complexity of the neural network, resulting in inefficiencies and the like. Aiming at the above problems, this paper studies the user's electricity consumption forecast based on the historical data of residential electricity consumption time series, and proposes a PCA-LSTM model based on deep neural network for the prediction of short-term time series data of user's research work. The model can effectively avoid the excessive dependence of data hypothesis in statistical methods, and avoid the disadvantages of LSTM network operation complexity. The PCA (principal component analysis) [14] algorithm is used to reduce the dimensionality of the data and extract relevant influencing factors. The main component in the optimization of the number of input variables, so as to achieve the effect of time series data prediction to effectively solve related problems.

2. Establishment of PCA-LSTM model based on deep neural network

2.1. PCA dimension reduction

When using the deep neural network to predict the short-term time series data of power consumption, the neural network can continuously train and learn through a large amount of historical data of user power consumption to determine the relationship between the input data and its parameters and achieve the ideal prediction. The effect is therefore much better than the traditional method. However, the prediction of short-term time series data on electricity consumption is not an easy problem because there are many factors that affect electricity consumption, such as geographical factors, weather, temperature, holidays, weekends/working days. If a lot of factors are used to make predictions, although it can reflect the trend of power consumption, the training convergence time of deep neural networks becomes very slow, resulting in inefficiency; if the influence factors are not considered, the historical data information will be Not enough expression, so we must consider the influencing factors but not all of them. At this time, we use PCA (principal component analysis) to extract the main components of the influencing factors. Because the raw data of the electricity consumption selected in this paper has multiple dimensions, the PCA dimension reduction is used to reduce the dimensionality of the original data, and the number of input variables is optimized to reduce the computational complexity of the neural network. The PCA principal component analysis method replaces the original multidimensional data with a low-dimensional comprehensive index by adopting a linear transformation to preserve the original data information to the greatest extent. The PCA operation flow is to de-average, calculate the covariance matrix and simultaneously calculate the eigenvalues and eigenvectors of the matrix, and keep the largest eigenvectors from small to large for eigenvalues, and transform the data into a new space constructed by eigenvectors. . The algorithm flow is:

Assume that the original sample data is represented as a matrix:

$$x=\begin{pmatrix} x_{11} & x_{12} & \cdots & x_{1p} \\ x_{21} & x_{22} & \cdots & x_{2p} \\ \vdots & \vdots & \vdots & \vdots \\ x_{n1} & x_{n2} & \cdots & x_{np} \end{pmatrix} \tag{1}$$

The first step: the original historical data needs to be standardized, and the calculation formula is:

$$x_{ij}=\frac{x_{ij}-x_j}{\sqrt{\operatorname{var}(x_j)}},(i=1,2,\ldots,n;\ j=1,2,\ldots,p) \tag{2}$$

among them:

$$x_j=\frac{1}{n}\sum_{i=1}^{n}x_{ij} \tag{3}$$

$$\operatorname{var}(x_j)=\frac{1}{n-1}\sum_{i=1}^{n}(x_{ij}-x_j)^2,(j=1,2,\ldots,p) \tag{4}$$

Step 2: Calculate a matrix of sample correlation coefficients for the electricity consumption data.

$$r=\begin{pmatrix} r_{11} & r_{12} & \cdots & r_{1p} \\ r_{21} & r_{22} & \cdots & r_{2p} \\ \vdots & \vdots & \vdots & \vdots \\ r_{n1} & r_{n2} & \cdots & r_{np} \end{pmatrix} \tag{5}$$

After the standardized raw data is still represented by x, the correlation coefficient of the standardized sample data is:

$$r_{ij} = \frac{1}{n-1}\sum_{i=1}^{n} x_{ij}x_{ij}(i,j=1,2,...,p) \tag{6}$$

Step 3: Use Jacobbian to find the eigenvalues of the correlation coefficient matrix r and the corresponding feature vector α :

$$\lambda_i, i = 1,2,..., p \tag{7}$$

$$\alpha_i = (\partial_{i1}, \partial_{i2},..., \partial_{ip}), (i = 1,2,..., p) \tag{8}$$

The fourth step: select the main component that meets the requirements and find the principal component expression. The sample data obtained by principal component analysis generally can obtain p principal components, but the information content of the p principal components is often different, and there will be differences between them. If the principal component corresponds to a large variance, then the corresponding amount of information is large, and vice versa. When actually selecting, it is usually chosen to reflect the main components of the original data information to a large extent. The basis of the selection is to accumulate and sum according to the contribution rate of each principal component, that is, the proportion of each principal component in the sum of all variances. The mathematical expression of the contribution rate is:

$$a_i = \lambda_i / \sum_{i=1}^{p} \lambda_i \tag{9}$$

The number of principal components to be selected at the end. That is, the determination of the value of k in the $F_1, F_2, F_3, \ldots, F_k$ is determined by the cumulative contribution rate $G(k)$ of the principal component. among them,

$$G(k) = \sum_{i=1}^{k} a_i \tag{10}$$

Usually, as long as the sum of the contribution rates reaches a certain value or more, it can be considered that the comprehensive indicators obtained can fully reflect the vast majority of information contained in the original sample data.

Step 5: Calculate the principal component $F_i = \partial_{1i}x_1 + \partial_{2i}x_2 + \ldots + \partial_{pi}x_p, i = 1,2,..., p$, the specific form can be as follows,

$$\begin{pmatrix} F_{11} & F_{12} & \cdots & F_{1k} \\ F_{21} & F_{22} & \cdots & F_{2k} \\ \vdots & \vdots & \vdots & \vdots \\ F_{n1} & F_{n2} & \cdots & F_{nk} \end{pmatrix} \tag{11}$$

Step 6: Calculate the principal component load, the formula is:

$$q_i = \sum_{i=1}^{p} \sqrt{\lambda_i \alpha_i}, i = 1,2,..., p \tag{12}$$

2.2. Long-term and short-term memory network (LSTM) model

Circulating neural network (RNN) [15] is a popular deep learning model in recent years, and has been successfully applied in many fields such as language recognition and text categorization. However, it also faces the problem of gradient explosion and gradient disappearance. The problem cannot be solved very well. Thus, the neural network model after optimization of the cyclic neural network-LSTM is obtained. The main change is the introduction of three types of gates at each index index position when remembering information: input gates, forgetting gates, and output gates. Through these three kinds of gate structures, the state information of the unit is controlled and maintained, the cell state information is stored and modified, and the information is selectively added or the existing information is deleted according to the result, thereby solving the long-term gradient of the RNNs. The problem of explosion and disappearance has also realized long-term memory and proved its superiority in a large number of experiments, so it is widely used in time series problems of long-term dependence. For the prediction problem mentioned in this paper is a typical timing problem, so the model is selected for training and prediction.

The specific internal structure of LSTM is shown in Figure 1:

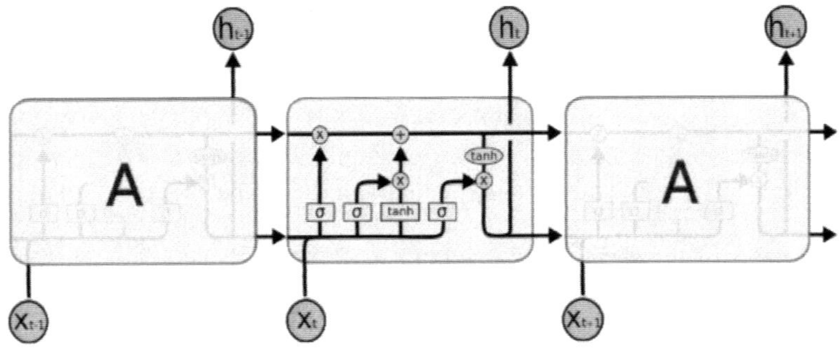

Figure 1. LSTM model structure

The steps for the LSTM to update the internal state during training are as follows:

In the first step, the LSTM needs to pass the forgetting gate to determine the information that the cell state needs to be discarded. The forgetting gate obtains a value between 0 and 1 according to the historical information of the previous state and the currently input information to represent the probability of forgetting the state of the upper layer of cells. That is to say, the forgetting gate is to control whether or not to forget, that is, to control whether the state of the upper layer hidden cells is forgotten with a certain probability. Using the activation function (generally the sigmoid function), The hidden state of the input previous sequence is denoted as h_{t-1} and the input of this sequence is recorded as x_t by which the output of the forgetting gate between [0.1] is recorded as $f(t)$; 0 represents all information of the discarded history state, and 1 represents all information about the historical status. Where x_t represents the input of t time, h_{t-1} represents the value of the hidden layer at $t-1$ time, W_f , U_f , and b_f represent the weight of the LSTM hidden layer in the forgetting gate, the weight of the current input, and the offset in the forgotten gate; $f(t)$ represents the output of the forgetting gate. The specific calculation is as follows:

$$f(t)=\sigma(W_f \cdot h_{t-1}+U_f \cdot x_t+b_f) \tag{13}$$

In the second step, the LSTM determines the update information through the input gate and places it in the cell. After the circulatory neural network has forgotten part of the previous state, it needs to supplement the latest memory from the current input, and the input gate is responsible for processing the input of the current sequence position to supplement the latest memory. The input gate is composed of two parts. The first part uses the sigmoid function to output as i_t , the second part uses the tanh function to output as a_t , and the two obtain the information updated to the cell state by the bitwise product, where W_i , U_i , W_a , U_a , b_i , b_a are the coefficients and offsets of the linear relationship. The specific calculation is as follows:

$$a_t=\tanh(W_a \cdot h_{t-1}+U_a \cdot x_t+b_a) \tag{14}$$

In the third step, the cell state is updated. After the first two "gates", the deletion and addition of the delivery information can be determined, that is, the update can be performed. Both the results of the Forgetting Gate and the results of the input gate will act on the Cell Status C_t and put the C_{t-1} update to the final cell state C_t, That is:

$$C_t = C_{t-1} * f_t + i_t * a_t \tag{15}$$

Finally, the result of the final cell output is determined by the output gate. First, the input information is determined by the sigmoid layer to determine the output of the input information, and then the cell state is processed by the tanh function. Finally, the final output is determined by bitwise multiplication with the previously obtained sigmoid value, and the specific calculation is as follows:

$$O_t = \sigma(W_o \cdot h_{t-1} + U_o \cdot x_t + b_o) \tag{16}$$

$$h_t = O_t * \tanh(C_t) \tag{17}$$

Long-term and short-term memory network (LSTM) models are effectively avoided by using very specialized structural design. The problems of gradient disappearance and explosion in the training of conventional cyclic neural network model (RNNs) can effectively avoid the use of historical sequence information. . In order to fully predict the power consumption short-term sequence data, the LSTM model is used to effectively predict the data based on the dimensionality reduction of the PCA.

3. PCA-LSTM model for time series data prediction

3.1. Data preprocessing

By observing the previous time series data prediction experiment, in the process of selecting this historical data and influencing factors, different types of influencing factors will be selected according to the principle of comprehensiveness. Because there are different dimensions in the data, the numerical difference is relatively large, and the input and output range of the nonlinear activation function in the neural network model is limited. In order to avoid linear representation of historical data, the data is compressed and the neural network is avoided. When the element appears saturated, it is necessary to standardize the data of the time series data so that the range of the range is in the same interval. The purpose of data standardization processing is to remove the influence of the data from the dimension and summarize the statistical distribution of the same sample.

In the short-term time series data of the user's electricity consumption, due to the different influencing factors between different user historical days, the value of the electricity consumption varies greatly. If the processing is directly processed, the prediction result will be inaccurate, so in order to improve the data prediction. The accuracy of the analysis requires normalization of the data for each dimension to reduce errors and improve accuracy. There are many methods of normalization, with min-max maximum and minimum normalization, that is, linear changes to the original data to map values to the [0-1] range; and Z-score standardization method. The normalization method used in this paper is the min-max method for data standardization. The formula is as follows:

$$X^* = \frac{X_i - x_{\min}}{x_{\max} - x_{\min}} \tag{18}$$

Among them, $x_{\min}$ is the smallest value in the original sample data sequence, $x_{\max}$ is the largest value in the original sample data, X^* is the data sequence that the neural network needs to input, and X_i is the original data sequence.

3.2. Using PCA to perform feature dimensionality reduction on training samples

As mentioned above, in the actual time series data prediction, the short-term time series data of power consumption often contains multiple influencing factors, resulting in a relatively large amount of information or even redundant information. If the training sample dimension of the deep neural network is too much, the corresponding network will become more complicated, which will lead to prolonged network training time, inefficient, and the prediction accuracy may also decrease accordingly. Subjective selection of the dimensions of the training samples will easily lead to the loss of important information, resulting in a decrease in the predictive performance of the neural network. If the original data samples are first subjected to dimensionality reduction, the primary response is selected with less loss information. The comprehensive dimension will reduce the computational complexity of the neural network, improve the training efficiency of the neural network, and greatly improve the prediction performance. In this paper, the PCA dimension reduction processing of the input short-term time series data is used to achieve data dimensionality reduction.

After reducing the dimensionality of the sample data PCA, we can use a small number of features to train and predict the deep neural network, which not only improves the accuracy of the experimental prediction but also improves the convergence speed of the neural network.

3.3. Training and testing of LSTM network model

The LSTM network model is used to build a deep learning framework. At the same time, the LSTM network model is built using the popular Tensorflow deep learning platform. There are many activation functions. The relu function is used in this paper.

First, some of the original power consumption sequence data samples are taken as the training set, and the remaining small power consumption time series data samples are used as the test set. At the end of the training, the weights and offsets of the LSTM network are saved, and the trained deep neural network model is used to test on the corresponding test set, to see how accurate the predictions is, and to compare the predictive effect fobtained by simply using the RNN and LSTM neural network models. effect.

The LSTM network model for short-term time series data prediction needs to determine the following parameters: Timestep time step, number of hidden layer layers, number of hidden layer neurons, number of iterations, learning rate, etc. In the process of training, the corresponding hyperparameters are corrected in real time according to the effect of the training, so as to achieve a perfect prediction effect.

4. Experiment and result analysis

4.1. Introduction to the data set

The data used in the short-term time series data forecasting of this electricity consumption is a user's electricity record from a total of 380 days from July 1, 2017 to July 15, 2018, and is mainly sampled by smart meters. The data taken include user ID, electricity date, total electricity consumption, "tip" time consumption, "peak" time consumption, "flat" time consumption, and "valley" time consumption. The total amount of electricity used by each user per day and the characteristics of the electricity consumption data during the "peak peak" period, as well as the influencing factors mainly include date, day of the week, holidays, temperature, weather, and so on.

Accurate analysis and prediction first need to process a large amount of raw data, which is different from the data preprocessing mentioned above. At this time, it mainly includes the combination of data, removal of invalid records, and removal of sparse records. The sample data are all from the real life data, all the acquisition process will be affected by external factors, there is "noise", and some of the user's part of the period of record data will be missing, resulting in a large error. In order to ensure the reliability and reality of the forecast, the users with complete data in the original data set and the data with total power consumption less than 50 are selected, because the survey found that the residential electricity demand does not exceed 50 kWh per day. Through filtering, a total of 4,456 users and 380 days of electricity usage data from July 1, 2017 to July 15, 2018 were selected. Among them, a part of the original data is selected as the training set for the training of the model, and the remaining data of the original data is used as a test set to verify the feasibility of the model.

4.2. Evaluation criteria for network models

In this paper, the evaluation criteria for the short-term time series data prediction results of users is the Root Mean Square Error, which reflects the performance of the model to control the absolute error. The calculation formula is:

$$RMSE = \sqrt{\frac{1}{n}\sum_{i=1}^{n}[p(i) - \hat{p}(i)]^2} \tag{19}$$

In the formula: $p(i)$ represents the actual value of the electricity consumption data, $\hat{p}(i)$ represents the predicted value of the user electricity consumption data; n represents the number of predicted verification data; i is the predicted point serial number. The smaller the $RMSE$, the better the prediction effect of the model.

4.3. Experimental results and comparative analysis

In this paper, the PCA-LSTM model based on deep neural network is used to predict the short-term time series data of the user's electricity consumption. In order to more intuitively express the prediction effect of the text model on the user's electricity consumption data, all users are selected from May to July 2018. A total of 60 days of electricity recorded data for prediction, and compared with SVM, RNN model, the comparison of the user's electricity consumption prediction effect and actual electricity consumption and the comparison between the predicted values and actual values of different models are shown in Figures 2, 3, and 4. Shown as follows:

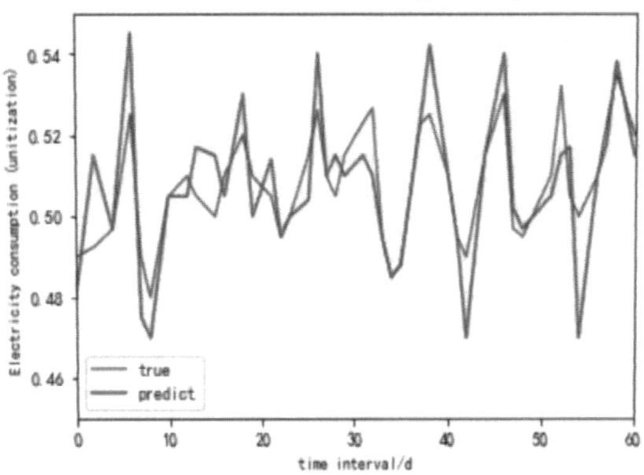

Figure 2. SVM model predicted value and actual value fit

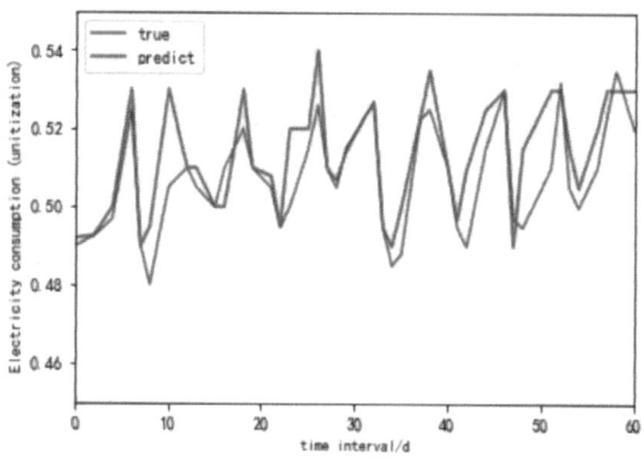

Figure 3. Comparison of predicted and actual values of RNN model

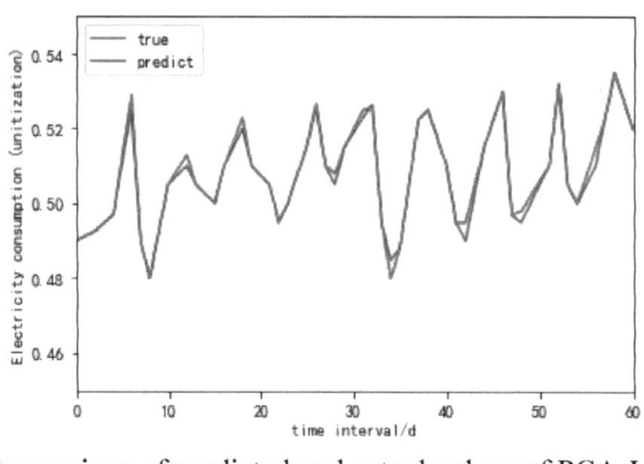

Figure 4. Comparison of predicted and actual values of PCA-LSTM model

In the above figure, we can visually see the prediction effects of different models. From the comparison between the predictions of the various time periods and the real values, we can see that the proposed model can make a good prediction for the user's electricity consumption time series data. The result also shows that the prediction method of this paper is effective, and the user short-term time series data can be well fitted.

Table 1. Comparison of experimental results between benchmark model and PCA-LSTM

Model	RMSE
HA	1.73
ARIMA	1.55
SVM	0.97
BP	0.72
RNN	0.64
PCA-LSTM	0.48

As shown in Table 1 above, the prediction effects of the deep neural network based PAC-LSTM model and the other five benchmark models used in this paper are given. By comparing the results of RMSE, the following conclusions can be drawn:

(1) Compared with other methods, the PCA-LSTM model based on deep neural network used in this paper achieves the best prediction effect and the best performance, so the model can better fit the user's electricity consumption time series data. ;

(2) The effect of the HA method is much lower than other prediction models, which is the worst method, because the method only uses the historical data average to predict the future, ignoring the characteristics of many data itself, making the error of the result larger;

(3) The prediction effect of the neural network model is better than the regression model as a whole, which also shows that the model used in this paper has a good prediction effect;

Through the comparative analysis of the above results, it is verified that the model proposed in this paper has high prediction accuracy and good prediction effect, and can better fit the time series data and power consumption behavior of user electricity consumption.

5. Conclusion

For the long-term and short-term memory network (LSTM) model, the characteristics of long-term and short-term dependence information of learning time series data and the excessive data dimension lead to the decline of neural network convergence speed. This paper proposes a PCA-LSTM model based on deep neural network for users. The short-term time series data of power consumption is used for prediction, and the prediction effect obtained by the model is simply compared with other models to highlight the feasibility of the model. By analyzing the problems of traditional neural network RNNs, and using the variant model LSTM based on the PCA dimension reduction, the required prediction effect is achieved. The experimental results show that the proposed prediction model not only improves the accuracy of time series data prediction, but also improves the convergence speed of LSTM network, and verifies the validity of the model.

In the future research, we will continue to explore the parameter selection problem of the network model, so how to conduct the adjustment is also the next research work.It is also necessary to consider expanding the data set and applying different data sets for training tests, thereby further improving the generalization ability of the model.

Acknowledgments

This work was financially supported by National Key Research and Development Plan 2017YFB400903.

References

[1] Wu J Z.The Driving and Current Situation of European Integrated Energy System Development[J].Automation of Electric Power Systems,2016,40(5):1-7.

[2] Luo J Z,Jin J H,Song A B,et al. Cloud computing: architecture and key technologies[J].Journal on Communications,2011,32(7):3-21.
[3] Weigend A S.Time series prediction: forecasting the future and understanding the past[M].Routledge,2018.
[4] BOX G E, P, JENKINS G M,REINSEL G C,et al.Time series analysis:Forecasting and control [M].New York:John Wiley & Sons,2015.
[5] Xiong Z B.Research on GDP Time Series Prediction Based on ARIMA and Neural Network Integration[J].Journal of Mathematical Statistics and Management,2011,30(2): 306-314.
[6] Langkvist M,Karlsson L,Loutfi A.A review of unsuper-vised feature learning and deep learning for time series modeling[J].Pattern Recognition Letters, 2014,42: 11-24.
[7] ZHANG Fan, DEB C,LEE S E, et al.Time series forecasting for building energy consumption using weighted Support Vector Regression with differential evolution optimization technique [J]. Energy and Buildings,2016,126:94-103.
[8] WONG F S .Time Series forecasting using backpropagation neural networks[J]. Neurocomputing,1991,2(4):147-159.
[9] WANG Z,SRINIVASAN R S.A review of artificial intelligence based building energy use prediction:contrasting the capabilities of single and ensemble prediction models[J].Renewable mand Sustainable Energy Reviews, 2017,75:796-808.
[10] Cheng X Q, Yan X L, Wang Y Z, et al.Overview of Big Data Systems and Analysis Techniques[J].Journal of Software,2014,25(9):1889-1908.
[11] CONNOR J,ATLAS L.Recurrent neural networks and time series prediction[C] //IJCNN-91-Seattle International Joint Conference.on Neural Networks. Seattle,WA,USA:IEEE,1991:301-306.
[12] HOCHREITER S, SCHMIDHUBER J. Long short-term memory [J].Neural computation, 1997,9(8):1735-1780.
[13] Lai G,Chang W C,Yang Y,et al. Modeling long-and short-term temporal patterns with deep neural networks[C]//The 41st International ACM SIGIR Conference on Research & Development in Information Retrieval.ACM,2018:95 — 104.
[14] YANG Kairui,MENG Fanrong,LIANG Zhizhen. Adaptively weighted PCA algorithm [J].Computer Engineering and Applications,2012,48(3):189-191.
[15] Hochreiter S,Bengio Y,Frasconi P,et al.Gradient flow in recurrent nets the difficulty of learning long-term dependencies[C].A Field Guide to Dynamical Recurrent Networks,Wiley-IEEE Press,2001:237-243.

Body-Boundary-Refined Human Parsing Network based on Fully Convolutional Network and Conditional Random Fields

Hanlin Zhao[a] and Xin Ma[b]

School of Control Science and Engineering, Shandong University, Jinan, China

[a]zhaohanlin@mail.sdu.edu.cn; [b]maxin@sdu.edu.cn

Abstract. Human parsing is still a big challenge in computer vision task to accurately label every body part in an image. Current methods of sematic segmentation mainly focus on dividing each independent part and ignore the structural priors of human body. However, the position relationships of different parts, especially which around the area of boundary, indicate their associations and also contribute to the semantic segmentation. In our work, we propose a body-boundary-refined part to refine the segmentation result of human part edge by simply utilizing the structure priors around the body boundary. It puts a penalty mechanism on wrong marginal pixels to improve segmentation performance around the area of body boundary. The network achieves competitive performance on the PASCAL-Parts-dataset and especially the area around the body boundary has been refined.

1. Introduction

Human parsing plays an increasingly significant role in many fields of computer vision like human behaviour analysis and human robot handovers. However, the semantic segmentation of human body part is still a challenging task, for every parts of human should be distributed a label with specific and pre-defined semantic information such as head and arms.

Fortunately, human parsing could be regarded as a problem of semantic segmentation. The majority of traditional approaches [1-9] based on the fully convolutional network (FCN) [1] mainly focus on addressing the problem of losing many useful spatial information caused by the operation of pooling or large convolutional stride in the classification network [10-12]. Some networks are favour of using encoder-decoder [3] [5] [8] structure to obtain efficient details from encoder network. And other researches are interested in the area of exploiting the context information [2] [4] [6] [7]. We expect our network could adjust more natural scene with the person images extracted from PASCAL Parts dataset where having more than one multi-scale persons. Therefore, we choose one of the most advanced FCN models called DeepLabv3 [2] as our baseline network with believe that human parsing still can be seen a semantic segmentation aimed to classify each pixel into correct label. However, there exist some differences between semantic segmentation and human parsing. For general semantic segmentation, a several of objects will be distributed everywhere in an image and we can't find obvious interrelationship between different categories. Human body parts have a closely positional relationship, which is driven us to exploit the connection in the pixel level.

Driven by the observation we select the conditional random field (CRF) to process the interaction between each pixel. However, the operation of conditional random field is always used for a post-processing [4] [9] structure followed by a fully convolutional network in most segmentation module because of the difficulties in the CRF training stage. Fortunately, a CRF-RNN structure that combines the strengths of convolutional neural network and conditional random field to formulate a deep network which can be trained end-to-end is introduced in this paper [13]. Although it is difficult to balance the training speed and the internal CRF parameters, we argue that a structural prior of human body such as CRF is beneficial for our task. Therefore, CRF-RNN is added in the human parsing network.

But, only structure of CRF not enough obtain all potential position relationship at least around the body's edge. For example, the potential correlation such that surrounding pixels' categories of head marginal pixels are likely to torso pixels or background pixels or head pixels. If prediction results of surrounding marginal pixels don't align with the structural of the human body, some punishment will be given to these wrong boundary pixels. To exploit these prior's relationships, we propose a structure named body-boundary-refined part to punish the wrong marginal pixels in the final loss function to refine segmentation result.

2. Related Work

2.1. Semantic segmentation

Semantic segmentation aims to label different objects with relative sematic labels in pixel level. Recently, FCN [1] has become the most successful basement model in semantic segmentation [2-7] [9] [14]. It replaces the fully connected layer with convolutional layer to reserve the spatial information of image to address pixel-level dense prediction and adopt skip connection to solve the problem of losing spatial information. A universal structure called encoder-decoder module [3] [7] [8] has been widely used and other successful networks [2] [6] think that it is useful to exploit multi-scale contextual information in semantic segmentation. Zhao et al. [6] propose a pyramid pooling module that is applied to harvest different sub-region representations to exploit global and local context information.

2.2. Human parsing

Recently, variant of FCN also has been successfully used in human parsing. Human Part Discovery [15] designs a like-FCN structure to obtain different human parts labels in two granularity levels. However, it misses out the structural priors of the human body. Look into Person (LIP) [16] not only contributes significantly to make a new benchmark, but also adopts a self-supervised structure-sensitive learning to give a self-supervise for the net. Qi et al. [17] performed a holistic and instance-level human parsing using conditional random field to allocate every part into every object with given detections and category-level segmentation module. HAZN [18] designs an efficient strategy of first zooming into objects and then zooming into parts to get a great performance. Other works like Chen et al. [19] introduce an attention strategy which allocates different weights according to different scales of objects into network to improve the performance in the small objects. In our work, the proposed body-boundary-refined part can utilize the structural of the human body to refine the body boundary in some extent.

2.3. Conditional Random Field

Currently, many works combine fully convolutional network with conditional random fields to improve the performance of semantic segmentation [4] [17] [20] via incorporating surrounding information with pairwise energy. More researches [13] [20] attempt to add CRFs into fully convolutional network to train end-to-end. Thanks to the great contribute of densecrf [21], it's more convenient to integrate fully connected CRFs with deep convolutional neural network in semantic segmentation via Mean-field approximate inference and efficient message passing. In the human parsing, the human body has many potential structural information could be used, such features could well solve by CRFs, so we incorporate the densecrf with deeplab to refine the prediction result.

3. Proposed Approach

Body-boundary-refined network consists of three components as shown in Figure 1. A variant fully convolutional network named DeeplabV3 [2] and a fully connected CRFs model named CRFasRNN [13] and our proposed model named body-boundary-refined part to refine the boundary of human body.

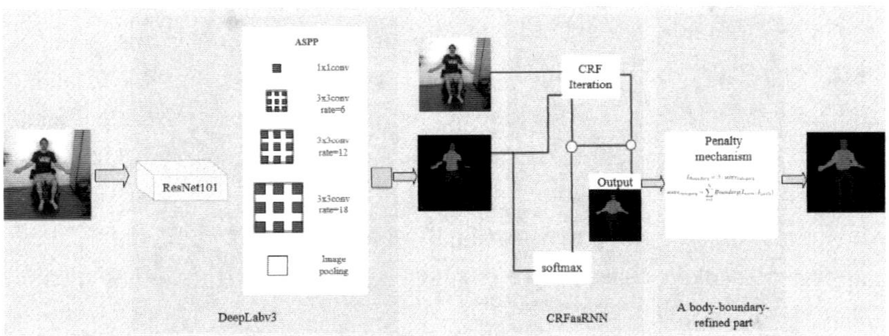

Figure 1. Body-boundary-refined network.

3.1. Deeplabv3 Network
For the task of semantic segmentation, there are two challenges in exploiting convolutional neural network. One problem is reduced feature resolution and another is multiple scales of objects in the image. We choose Deeplabv3 as baseline network to solve the case of having multi-scale persons.

3.2. Fully Connected Conditional Random Fields
Conditional random field, which can establish relationships between mutual independence pixels, is a probability graph model with Markov property. In our work, five labels such as head, torso, arms, legs, and background should be given in an image. Although this task is similar with semantic segmentation, the special structural relationship is unique in the human segmentation. Thus, human body image should contain so many useful priors that we can't ignore. Therefore, CRFs is followed by the deeplabv3 to represent structural priors of human body.

3.3. Body-Boundary-Refined Part
When we only use deeplabv3 to segment human sematic body part, some terrible results appear that these connected pixels between head part and torso part have disappeared. Therefore, we realise that it is significant to encode the structural priors of human body part especially around boundary into network. Motivated by this intuition, we design a body-boundary-refined part which includes penalty weight(β) and category score ($score_{category}$, as in equation (2)) to improve the importance of human body boundary. We put the $L_{boundary}$ (as in equation (1)) as a small part of total loss to punish the terrible boundary results of prediction. In addition, in order to conveniently compute the punishment, we have fixed the penalty weight β as 0.01 in our work.

$$L_{boundary} = \beta \bullet score_{category} \tag{1}$$

$$score_{category} = \sum_{i=1}^{N} Boundary\left(I_{norm}, I_{(pre)i}\right) \tag{2}$$

The category score (as in equation (2)) represents that how much punishment network would put into the total loss. Fortunately, these four categories have some fixed position relationship especially around the human body boundary. For example, if boundary pixels of head have been predicted by network, obviously we can infer categories of surrounding pixels. In another words, the eight neighborhood of boundary pixels in prediction represent the degree of good or bad about structural information. That is, for example, if there is a boundary pixel connected with head and torso, then head pixels and torso pixels will be found in the eight neighborhoods of this pixel, thus the eight neighborhood of boundary pixels in prediction will represent the structural information about human body. Therefore, some norms (I_{norm}, as in equation (3)) are defined to present the structural priors of human body. Let $L=\{l_0,l_1,l_2,l_3,l_4\}$ be the labels of background, head, torso, arms, and legs and $X=\{x_1,x_2,...x_N\}$ be the boundary pixels. Let $H\in\{\eta_0,\eta_1,\eta_2,\eta_3,\eta_4,\eta_5,\eta_6,\eta_7,\eta_8\}$ be the categories of every pixels in the eight neighborhood and η_0 be the boundary pixels in the central position of eight neighborhood. Where $P\{\eta\in\{l_0,l_1,l_2\}\mid X=l_1\}$ means the probability of appearing background or head or torso pixels is larger than other situations around the head marginal pixels. Similarly, $P\{\eta\in\{l_0,l_2,l_3\}\mid X=l_3\}$ means the probability of appearing background or torso or arms pixels is larger than other situations around the arms marginal pixels and $P\{\eta\in\{l_0,l_2,l_4\}\mid X=l_4\}$ means the probability of appearing background or torso or legs pixels is larger than other situations around the legs marginal pixels.

$$I_{norm}=\begin{cases}P\{\eta\in\{l_0,l_1,l_2\}|X=l_1\}\\ P\{\eta\in\{l_0,l_2,l_3\}|X=l_3\}\,\eta\in H\\ P\{\eta\in\{l_0,l_2,l_4\}|X=l_4\}\end{cases} \tag{3}$$

After acquiring the pixels of boundary in the prediction results, the norm is performed to decide whether giving punishment to the total loss. As in equation (4), where $Boundary\left(I_{norm},I_{(pre)i}\right)$ means that if the boundary pixel in the prediction result is satisfied with the norm, then zero punishment will give to the loss. However, the punishment should put to the loss when the prediction result breaks the norms.

$$Boundary\left(I_{norm},I_{(pre)i}\right)=\begin{cases}0, I_{norm}=I_{(pre)i}\\ 1, I_{norm}\neq I_{(pre)i}\end{cases} \tag{4}$$

4. Experiments

We evaluate the performance of our network on the PASCAL Parts dataset and some images from Freiburg Sitting People dataset [15] to visualize the segmentation effect of our network.

4.1. Experimental Settings

PASCAL Parts dataset: In our experiment, we extract 3589 images which contain human body by key annotation word of person from original PASCAL Parts dataset and fuse 24 original human parts classes into 4 classes to evaluate our network. Then, we randomly divide dataset into 80% training and 20% testing.

Metrics: In our experiment, pixel accuracy and intersection over union are used to evaluate performance of our network. The intersection over union (IOU) is also used to balance quantitative gap from background and foreground pixels. Confusion matrix is used to compute the mean IOU and IOU of every classes.

Training: We use the pre-trained model from Resnet101 and the scale of input images is fixed as 300x300 for training. Two types of augmentations to our training data is mirroring and cropping. Polynomial decay is adopted in the setting of learning rate. The initial learning rate is 7e-3 and the end learning rate is 1e-4. After training Deeplabv3 for 80 epochs we add the CRF and body-boundary-refined part into our end-to-end network by the trained deeplabv3 to increase the training speed.

4.2. Results and Comparisons

In our task, only four labels are divided into human body parts and there exist few same works with us. Therefore, we compare FCN-8s [1] and Deeplabv3 [2] with our network in the same experiment environment to show the improvement of our network. Table 1 show the mean intersection over union (IOU) and pixel accuracy of FCN-8s, Deeplabv3 and Deeplabv3+CRF in the person parts dataset extracted from PASCAL Parts dataset by us with four parts labels and background label. We first train FCN-8s [1] with VGG16 [10] extracting feature network in the same dataset getting bad mean IOU result. Then when using Deeplabv3 [2] with ResNet101 [11] extracting feature network, we obtain good mean IOU result. However, many details are lost especially in the small area like hand and foot. According to the third column results of Figure 2, human body parts are refined by adding CRF into Deeplabv3 and the mean-IOU is improved about 1.5%. Compared with previous model, although our network obtains a little improvement in the mean-IOU ranging from 58.72% to 60.64%, the segmenting effect on the body boundary like the area of arms or head or legs is greatly improved as the fourth column shown, especially in the joints of the two parts. The refined inference image and a little improvement in mean IOU have confirmed the effectiveness of our body-boundary-refined part.

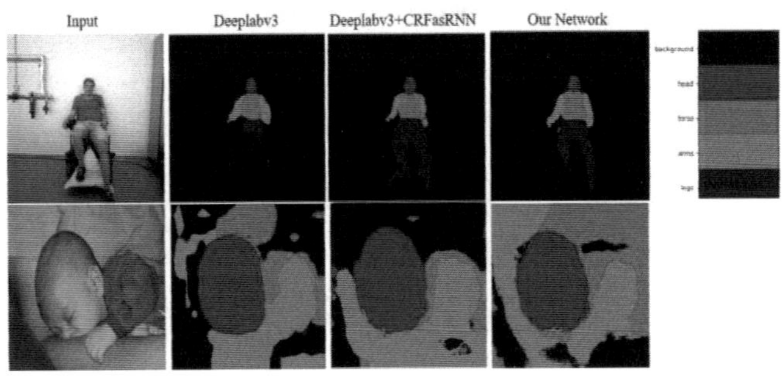

Figure 2. Qualitative results on the PASCAL dataset and the Freiburg sitting people dataset.

Table 1. Performance comparison in terms of mean IOU and pixel accuracy on the validations.

Method	Head	Torso	Arms	Legs	Bkg	mIOU	PA
FCN-8s [1]	41.31	17.98	3.61	0.02	87.87	30.16	86.92
Deeplabv3 [2]	71.21	54.65	37.98	32.90	89.30	57.21	88.17
Deeplabv3+CRFas RNN	75.54	54.26	42.99	30.96	89.89	58.72	89.18
Our network	**75.64**	**58.55**	**46.42**	**32.19**	**90.40**	**60.64**	**89.54**

5. Conclusion

In this paper, we present a novel body-boundary-refined part based on semantic segmentation and fully connected CRF to improve the performance of human parsing. Our approach applies Deeplabv3 as our baseline network and fuses denseCRFs into Deeplabv3. To exploit more structural priors of human body part particularly in the area of boundary, we introduce a new body-boundary-refined part for human parsing task, including 3,589 images extracted from PASCAL Parts dataset with pixel-wise annotations on four semantic part labels. Experimental results on the dataset demonstrate that our proposed approach refine boundary of human, and achieves a good performance for the human parsing.

Acknowledgments

This work was supported in part by the Chinese National Key Research and Development Plan (2018YFB1305803), Chinese National Natural Science Foundation (61673245), Chinese National Programs for High Technology Research and Development (2015AA042307).

References

[1] Long, Jonathan and Shelhamer, Evan and Darrell, Trevor. Fully convolutional networks for semantic segmentation. Proceedings of the IEEE conference on computer vision and pattern recognition, 2015:3431--3440.

[2] Chen, L. C., Papandreou, G., Schroff, F., Adam, H. Rethinking atrous convolution for semantic image segmentation. arXiv preprint arXiv:1706.05587, 2017.
[3] Badrinarayanan V, Kendall A, Cipolla R. Segnet: A deep convolutional encoder-decoder architecture for image segmentation [J]. IEEE transactions on pattern analysis and machine intelligence, 2017, 39(12): 2481-2495.
[4] Chen, L. C., Papandreou, G., Kokkinos, I., Murphy, K., Yuille, A. L. Deeplab: Semantic image segmentation with deep convolutional nets, atrous convolution, and fully connected crfs. IEEE transactions on pattern analysis and machine intelligence, 40(4): 834-848, 2018.
[5] Lin, G., Milan, A., Shen, C., Reid, I. D. RefineNet: Multi-path Refinement Networks for High-Resolution Semantic Segmentation. CVPR. Vol. 1. No. 2, 2017.
[6] Zhao, Hengshuang and Shi, Jianping and Qi, Xiaojuan and Wang, Xiaogang and Jia, Jiaya. Pyramid scene parsing network. IEEE Conf. on Computer Vision and Pattern Recognition (CVPR), 2017:2881--2890.
[7] Chen L C, Zhu Y, Papandreou G, et al. Encoder-decoder with atrous separable convolution for semantic image segmentation[C]. Proceedings of the European Conference on Computer Vision (ECCV). 2018: 801-818.
[8] Ronneberger, Olaf and Fischer, Philipp and Brox, Thomas. U-net: Convolutional networks for biomedical image segmentation. International Conference on Medical image computing and computer-assisted intervention, 2015:234--241.
[9] Chen L C, Papandreou G, Kokkinos I, et al. Semantic image segmentation with deep convolutional nets and fully connected crfs[J]. arXiv preprint arXiv:1412.7062, 2014.
[10] K. Simonyan and A. Zisserman, Very deep convolutional networks for large-scale image recognition. International Conference on Learning Representations, 2015.
[11] He, Kaiming and Zhang, Xiangyu and Ren, Shaoqing and Sun, Jian. Deep residual learning for image recognition. Proceedings of the IEEE conference on computer vision and pattern recognition, 2016:770--778.
[12] Szegedy C, Liu W, Jia Y, et al. Going deeper with convolutions[C]. Proceedings of the IEEE conference on computer vision and pattern recognition. 2015: 1-9.
[13] Zheng S, Jayasumana S, Romera-Paredes B, et al. Conditional random fields as recurrent neural networks[C]. Proceedings of the IEEE international conference on computer vision. 2015: 1529-1537.
[14] Yu, Fisher, and Vladlen Koltun. Multi-scale context aggregation by dilated convolutions. arXiv preprint arXiv:1511.07122, 2015.
[15] Oliveira, G. L., Valada, A., Bollen, C., Burgard, W., Brox, T. Deep learning for human part discovery in images. IEEE International Conference, 2016:1634-1641.
[16] Gong, Ke and Liang, Xiaodan and Zhang, Dongyu and Shen, Xiaohui and Lin, Liang. Look into Person: Self-Supervised Structure-Sensitive Learning and a New Benchmark for Human Parsing. CVPR, 2017.
[17] Li, Qizhu and Arnab, Anurag and Torr, Philip HS. Holistic, Instance-level Human Parsing. arXiv preprint arXiv:1709.03612, 2017.
[18] Xia, Fangting and Wang, Peng and Chen, Liang-Chieh and Yuille, Alan L. Zoom better to see clearer: Human and object parsing with hierarchical auto-zoom net. European Conference on Computer Vision, 2016:648--663.
[19] Chen, Liang-Chieh and Yang, Yi and Wang, Jiang and Xu, Wei and Yuille, Alan L. Attention to scale: Scale-aware semantic image segmentation. Proceedings of the IEEE conference on computer vision and pattern recognition, 2016:3640--3649.
[20] Teichmann, Marvin TT and Cipolla, Roberto. Convolutional CRFs for Semantic Segmentation. arXiv preprint arXiv:1805.04777, 2018.
[21] Krähenbühl, Philipp and Koltun, Vladlen. Efficient inference in fully connected crfs with gaussian edge potentials. Advances in neural information processing systems, 2011:109--117.

Artificial Intelligence Promotes the Evolution of English Writing Evaluation Model

Zheqian Su[1], Liu Miao[1] and Jiang Man[2,*]

[1] Changchun University of Chinese Medicine Foreign Language Department, Changchun, 130117

[2] Changchun University of Chinese Medicine Planning and Finance Division, Changchun, 130117

[*] 18146034@qq.com

Abstract: On the background of "Internet +" era, the domestic higher education is showing the trend of artificial intelligence. The reliability and scientificity of computer intelligent evaluation are further carried out, and the mode of intelligent evaluation and data analysis in optimizing the precise teaching of English writing is explored, which can lay a foundation for the large-scale use of the technology. Based on data-driven theory, this study further analyzed the role of AI in promoting in-depth learning by comparing AI writing review model with manual review model.

Key words: Artificial Intelligence; English Writing; Evaluation Model

1. Introduction

The research on second language learning tools has a long history, from the initial phonograph English phonetics learning to radio news, from the promotion of cassette recorders in traditional English classrooms to situational writing teaching in bilingual laboratories[1], from the application of multimedia technology in English writing to AI-assisted reading in the context of mobile-interconnected era,the ever-changing information technology plays a revolutionary role in the process of second language acquisition, which also makes the mode of second language writing more and more intellectualized and humanized.

2. Research Background

2.1. Concepts of "Artificial Intelligence"

Artificial Intelligence (AI) is an interdisciplinary subject that simulates human intelligence. It mainly simulates and reproduces human thinking process[2]. The term "artificial intelligence" was first introduced at the Dartmouth Society in 1956. Four scholars, John McCarthy, Marvin Minsky, Allen Newell and Herbert Simon, then defined "Artificial Intelligence" as a discipline. Its main task is to establish the theory of intelligent information processing, and then design some computing systems that can show some approximate human intelligent behavior.

2.2. Application of Artificial Intelligence in English Writing Review

For more than half a century, Artificial Intelligence (AI) has been widely used in all parts of the world and in various fields, especially in second language intelligent writing and evaluation. The application of AI in the field of writing can be divided into four stages: one is to make courseware in high-level language; the other is to make courseware with writing system as work; the third is to apply multimedia network information technology in a large area; and the fourth is to combine expert system with AI program application [3].

The application of AI in the field of writing review has also experienced the following three stages: the first stage focuses on the development and application of the primary intelligent evaluation system. In the 1960s, Ellis Page and others in the United States developed the world's first automatic Writing Evaluation (AWE) system, Page Essay Grade (PEG), which only provided the score of the composition without any other form of feedback. The second stage focuses on the intermediate intelligent feedback. In the 1990s, on the basis of the AWE system, there appeared a feedback evaluation system featuring scanning, evaluation, diagnosis and scoring. However, due to the lack of computer corpus and other factors, there were many problems in the surface of feedback information [4]. The third stage is the advanced intelligent network evaluation system. At present, the Automated Essay Grading (AEG) is widely used in the world. The system plays an important role in the evaluation of TOEFL examinations as well as the manual evaluation. In addition, Writing Roadmap developed by McLaugher, Peer Mark system launched by Turnitin, Calibrated Peer Review system of the University of California, Peerceptiv system developed by the University of Pittsburgh and Peer Scholar system developed by the University of Toronto in Canada have been widely used in the process of English writing review[5]. This kind of review system is based on big data technology, which can comprehensively evaluate the author's linguistic skills, and at the same time examine the author's cognitive ability, and provide real-time feedback [6].

In China, the development and application of AI in the second language writing evaluation system started relatively late, until the end of 2013, China officially ushered in the era of 4G network. According to "the China Internet Education Trend Report 2013-2014", the use rate of mobile learning among the major groups was more than 80%. By the end of 2014, the total number of educational APPs in China had exceeded 70,000. With the opening of mobile-interconnected 4G, the application of artificial intelligence in second language learning has been gradually promoted. Unlike the traditional manual marking, the real-time marking and feedback system for compositions embedded in mobile phones has been recognized by teachers'universities. Take "Correcting Network (Sentence Cool)" as an example, it is based on corpus and cloud computing and trains scoring engine. It automatically corrects students' compositions according to the scoring standards of CET-4 ,CET-6, National Matriculation Test and High school Entrance Examination, gives the scoring and gives the scoring in grammar and vocabulary. Feedback hints should be given where the expression is not standardized and suggestions for students'revision should be given.

3. Review of Domestic and Foreign Research

3.1. Research Trends at Home and Abroad

By searching the keywords of "Artificial Intelligence" and "Writing" on CNKI, we found that 285 articles were collected in the whole network, especially the most concentrated ones published after 2013, showing an upward trend as follows: 2013 (3 articles), 2014 (7 articles), 2015 (22 articles), 2016 (34 articles), 2017 (71 articles), 2018 (102 articles). A deep search of the subject words of "Artificial Intelligence" and "English Writing" reveals that a total of 12 articles are included in CNKI.

Through searching the Scopus English academic literature database for topics, abstracts and keywords "AI" and "writing", 50 English academic documents were collected from 2000 to December 2018. The overall research shows an upward trend, including 2013 (1 article), 2014 (6 articles), 2015 (2 articles), 2016 (3 articles), 2017 (7 articles), 2018 (12 articles). Through visual subject analysis, it is found that "Artificial Intelligence News Writing" or "Robot News Writing" occupies the mainstream, from which we can see that the news media have begun to lay out "Artificial Intelligence Robot" instead of "Artificial" in the industry for news reporting.

From the perspective of current research at home and abroad, the research on the application of artificial intelligence (AI) in second language writing mainly focuses on the following three directions: the relationship between AI and the psychological dynamics of second language writers, the relationship between AI and the achievement of second language writing, and the relationship between AI and the teaching of second language writing.

3.2. A Study on the Relationship between AI and the Psychological Dynamics of Second Language Authors

Some domestic scholars have done some research on teaching practice by taking the "Correcting Network" as an example. It is found that AI system can automatically improve information updating and pushing according to users' usage characteristics[7]. At the same time, based on large data theory and corpus construction, this kind of intelligent marking system can complete multi-dimensional quality feedback of vocabulary, grammar and content for L2 writers' individualized writing, thereby eliminating L2 writers' quality feedback. The author's anxiety and conflict in the process of writing (Tang Jinlan et al., 2012; Wang Ying et al., 2012).

Hua Lulu, Chen Lin and Sun Mengmeng(2017) believe that AI system can provide individualized and situational learning resources for L2 idiom writers, avoid the uniformity of feedback information in the process of reviewing, and play an effective role in stimulating the author's learning motivation in the process of L2 writing [8]. Other domestic scholars believe that AI assessment system can stimulate students' potential interest in English writing teaching and promote second language writers to shift from shallow learning to deep learning [9]. At the same time, AI system based on cloud computing and speech recognition technology can also monitor the changes of the ability and mood of second language writers, and provide teachers with first-hand learning analysis data by generating relevant data and graphs [10].

Some foreign scholars have deconstructed AI review system from the perspective of cognitive psychology. AI relies on big data to make the traditional static review process become a dynamic information intensive feedback process. Through monitoring the author's writing ability and emotional changes, the diachronic data tracking is completed[11]. Schunn et al. (2016) pointed out that the current North American Writing Peer Assessment Computer Aided System (CAIS) can provide real-time feedback on the writing motivation and quality of L2 writers, and complete diachronic information interaction and follow-up survey by combining background database.

Some scholars have found that the current AI review system is not perfect. They believe that some online English composition review websites only provide vocabulary and grammar improvement suggestions in information feedback, but do not provide relevant examples. To some extent, this will cause secondary anxiety for second language writers (Shi Xiaoling, 2012; Jiang Yan, Ma Wulin, 2013; Zhang Li, 20 13).

3.3. The Relationship between AI and Second Language Writing Achievement

At present, AI rating system is different from traditional manual scoring and computer closed scoring. It evaluates and analyses students' second language writing in an open form. The effective application of AI rating system is positively correlated with writing achievement and writing motivation[12]. Huang Jing & He Huaqing (2018) found that AI intelligent feedback had a positive effect on students'English writing quality, especially in vocabulary[13].

3.4. The Relationship between AI and Second Language Writing Teaching

Scholars at home and abroad generally believe that traditional second language writing is affected by the mother tongue environment to a certain extent, which is often neglected by manual review. AI evaluation system can evaluate writing ability according to the social, linguistic and cultural environment in which the author lives[14-15], which makes the teaching of second language writing more humanized and provides a more effective communication bridge between teachers and students[16]. Dikli & Bleyle (2014) conducted a study of AI reviews and found that L2 writers have a positive attitude towards the feedback provided by AI system. According to Siddhartha Ghosh (2019), AI assessment systems such as AEG can correct the target language deviation based on the cognitive characteristics of second language learners, avoid the influence of mother tongue on second language writing and provide more authentic corpus usage.

Some domestic scholars have compared the differences between traditional English composition marking and AI marking: traditional manual marking occupies a large amount of teachers' personal time, and can not make personalized learning plans according to individual problems in the process of marking; AI marking system makes teachers more effective in the teaching process, and teachers can target teaching according to the data summarized by the intelligent system. Learning intervention to achieve the goal of accurate teaching [17]. Through empirical research, Wu Yong and Zhang Wenxia (2016) found that L2 learners are more inclined to adopt teacher feedback. The feedback provided by AI is only for reference. Teacher feedback combined with AI feedback can help stimulate learners' autonomous learning consciousness.

From the point of view of domestic and foreign research, the research on the application of AI evaluation system in writing only focuses on the two methods of questionnaire survey and interview, few studies involve text analysis (Huang Jing, Zhang Wenxia, 2014; Wu Yong, Zhang Wenxia, 2016).

4. An Empirical Study of AI Evaluation System in English Writing

Second language writing is a complex psychological cognitive process, which is influenced by many factors. For second language idiom writers, writing anxiety is a negative emotional experience in the process of second language writing [18]. It has a significant negative correlation with writing ability and writing achievement[19]. Nowadays, based on big data and cloud computing, AI Intelligent Writing Review System can control the second language writers' learning style, cognitive tendency, learning strategies, motivation types and humanistic accomplishment. It not only reduces the writing anxiety of second language learners to a certain extent, but also stimulates the writers' writing interest and motivation [20]. In order to verify the effectiveness of AI assessment system in English writing, this study, based on data-driven theory, adopts the method of stratified cluster random sampling to select 2017 undergraduates (N=200) from a Medical University in China. In the process of College English writing, the experimental group adopted the marking system of AI, while the control group adopted the traditional form of teacher's manual marking. The study lasted for one semester and 10 writing exercises were conducted. There was no difference between the experimental group and the control group in terms of writing topics, writing time and writing requirements. The full score of the composition was 15 points. The evaluation process followed the unified scoring criteria of CET-4. In this study, FLCAS and the self-designed College English Writing Competence Scale were used to measure the two scales. The Cronbach coefficient of the former is 0.87 and the Cronbach coefficient of the latter is 0.79, which shows that the two scales have good internal consistency. The scoring criteria were Likert 5-point scale, and the data were analyzed by SPSS 22.0.

Table 1: Independent sample t-test of English writing achievement, writing anxiety and writing ability between the experimental group and the control group in the post-test

	Rank	Number	Mean value	Standard Deviation	t	Sig.
English Writing Achievements	Experimental Group	100	10.56	3.82	5.49**	0.031
	Control Group	100	8.94	2.77		
English Writing Anxiety	Experimental Group	100	3.06	0.21	7.17***	0.047
	Control Group	100	4.12	0.36		
English Writing Ability	Experimental Group	100	4.22	0.31	4.96**	0.019
	Control Group	100	3.54	0.46		

P<0.05;**P<0.01;P<0.001**

Before the beginning of the experiment, two groups of subjects were tested on their writing performance, writing ability and writing anxiety. It was found that there was no significant difference between the two groups (p>0.05). After a semester's teaching experiment, independent sample t-test showed as in table 1 that the English writing ability and writing achievement of the experimental group were significantly higher than those of the control group (p<0.01), while the writing anxiety of the experimental group was significantly lower than that of the control group (p<0.001).

Through paired sample t-test (table 2), the writing performance of the experimental group in the post-test was significantly higher than that in the pre-test (p< 0.01), and the writing ability of the experimental group was also significantly higher (p<0.01). Growth (p<0.05), the writing anxiety of the experimental group in the post-test was significantly lower than that of the pre-test (p<0.01).

Table 2: T-test of paired samples of English writing achievement, writing anxiety and writing ability between the experimental group and the control group in the pre-and post-test

	Rank	Mean value	Standard Deviation	t	Sig.
English Writing Achievements	**Experimental Group**				
	pretesting	8.24	3.09	5.77**	0.000
	post-test	10.56	3.82		
	Control Group				
	pretesting	8.13	2.61	1.76	0.036
	post-test	8.94	2.77		
English Writing Anxiety	**Experimental Group**				
	pretesting	3.45	0.44	6.31**	0.026
	post-test	3.06	0.21		
	Control Group				
	pretesting	3.42	0.41	1.91	0.269
	post-test	4.12	0.36		
English Writing Ability	**Experimental Group**				
	pretesting	3.24	0.70	4.37**	0.017
	post-test	4.22	0.31		
	Control Group				
	pretesting	2.97	2.61		
	post-test	3.54	0.46	1.83	0.176

P<0.05;**P<0.01;P<0.001**

Pearson correlation analysis shows in table 3 that the effectiveness of the use of corrective web is positively correlated with English writing achievement and writing ability (p<0.01), negatively correlated with writing anxiety (p<0.01), and positively correlated with writing achievement (p< 0.05), writing ability (p< 0.05), and has negative correlation with writing anxiety (p<0.01); the sub-dimension of feedback adoption was significantly positively correlated with English writing achievement and writing ability (p<0.01); and the sub-dimension of convenience was only positively correlated with English writing achievement (p< 0.05).

Table 3: Pearson correlation analysis of the effectiveness of the use of marking network with English writing achievement, writing anxiety and writing ability

	Effectiveness of Net Usage	Duration	Feedback to Adopt	Convenience of Use
English Writing Achievements	0.937**	0.204*	0.547**	0.202*
English Writing Anxiety	-0.238**	-0.229**	-0.166	-0.089
English Writing Ability	0.890**	0.216*	0.460**	0.126

P<0.05;**P<0.01;P<0.001**

Therefore, it can be seen from the analysis that the effective use of the marking network promotes English writing achievement and writing ability, and also inhibits writing anxiety. The longer the marking network is used, the greater the improvement of students' English writing achievement and writing ability, and the lower their writing anxiety to a certain extent. The more they are adopted, the better their writing performance and writing ability will be. The convenience of using the marking network can also promote the improvement of writing performance to a certain extent.

5. Epilogue

With the further popularization of artificial intelligence in the field of education, College English writing teaching has also ushered in the "two-way liberation" of teachers and students: on the one hand, teachers release more time through AI evaluation system, which can better observe students' learning process and characteristics, formulate more detailed analysis of learning conditions and feedback of opinions; on the other hand, students can continuously improve through AI evaluation system. Self-writing ability, through machine review and feedback, finds the joy and confidence of writing in the process of repeated revision. Introducing artificial intelligence into college English writing teaching can not only deeply integrate teaching and learning, create a new mode of second language writing, but also stimulate the internal motivation of second language learners, fundamentally improve the quality of English teaching and effectively release the power of information technology.

ACKNOWLEDGMENTS

Our thanks to the projects supported by Research Project of the National Administration of Traditional Chinese Medicine Higher Education "Implementing University Study and Research": Research on the Theory and Practice of the Teaching Reform of "New Model of Apprenticeship System" in Colleges and Universities of Chinese Medicine (No. GZ-10);Scientific Research Project of Jilin Vocational and Technical Education Association in 2018: Study on the "New Model of Apprenticeship System" in Chinese Medicine Vocational Colleges in Jilin Province (No. 2018XHZ024); Project of Social Science Foundation of Jilin Province in 2019: Investigation on Cultural Identity of Traditional Chinese Medicine in Jilin Province and Study on Promotion Path（No.2019B89）; 2018 Changchun University of Chinese Medicine curriculum thinking education and teaching reform project.

References

[1] Hu Jiasheng, Jin Yan. Theoretical and Practical Research on the Integration of Educational Technology and Foreign Language Curriculum [J]. Audiovisual Education in China, 2015 (4): 114-120.

[2] Luckin R, Holmes W. Intelligence Unleashed: An argument for AI in Education [EB/OL].https://www.pearson.com/news/blogs/CompanyBlog/2016/03/intelligence-unleashed-an-argument-for-ai-in-education.html, 2017-07-20.

[3] He Gaoda, Luo Zhongmin. Application of Artificial Intelligence in Foreign Language Teaching [J]. Foreign Language Audiovisual Teaching, 2008 (5): 74-80.

[4] Stevenson, M., Phakiti, A. The Effects of Computer-generated Feedback on the Quality of Writing [J]. Assessing Writing, 2014 (19): 51-65.

[5] A study on the mutual evaluation system of Christian D. Schunn. Peerceptiv to assist students in English writing [J]. Foreign Language Audiovisual Teaching, 2016 (8): 38-44.

[6] Siddhartha Ghosh.Online automated essay grading system as a web-based learning (WBL) tool in engineering education [M].Web-based Engineering Education: Critical Design and Effective Tools, 2010:53-62.

[7] Yang Xiaoqiong, Dai Yunqian. Practical Research on the Teaching Model of College English Self-directed Writing Based on Correction Network [J]. Foreign Language Audiovisual Teaching, 2015 (2): 17-23.

[8] Hua Lulu, Chen Lin, Sun Mengmeng. Research on Artificial Intelligence Promoting English

Learning Change [J]. Modern Distance Education, 2017 (6): 27-31.

[9] Zhang Qi. A Review of the Research on E-Learning Environmental Deep Learning in China [J]. Modern Distance Education, 2014 (5): 60-67.

[10] Liu Qingtang, Mao Gang, Yang Lin, Cheng Yun. Development and Prospect of Intelligent Teaching Technology [J]. China Audiovisual Education, 2016 (6): 8-15.

[11] Wardle, E. & Roozen, K. Addressing the Complexity of Writing Development: Toward an Ecological Model of Assessment [J]. Assessing Writing, 2012 (2): 106-119.

[12] Yan Zhiming, Tang Xia Xia, Qin Xuan, Zhang Fei, Duan Yuanmei. The connotation, key technologies and application trend of educational artificial intelligence (EAI) -- Analysis of reports of "Preparing for the Future of Artificial Intelligence" and "National Strategic Plan for Artificial Intelligence Research and Development" [J]. Journal of Distance Education, 2017 (1): 26-35.

[13] Huang Jing, He Huaqing. Research on the influence of human-computer feedback on students'writing behavior [j]. Foreign Language Audiovisual Teaching, 2018 (12): 19-24.

[14] Behizadeh, N. & Engelhard Jr. G. Historical View of the Influences of Measurement and Writing Theories on the Practice of Writing Assessment in the United States [J]. Assessing Writing, 2011 (3): 189-211.

[15] Slomp, D. H. Challenges in Assessing the Development of Writing Ability: Theories, Constructs and Methods [J]. Assessing Writing, 2012 (2): 81-91.

[16] Deane, P. On the Relation between Automated Essay Scoring and Modern Views of the Writing Construct [J]. Assessing Writing, 2013 (1): 7-24.

[17] Gan Ronghui, He Gaoda. Learning Analysis and Foreign Language Teaching Research Prospects in the Big Data Era [J]. Foreign Language Audiovisual Teaching, 2016 (3): 40-45.

[18] Wu Yuhong, Gu Satellite. An Empirical Study on Cooperative Learning to Reduce English Writing Anxiety of Non-English Majors [J]. Foreign Language and Foreign Language Teaching, 2011(6):51-55.

[19] Guo Jidong. Effects of Writing Self-concept and Self-efficacy on College Students' English Writing Achievements: Mediating Role of Writing Anxiety [J]. Journal of Foreign Languages, 2018 (2): 69-74.

[20] Wang Zhe, Zhang Yue. Big Data Approach and Innovative Research in Foreign Language Teaching: A Case Study of Millions of English Writing with the Same Topic in 2015 [J]. Foreign Language Audiovisual Teaching, 2015 (5): 3-8.

An Improved Surrogate Based Optimization Method for Expensive Black-box Problems

Pengcheng Ye[1, 2 *] **and Guang Pan**[1, 2]

[1]School of Marine Science and Technology, Northwestern Polytechnical University, Xi'an 710072, China
[2]Key Laboratory for Unmanned Underwater Vehicle, Northwestern Polytechnical University, Xi'an, Shaanxi 710072, China

[*]yepengcheng@nwpu.edu.cn

Abstract. For expensive black-box problems, surrogate modelling techniques are generally used to decrease the computational source. In this study, an improved surrogate based optimization (SBO) method is presented to solve the real-world engineering applications with expensive black-box objective responses. An optimized ensemble of surrogates combing three typical surrogate modelling techniques is adapted to efficiently predict the objective response. Meanwhile, the hierarchical design space reduction (HSR) strategy is employed for obtaining the smaller design subspace for improving the optimization efficiency. During the search, all test problems are considered as the real-world engineering applications whereas the actual global optima as well as the function characteristics are unknown in advance. The results show that the proposed method is superior in identifying the global optimum.

1. Introduction

The rapid progress of the engineering applications like underwater vehicles has improved the significance for optimal design of the related industrial products. Nevertheless, the classical optimization methods like ant colony algorithm, genetic algorithm (GA) generally require plenty of computationally expensive function evaluations to identify the acceptable optima. To overcome the computational difficulty, SBO approaches are broadly employed for accelerating the search process. A new SBO method called SEUMRE is introduced to handle the engineering design optimization problems in literature [1]. A metamodel-based optimization approach named HAM which can adaptive select suitable metamodelling techniques is presented in literature [2]. Shi et al. use the radial basis functions (RBF) method for handling the satellite constellation system problem [3]. Sun et al. use the efficient global optimization (EGO) method to obtain the higher maximum LDR for underwater vehicles [4]. However, these studies remain unclear which surrogate modeling technique is most suitable for a given problem.

To achieve the better optimal results and reducing the number of function evaluations (NFE), ensemble of surrogates has been recently drawn considerable attention and proven a promising method for various optimization problems. Goel et al. present a novel weight factors selection method selected weight factors using a given global error measure [5]. The corresponding weights are calculated by solving a optimization problem in literature [6]. Lee and Choi use the prediction point of interest to obtain the pointwise ensemble of surrogates [7].

Motivated by the authors' previous study [8], an adaptive and improved surrogate based optimization method using the ensemble of surrogates and HSR strategy (ISGO-HSR) is studied to enhance the efficiency and accuracy for solving the expensive black-box problems where the actual global optima are unknown beforehand. A new Latin hypercube type sampling method called TPSLE [9] is employed to generate the initial sample points to build the single and ensemble of surrogate models. Moreover, three various search subspaces: important local space (ILS), original global space (OGS) and promising joint space (PJS) are computed by HSR strategy. New points are supplemented to the sample set for sequentially improving the accuracy of prediction models. Tested using some representative benchmark optimization problems [10], ISGO-HSR method shows improved in identifying the global optimum.

2. ISGO-HSR method

An ensemble of surrogates based optimization approach using the HSR strategy is proposed to handle the complicated real-word engineering applications in the authors' previous work [8]. The ensemble of surrogates can be obtained by dealing with the optimization formulation below

$$\begin{aligned} &Find \quad \alpha, \beta \\ &\min \quad E_e = \sqrt{\frac{1}{N}\sum_{k=1}^{N}\left(y(\boldsymbol{x}_k)-\sum_{i=1}^{s} w_i \cdot \tilde{y}_i^{(-k)}(\boldsymbol{x}_k)\right)^2} \\ &\quad w_i = w_i^* \Big/ \sum_{j=1}^{s} w_j^*,\ w_i^* = \left(E_i + \alpha \bar{E}\right)^{\beta}, \\ &\quad \bar{E} = \frac{1}{s}\sum_{i=1}^{s} E_i, \\ &\quad E_i = \sqrt{\frac{1}{N}\sum_{k=1}^{N}\left(y(\boldsymbol{x}_k)-\tilde{y}_i^{(-k)}(\boldsymbol{x}_k)\right)^2}, \\ &s.t. \quad E_i + \alpha\bar{E} > 0,\ \alpha < 1,\ \beta < 0 \end{aligned} \tag{1}$$

where E_i and E_e are the prediction sum of squares of root mean square ($\mathrm{PRESS_{RMS}}$) of the *ith* surrogate and the ensemble of surrogates separately, $\bar{E}$ represents the average value of all surrogates' $\mathrm{PRESS_{RMS}}$, N indicates the number of training points, $\boldsymbol{x}_k$ represents the *kth* sample point, $y(\boldsymbol{x}_k)$ and $\tilde{y}_e(\boldsymbol{x}_k)$ denote the true response and relevant prediction response from an ensemble of surrogates at the *kth* sample point respectively, $\tilde{y}_i^{(-k)}(\boldsymbol{x}_k)$ indicates the approximated response of the *ith* single surrogate achieved by leave one out cross validation method. HSR strategy is conducted to identify the design subspaces OGS, PJS and ILS during each three iterations. In fact, the design subspaces PJS and ILS will speed up the convergence, and more promising points will be generated meat the actual optimal point. The detail description on OGS, PJS and ILS can be found in literatures [8].

It's noted that the convergence criterion in this work is different from the authors' previous paper [8]. In the authors' previous paper, the actual global optima of all test examples are assumed to be known in advance. The convergence criterion is formulated as follows

$$\begin{cases} \left|\dfrac{f_{opt}-f_{min}}{f_{min}}\right| < Tol, & if\ f_{min} \neq 0 \\ f_{opt} < Tol, & if\ f_{min} = 0 \end{cases} \tag{2}$$

where f_{min} indicates the real global optimum, f_{opt} represents the current best solution, Tol is selected to judge whether the present best result satisfy the demand. The convergence criterion abovementioned is suitable for the real-world applications where the function characteristics are sufficiently gained. However, the global minima of most engineering design optimization problems like aerospace designs are unknown beforehand. Therefore, the actual global optima of all test examples are supposed to be undiscovered in this study. The convergence criterion is changed as follows

$$\left|\bar{F}_{i+1} - \bar{F}_i\right| \le \varepsilon, \qquad \bar{F}_i = \frac{\sum_{j=1}^{5} f_j}{5} \tag{3}$$

where f_j indicates the *jth* minimum prediction value, ε represents an allowable error determined by the researcher. If the improvement from the average of five minimum function values becomes negligible, the current best result is considered to be acceptable. This convergence criterion offers an efficient way to solve the expensive black-box problems.

The ISGO-HSR method improved from an existing ESGO-HSR approach is presented to well handle the most real-world applications without knowing the actual global optima. The global minima of all test examples are assumed to be undiscovered in this paper. More details of the ESGO-HSR approach can be found in the authors' previous paper [8].

3. Results and discussions

The accuracy and efficiency of ISGO-HSR method are tested using six well-known benchmark optimization problems SC, GP, RB, HN6, TR10, F16 formulated below. All problems are considered as the expensive black-box problems whereas the actual global optima as well as the function characteristics are unknown in advance. All design parameters in ISGO-HSR method which can refer to the literature [8] are the same for all problems. Furthermore, ε is set to be 0.001 for all test problems.

(1) Six-hump Camel-Back function (SC), $n=2$

$$f(x) = 4x_1^2 - 2.1x_1^4 + \frac{x_1^6}{3} + x_1x_2 - 4x_2^2 + 4x_2^4, x \in [-2, 2], f_{min} = -1.0316 \tag{4}$$

(2) Goldstein and Price function (GP), $n=2$

$$f(x) = \left[1 + (x_1 + x_2 + 1)^2 (19 - 14x_1 + 3x_1^2 - 14x_2 + 6x_1x_2 + 3x_2^2)\right] * \left[30 + (2x_1 - 3x_2)^2 (18 - 32x_1 + 12x_1^2 + 48x_2 - 36x_1x_2 + 27x_2^2)\right], x \in [-2, 2], f_{min} = 3 \tag{5}$$

(3) Rosenbrock function (RB), $n=2$

$$f(x) = 100(x_2 - x_1^2)^2 + (x_1 - 1)^2, x \in [-2, 2], f_{min} = 0 \tag{6}$$

(4) Hartman function (HN6), $n=6$

$$f(x) = -\sum_{i=1}^{4} c_i \exp\left[-\sum_{j=1}^{n} \alpha_{ij}(x_j - p_{ij})^2\right], i = 1, 2, \ldots, n, x \in [0, 1], f_{min} = -3.3224 \tag{7}$$

$$[a_{ij}] = \begin{bmatrix} 10 & 3 & 17 & 3.5 & 1.7 & 8 \\ 0.05 & 10 & 17 & 0.1 & 8 & 14 \\ 3 & 3.5 & 1.7 & 10 & 17 & 8 \\ 17 & 8 & 0.05 & 10 & 0.1 & 14 \end{bmatrix},$$

$$[p_{ij}] = \begin{bmatrix} 1312 & 1696 & 5569 & 124 & 8283 & 5886 \\ 2329 & 4135 & 8307 & 3736 & 1004 & 9991 \\ 2348 & 1451 & 3522 & 2883 & 3047 & 6650 \\ 4047 & 8828 & 8732 & 5743 & 1091 & 381 \end{bmatrix} * 10^{-4}$$

(5) Trid function (TR10), $n=10$

$$f(x) = \sum_{i=1}^{n}(x_i - 1)^2 - \sum_{i=2}^{n} x_i x_{i-1}, x \in [-100, 100], f_{min} = -210 \tag{8}$$

(6) A function of 16 variables (F16), $n=16$

$$f(x)=\sum_{i=1}^{n}\sum_{j=1}^{n}\alpha_{ij}\left(x_i^2+x_i+1\right)\left(x_j^2+x_j+1\right),\ i=1,\ 2,\ \ldots,\ n,\ x\in[-1,\ 1],\ f_{min}=25.8750 \tag{9}$$

$$\left[a_{ij}\right]=\begin{bmatrix}
1&0&0&1&0&0&1&1&0&0&0&0&0&0&0&1\\
0&1&1&0&0&0&1&0&0&1&0&0&0&0&0&0\\
0&0&1&0&0&0&1&0&1&1&0&0&0&1&0&0\\
0&0&0&1&0&0&1&0&0&0&1&0&0&0&1&0\\
0&0&0&0&1&1&0&0&0&1&0&1&0&0&0&1\\
0&0&0&0&0&1&0&1&0&0&0&0&0&0&1&0\\
0&0&0&0&0&0&1&0&0&0&1&0&1&0&0&0\\
0&0&0&0&0&0&0&1&0&1&0&0&0&0&1&0\\
0&0&0&0&0&0&0&0&1&0&0&1&0&0&0&1\\
0&0&0&0&0&0&0&0&0&1&0&0&0&1&0&0\\
0&0&0&0&0&0&0&0&0&0&1&0&1&0&0&0\\
0&0&0&0&0&0&0&0&0&0&0&1&0&1&0&0\\
0&0&0&0&0&0&0&0&0&0&0&0&1&1&0&0\\
0&0&0&0&0&0&0&0&0&0&0&0&0&1&0&0\\
0&0&0&0&0&0&0&0&0&0&0&0&0&0&1&0\\
0&0&0&0&0&0&0&0&0&0&0&0&0&0&0&1
\end{bmatrix}$$

The computational cost is generally direct proportion to the NFE for computation-intensive black-box problems. Thus, NFE is used for representing the search efficiency. Meanwhile, the optimal result f_{opt} obtained is also employed to indicate the optimization quality and accuracy. Ten runs are conducted for all test examples for eliminating the random variation. The optimal results are compared with two recently introduced SBGO approaches SEUMRE and HAM. The collected mean and scope values from the optimal results f_{opt} and NFE are recorded in Tables 1 and 2. The sign > means that at least one of the experiments fails to meet the convergence criteria expressed in Eq. (6) within 300 function evaluations. And, the values given in the parentheses indicate the amount of failures. The better mean values of f_{opt} and smaller NFE are highlighted with bold. Through analysing the test results in Tables 1 and 2, we can find that:

(1) ISGO-HSR method can successfully achieve the acceptable optimal solutions for all test examples without knowing the actual global optima in advance.

(2) ISGO-HSR method uses the least number of NFE for most problems. SEUMRE and HAM can only fast catch the global optima for simple problems as GP and RB. However, they require largest NFE for complicated problems including HN6, TR10 and F16, even fail to find the acceptable global optima.

(3) ISGO-HSR method performs the superior capability for all test experiments. It's more efficient and accurate than the comparative methods SEUMRE and HAM.

In ISGO-HSR method, the ensemble of surrogates is used to decrease the required evaluations of the expensive black-box functions. Meanwhile, the HSR strategy is used to accelerate the search in a smaller and more promising subspace. Therefore, the success rate of the optimization method ISGO-HSR is more superior than the comparative methods SEUMRE and HAM. The recently presented ISGO-HSR method shows satisfactory performance From the efficiency and accuracy perspectives.

4. Conclusions

An improved surrogate based optimization method ISGO-HSR has been presented to efficiently handle the expensive black-box problems in this study. The primary contributions are summarized as below

(1) Introduce an improved surrogate-based optimization method ISGO-HSR that can solve the expensive black-box problems where the function characteristics are unknown in advance.

(2) The test results demonstrate that the proposed optimization method ISGO-HSR is efficient and superior in handling the expensive black-box problems.

Table 1. Comparative results f_{opt} obtained by all optimization methods

Fun.	SEUMRE		HAM		ISGO-HSR	
SC	[-1.0316, -1.0284]	-1.0309	[-1.0316, -1.0256]	-1.0309	[-1.0316, -1.0290]	**-1.0313**
GP	[3.0000, 3.0104]	3.0012	[3.0000, 3.0021]	**3.0003**	[3.0000, 3.0203]	3.0040
RB	[0.0000, 0.0395]	0.0058	[0.0000, 0.0016]	0.0003	[0.0000, 0.0007]	**0.0002**
HN6	[-3.3221, -2.6475]	-3.1555	[-3.3223, -3.1622]	-3.2492	[-3.3223, -3.3191]	**-3.3216**
TR10	[911.16, 2723.58]	1883.87	[-209.19, -77.32]	-131.07	[-210.00, -210.00]	**-210.00**
F16	[28.4079, 32.2758]	30.4832	[25.9053, 26.7403]	26.1413	[25.8758, 26.0356]	**25.9071**

Table 2. Comparative results NFE obtained by all optimization methods

Fun.	SEUMRE		HAM		ISGO-HSR	
SC	[38, 69]	55.5	[39, 71]	55.8	[26, 66]	**47.5**
GP	[104, 146]	113.1	[150, 224]	180.4	[89, 168]	**111.5**
RB	[36, 118]	74.8	[56, 120]	80.7	[30, 105]	**56.7**
HN6	[36, 160]	118.9	[76, 202]	156.7	[70, 91]	**77.5**
TR10	[>300, >300]	>300(10)	[>300, >300]	>300(10)	[162, 186]	**171.7**
F16	[258, >300]	>296.0(8)	[>300, >300]	>300(10)	[83, 110]	**96.0**

Acknowledgments

This research was supported by the National Science Foundation of China (Grant No. 61803306).

References

[1] Younis A, Dong Z 2010 Metamodeling and search using space exploration and unimodal region elimination for design optimization *Engineering Optimization* **42** 517-533

[2] Gu J, Li G, Dong Z 2012 Hybrid and adaptive meta-model-based global optimization *Engineering Optimization* **44** 87-104

[3] Shi R, Liu L, Long T Wu Y, Wang G 2018 Multidisciplinary modeling and surrogate assisted optimization for satellite constellation systems *Structural and Multidisciplinary Optimization* **58** 2173-2188

[4] Sun C, Song B and Wang P 2015 Parametric geometric model and shape optimization of an underwater glider with blended-wing-body *International Journal of Naval Architecture and Ocean Engineering* **7** 995-1006

[5] Goel T, Haftka RT, Shyy W Queipo N 2007 Ensemble of surrogates *Structural and Multidisciplinary Optimization* **33** 199-216

[6] Acar E, Rais-Rohani M 2009 Ensemble of metamodels with optimized weight factors *Structural and Multidisciplinary Optimization* **37** 279-294

[7] Lee Y, Choi D-H 2014 Pointwise ensemble of meta-models using *v* nearest points cross-validation *Structural and Multidisciplinary Optimization* **50** 383-394

[8] Ye P, Pan G and Dong Z 2018 Ensemble of surrogates based global optimization methods using hierarchical design space reduction *Structural and Multidisciplinary Optimization* **58** 537-554

[9] Pan G, Ye P and Wang P 2014 A novel latin hypercube algorithm via translational propagation *Scientific World Journal* **2014** 163949 https://doi.org/10.1155/2014/163949

[10] Dong H, Song B, Dong Z Wang P 2016 Multi-start space reduction (MSSR) surrogate-based global optimization method *Structural and Multidisciplinary Optimization* **54** 907-926

A Hybrid Meta-Heuristic Solution to Operation Optimization of Hot Oil Pipeline

Li Bo[1,a], He Miao[2,b], Kang Yang[1,c], Shi Haoyu[1,d],Zhao Jiali[1,e] and Liu Yan[3,f]

[1] PetroChina Pipeline R&D Center. No. 51 Jinguang Road, Guangyang District, Langfang, Hebei
[2] Technology Service Center of PetroChina Pipeline Company. No. 18 Sihai Road, Guangyang District, Langfang, Hebei
[3] PetroChina Pipeline Company. No. 408 Xinkai Road, Guangyang District, Langfang, Hebei

[a]libocolby@yeah.net; [b]hemiao08@petrochina.com.cn; [c]kangyang@petrochina.com.cn; [d]shihaoyu@petrochina.com.cn; [e]zhaojiali@petrochina.com.cn; [f]sdliuyan@petrochina.com.cn

Abstract. In the process of pipeline transportation, the temperature and pressure of waxy and high viscosity crude oil are decreasing. In order to complete the pipeline transportation task, stations are usually set at intervals to overcome the pressure loss caused by friction and impact in the flow and the temperature loss caused by the radial temperature difference. The main equipment of the station is pumps and heaters. The operating fee of pumps and heaters is the core costs of hot oil pipelines, so it is necessary to reasonably optimize the pump and heater operating schemes. Because the hydraulic and thermodynamic of the hot oil pipeline are coupled with each other, and there are a lot of nonlinear constraints and integer variables, this optimization problem is a kind of mixed integer nonlinear programming, which is difficult to solve by using the traditional optimization method. In this paper, a hybrid meta-heuristic algorithm based on particle swarm and pattern search is proposed, which can deal with the problem effectively. The operational cost of the optimized scheme is much lower than that of the actual operating scheme. So this method can provide guidance for the cost reduction and efficiency improvement of the hot oil pipeline.

1. Introduction

There are four modes of crude oil transportation: highway, railway, waterway and pipeline. Pipeline transportation is the most ideal and main way of crude oil transportation, which integrates various factors such as economy, transportation capacity and oil loss. The task of long distance pipelines is to transport crude oil from oil fields in the upstream to refineries or wharfs in the downstream. There are two energy losses in the process of hot oil transportation, namely, friction loss and heat dissipation loss. Therefore, it is necessary to provide energy for oil flow from two aspects, heat energy from heating station and pressure energy from the pumping station.

Domestic crude oil mostly belongs to the type of waxy and high viscosity oil, and its fluidity at room temperature is very poor. In order to maintain the fluidity, the cost of energy consumption is usually high. The electric charge of the pumping station and the fuel cost of the heating station is the main operating cost composition of the hot oil pipeline. And these two costs are coupled. Increasing electricity cost of the pumping station means that the temperature rise of the pump is higher, and the fuel cost of the heating station is reduced. Increasing the fuel cost of the heating station means that the viscosity of the oil decreases and the friction loss decreases. Then the corresponding electricity cost of the pumping station is reduced. How to control electricity and fuel charges reasonably and make the lowest total operating cost of hot oil pipeline is a huge problem faced by the pipeline operation management department[1].

In order to solve the problem of hot oil pipeline operation optimization (hereinafter referred to as HOPOO), the optimization model of HOPOO is established and a hybrid meta-heuristic algorithm is designed based on it.

2. Mathematical models

As is illustrated in Figure 1 and Figure 2, the HOPOO model is mainly composed of stations and pipelines. Oil flow can be injected or distributed in the station. Stations are divided into pumping stations (only contain pumps), heating stations (only contain heaters) and heating & pumping stations (contain both pumps and heaters). Pumps in stations are usually connected in series, while heating furnaces are usually connected in parallel. The pumps in the station are usually centrifugal pumps with constant speed, and some stations are equipped with one speed-regulating pump. Stations are often equipped with pressure regulating valves in order to prevent the outlet pressure of the station is too high. [2]

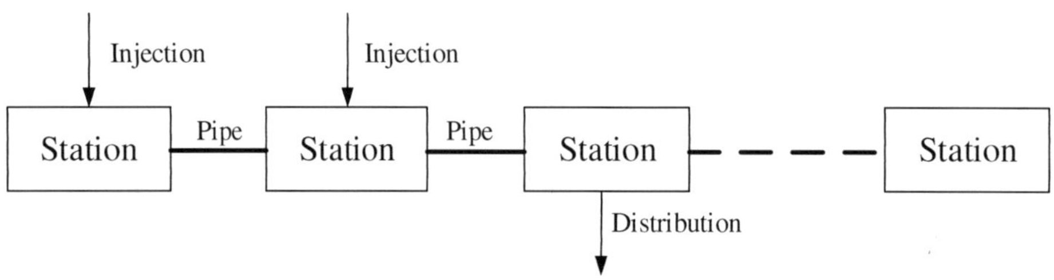

Figure 1. The pipeline structure chart

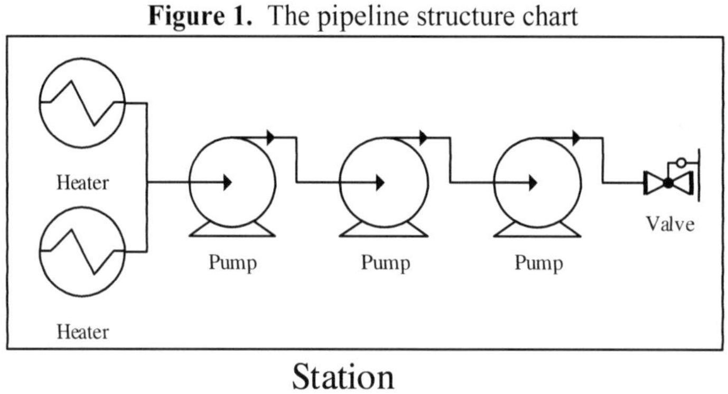

Figure2. The station structure chart

The HOPOO problem can be described as calculating the optimal operation scheme of pumps and heaters for each station with the aim of the lowest total operating cost.

2.1. Decision variables

Let ΔH_i represent the effective pressure head provided by the station i, let ΔT_i represent the temperature rise provided by the station i. For pumping stations $\Delta T_i = 0$, and for heating stations ΔH_i=0. ΔH_i and ΔT_i are the decision variables for HOPOO problem, $i = 1,2,...m$. Here m stands for the number of stations. ΔH_i and ΔT_i must be within a certain range, that is, $\Delta H_i \in \left[0, \Delta H_i^{\max}\right]$ and $\Delta T_i \in \left[0, \Delta T_i^{\max}\right]$.

2.2. Objective function

When ΔH_i is given, the optimal pump scheme of the station i can be obtained by enumeration method. The optimal pump scheme of the station i includes k_i^c -the opening number of constant speed pumps, h_i^c -the pressure head provided by the constant speed pump, h_i^r -the pressure head provided by the speed regulating pump, n_i -the rotate speed of the speed regulating pump, h_i^t -the pressure head loss of the regulating valve, Q_i -the flow rate of station i. The relationship between the above parameters is expressed by formula 1-3.

$$\Delta H_i = k_i^c \cdot h_i^c + h_i^r - h_i^t \tag{1}$$

$$h_i^c = a_{i,1} - b_{i,1} Q_i^{1.75} \tag{2}$$

$$h_i^r = a_{i,2} n_i^2 - b_{i,2} n_i^{0.25} Q_i^{1.75} \tag{3}$$

$a_{i,1}$, $b_{i,1}$, $a_{i,2}$, $b_{i,2}$ can be obtained by least square regression of the measured pressure head, flow rate and rotate speed.Or the relationship between k_i^c, h_i^r and Q_i can be directly fitted by using actual data through machine learning.

The operating cost of station i can be expressed by formula 4-6

$$F_i = e_i^p \left(k_i^c p_i^c + p_i^r\right)\Delta t + e_i^h \Delta t \cdot \rho Q_i c \cdot \Delta T_i / E / \eta_i \tag{4}$$

$$p_i^c = a_{i,3} Q_i^{b_{i,3}} + c_{i,3} \tag{5}$$

$$p_i^r = a_{i,4} + b_{i,4} n_i + c_{i,4} Q_i + d_{i,4} n_i^2 + e_{i,4} n_i Q_i + f_{i,4} Q_i^2 \tag{6}$$

$a_{i,3}$, $b_{i,3}$, $c_{i,3}$, $a_{i,4}$, $b_{i,4}$, $c_{i,4}$, $d_{i,4}$, $e_{i,4}$, $f_{i,4}$ can be obtained by least square regression of the measured power, flow rate and rotate speed.

e_i^p -electricity fees of station i; Δt - running time; e_i^h -fuel costs of station i; ρ -crude oil density; c -specific heat capacity of crude oil; E - heat value of fuel; η_i -efficiency of heaters in station i.

The objective function of HOPOO is:

$$\min\left(\sum_{i=1}^{m} F_i\right) \tag{7}$$

2.3. Constraints

Constraints of HOPOO are as follows.

The rotate speed of the speed regulating pump must be within a certain range:

$$n_i^{\min} \le n_i \le n_i^{\max} \tag{8}$$

The outlet pressure of the station must be below the maximum allowable pressure and the inlet pressure of the station must be higher than the specified pressure:

$$\rho g H_i^O \le P_i^{\max} \tag{9}$$

$$P_i^{\min} \le \rho g H_i^I \tag{10}$$

When the oil flow is heated to a certain temperature, it enters the pipe and continues to dissipate heat to the surrounding medium, which reduces the oil flow temperature. The temperature distribution of the pipeline along the path can be expressed in Formula 11-12.

$$T_{i+1}^I = T_0 + \left(T_i^O - T_0\right) e^{-a_i L_i} \tag{11}$$

$$a_i = \frac{K_i \pi D_i}{\rho Q_i c} \tag{12}$$

K_i -the overall heat transfer coefficient, which can be calculated by actual operation data; D_i -pipe diameter; T_0 -temperature of soil. T_i^O -the outlet temperature of station *i*; T_i^I -the inlet temperature of station *i; L_i* -pipe length.

The loss of pressure energy in pipeline transportation mainly includes two parts: one is the elevation energy loss caused by topographic fluctuation, it is a fixed value that does not change with the flow rate; the other is the energy loss caused by friction and impact of oil flow in the pipeline. The pipeline pressure head loss can be calculated in Formula 13-14. It should be pointed out that the pressure at any point in the pipe should not be less than 0.

$$H_i^O - H_{i+1}^I = f_i L_i + Z_i - Z_{i+1} \tag{13}$$

$$f_i = 0.0246 \frac{Q_i^{1.75} \nu^{0.25}}{D_i^{4.75}} \tag{14}$$

Z_i -the elevation of station *i*; H_i^O -the outlet pressure head of station *i*; H_i^I -the inlet pressure head of station *i*; ν -oil viscosity which can be calculated by a polynomial equations about T_a (the average temperature of oil), seen in Fig.3.

$$T_a = \frac{1}{3} T_i^O + \frac{2}{3} T_{i+1}^I \tag{15}$$

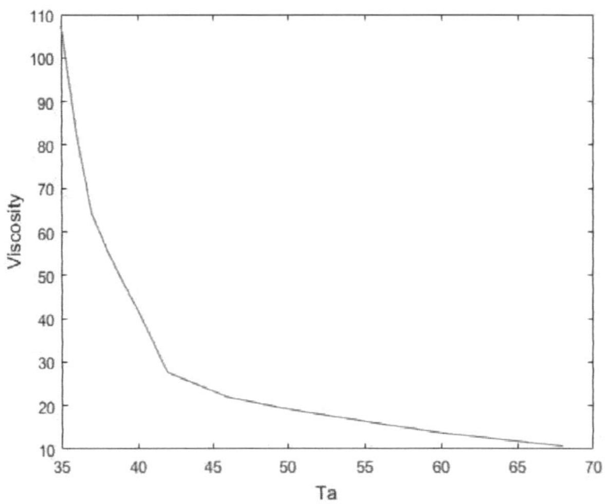

Figure 3. The relationship between viscosity and Ta

Formula 11-15 can be described as hydraulic and thermal simulation of the pipeline. In order to ensure the accuracy of the calculation, it is usually necessary to divide the pipe into small sections for hydraulic and thermal simulation.

The relationship between the pipeline and the station can be described by formula 16-17

$$\Delta H_i = H_i^O - H_i^I \tag{16}$$

$$\Delta T_i = T_i^O - T_i^I - \Delta T_i^P \tag{17}$$

ΔT_i^P -the temperature rise caused by the rotation of the pump, it is related to the flow rate and the pump speed or you can approximate $\Delta T_i^P \approx 0.5 k_i^c$

2.4. A brief summary

To sum up, the HOPOO model can be described as a mixed integer non-linear programming(MINLP). When experts or scholars solved this kind of problem, they often had to make a series of simplifications or assumptions, which would cause the optimization result to deviate from the actual operating situation to some extent. In addition to the difficulty of MINLP problem itself, multiple parameters or conditions are coupled with each other in the HOPOO model, which greatly increases the difficulty of algorithm design.

3. A hybrid Meta-Heuristic algorithm

In the engineering practice of operations research, in addition to precise solving algorithms(such as dynamic programming, linear programming, integer programming, branch and bound method, etc), various intelligent optimization algorithms of heuristics or meta-heuristics[3](such as genetic algorithm, simulated annealing, particle swarm, tabu search[4], etc) are more and more widely used. These intelligent optimization algorithms have good adaptability to nonlinearity and nonconvexity, and have good reliability and robustness. The algorithm structure is relatively simple, and some algorithms have inherent parallelism. However, in the face of large and complex optimization problems, the convergence rate of intelligent optimization algorithms significantly decrease, and sometimes they can only converge to a poor local optimal solution.

To solve the HOPOO problem, a hybrid meta-heuristic optimization algorithm is designed. The algorithm is divided into two stages. In the first stage, the intelligent optimization algorithm is used to get the solution of the original problem(It doesn't matter if this is just a poor local optimal solution); In the second stage, the local search method is used to further optimize the solution of the first stage. The second stage will help us get out of the local optimal solution that intelligent optimization algorithm falls into.

After algorithm testing, particle swarm optimization[5](PSO) is adopted in the first stage, and pattern search algorithm is adopted in the second stage. The algorithm flow chart is shown in Figure 4.

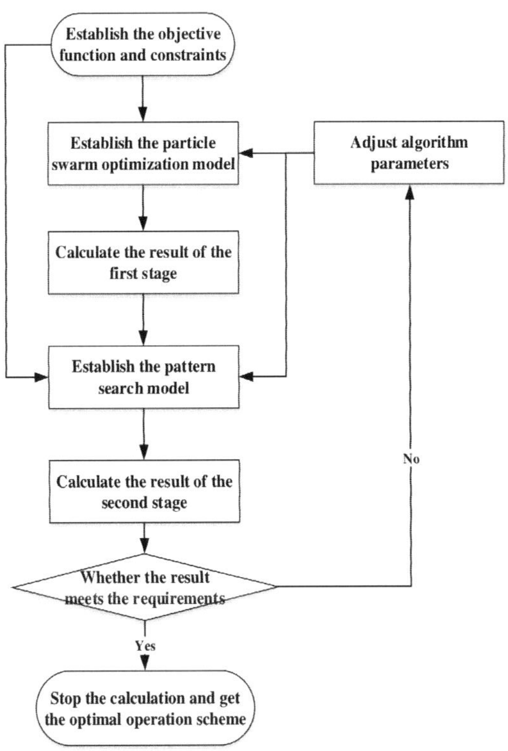

Figure 4. Flow chart of the hybrid meta-heuristic algorithm

A basic variant of the PSO algorithm works by having a population(called a swarm) of candidate solutions(called particles). These particles are moved around in the search-space according to a few simple formulae. The movements of the particles are guided by their own best known position in the search-space as well as the entire swarm's best known position. When improved positions are being discovered these will then come to guide the movements of the swarm. The process is repeated and by doing so it is hoped, but not guaranteed, that a satisfactory solution will eventually be discovered. The Outline of the algorithm is as follows:

- The particle swarm algorithm begins by creating the initial particles, and assigning them initial velocities.
- It evaluates the objective function at each particle location, and determines the best (lowest) function value and the best location.
- It chooses new velocities, based on the current velocity, the particles' individual best locations, and the best locations of their neighbors.
- It then iteratively updates the particle locations (the new location is the old one plus the velocity, modified to keep particles within bounds), velocities, and neighbors.
- Iterations proceed until the algorithm reaches a stopping criterion.

Pattern search[6] is a family of numerical optimization methods that do not require a gradient. As a result, it can be used on functions that are not continuous or differentiable. One such pattern search method is convergence, which is based on the theory of positive bases. Optimization attempts to find the best match in a multidimensional analysis space of possibilities. Starting from the initial base point, the algorithm alternately implements two kinds of search: axial search and pattern search. The axial search is carried out along coordinate axes in turn to determine the new base point and the direction conducive to the descent of the function value. The pattern search is carried out along the line of two adjacent base points in an attempt to make the value of the function decline faster.

4. Numerical experiment

We applied this algorithm to a practical pipeline. There are 10 stations in the pipeline, the total length of which is about 550 kilometers. The test conditions are consistent with two actual operation conditions. The difference between two operating conditions is mainly in pipeline transportation capacity. The 1st condition has a larger capacity, while the 2nd has a smaller one. The total heat transfer coefficient and pipe friction are checked by the actual operation data before the hydraulic and thermal calculation of the pipeline. Because the information of pipeline, station and equipment is too large and complex, this paper does not list all the model parameters. The station configuration of the test pipeline is shown in Table 1.

Table 1. Station configuration

Station	Type	Number of pumps	Number of Heaters	Injection/ distribution	Speed regulating pump
1	heating & pumping	3	5	Injection	no
2	heating	0	2	Injection	no
3	heating & pumping	2	2	none	yes
4	none	0	0	Injection	no
5	heating & pumping	3	2	none	no
6	heating & pumping	2	2	Distribution	yes
7	heating & pumping	2	3	Distribution	yes
8	heating & pumping	2	2	Injection	yes
9	heating	0	3	none	no
10	none	0	0	none	no

Figure 5 and Figure 7 are iterations of the two stages under two different operating conditions. It can be seen that under the 1st operating condition, the optimization space of the second stage is very limited, and only 0.4% is optimized on the basis of the results of the 1st stage. This is because in the 1st operating condition pipeline transportation capacity is large. At this time, the optimization space of the entire pipeline is very limited, so the results of the first stage are often good enough. While the capacity is small, in the second-stage pattern search method can further optimize the results of PSO by 2.6%.

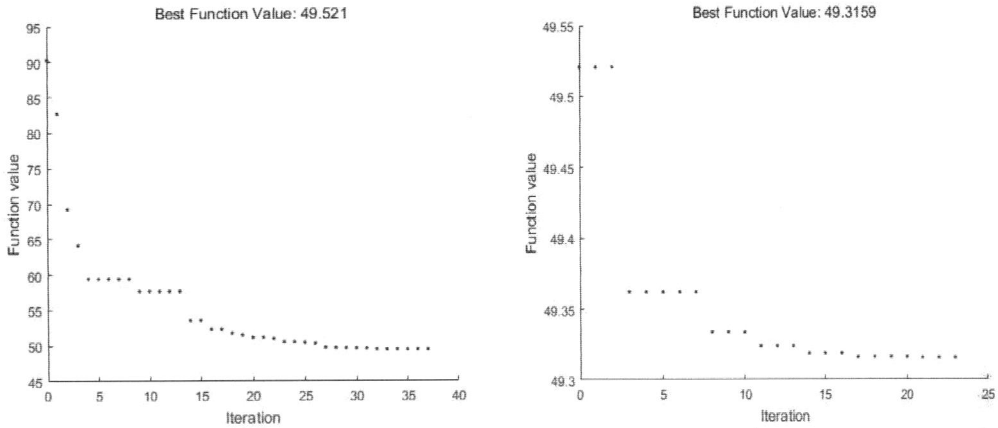

Figure 5. Iterative process of two-stage optimization under the 1st operating condition

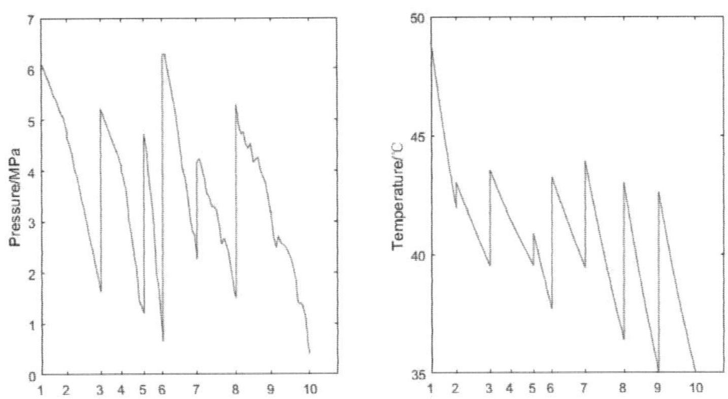

Figure 6. Hydraulic and thermal distribution of the optimal scheme under the 1st operating condition

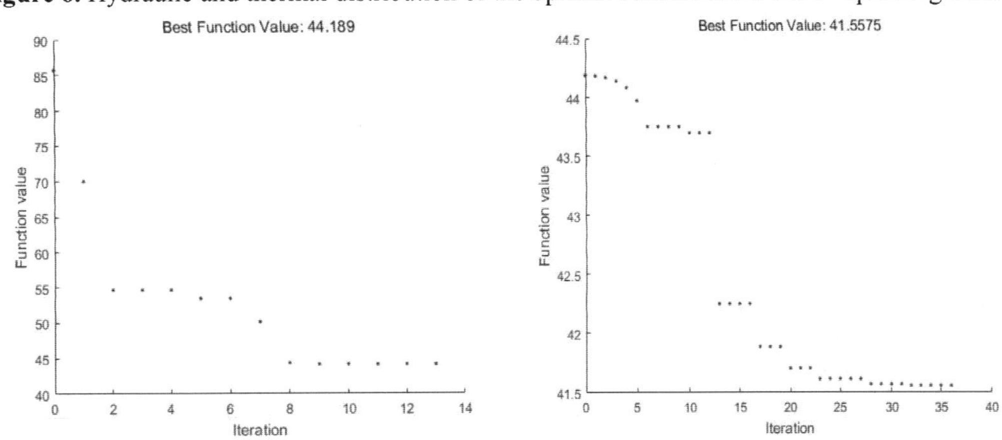

Figure 7. Iterative process of two-level optimization under the 2nd operating condition

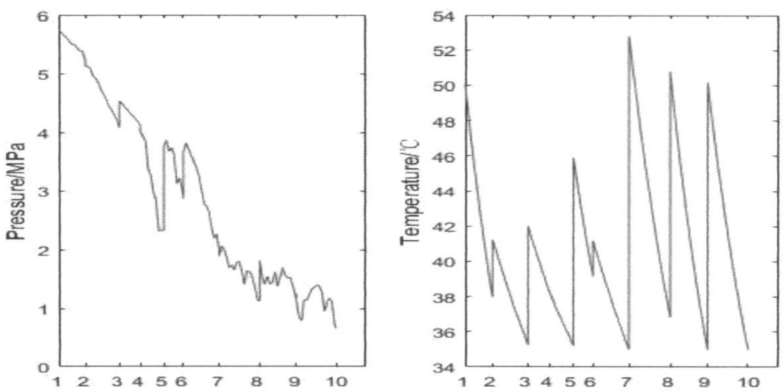

Figure 8. Hydraulic and thermal distribution of the optimal scheme under the 2nd operating condition

Figure 6 and Figure 8 are the hydraulic and thermal distribution of the optimal scheme under two different operating conditions. By contrast, we can see that at low throughput, we can operate fewer pumps, but we cannot reduce the use of heaters. From Formula 11-15, we can deduce that friction loss decreases with the reduction of pipeline transportation capacity, but at the same time the temperature at the end of the pipe decreases sharply, which means that we need to use heaters to raise the temperature higher.

5. Conclusion

In the two test examples, as shown in Table 2, the optimized operation cost is 6.06% and 9.26% lower than the actual operation cost. This means not only a reduction in the cost of tens of thousands yuan per day, but also less pollutant emissions. At the same time, it can provide a more scientific and convenient programming tool for pipeline dispatchers. Using this method, the optimal operating scheme can be obtained in a few minutes, which will save a lot of manpower and time costs. All of these are of great significance to pipeline management organizations.

Table 2. Comparison of actual cost and optimized cost

Operating condition	Actual operating cost 10^4 yuan/day	Optimal operating cost 10^4 yuan/day	Optimizing space
1	52.5	49.3159	6.06%
2	45.8	41.5575	9.26%

The HOPOO model and the corresponding algorithm in this paper are all under steady-state conditions, which means that the parameters of pipelines and equipments are in a stable state which does not change with time. The transient optimization of hot oil pipeline is far more difficult, which will depend on more intelligent methods, such as reinforcement learning, deep learning and so on.

Reference

[1] Wang, Yu, et al. A survey on oil/gas pipeline optimization: Problems, methods and challenges. Proceedings of 2012 IEEE International Conference on Service Operations and Logistics, and Informatics. IEEE, 2012.
[2] de la Cruz, Jesús Manuel, et al. Multiobjective optimization of the transport in oil pipelines networks. EFTA 2003. 2003 IEEE Conference on Emerging Technologies and Factory Automation. Proceedings (Cat. No. 03TH8696). Vol. 1. IEEE, 2003.
[3] Kang S L, Zong W G. A new meta-heuristic algorithm for continuous engineering optimization: harmony search theory and practice[J]. Computer Methods in Applied Mechanics & Engineering, 2005, 194(36):3902-3933.

[4] Glover F, Laguna M. Tabu Search[M]// Metaheuristic Procedures for Training Neutral Networks. 2006.
[5] Kennedy, J. (2010). Particle swarm optimization. Encyclopedia of machine learning, 760-766.
[6] Lewis R M, Torczon V. Pattern Search Algorithms For Bound Constrained Minimization[J]. Siam Journal on Optimization, 1999, 9(4):1082-1099.

Hybrid Algorithm for Solving Traveling Salesman Problem

Ping Zhao*and Degang Xu

School of Information Science and Engineering, Henan University of Technology, Zhengzhou, P. R China

*Corresponding author E-mail: zhaoping_xyz@163.com

Abstract. The basic genetic algorithm has the disadvantages of falling into local optimum and slow convergence. To solve this problem, a hybrid algorithm combining simulated annealing strategy is proposed. The cooling process in simulated annealing is used to complete the iterative process in the hybrid algorithm. The algorithm is used to solve the traveling salesman problem. The results show that the convergence speed and accuracy of the hybrid algorithm is significantly better than the basic genetic algorithm.

1. Introduction

The Genetic Algorithm (GA) are global optimization methods a which founded by Holland[1], University of Michigan, and GA is a random search algorithm by simulating the natural selection and natural genetic mechanisms of the biological world. However, genetic algorithm has the disadvantages of premature convergence and local extremum [2]. There are many problems in solving the Traveling Salesman Problem (TSP). Therefore, many scholars have proposed various improved algorithms to solve the TSP problem. Sun et al [3] proposed an indefinite period traveling salesman problem, and used exact as well as heuristic procedures for solving TSP. The computational experiments showed that the proposed method produced high quality solutions for given problems. Zhang X L[4] uses heuristic information to reduce the cross-edges in the travel route, which has great advantages in solving such problems with the mutation operator in the genetic algorithm. Venkatesh et al [5] proposed an artificial bee colony algorithm for the generalized covering TSP. The computational experiments showed the effectiveness of the approach over all the other state-of-the-art approaches in terms of solution quality. Liu F[6] uses multiple non-enhanced mutation operators to adjust the individual's mutation strategy according to the individual's evolutionary characteristics, which can not only maintain individual diversity., while increasing the speed of convergence. The ant colony algorithm and particle swarm optimization algorithm are combined[7] to solve the problem. The experimental results show that the algorithm has good convergence performance. T Srinivas Rao [8] uses ant colony optimization algorithm to design the visit route. The results show that the algorithm performs well in selecting the shortest route.

In this paper, the basic genetic algorithm solves the TSP problem with slow convergence speed. The satisfactory travel route is not obtained. The paper proposed a hybrid algorithm, that is the combination of genetic algorithm and simulated annealing strategy [9], and the effectiveness of the hybrid algorithm is verified.

2. Traveling Salesman Problem

The TSP is one of the most widely studied NP-hard combinatorial optimization problem which cannot be solved exactly in polynomial time. This problem is, for a given set of n cities , to determine a minimum distance circuit(Hamiltonian circuit or cycle) passing through each vertex onc and only once. In a classical two-dimensional TSP, TSP can be represented as graph G={V, E}, where V=1,2…,*n* ;*n* is the set of nodes and E is the set of edges. A salesman has to travel n cities at minimum cost.In this tour, salesman starts from a city, visits all the cities exactly once and comes to the starting city using minimum cost. Let d_{ij} be the distance for traveling from *i*-th city to *j*-th city. Then the model is mathematically formulated as

Minimize

$$Z=\sum_{i=1}^{n}\sum_{j=1}^{n}d_{ij} \quad (1)$$

Determine

$$d_{ij} \quad i=1,2,\ldots,n \quad j=1,2,\ldots n \quad (2)$$

Subject to

$$\sum_{i=1}^{n}d_{ij}=1,\ j=1,2,\ldots,n \quad (3)$$

$$\sum_{j=1}^{n}d_{ij}=1,\ i=1,2,\ldots,n \quad (4)$$

where d_{ij}=1 if the salesman travels from city-*i* to city-*j*, otherwise d_{ij}=0.

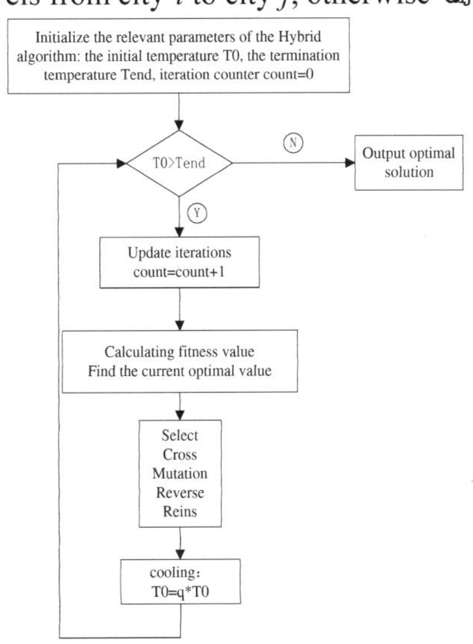

Figure 1. Hybrid algorithm flow chart

3. Steps of the hybrid algorithm

The algorithm of fusion simulated annealing is based on the basic genetic algorithm. The annealing process of the simulated annealing algorithm is integrated into the genetic algorithm. The mixture of the two algorithms is used to optimize the algorithm and improve the search ability of the GA. The execution process of this algorithm is shown in Figure 1.

3.1. Implementation of genetic algorithm

Genetic algorithm is a randomized search method that draws on the natural selection of the biological world and the natural selection mechanism [10]. The main advantage of the algorithm is the random selection mechanism. There is no special requirement for the search space in the process of solving the problem. It has the advantages of simple operation and fast convergence, and has strong competitiveness in solving the path problem.

The main feature of genetic algorithm is the group search strategy and the information exchange between individuals in the group. It is actually simulating the overall learning process from the individual to group. Each individual corresponds to the research problem' solution when simulating the algorithm. The genetic algorithm can eliminate the individuals with low fitness by selecting operator, cross operator and mutation operators, so that the group evolves to the best region in the search space, and finally generates chromosomes that correspond to the optimization goals.

In the process of solving the problem, the main workflow of the standard genetic algorithm is as follows:

(1) Chromosome coding

There are many ways to encode chromosomes. In this paper, each city is coded by natural number coding. The city point is used as the gene, and the chromosome formed according to the order of traveler's visit is the solution to the problem.

(2) Population initialization

When chromosome coding is completed, the number of populations needs to be set according to the size of the city, and then an individual is randomly selected from the population as the initial solution to the problem.

(3) fitness function

In the process of the salesman access to the city route, the main cost is the distance, and the fitness function is the reciprocal of the total distance after visiting *n* cities.

$$f=\frac{1}{Z} \tag{5}$$

The goal of TSP optimization is to select chromosomes with the largest fitness function, which correspond to shorter route distance, that is the better chromosome quality.

(4) Select operation

The selection operation is to select a certain number of individuals from the old population to the new group according to a certain probability, with the fitness value of the individual. which can make the individuals with large fitness values remain as much as possible, thereby improving the performance of the algorithm.

a: Calculate the probability that each individual will occupy the entire population:

$$f_i=\frac{F_i}{\sum_{i=1}^{n} f_i} \tag{6}$$

b: For the *i*-th individual, calculate the cumulative fitness from the first individual to the *i*-th individual:

$$p_i=\sum_{i=1}^{n} f_i \tag{7}$$

c: Generate a random number between [0, 1] and select the individual according to a certain ratio.

(5) Cross operation

The purpose of the crossover operation is to combine new individuals so that a more efficient search can be performed. There are many types of crossover operators, such as single-point crossover, double-point crossover, multi-point crossover for binary coding, partial mapping crossover, sequential crossover, etc. for natural number coding. In this paper, the cross operation is partial mapping intersection. Firstly, the parent samples are grouped in pairs, and each group is crossed according to the same process. Take the number of cities as 10 as an example:

(6) Mutation operation

The mutation operation is based on randomly generating two random integer rand1 and rand2 between [1,n] in the selected individuals, and the two points are changed to make them mutate.

(7) Evolutionary reverse operation

In order to further improve the local search ability of the genetic algorithm, after the traditional selection, crossover and mutation, the evolutionary reverse operation is introduced into the algorithm optimization. It is necessary to judge whether the fitness value is improved according to the reversal. If fitness value better than before, the reverse result is valid.

3.2. Implementation of simulated annealing algorithm

The simulated annealing algorithm is a general optimization algorithm. It's annealing process is mainly composed of heating process, isothermal process and cooling process. The implementation process is

a: Control parameters

The main control parameters of the simulated annealing algorithm are initial temperature T_0, cooling rate q, ending temperature T_{end}, and chain length L.

b: Generation of initial solution

For TSP problem with n cities, the solution to the problem is a sort in 1-n full sorting, where each number corresponds to the city number according to the encoding rules. Using a random selection method, select an individual from the population as the initial solution to the problem.

c: Solution transformation to generate a new solution

A new route is generated using the two-neighbor transform method for the currently obtained solution. The location to be transformed

d: state acceptance function

Using the Metropolis criterion as a state acceptance function, the probability of accepting a new solution is

$$P=\begin{cases}1, df<0\\ \exp\left(-\frac{df}{T}\right), df\geq 0\end{cases} \tag{8}$$

If $df<0$, the new path is accepted with probability 1; otherwise the new path is accepted with the probability $\exp\left(-\frac{df}{T}\right)$.

The advantage of the Metropolis acceptance criterion is that it can accept the optimization solution with a certain probability, avoiding the local optimization as much as possible, and finally obtained the global or approximate global optimal solution of the system.

e: Stop rule

After several successive cooling, the iteration is stopped when the initial temperature is less than the termination temperature.

The simulated annealing algorithm has certain advantages in solving the TSP problem and is very competitive in solving the optimization problem. Aiming at the shortcomings of genetic algorithm in local search ability, the paper combines genetic algorithm and simulated annealing algorithm to enhance the local search ability of the algorithm. The algorithm only needs to integrate the simulated annealing algorithm based on genetic algorithm. In this paper, the degenerative process of the simulated annealing algorithm is mainly used as the iteration number of the hybrid algorithm, and the optimal solution is obtained.

4. Simulation results analysis

According to the actual problem, the traveling salesman problem contains 31 cities to visit, and the visit route is reasonably arranged to minimize the total access distance. In this paper, the mixed algorithm computer program of this problem is compiled and experimentally calculated using MATLAB language.

IOP Conf. Series: Materials Science and Engineering **646** (2019) 012032 doi:10.1088/1757-899X/646/1/012032

Table 1. Comparison of calculation results between hybrid algorithm, genetic algorithm and simulated annealing algorithm

calculation order	hybrid algorithm	simulated annealing algorithm	genetic algorithm
1	15511.2368	16934.921	15470.7389
2	15402.3419	16476.5901	15622.0458
3	15433.4199	16811.7854	16914.8436
4	15470.2901	17674.6774	15593.6963
5	15414.5194	16098.364	15990.0235
6	15385.4258	16777.4279	16007.0926
7	15416.5038	16920.723	15511.2368
8	15433.4199	16779.8239	16445.8346
9	15514.5104	17079.4438	15834.57
10	15450.5079	17054.7815	15983.4004
11	15591.2485	16869.2385	16174.814
12	15538.1899	16911.5293	16189.3939
13	15437.0016	17130.2279	15476.8498
14	15487.2062	16516.5621	15551.6628
15	15506.2786	16914.9322	15975.8032
16	15402.3419	16892.7235	15437.0016
17	15699.9886	16571.8668	16207.7268
18	15874.316	16669.738	15397.4314
19	15512.8383	16384.737	16227.8195
20	15453.8228	17083.0706	16522.2833
21	16802.3777	17017.0281	15849.924
22	15699.9886	16464.0487	16214.731
23	15450.5079	16462.4949	15531.8806
24	15492.7102	15906.4685	15437.0016
25	15582.485	16668.149	16035.5721
26	15900.2912	15620.6433	15930.0867
27	15646.9416	16113.4146	15454.082
28	15480.4395	15846.2795	15683.0725
29	15495.8921	16159.6573	15581.7823
30	15385.4258	15618.8662	15612.5596

Table 2. Average distance and optimal distance of the three algorithms

Compare content	hybrid algorithm	simulated annealing algorithm	genetic algorithm
Average distance (km)	15562.42	16614.34	15862.17
Optimal solution (km)	15385.4258	15618.8662	15397.4314
times of obtaining the optimal solution	2	1	1

According to the characteristics of such problems, the main parameters used by the author in solving the problem are: population size 100, evolutionary algebra 270, generation gap 0.9, crossover probability 0.9, mutation probability 0.05, initial temperature 1000. The termination temperature was 1e-3 and the cooling rate was 0.95. In order to facilitate the comparison, the author also solved the problem 30 times with genetic algorithm and simulated annealing algorithm respectively. Under the premise that the number of searches for each solution is 270 times, the comparison of the calculation results of the three algorithms is shown in Table 1 and Table. 2.

It can be seen from Table 1 and Table 2 that the solution to the problem is solved by the hybrid algorithm. The average value of the access route is 15562.42km, and the shortest access route distance is 15385.4258km, which is better than genetic algorithm and simulated annealing algorithm. The algorithm and simulated annealing algorithm have higher computational efficiency.

It can be seen from the above experimental calculation that by combining the genetic algorithm and the simulated annealing algorithm to construct a hybrid algorithm for solving the traveling salesman problem, the convergence speed and the optimization speed of the genetic algorithm can be overcome to some extent, so that it can be better. Calculation results of genetic algorithm and simulated annealing algorithm. At the same time, the hybrid algorithm also has the characteristics of high computational efficiency and stable calculation results.

5. Conclusion

This paper discusses the mathematical model of traveling salesman problem and the detailed implementation process of genetic algorithm and simulated annealing algorithm. It points out that genetic algorithm has the disadvantages of poor local search ability and easy to produce premature phenomenon, and incorporates simulated annealing in the process of genetic algorithm search process. The new algorithm takes into account both local and global aspects, overcoming the shortcomings of genetic algorithm. The experimental results show that the hybrid genetic algorithm for TSP can overcome the shortcomings of the local search ability of the genetic algorithm to solve the problem and the shortcomings of the simulated annealing algorithm in the global search ability to a certain extent, which is better than the simulated annealing algorithm. And the calculation results are better than genetic algorithm and simulated annealing algorithm. At the same time, the hybrid algorithm also has the characteristics of high computational efficiency and stable calculation results, which shows its good optimization performance. The hybrid algorithm to solving multi-objective optimization problems is a further research direction.

References

[1] Holland J H .1992. Adaptation in natural and artificial system[M].*Ann Arbor: The University of Michigan Press*

[2] Dong G , Guo W W and Tickle K .2012. Solving the traveling salesman problem using cooperative genetic ant systems[J]. *Expert Systems with Applications*, **39(5)**:5006-5011.

[3] Lei Sun, Mark H. Karwan, Moustapha Diaby. 2019.The indefinite period traveling salesman problem[J]. *European Journal of Operational Research*,**270(3)**:1171-1181.

[4] Zhang X L, Zuo G C and Yang J.2010. Solving Travelling Salesman Problem by Genetic Algorithms with Heuristic Mutation Stratgey[J]. *Computer Applications and Software*, **27(3)**:237-240.

[5] Venkatesh Pandiri, Alok Singh.2019.An artificial bee colony algorithm with variable degree of perturbation for the generalized covering traveling salesman problem[J].*Applied Soft Computing*, **78:**481-495.

[6] Liu F, Lv S H and Zhao Z H.2011. Study of hybrid genetic algorithm based on multi-step reinforcement mutation operator[J]. *Computer Engineering and Applications*, **47(29)**:46-48.

[7] Aditi Khanra , Manas Kumar Maiti and Manoranjan Maiti .2015. Profit maximization of TSP through a hybrid algorithm[J]. *Computers & Industrial Engineering*,**88**:229-236.

[8] T Srinivas Rao.2018.An Ant Colony TSP to Evaluate the Performance of Supply Chain Network[J].*Materials Today: Proceedings*,**5**:13177-13180.

[9] Bertsimas D , Tsitsiklis J .1993. Simulated Annealing[J]. *Statistical Science*, **8(1)**:10-15.

[10] Chen G L, Wang X F, Zhuang Z Q, et al. Genetic Algorithms and Its Application[M]. *POSTS & TELECOM PRESS*, 1999.

IOP Conf. Series: Materials Science and Engineering **646** (2019) 012033 doi:10.1088/1757-899X/646/1/012033

Optimizing the Perimeter Control for Urban Traffic Sub-region by BP Neural Network

Wenchao Liu and Wenxing Zhu*

Shandong University, Jinan, China

*zhuwenxing@sdu.edu.cn

Abstract. In this paper, we improve the effect of perimeter control for Urban Traffic Sub-region by optimizing signal timing in traffic sub-region. Firstly, we verify the influence of different signal timing on the shape of macroscopic fundamental diagram(MFD) by simulation. Then, we use BP neural network to model the mapping relationship between signal timing and MFD maximum volume(saturated traffic volume of sub-region). In perimeter control based on MFD, the larger the saturated traffic volume, the more vehicles passing through the sub-area. When the output of the BP neural network model reaches the maximum value, the signal timing of the road network is the best one. Comparing the perimeter control effects of different signal timings, we find that the larger the saturation traffic, the larger the traffic in the boundary control and the smaller the vehicle delay.

1. Introduction

With the advancement of urbanization and the increase of car ownership, single point control and main stem control can no longer solve the increasingly serious traffic congestion problem. In view of the over-saturated state of urban regional traffic, the perimeter control strategy has a good adjustment effect. Perimeter control strategy is a kind of urban traffic sub-region control strategy based on macroscopic fundamental diagram(MFD). It can improve the traffic situation in sub-region by adjusting the number of vehicles entering and leaving sub-region to maximize the traffic volume. MFD can intuitively describe the relationship among average traffic flow, average density and average speed in homogeneous traffic network, which provides a theoretical basis for macro-network regulation[1-3].

In 2007, Daganzo proved the existence of MFD through theory, and pointed out that MFD is the inherent attribute of road network and does not depend on OD distribution of road network[4]. Buisson et al. used French highway and urban road traffic data to prove that heterogeneity has a greater impact on the shape of macro-basic maps[5]. Geroliminis et al. found that MFD also exists in inhomogeneous networks, and the shape of MFD is affected by signal control[6]. The shape of MFD determines the maximum traffic volume that can be reached in the sub-region, i.e. saturated traffic volume. That is to say, different signal timing correspond to different saturated traffic volumes for the same traffic sub-area.

In order to obtain the maximum saturated volume, we use BP neural network to model the saturated traffic volume in the sub-region with the input of signal timing. We establish three BP neural network models with one hidden layer, two hidden layers and three hidden layers, and the validity of the models were compared and analyzed. The BP neural network with three hidden layers is used to find out the maximum saturated traffic volume in the sub-region and the corresponding signal timing scheme, that is, the optimal signal timing. Bang-Bang perimeter control is carried out under the optimal signal timing, and the results are compared with those under other non-optimal signal timing. The results show that the effect of perimeter control can be optimized by maximizing saturated traffic volume.

2. The Influence of Signal Control on MFD

This paper is based on the simulation data of VISSIM, which is a simulation software suitable for macro traffic simulation. We build a 4×4 road network, As shown in the Figure 1. All traffic signals are fixed with a period of 100 seconds, in which the main road adopts three phase signal control and the rest is two phase signal control. So the dimension of the signal timing vector is 5.

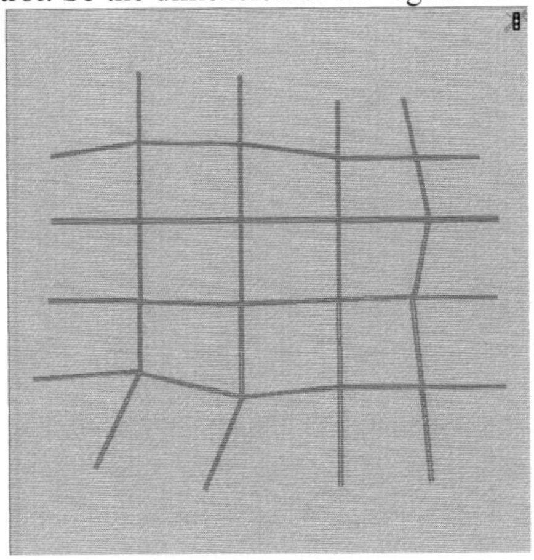

Figure 1. Sub-area network

We use different signal timing to simulate traffic and draw MFD of road network. The simulation results and MFD curves are shown in the Figure 2. Comparing these figures to determine whether different signal timing will affect the shape of MFD.

The red curve in the graphs represent the relationship between the cumulative number of vehicles in the road network and the traffic volume of the sub-region. It is fitted by a cubic function. When the state of the road network is at the vertex of the curve, the road network is saturated[7]. The cumulative number of vehicles corresponding to saturated state is the critical number of vehicles, and the corresponding traffic volume is the saturated traffic volume. As can be seen from Figure 2, the MFDs are different under each signal timing. They have different saturated traffic volume and critical cumulative number of vehicles. In order to determine the mapping relationship between signal timing and saturated traffic volume, we use BP neural network to model saturated traffic volume.

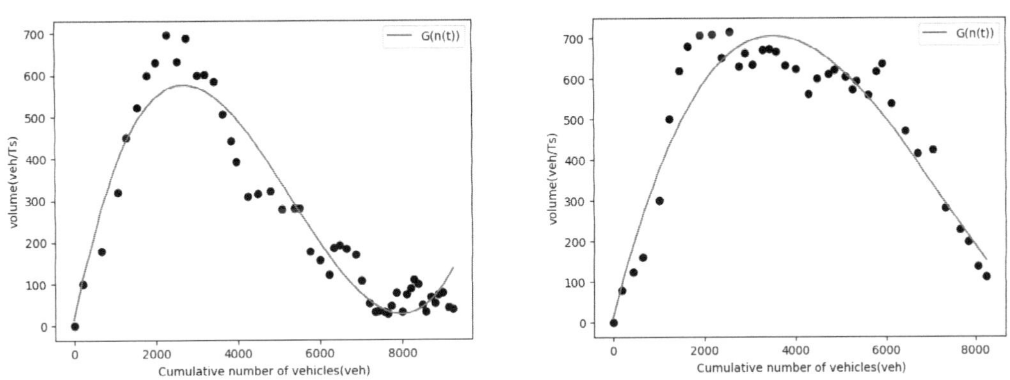

(a) (b)

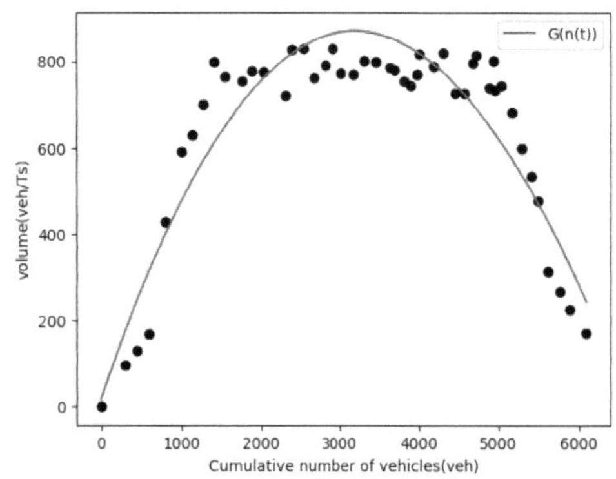

(c)

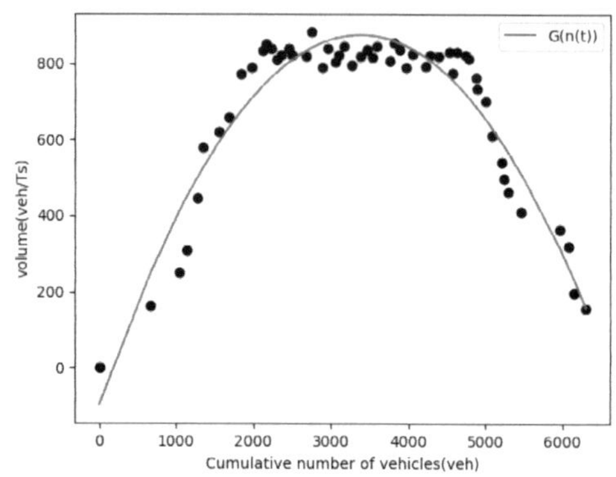

(d)

Figure 2. macroscopic fundamental diagram

3. Saturated Traffic Volume Modeling in Traffic Subarea

Using BP neural network to model the peak flow of MFD, i.e. saturated traffic volume of the sub-region, and the input of the model is signal timing. We establish three BP neural network models with one hidden layer, two hidden layers and three hidden layers, and the validity of the models were compared and analyzed.

3.1. BP Neural Network Model

Back propagation neural network is a concept proposed by Rumelhart and McClelland in 1986. BP neural network can learn and store the nonlinear mapping relationship between input and output. Here we use it to model the saturated traffic volume. The input of the model is the signal timing vector of the sub-region. It modifies the weights and thresholds of the network through the reverse propagation of errors, so that the errors between the predicted values and the real values are continuously reduced. When the error reaches the set value or the number of iterations reaches the maximum value, the training stops and the modeling is completed. The structure of BP neural network is shown in the Figure 3.

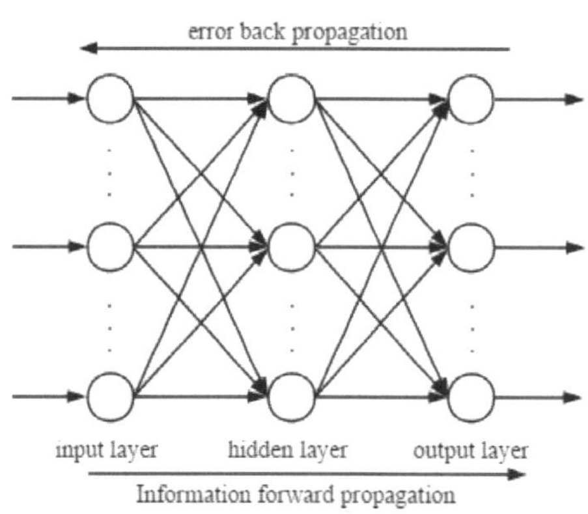

Figure 3. The structure of BP neural network

In the training process of neural network, the first is the forward transmission of information. In this process, the input vectors are computed layer by layer to get the output[8]. Taking a neuron as an example, its transmission process is as follows:

$$a_j = f(\sum_{i} x_i w_{ij} + b_j) \tag{1}$$

where x_i and a_j are the input of the neural network and the output of the hidden layer neuron j, respectively; w_{ij} is the weight of x_i to a_j, and b_j is the bias; f is an activation function. In the process of error back propagation, we use loss function to measure the error between predicted value and real value, and use gradient descent method to update the weights of the network, which makes the predicted value of the network as close as possible to the real value. As iteration proceeds, the value of loss function decreases. The loss function used in this modeling is MSE (mean square error).

$$MSE = \frac{1}{m}\sum_{i=1}^{m}(y_{pi} - y_{ti})^2 \tag{2}$$

where MSE and MAE are mean square error and mean absolute error respectively; m is the number of output nodes; y_{pi} is the predicted value of BP neural network and y_{ti} is the true value.

The modeling steps are as follows[9-11]:

(1) Determine the network structure. The number of neurons in input layer and output layer is determined according to the dimension of input and output. The number of neurons in hidden layer is determined by heuristic method.

(2) Initialize the network. Initialization of weights and thresholds in networks.

(3) Forward transmission of information. The input vector is computed layer by layer to get the predicted value.

(4) Reverse propagation of errors. The errors are computed by gradient descent method to get the change of weights, which are updated from the end of the network.

(5) Determine whether the termination conditions are met. The termination condition is that the number of iterations reaches the upper limit or the error is small enough. If the condition is satisfied, the modeling stops, otherwise it goes back to step (3).

3.2. Analysis of Modeling Result

After testing, it is found that the BP neural network with one hidden layer, two hidden layers and three hidden layers has the best prediction effect when its structure is 5-12-1, 5-12-5-1 and 5-12-1-4-1, respectively. Three models are tested with simulation data, and the test results are shown in the Figure 4.

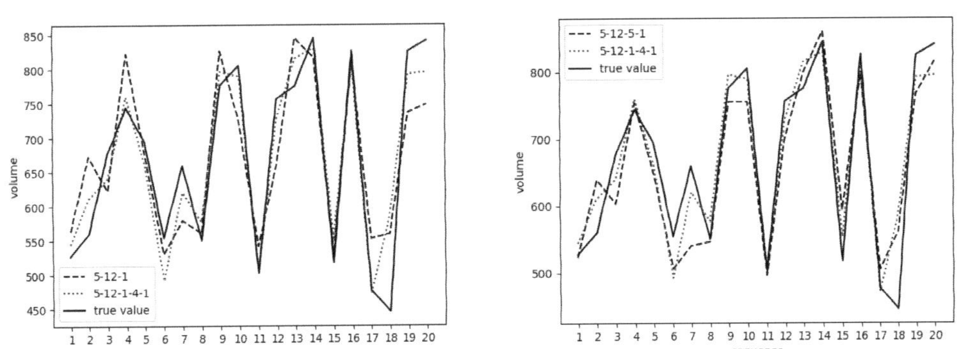

(a) (b)

Figure 4. Model validity comparison

According to the figures, the prediction result of BP neural network of 5-12-1-4-1 is the closest to the real value, which is more reliable than the other two models. So we use BP neural network model with three hidden layers to find the best signal timing vector to maximize the saturated traffic volume. Here we can use genetic algorithm to optimize or traversal search. After optimization, it is found that the optimal timing vector is [35, 13, 43, 51, 43], the corresponding maximum saturated traffic volume is 877 veh/Ts, and Ts is the sampling period.

4. Optimized boundary control

Perimeter control is a regional vehicle number control technology based on MFD. When the number of vehicles in the area exceeds the critical number of vehicles, the number of vehicles entering the sub-area will be reduced through the signal lights at the boundary intersection. When the number of vehicles in the area is less than the critical number of vehicles, the number of vehicles entering the sub-area will be increased. In this way, the number of vehicles in the area can be maintained near the critical value, and the traffic volume in the area can also be maintained near the saturated traffic volume. Therefore, during the implementation of boundary control, the larger the saturated traffic volume, the more vehicles passing through the road network. In the third chapter, we use the BP neural network model to find the signal timing which maximizes the saturated traffic volume is [35, 13, 43, 51, 43].

We use Bang-Bang control strategy to verify whether the optimized signal timing vector can improve the effect of boundary control. The signal timing vector and its corresponding saturated traffic volume are shown in the Table 1. The first group in the table is the optimal timing vector and the maximum saturated traffic volume.

Table 1. signal timing scheme

signal timing vector(s)	*saturated traffic volume(veh)*
[35,13,43,51,43]	877
[30,17,44,56,38]	810
[38,30,23,55,39]	695
[40,39,12,58,36]	511

The effect of the Bang-Bang control under four timing vectors is shown in the Figure 5. Figure 5.(a) shows the variation of sub-area flow with simulation time in each signal cycle, and Figure5.(b) shows the variation of vehicle average delay with simulation time in each signal cycle.

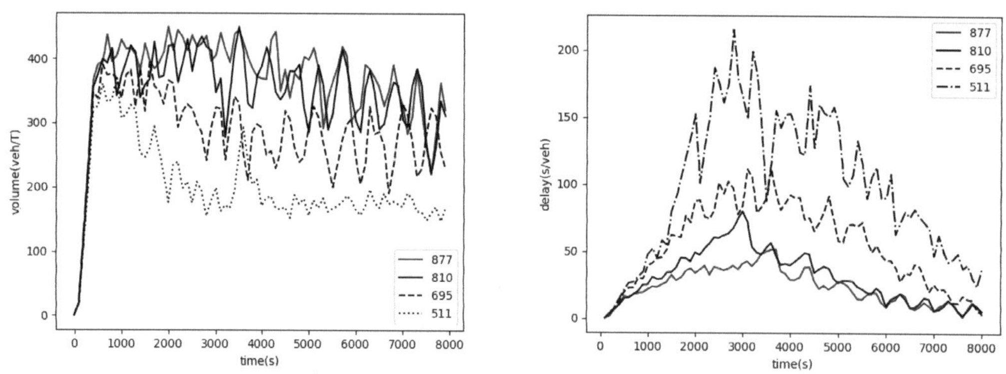

(a) (b)

Figure 5. Contrast of boundary control effect

As can be seen from Figure 5, compared with the Bang-Bang control under the other three signal timing vectors, the traffic volume in the Bang-Bang control under the optimized signal timing vector is larger and the vehicle delay is smaller. This proves that the optimized sub-region signal timing vector can improve the effect of boundary control based on MFD.

5. Conclusion

In this paper, we use BP neural network to model the saturated traffic volume in the sub-region with the input of signal timing. The simulation data show that the BP neural network with three hidden layers can well reflect the mapping relationship between signal timing and saturated traffic volume. With the help of BP neural network model, we get the signal timing which maximizes the saturated traffic volume, and call it the optimal signal timing. It is found that compared with other signal timing, the optimal signal timing can increase the traffic volume of sub-region and reduce the delay of sub-region.

References

[1] Gan Q J 2014 Macroscopic modeling and analysis of urban vehicular traffic *J.* Dissertations & Theses - Gradworks
[2] Knoop V L, Hoogendoorn S P and Van Lint J W C 2013 The impact of traffic dynamics on macroscopic fundamental diagram (poster) *J.* Civil Engineering & Geosciences pp 236-250
[3] Shraiber A, Haddad J 2014 Robust control design for a perimeter traffic flow controller at an urban region *C.* Control Conference
[4] Dangzo C F 2007 Urban gridlock:Macroscopic modeling and mitigation approaches *J.* Transportation Research Part B pp 0-62
[5] Buisson C, Ladier C 2009 Exploring the Impact of Homogeneity of Traffic Measurements on the Existence of Macroscopic Fundamental Diagrams *J.* Transportation Research Record: Journal of the Transportation Research Board pp 123-136
[6] Ludovic L, Nikolas G 2013 Estimating MFDs in simple networks with route choice *J.* Transportation Research Part B pp 468-484
[7] Wu C Y, Li M, Jiang R, Hao Q Y and Hu M B 2018 Perimeter control for urban traffic system based on macroscopic fundamental diagram J. Physica A pp 231-242
[8] Sadeghi B H M 2000 A BP-neural network predictor model for plastic injection molding process *J.* Journal of Materials Processing Technology pp 411-416
[9] Xiao Z, Ye S J and Zhong B 2009 BP neural network with rough set for short term load forecasting *J.* Expert Systems with Applications pp 273-279
[10] Jin W, Li Z J and Wei L S 2000 The improvements of BP neural network learning algorithm C. 5th International Conference
[11] Baoan Y, Hai J I 2001 A Study of Commercial Bank Loans Risk Early Warning Based on BP Neural Network *J.* Systems Engineering-theory & Practice

Ground-based Multi-platform Point Clouds Registration

Yuan Gao*and Mingchu Li

School of Software Technology, Dalian University of Technology, Dalian 116620, China;

* Correspondence: 937018401@qq.com; ssdut@dlut.edu.cn

Abstract: Laser scanning system provides an efficient solution to rapidly acquire 3D information of large-scale scenes. Point clouds collected by laser scanning systems contain numerous objects with significant disparities in size, complicated and incomplete structures, holes, varied point densities, and huge data volumes, raising great challenges for automated point clouds registration, segmentation, and object detection. The dissertation presents a hierarchical merging based multi-platform point clouds registration algorithm to align MLS point clouds and unordered TLS point clouds from various scenes and validates its performance on nine challenging datasets. The algorithm improves the efficiency and accuracy of point cloud registration and enhances the registration ability of the algorithm for low-overlap and high-symmetry point clouds.

Key words: Point clouds, automatic registration, Urban scene mapping

1. Introduction

When the mobile laser scanning system (MLS) acquires the three-dimensional data of the ground objects around the road, the data of the three-dimensional target in the scene is incomplete due to the mutual occlusion between the targets, which seriously affects the accuracy of target recognition and model reconstruction. In practical applications, MLS is used to collect large-scale cities or high-speed scene point clouds, and then TLS is used to supplement the key areas and data holes to compensate for the lack of data coverage and scene representation of a single laser scanning system. There are spatial benchmarks, observation angles, and data quality differences between ground-based multi-platform laser point clouds, which poses a huge challenge to ground-based multi-platform laser point cloud automated registration.

Aiming at the low efficiency of existing ground-based multi-platform point cloud registration algorithm, difficulty in solving low overlap registration, lack of MLS and TLS point cloud registration theory research, this paper proposes a fast and robust ground-based multi-platform laser point. Ground-based Multi-platform Point Clouds Hierarchical Registration (GMPCHR), which realizes multi-view TLS point cloud registration under unified theoretical framework and multi-view TLS and MLS point cloud registration, improving point cloud matching. The quasi-efficiency, accuracy, and scope of application enhance the registration ability of the algorithm for low-overlap and high-symmetry point clouds.

2. Related work

When MLS obtains the 3D data of the ground objects around the road, the data of the 3D target in the scene is incomplete due to mutual occlusion between the targets, which seriously affects the accuracy of target recognition and model reconstruction, and limits the actual efficiency of the laser scanning system. In practical applications, MLS is used to collect large-scale cities or high-speed scene point clouds, and then TLS is used to compensate key areas and data holes to make up for the lack of data coverage and scene expression of a single laser scanning system. There are spatial benchmarks, observational perspectives, and data quality differences between the MLS point cloud and the TLS point cloud, which poses a challenge for automatic registration of MLS and TLS point clouds.

In order to reduce the computational complexity and improve the efficiency of point cloud registration, scholars propose a multi-view cloud registration strategy based on shape growth ([1], [2]). The multi-view cloud registration method based on shape growth first selects the point cloud with the largest surface area or the maximum number of points as the seed point cloud, then pairs the seed point cloud and other point clouds, and finally merges with the seed point cloud. There is a point cloud with enough pairs of features of the same name. Iterate through the above process until all point clouds are merged. Compared with the algorithm based on minimum spanning tree and global energy optimization, the algorithm has lower computational complexity and higher computational efficiency. However, it has higher requirements on point cloud overlap and cannot solve low-overlap point cloud registration.

Yu et al proposed a multi-scale ICP method to achieve non-rigid registration of overlapping MLS point clouds [3]. The algorithm first detects the features of each feature in the point cloud, and divides them into five levels according to the size of the feature. Then, according to the order of size from large to small, the features of each layer are used as the input of the ICP algorithm to carry out the point cloud registration; finally, the rigid transformation combination of multiple cubic spline control points is used to express the non-rigid transformation model. Yang et al firstly use FPFH and semantic feature points as feature descriptors to perform exhaustive pairwise registration, and then calculate the number of the same name features between point clouds as weights to construct an undirected weighted graph. Finally, the method implements the conversion of any point cloud to the reference point cloud by calculating the minimum spanning tree of the graph [4]. Zai et al fused the feature similarity and geometric consistency to construct the global energy equation, and transformed the problem of point cloud registration into the problem of energy optimization; then used the game theory method to optimize the solution to obtain the global optimal feature set of the same name [5]. This method improves the accuracy and recall rate of feature recognition of the same name, but the computational complexity is high and it is difficult to apply to large-scale point cloud registration. Yan et al first segmented the on-board laser point cloud according to the trajectory and removed its ground point cloud; then used the point cloud three-dimensional bounding box to find the overlapping point cloud and used the improved ICP method for registration. Overlapping point clouds; finally, using the registration parameters to reverse the positioning and attitude error of the on-board laser scanning system [6]. When there is a large deviation between overlapping MLS point clouds, the ICP algorithm will converge to the local extremum, resulting in registration failure.

3. Method

3.1. Data preprocessing

In this paper, the point density in the three-dimensional neighborhood is used to remove the abnormal points in the point cloud, and the remaining point clouds are subjected to regular down-sampling. Then, along the direction of the vehicle's trajectory, the point cloud is divided into a series of local point clouds in a XY plane according to a certain width, length and overlap.

3.2. Point cloud multi-level feature calculation

This paper calculates the local feature descriptors of key points in the laser point cloud and the point cloud global feature descriptors of the point cloud block. This paper develops a point cloud overall aggregation descriptor, which realizes the accurate characterization of the overall characteristics of the laser point cloud by encoding the BSC descriptors of all key points in the point cloud. The algorithm first randomly selects the BSC descriptors of N key points from all laser point clouds, and uses the K-means algorithm to cluster them into K categories, and uses the K-type center as a visual word to obtain a visual dictionary. Then, calculate the laser separately. The sum of the BSC descriptors of all key points in the point cloud to the residuals of their corresponding visual words; finally, the residuals of all the visual words are concatenated and normalized to be the global aggregate descriptor (GAD) of the point cloud.

3.3.Point cloud near-neighbour (overlapping) structure diagram fast construction

The relative position between the ground-based multi-platform laser point cloud registration midpoint clouds is unknown, increasing the complexity of point cloud registration. In order to estimate the relative position between the point clouds, most existing algorithms first perform an exhaustive pairwise registration of the point clouds in the scene, and then filter the point clouds with large overlap by pre-defined registration indicators. Such methods are computationally time consuming and difficult to apply to ground-based multi-platform laser point cloud registration for large scenes. This paper develops an adjacent (overlapping) point cloud efficient indexing method based on point cloud overall aggregation descriptor similarity, which realizes the rapid construction of point cloud neighbor (overlap) structure relationship diagram, which reduces the complexity of ground-based multi-platform laser point cloud registration algorithm.

The two-point cloud with large overlap contains more BSC descriptors with the same name, so its corresponding point cloud overall aggregation descriptor GAD has greater similarity. Specifically, the similar components and dissimilar components in the GAD descriptors of adjacent two-point clouds are derived from their overlapping and non-overlapping regions, respectively. Based on the above considerations, this paper first calculates the Manhattan distance of K corresponding components between GAD features, and reorders them according to the distance from small to large; then calculates the similarity between features and calculates the overlap between point clouds.

In order to verify the relationship between GAD feature similarity and its corresponding point cloud overlap, this paper uses linear regression analysis to calculate the correlation between them. For any point cloud, first calculate the overlap degree between the point cloud and other point clouds and the corresponding GAD feature similarity; then use the degree of overlap as the abscissa, and the normalized similarity as the ordinate to form the data point; finally, the method of squares fits the regression equation and calculates the deterministic coefficient R^2 of the equation.

3.4. The best registration point cloud pair calculation considering the maximum overlap degree

The two point clouds with large overlap have more similar features. We use the overall aggregation of the point cloud to describe the sub-similarity to indirectly reflect the overlap between the point clouds, and preferentially match the point cloud with the most similarity of the point cloud GAD. We use a two-two-point cloud registration algorithm that combines BSC to describe sub-similarity, geometric consistency and visual compatibility, and register the two sets of point clouds with the largest aggregate similarity in the overall aggregation descriptor, including the BSC descriptor matching of the same name. There are four main steps to improve geometric consistency, such as error matching culling, space conversion parameter calculation, and visual compatibility testing. Finally, we merge the source point cloud and the target point cloud that satisfy the visual compatibility into a new point cloud and calculate its BSC descriptor sub-set. We use the BSC to describe the sub-sets and incremental merges to obtain the BSC descriptor sub-sets of the new point cloud, thus implementing point cloud fusion.

4. Experimental results and analysis

4.1. Multi-view TLS point cloud registration results

Figure 1 shows the overall and detailed results of the multi-view TLS point cloud registration at 32 stations of Wuhan Longquanshan Park. Figure 1a and b are the details of one of the building registrations and the station location of the scanning station, point clouds of different scanning stations are represented by different colors.

(a) (b)

Figure 1. Overall and details of multi-view TLS point cloud registration at 32 stations of Wuhan Longquanshan Park.

4.2. Multi-view TLS and MLS point cloud registration results
Figure 2 shows the overall and detailed results of the algorithm's MLS point cloud and 5-station multi-view TLS point cloud registration at the School of Geographical Sciences. Figure 2 shows the results of MLS point cloud and multi-view TLS point cloud registration, as well as the position of the multi-view TLS point cloud station. The MLS point cloud is shown in gray, and the TLS point clouds of different scanning stations are represented by different colors.

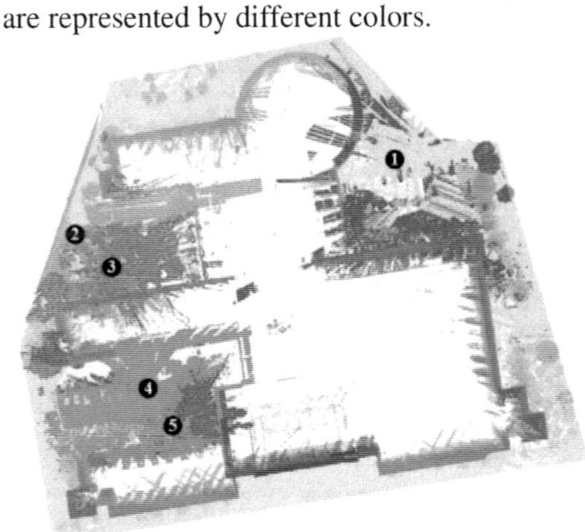

Figure 2. Results of MLS point cloud and multi-view TLS point cloud registration

4.3. Registration Accuracy Quantitative Evaluation and Analysis
In this paper, the minimum, maximum, mean and root-mean-square errors of the rotation angle error and translation error, as well as the accuracy of registration, are used to quantitatively evaluate the performance of the proposed method. In order to obtain the true value of the point cloud registration, this paper first uses the manually labeled pair of points with the same name to perform rough point cloud registration, and then uses the multi-view ICP algorithm for fine registration. The experimental results show that the ground-based multi-platform laser point cloud registration method achieves excellent performance in multi-view point cloud registration, multi-view TLS, and MLS point cloud

registration in different scenes. The average rotation angle error is less than 0.1 degree. The translation error is less than 0.1 meters, which can meet the three-dimensional target extraction and three-dimensional reconstruction accuracy requirements.

4.4. Quantitative evaluation and analysis of time efficiency

This section quantitatively evaluates the complexity of the registration algorithm, the time spent for each key step, and the total elapsed time. Experimental results show that the multi-platform laser point cloud registration method proposed in this paper achieves high computational efficiency in multi-view TLS point cloud registration, multi-view TLS, and MLS point cloud registration in different scenarios. In particular, it is worth noting that the number of two-by-two registrations in this paper is obviously less than the number of exhaustive registrations. It takes only half an hour to automate the registration of TLS points in 32 stations of Wuhan Longquanshan Park, significantly improving the computational efficiency.

4.5. Experimental comparison and discussion

In order to further analyze the performance of the GMPCR algorithm, the paper compares the current multi-view TLS point cloud registration algorithms from four aspects including accuracy of registration, rotation angle error, translation error, and algorithm time-consuming. As shown in Table 1. At the same time, in order to verify the impact of local feature descriptors and multi-view matching strategies on the performance of the registration algorithm, this paper also analyzes the performance of variants GMPCR* and GMPCR# of the GMPCR algorithm. The GMPCR# and GMPCR* use the multi-view TLS point cloud matching strategy based on the minimum spanning tree and the multi-view TLS point cloud matching strategy based on shape growth instead of the hierarchical registration strategy in this paper, the other steps are consistent with the GMPCR algorithm.

Table 1. Algorithm performance comparison

Data set		SRR (%)	Rotation angle error $\boldsymbol{e}^r_{r,i}$ (mdeg)		Translation error $\boldsymbol{e}^t_{r,i}$ (mm)		operation hours (min)
			Ave	RMSE	Ave	RMSE	
Wuhan Longquan shan Park	Weber et al., 2015	100.0	49.1	12.4	43.3	7.2	361.6
	GMPCR*	100.0	48.6	9.8	42.9	8.1	177.8
	Guo et al., 2014	100.0	44.5	13.7	37.2	9.4	186.2
	GMPCR#	100.0	44.7	13.6	36.8	9.8	94.6
	GMPCR	100.0	**40.2**	9.2	**31.1**	8.0	**39.5**

5. Conclusions

Aiming at the problems of the existing ground-based multi-platform point cloud registration algorithm, such as low efficiency, difficulty in resolving low overlap registration, lack of theoretical research on MLS and TLS point cloud registration, etc. This paper presents a fast, robust method for automatic registration of ground-based multi-platform LiDAR point clouds, which achieves a multi-vision TLS point cloud registration and multi-view TLS and MLS point cloud registration under a unified theoretical framework. The results of comprehensive comparison experiments show that the proposed method for multi-platform LiDAR point cloud registration improves the efficiency, accuracy, and scope of point cloud registration, and enhances the ability of the algorithm to register with low overlap and high symmetry point clouds. In the future work, the ground-based multi-platform LiDAR point cloud registration method will be further optimized and expanded, including: closed-loop automated detection and redistribution of registration errors based on closed-loop constraints; further improve the theoretical system of multi-platform point cloud registration, and expand this method for multiplatform laser point cloud registration such as TLS, MLS and ALS.

References

[1] Mian A, Bennamoun M, Owens R A. 2006a. A novel representation and feature matching algorithm for automatic pairwise registration of range images[J]. *International Journal of Computer Vision*, **66(1)**: 19-40.

[2] Guo Y, Bennamoun M, Sohel F, et al. 2014. 3D object recognition in cluttered scenes with local surface features: A survey [J]. *IEEE Transactions on Pattern Analysis and Machine Intelligence*, **36(11)**:2270–2287.

[3] Yu F, Xiao J, Funkhouser T. 2015. Semantic alignment of LiDAR data at city scale[C]. *IEEE Conference on Computer Vision and Pattern Recognition*, 722-731.

[4] Yang B, Dong Z, Liang F, et al. 2016. Automatic registration of large-scale urban scene point clouds based on semantic feature points[J]. *ISPRS Journal of Photogrammetry and Remote Sensing*, **113**: 43-58.

[5] Zai D, Li J, Guo Y, et al. 2017. Pairwise registration of TLS point clouds using covariance descriptors and a non-cooperative game[J]. *ISPRS Journal of Photogrammetry and Remote Sensing*, **134**: 15-29.

[6] Yan L, Tan J, Liu H, et al. 2018. Automatic non-rigid registration of multi-strip point clouds from mobile laser scanning systems[J]. *International Journal of Remote Sensing*, **39(6)**: 1713-1728.

On-line Data Analysis and Optimization of Crankshaft Dynamic Balance

Chuanhong Zhou[1,a], Lei Ding[1,b], Lianghuang Lin[1,c]

[1]School of Mechatronic Engineering and Automation, Shanghai Key Laboratory of Intelligent Manufacturing and Robotics Shanghai University, Shanghai, China

E-mail: [a]13817297415@139.com; [b]18817834919@163.com; [c]980825296@qq.com

Abstract. The crankshaft is a high-speed rotating component that is widely used in many fields. Dynamic balance of the crankshaft is an important process stage that affects the quality of the crankshaft. In order to optimize the quality of crankshaft dynamic balance machining, this paper illustrates the importance of the initial unbalance of crankshaft to the dynamic balance machining of crankshaft. For the initial imbalance distribution of crankshaft, the improved mean method, gradient descent method and mean shift algorithm are used to calculate the machining center. The position of the hole realizes the crankshaft imbalance correction with different effects and reduces the processing strength of the crankshaft dynamic balance phase. The study is applicable to the machining of crankshafts in geometric centering processes in batches.

Keywords: crankshaft dynamic balance, online control, mean method, gradient descent method, mean shift algorithm

1. Introduction

The crankshaft acts as a rotor and, in the working state, mainly outputs power by rotating motion. Rotor unbalance is a common phenomenon in the rotor. The crankshaft blank produced by the same mold has similar axial mass distribution, crack defects and shaft deformation. As a result, the uneven distribution of crankshaft blanks in the same batch presents a certain deviation trend [1]. The crankshaft roughing stage of the geometric centering process determines a pair of center holes for the machining of the entire batch of crankshafts, thereby determining the initial unbalance of the crankshaft. The initial unbalance is continuously reduced during crankshaft roughing and finishing, and finally the final balance of the crankshaft is used to control the dynamic imbalance within the allowable range. If the unbalanced offset trend can be grasped by sampling analysis, and the better central hole position is found, the unbalance amount entering the final balance of the crankshaft can be as small as possible, so that the crankshaft dynamic balance processing quality can be improved, the crankshaft quality can be improved, and the production cost can be reduced.

2. Crankshaft dynamic balance online quality control

The online control of the crankshaft machining is carried out, and the position of the central hole is adjusted one or more times during the crankshaft machining process to optimize the dynamic balance machining effect of the crankshaft. By collecting the dynamic balance test data of the partially processed crankshaft, the center hole position is analyzed and calculated, and the result is fed back to

IOP Conf. Series: Materials Science and Engineering **646** (2019) 012035 doi:10.1088/1757-899X/646/1/012035

the machining. process, so that the subsequent machining effect is improved. The application of reasonable online control technology can further improve the processing quality in the final equilibrium stage, which is conducive to actual production. During the crankshaft machining process, the initial unbalance amount is a dynamic change [2], which is shown in Figure.1, which is U_0 in the roughing process, U_1 in the roughing process, and U_2 in the finishing process. U_2 is necessary for the final equilibrium phase. U_2 data analysis is the best and no additional acquisition is required. The crankshaft online system operating mode can be represented by Figure.2:

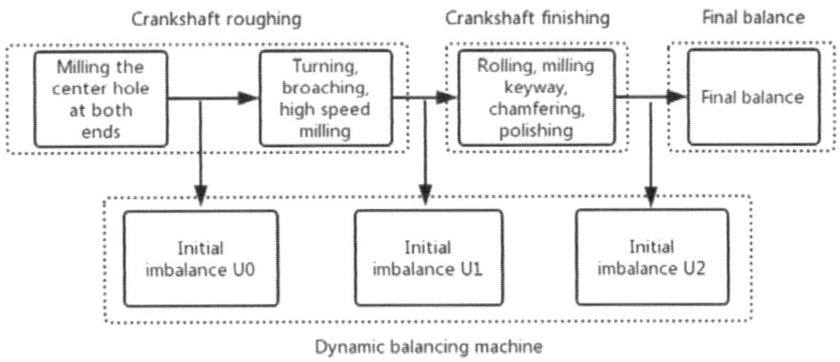

Figure 1. Initial imbalance of the crankshaft

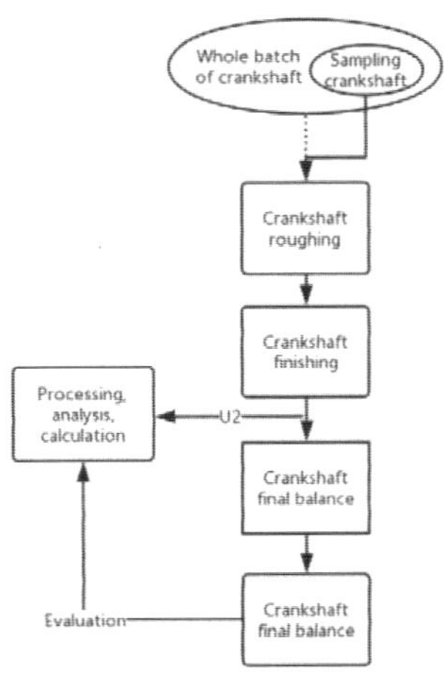

Figure 2. Crankshaft online quality control mode

3. Crankshaft dynamic balance analysis and optimization

3.1 Crankshaft dynamic imbalance distribution trend

The center hole position of the traditional geometric centering process is centered at the geometric center or the mean center of the sampled crankshaft [3]. The data points in the standard coordinates are used to describe the offset of the unbalance on the end face of the crankshaft. The mean center of the sample number n is calculated as:

$$\begin{cases} x = \sum_{i=1}^{n} x_i/n \; (n = 1,2 \ldots, n) \\ y = \sum_{i=1}^{n} y_i/n \; (n = 1,2 \ldots, n) \end{cases} \tag{1}$$

The result is greatly affected by the randomness of sampling. It is necessary to ensure that the number of samples is sufficient to stabilize the results; and the data does not exclude sample points that deviate from the expected value, and the accuracy of the results is insufficient. Figure 3. shows the frequency distribution of the imbalance of the entire batch of crankshafts on the two end faces.

The mean method is improved, and data noise filtering is added [4] to enhance the stability and accuracy of the results. Round with the mean center(O_x, O_y), R is the maximum dynamic balance offset, the radius of the R_f split circle, and k is the filter strength:

$$(x - O_x)^2 + (y - O_y)^2 = {R_f}^2 \quad (2)$$

$$R_f = (1 - k) \cdot R \qquad (0 \le k < 1) \quad (3)$$

The mean center is recalculated according to the data points in the circle, and the updated mean center has a certain offset from the original mean center. As shown in the histogram, when the filter intensity k=0.5, As shown in Figure 4, the mean center of the k value is taken as 0.5. the data points with a large offset of about 5% can be filtered, the calculated mean center is represented by a green dot, and the semi-transparent area indicates the range of the noise data; Bit represents the old mean center.

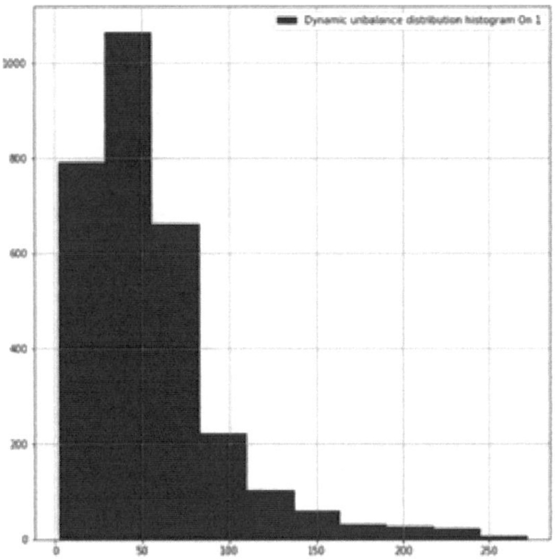

Figure 3. The frequency distribution of the imbalance of the entire batch of crankshafts on the two end faces

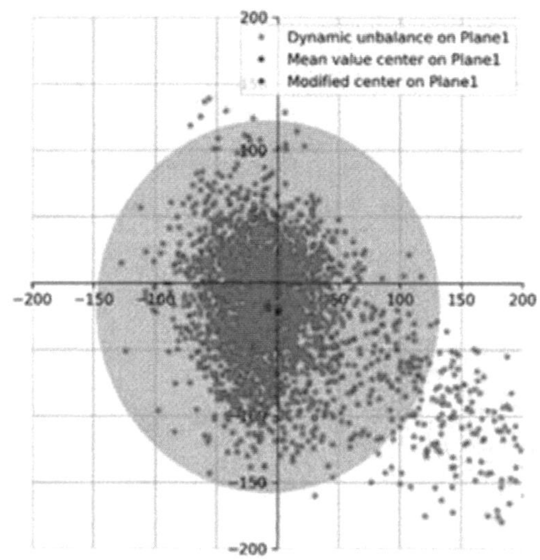

Figure 4. The mean center of the k value is taken as 0.5

3.2 Dynamic balance analysis based on gradient descent method

To minimize the machining of the crankshaft balance and the shortest machining time, the gradient hole method can be used to calculate the center hole position[5]. The gradient essence is a vector, and the function has the largest rate of change along the gradient direction. In the sampled data points, the sum of the Euclidean distances of the central hole and all data points can represent the sum of the unbalanced quantities of the sampled samples[6]. The loss function of the crankshaft dynamic balance problem can be defined as:

$$\min f(x, y) = \min \sum_{i=1}^{n} \sqrt{(x_i - x)^2 + (y_i - y)^2} \quad (4)$$

Then the gradient can be expressed as the gradient descent method uses the negative gradient direction as the direction of the iteration:

$$\nabla f(x, y) = \left(\sum_{i=1}^{n} \frac{x - x_i}{\sqrt{(x - x_i)^2 + (y - y_i)^2}}, \sum_{i=1}^{n} \frac{y - y_i}{\sqrt{(x - x_i)^2 + (y - y_i)^2}}\right)^T \quad (5)$$

The equation of the gradient descent method can be expressed as follows, where θ_i represents the current point and the updated point, α represents the step size, and $J(\theta)$ represents the loss function:

$$\theta_i = \theta_i - \alpha \frac{\partial}{\partial \theta_i} J(\theta) \quad (6)$$

The unbalanced amount of the entire batch of crankshafts is represented by a scatter t[7], the forward path of the iterative result is indicated by a black trajectory, and the central hole position is identified by a blue dot. The iterative process and results on both ends of the crankshaft are shown in Figure. 5:

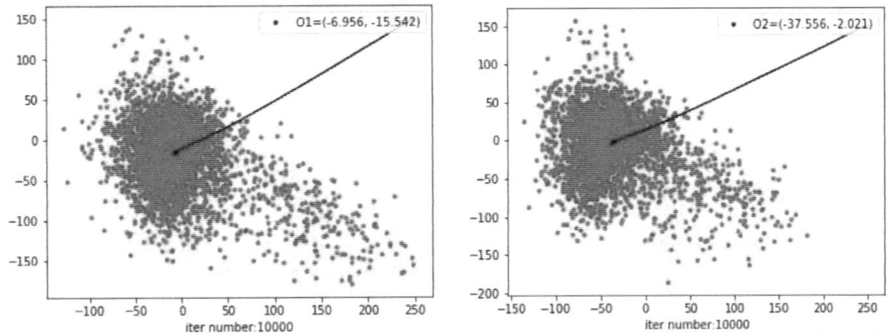

Figure.5 Process and results of calculating the center hole position by the gradient descent method

Since the gradient descent method is a global search algorithm, it is necessary to increase the number of samples to reduce the influence of sampling randomness on the results. After repeated experiments, the number of samples reached 200 or above can make the calculation results better. As shown in table 1.only some calculation results are listed. Generally speaking, the gradient descent method has smaller $\overline{D}$ and larger ΔD than the improved averaging method. It gets a smaller amount of imbalance and the correction is better:

Table 1. Comparison of calculation results of A end face sampling data

Magnitude	Improved mean algorithm	Value of $\overline{D}$ [g*cm]	Value of ΔD [g*cm]	Gradient descent center	Value of $\overline{D}$ [g*cm]	Value of ΔD [g*cm]
100	(-8.908，-20.154)	54.674	6717.1	(-8.74，-20.758)	54.715	6594.9
100	(-8.156，-15.729)	54.481	7292.2	(-8.806，-12.81)	54.567	7035.9
200	(-9.286，-18.234)	54.576	7009.1	(-7.59，-15.806)	54.471	7322.0
200	(-5.103，-18.163)	54.559	7059.8	(-6.432，-15.412)	54.470	7325.0
300	(-9.994，-15.549)	54.557	7065.7	(-7.035，-14.628)	54.474	7313.1
300	(-8.776，-16.672)	54.508	7211.7	(-6.130，-16.662)	54.485	7280.3

3.3 Dynamic balance analysis of crankshaft based on mean shift algorithm

To obtain a larger number of small unbalanced crankshafts and improve the first pass rate of crankshaft machining, the center shift position can be calculated using the mean shift algorithm to improve the distribution of crankshaft unbalance.

In the N-dimensional space, there are k sample points in the high-dimensional sphere region S_h with bandwidth h, and the basic vector form of the mean offset is:

$$M_h(x) = \frac{1}{k}\sum_{x_i \in S_h}(x_i - x) \tag{7}$$

The Gaussian kernel function is introduced to participate in the calculation, so that the contribution of the data points closer to the current point is larger at each iteration, and the contribution of the data points farther from the current point is smaller.

$$N(x) = \frac{1}{\sqrt{2\pi}h} e^{\frac{-x^2}{2h^2}} \tag{8}$$

The vector of the mean shift can be expressed as:

$$M_h(x) = \frac{\sum_{i=1}^{n} G_H(x_i - x) w(x_i)(x_i - x)}{\sum_{i=1}^{n} G_H(x_i - x) w(x_i)} \tag{9}$$

Where $w(x_i)$ is greater than 0, indicating the weight of each sample; H is the bandwidth matrix:

$$G_H(x_i - x) = |H|^{-\frac{1}{2}} G(H^{-\frac{1}{2}}(x_i - x)) \tag{10}$$

$$H = \begin{pmatrix} h_1^2 & 0 & \dots & 0 \\ 0 & h_2^2 & \dots & 0 \\ \dots & \dots & & \dots \\ 0 & 0 & \dots & h_d^2 \end{pmatrix}_{d\times d}$$

The averaging offset vector can be expressed as:

$$M_h(x) = \frac{\sum_{i=1}^{n} G\left(\frac{x_i - x}{h_i}\right) w(x_i)(x_i - x)}{\sum_{i=1}^{n} G\left(\frac{x_i - x}{h_i}\right) w(x_i)} \tag{11}$$

By defining the kernel function and the mean shift function for the iterative update calculation; setting the minimum iteration distance to the most iterative convergence condition; setting the bandwidth adjustment sample point range, you need to take h>10 to meet the dynamic balance analysis, h too small The result is too many sorting categories to determine the center hole position. The results of the analysis and calculation show that the green dot in the scatter plot indicates the position of the center hole, and the orange dot is the data noise, As shown in Figure.6, we can see that the points around the density center are densely distributed:

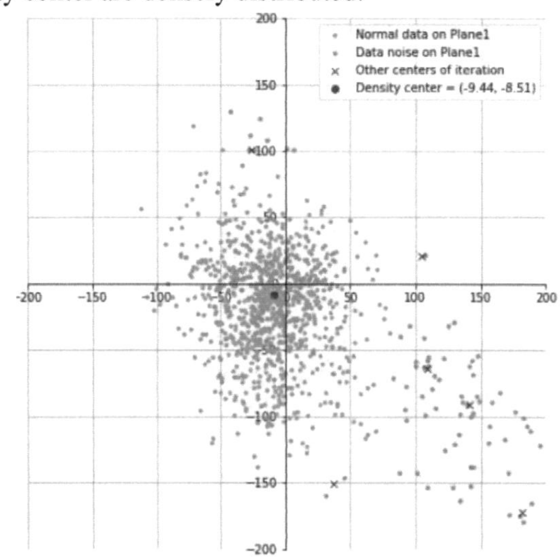

Figure 6. Mean Shift Algorithm Results at 1000 Samples

Since the dynamic balance of the entire batch of crankshafts cannot be obtained in practice, the training algorithm requires a relatively high number of data samples, so the product with a quantity of 200 is evaluated for results. The center hole position of each algorithm is shown in Table 2.

Table 2. Comparison of results at both ends of the crankshaft with a sample number of 200

	Center of the A end face	$\bar{D}$ value of the A end face [g*cm]	Center of the B end	$\bar{D}$ value of the B end face [g*cm]
Center of the mean method	(1.316,-21.642)	55.421	(-30.967,-7.403)	52.961
Center of k=0.5 mean method	(-5.103,-18.163)	54.559	(-38.458,-2.616)	52.374
Center of gradient descent	(-6.432,-15.412)	54.470	(-35.722,-1.346)	52.396
Center of mean shift method	(-7.56,-12.45)	54.553	(-31.05,-0.04)	52.739

The mean shift algorithm needs to train the algorithm model, so the number of samples should be increased appropriately. After repeated experiments, the $\bar{D}$ and ΔD of the center hole calculated by the mean shift algorithm are not more advantageous than the previous two algorithms, but still have a good correction effect; the result of the algorithm can generally obtain a larger number of imbalances at 50g.Crankshaft within cm, improve the first pass rate of the crankshaft, only listed some results, In Figure 7.,the gray figure in the histogram corresponds to the geometric center of the unbalanced frequency distribution, the yellow is the result of the improved mean method, the green is the result of the gradient descent method, blue The result of the mean shift algorithm. Take the results of one of the end faces and plot the frequency distribution as shown in figure 7.:

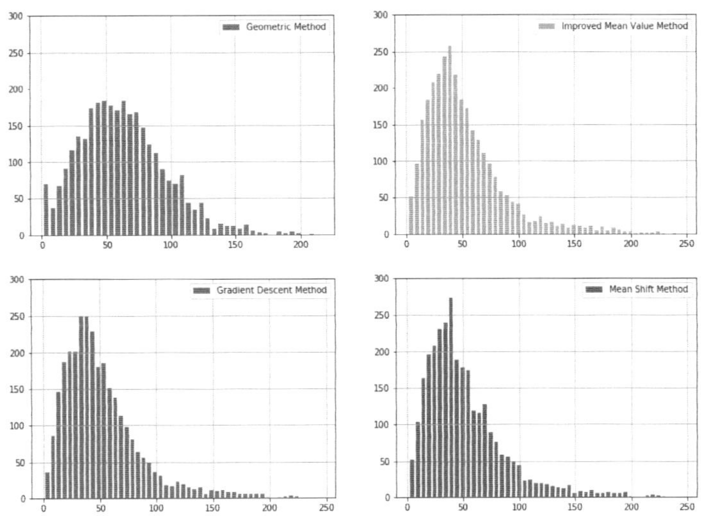

Figure 7. The distribution of the frequency of the entire batch of crankshafts corresponding to the number of samples being 200

4. Conclusions

(a) The mean shift algorithm can obtain better unbalance correction effect, but it is not as good as the former two; it is suitable for a slightly larger number of samples, which requires the central hole to make the crankshaft processing the highest pass rate, and the crankshaft with the highest number of high-quality crankshafts Balance processing.

(b) Gradient descent method can obtain the best unbalance quantity correction effect; it is suitable for crankshaft dynamic balance machining with a small number of samples, minimum required dynamic unbalance, minimum workload and shortest processing time.

Acknowledgment

Thanks to my teacher for your guidance and advice and thank Shanghai Key Laboratory of Intelligent Manufacturing and Robotics for assistance with our work.

References

[1] Zhu Zhengde. A little enlightenment on the development of crankshaft roughing manufacturing technology[J]. Diesel Engine Design and Manufacture, 2010, 16(02): 1-4+43.

[2] Zhu Zhengde. Analysis of the Selection of Geometric Centering and Mass Centering in Crankshaft Machining[J]. Automobiles & Parts, 2009(28): 26-28

[3] Wei G. Hybrid Harmonic Averaging Operator and its Application to Decision Making. [C] Iita International Conference on Services Science. IEEE, 2009.

[4] Aydore S, Thirion B, Grisel O, et al. Using Feature Grouping as a Stochastic Regularizer for High-Dimensional Noisy Data [J]. 2018
[5] Barzilai J, Borwein J M . Two-Point Step Size Gradient Methods [J]. IMA Journal of Numerical Analysis, 1988, 8(1): 141-148
[6] Kiwiel KC.Convergence and Efficiency of Subgradient Methods for Quasiconvex Minimization [J]. Mathematical Programming, 2001, 90(1): 1-25
[7] Sutton RS , Maei H R , Precup D, et al. Fast Gradient-descent Methods for Temporal-difference Learning with Linear Function Approximation [C]. Danyluk Et. 2009

Research and Design of ERP System for Small and Medium-sized Enterprises under Great Intelligence Mobile Cloud

Yong Wen*

Department of finance, GuangDong University of Science & Technology, Dongguan, Guangdong, China

*34235485@qq.com

Abstract. Based on the information management needs of SMEs, the article adopts the rapid prototyping method, designs the overall framework of the cloud ERP system, maps out the functional modules, adopts the B/S architecture, uses Java as the development tool, and SQL Server as the background server. The cloud computing and APP technologies are used to realize instant access between the PC and the mobile terminal, and the cloud ERP system with relatively complete system functions, simple interface and mobile operation is realized.The system incorporates new technologies such as big data analytics, artificial intelligence, machine learning and data mining to enable it to realize intelligent financial analysis, intelligent financial forecasting and intelligent financial decisions.

1. Introduction

At present, the mainstream ERP system in the market is designed and developed based on the management needs of large and medium-sized enterprises. The functions are large and complete, the interface is complex, the business process is cumbersome, the internal control cost is high, and the project implementation is difficult [1];The system design lacks experienced consultants and business experts to participate deeply, and can not fully meet the business needs; the system is not well integrated into cloud computing and mobile Internet technology, and information transmission is lagging [2-3].The design of large ERP systems does not quite match the business management processes of SMEs.Therefore, the development of cloud ERP system for small and medium-sized enterprises is a top priority and has broad application prospects.The system complies with relevant regulations such as the Accounting Standards for Business Enterprises, the latest tax law, and the business processes of SMEs, tailored to the needs of customers;The system utilizes technologies such as cloud computing and APP to realize instant communication between the PC and the mobile terminal, realizing real-time management and control of the enterprise;Senior CPAs participate in system development and business process reengineering, optimize management, make the system light and flexible, intuitive interface, simple operation, and meet the needs of enterprise management for related stakeholders and enterprise managers.Under the premise of following internal control, the system will properly post the business review link, which not only ensures the immediate processing of the business, but also protects the enterprise's asset security integrity and the authenticity of accounting information, ensuring the orderly and efficient operation of the business activities.

With the development of cloud computing and mobile Internet technologies, and the continuous improvement of enterprise management information systems, the enterprise management model is developing towards digitalization and platformization [4].The core part of enterprise management informationization is still financial management informationization [5].The finance department must leverage the cloud ERP platform to achieve real-time control of enterprise management.The system realizes the seamless connection between cloud financial management system and cloud ERP, OA and other systems, standardizes the front-end business processing flow, completes automatic data collection and real-time processing through multiple ways, helps SMEs reduce operational risks and improve management efficiency.

2. System analysis

Due to the limited space, the following mainly selects the financial management system in the cloud ERP as the research object, and briefly introduces the process of system analysis, system design, system implementation and testing.

2.1. The goal of the new system

Through the combing and optimization of the company's business, it forms a unified management standard, reconstructs the business management platform, and gradually realizes the interconnection of management data, providing timely and accurate data support for the company's operation and decision-making.

2.1.1. Organizational management architecture using group + branch. Focusing on the company's business, we will build an internal interactive communication platform to form a consistent network working environment. Through the platform, employees of member companies can centrally collect, publish and share internal information and promote cultural integration of enterprises.

2.1.2.Realize the standardized management. Follow scientific management methods and management standards, formulate enterprise management processes, and set up necessary control links to ensure smooth and consistent processes.

2.1.3. Building a holistic learning organization. Improve the reuse efficiency of internal knowledge, and form the knowledge replication ability and innovation ability of enterprises.

2.1.4. Support for large clusters. That is, deployment mode that supports load balancing.

2.1.5.The information flow and the capital flow are effectively combined. The financial comprehensive monitoring of the company's information flow, the pre-judgment of the financial risk management and control, and the realization of standardized management [6].

2.1.6. Using mobile internet technology. Support mobile phone camera, use FRID smart tag and bar code for information control, intelligent entry of accounting vouchers, automatic accounting. Realize electronic invoicing, online auditing, cloud reimbursement, payment, account books, and reports in one go.

2.1.7. Real-time tracking management. Real-time inventory management, intelligent purchasing, and immediate costing. With dynamic price management function, real-time view of cost and gross profit, adjust commodity prices according to local market conditions.

2.1.8. Establish network-based resource sharing and form a resource chain for diversified industries. Realize data docking and interconnection between other systems and financial systems.

2.1.9. Support WeChat function. The system should be able to access the corporate WeChat public number to meet the development of the WeChat public platform.

2.1.10. Support mobile phone operation function. Develop an Android-based mobile APP platform, so users can log in to the platform anytime and anywhere as long as they have access to the Internet.

2.2.System Requirements Analysis

Identify the specific needs of users, combined with the actual business level of the enterprise, determine the functions that the system should achieve and some characteristics based on it, including the operating environment of the system and evaluate the development potential of the system [7].

*2.2.1. Functional requirements analysis.*After analysis, the system requires the following functions: support e-commerce platform, online and offline inventory integration, financial business integration; realize various functions of cloud financial management, mobile query, modify, delete, add data is convenient,system stability and reliability.

*2.2.2. System performance requirements analysis.*The performance requirements of software products include the following aspects: integrity, adaptability, maintainability, normativeness, security, reliability, and practicability.Good safety, reliability and practicality are the advantages of this system.

2.2.2.1. Safety. It includes three aspects.First, network security: Cloud ERP protects data privacy through encryption algorithms, key technologies and other authentication methods, encrypts data transmission, storage and processing, conceals information, and protects data security. The hardware firewall is used to effectively prevent attacks from the network and the destruction of the computer system, so that the information system is stable, efficient, and safe.

The second is the security of the information system: to ensure that the business process and authority are clearly defined; the system uses the authorized use mode to prevent unauthorized users or terminals from accessing; effectively prevents users from entering illegal data to ensure data integrity; establish a system authentication module to block illegal Plug-in and program implantation; use necessary transaction processing technology to prevent abnormal business process data caused by network problems, crashes, etc.; establish system logs, form process data for key business operations, in order to cause operational errors, etc, and recover accordingly[8].

2.2.2.2. Reliability. The system should have a certain fault tolerance. When the user enters illegal data, the system must perform necessary automatic updates and error prompts to prevent data loss caused by abnormalities.

2.2.2.3. Practicality. System interface development uses the J2EE specification. For the program of the human-computer interaction part, the data operation guiding function is provided to enable the user to quickly complete the instruction and data entry. In principle, the user's common interface and function require the system to give a correct response within 3 seconds after the user issues the corresponding command.The system should provide convenient operation, generalized modules, high practicality and strong scalability, and provide good response for data collection, data verification, report presentation and other aspects. Bringing data quantification and real-time monitoring to project management, providing good financial flow management for financial information, and providing systematic data support for company decision-making.

In addition, the operating and maintenance costs of the system should be fully considered, the dependence on hardware and third parties should be reduced, and the cost of system operation should be reduced.

2.2.3. Business process analysis. The system service flow chart is shown in Figure 1.

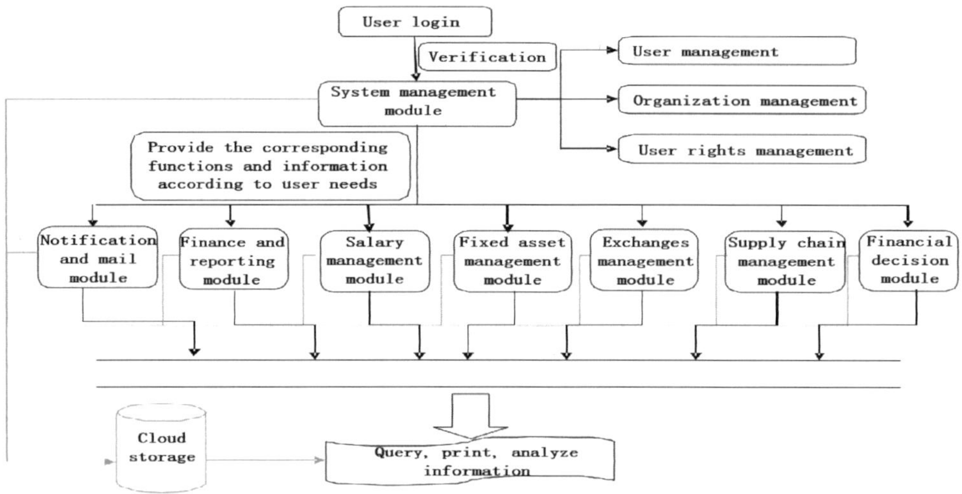

Figure 1. System service flow chart.

2.2.4. Hierarchical logic model.

2.2.4.1. First layer logical model. The user inputs the data into the cloud financial management system, and after the system processes, automatically generates financial information feedback to the user.

2.2.4.2. The second layer of logical model. As shown in Figure 2.

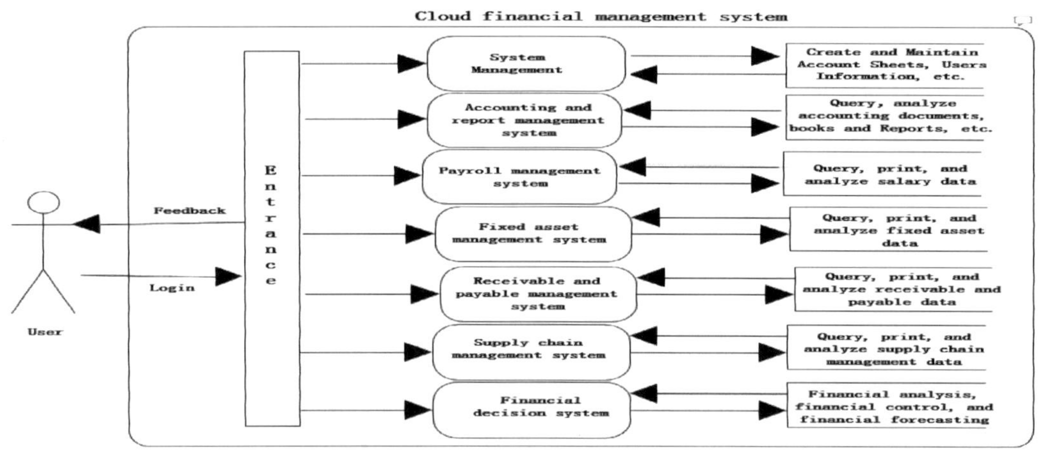

Figure 2. Layer 2 logical model.

3. System design

3.1. Design principles and ideas

System design principles:System coding must follow the design documentation, conform to the design philosophy, and adhere to a uniform coding style. The development language of this system is Java language, and the platform chooses eclipse3.5 SDK. The database selects the SQL Server database.

3.2. Overall architecture design

According to the previous analysis of the overall needs, the overall architecture of the system adopts the B/S structure.

3.3. Functional module architecture design

The main functional modules of the cloud financial management system are: general ledger and report function module, salary management function module, fixed asset management function module, transaction management function module, supply chain management function module, and financial decision function module.

3.4. Key technology solutions

There are several other methods for data backup and restoration in this system data maintenance module: one is SQL Back database automatic backup tool SQLBackupAndFTP; the other is to use SQL program for backup, which is divided into full backup, incremental backup and log Backup; third is to directly copy the data file and related configuration files.

3.5. Functional module design

The following takes the knowledge base management sub-module in the financial decision system as an example to introduce the function, input, output and processing flow of the module.

3.5.1. Functional description. This module is used to enter, edit, and manage company systems, financial systems, accounting standards, tax laws, financial management, production processes, system operations, design drawings, design tools, business data, design standards, competitors, business partner descriptions and more. Help the business staff to solve all kinds of confusion in their work in a question and answer form.

3.5.2. Input Output. Input items: query keywords, time, company rules and regulations, accounting standards, company financial management system, tax regulations, system operations, design drawings, design tools, business data, competitors, business partners, design standards and other related help documents .Output items: query and print various questions and answers.

3.5.3. Process. The knowledge base processing flow is divided into five stages. The first stage is knowledge collection, including: collecting various knowledge from business management systems, office management systems, production management systems, existing knowledge stocks, and external systems. The second stage is knowledge storage, which organizes the collected knowledge. The third stage is knowledge management, including: knowledge encryption, process management, classification management, operation tracking, and system warning. The fourth stage is the use of knowledge, such as: knowledge discovery, knowledge retrieval, knowledge push, knowledge query, knowledge download, and so on.The fifth stage is knowledge innovation, which realizes knowledge innovation and value enhancement through knowledge portal, knowledge attention, expert comment, knowledge conversion, etc., and then forms new knowledge in the knowledge base. Thus, the knowledge base forms a closed loop that is constantly updated and developed.

3.6. Database design

The database adopts cloud deployment methods such as distributed technology, virtual technology, and ESB enterprise service bus. It is the focus of the financial management system. Therefore, a well-designed database is one of the important tasks in the overall design.

3.6.1. Database design. In the requirements phase, you must determine the database structure, such as the database, database table structure, and so on. The data table structure is a prerequisite for the data window. In the design of the database, the database specification is usually designed to achieve a three-paradigm.It mainly includes the determination of the data table fields and the determination of the relationship between the data tables.

3.6.2. Database security design. The security of the system is closely related to the input cost and design structure of the system. Therefore, to develop a technology, security strategy and effective program that meets the actual operational needs must consider a variety of factors.

3.6.2.1. Checking from the database portal. At the time of data entry, a validity check is performed, which often requires control of the entry of the static table to prevent "sickness from entering the mouth."

3.6.2.2. Multiple protection with SQL Server. SQL Server security protection is used to ensure data security. SQL Server servers are divided into four security levels: Microsoft SQL Server object security control, database security control, advanced server platform security control, and user-assisted software control.Each level of security control is a layer of firewall to prevent the destruction of illegal users.

4. System implementation and testing

Follow the system requirements analysis and system design ideas to build a new system platform, design programs and coding, build cloud data warehouse, system testing and debugging, system trial run, system acceptance, and write the documentation of each stage, including: system requirements analysis manual , system design specification, system platform and hardware manual, database source mode list, program flow chart and source program list, system use and maintenance manual, etc.Due to the limited space, these processes are not repeated here.

5. Summary and outlook

In the new era of mass entrepreneurship and innovation, the article is based on the Guangdong-Hong Kong-Macao Greater Bay Area, focusing on the long-term, small ERP system located in the front-end financial information collection of SMEs, seamlessly connected with the OA office system, with professional accountants escorting, its help Enterprises realize informationization. The system uses Java as the front-end development tool and SQL Sever as the back-end database to organize the system development and smooth implementation according to the principles of software engineering.
Due to lack of experience and time constraints, only some functions of this ERP system have been put into the cloud, in the future the cloud computing and ERP are better integrated, and the functions of the system are continuously expanded and improved. The graphic design software such as PS, AI, ID, FL is used. The system interface is more beautiful and more intuitive, and the security technology of the cloud platform is improved, and the security design of the cloud storage in the front end, the background and the data transmission is improved, so that the system is more secure and reliable.

Acknowledgement

[Fund Project] Characteristic Innovation Project of “Innovation and Strong School Project” of Higher Education in 2017 in Guangdong Province (Project No.: 2017WTSCX125).

References

[1] Jiang C 2018 China. D. Emotion cognition research based on the analysis of EEG and ERP signals.Taiyuan: Taiyuan University of Technology Library.
[2] Zuxiang X 2018 China. D. The Implementation and Application of SAP System MM Module for Auto Parts Enterprise.Jinan: Shandong University Library.
[3] Hua F 2018 China. J. Enterprise intelligence financial innovation based on intelligent manufacturing environment.Communication of Finance and Accounting.**29**(2018)p64
[4] Michael O 2004 Pump Up SQL Server 2000. Windows & Net Magazine (New York: Windows & .NET Magazine Press)pp 116-118
[5] Yong W 2013 China. J. Research on the Status Quo and Countermeasures of Financial Management Informationization in Private Enterprises.Communication of Finance and Accounting.**11**(2013)pp 61-64
[6] Ivar J,Grady B and Rumbaugh J 2004 USA.M.The Unified Sotfware Development Process (New York: Addison-Wesley Professional Press)
[7] Dhar A 2003 USA.J. Automating SQL Server with SQL DMO.Windows Developer Network.**14**(2003)pp 260-261
[8] Yu L 2018 China. D. Research of CL Company Business Process Optimization Based on ERP Application.Lanzhou: Lanzhou University Library.

A Hybrid Genetic Algorithm for the Distributed Permutation Flowshop Scheduling Problem with Sequence-Dependent Setup Times

Jiangping Huang[1], Quanke Pan[1,*] and Qingda Chen[2]

[1] School of Mechanical and Electrical Engineering and Automation, Shanghai University, Shanghai, China
[2] State Key Laboratory of Synthetical Automation for Process Industries, Northeastern University, Shenyang, China

[*]Email: panquanke@shu.edu.cn

Abstract. The distributed permutation flowshop scheduling problem (DPFSP) has attracted many researchers' attention in recent years. In this paper, we extend the DPFSP by considering the sequence-dependent setup time (SDST). A new hybrid genetic algorithm (HGA) for the DPFSP with the SDST (SDST/DPFSP) is presented to minimize the maximum of the completion time. At first, a new population initialization is proposed. And then, the newly-designed operators are described in details, and we also introduce the mutation rate and the crossover rate to balance the mutation operator and the crossover operator. To further improve the obtained solution, a new local search method is developed. At last, the orthogonal experimental design is applied to adjust the parameters in the HGA, and a comprehensive computational campaign based on the 135 instances demonstrates the effectiveness of the proposed HGA for the SDST/DPFSP.

1. Introduction

The DPFSP, which is a kind of extension of the PFSP, has been an active research field and has impact on many manufacturing industries, such as the compact strip production in the Steel manufacturing industry [1]. In the DPFSP, there exists f identical factories, each of which is a flowshop with m machines and capable of processing all of the jobs independently. The scheduling task of the DPFSP is determining the assignment of the jobs to a certain factory and the processed order of the jobs in a factory. In addition, there are always some operations between the processing of two consecutive jobs such as machine-cleaning, tool-replacing, job-transporting, and so on, which can waste some time called the setup time. Furthermore, the setup time of one job is not only determined by the machine processing it but also by the job processed immediately before it, that is to say, the setup time is sequence-dependent. Therefore, to make the problem studied in the paper more practical, we consider the DPFSP with the SDST to minimize the maximum completion time. To best of our knowledge, there is few literatures of the SDST/DPFSP. Therefore, we will show the related literatures about the DPFSP and the PFSP with the SDST (SDST/PFSP).

The DPFSP is first proposed by Naderi and Ruiz in [2] with the objective of minimizing the makespan. In the following years, many researches have been done to it, and the related effective heuristics and metaheuristics has been developed to address different objectives. Pan, Gao and Wang present three constructive heuristics and four metaheuristics in [3]. And later, Ruiz, Pan and Naderi propose simple iterated greedy algorithms to minimize the makespan among all factories in [4]. In [5], Pan, Gao and Li address a novel distributed assembly permutation flowshop scheduling problem with three constructive heuristics, two variable neighborhood search methods, and an iterated greedy algorithm. An artificial chemical reaction optimization (CRO), a hybrid discrete cuckoo search (HDCS) algorithm and a new heuristic are proposed in [6][7] and [8] respectively, and the genetic algorithm (GA) is also applied to the DPFSP in [9][10], which has been proved an effective method for the DPFSP. In addition, some other algorithms such as the modified iterated greedy algorithm, the efficient tabu search algorithm are developed in [11][12].

The corresponding literatures of the SDST/PFSP are abundant as well, and we will list as follows. Mirabi [13] presents an ant colony optimization (ACO) algorithm. Rios-Mercado and Bard develop a branch-and-cut algorithm and two different mathematical formulations in [14], and later, they further develop the mentioned algorithms in [15]. Three heuristics, a memetic algorithm, a genetic algorithm and a NEH heuristic are compared in [16] by Kaweegitbundit. Vanchipura and Rajesh develop two constructive heuristic algorithms in [17] and a neighbourhood search known as variable neighbourhood descent (VND) to further improve the two constructive heuristics in [18]. Four iterated local search (ILS) algorithms are constructed in [19] by Wang and Dong. The migrating birds optimization metaheuristic (MBO) is applied to both [20] and [21].

As shown above, the DPFSP and the SDST are both important, and some researches have been done to them respectively, while there is few literature about the SDST/DPFSP. Therefore, we study the SDST/DPFSP in the paper. According to [2], the DPFSP is a NP-Complete problem, and without doubt, the SDST/DPFSP is a NP-Complete problem as well. Therefore, the algorithms applied to the DPFSP are also suitable for the SDST/DPFSP. With the extensive application of the GA, we apply it to the SDST/DPFSP as well. To further improve the quality of the solution we have obtained, we explore the GA with local search, which is called the HGA in the paper.

procedure Initialization of one individual
//how to create an individual according to the makespan
  generate a job permutation randomly
  for $i := 1$ **to** n **do** //where n is the number of the jobs
    calculate the $C_{\max}$ of each factory
    find factory $F_{\min}$ with the minimized $C_{\max}$
    place job J_i to factory $F_{\min}$
  endfor

Figure 1. The pseudo codes of the initialization of an individual.

procedure mutation
  create a random number k
  for $i := 1$ **to** k **do**
    select a factory F_{rand} randomly
    select a job J_{rand} randomly from factory F_{rand}
    replace job J_{rand} to the best position in all the factories
  endfor

Figure 2. The pseudo codes of the mutation operator.

2. Problem description

The SDST/DPFSP can be described as follows. A set of n jobs, $J = \{J_1, J_2, \mathrm{K}, J_n\}$, can be processed in f identical factories, $F = \{F_1, F_2, \mathrm{K}, F_f\}$, each of which can process one job J_j, ($J_j \in J$) independently just like an independent flowshop, and all of them are with m machines, $M = \{M_1, M_2, \mathrm{K}, M_m\}$, respectively. All jobs are processed in the same production routine that starts from the first machine and ends at the last machine with no machine being skipped. The processing time of J_j on machine M_i ($M_i \in M$) is denoted by $p_{j,i}$, the SDST of J_j can be described as $S_{i,j',j}$ on the condition that job J_j is not the first job processed on machine M_i and the job immediately processed before it is job $J_{j'}$ ($J_{j'} \in J$), if the job J_j is the first processed job, the SDST of J_j on machine M_i is denoted as $S_{i,j,j}$. The objective of the SDST/DPFSP in the paper is to find a permutation to minimize the makespan ($C_{\max}$).

3. A hybrid genetic algorithm

In this section, we present the HGA for the SDST/DPFSP with the objective of minimizing the makespan. We will describe the solution representation, population initialization, the design of the operators and the local search method in details and give the pseudo codes.

3.1. Solution representation

We adopt the job-based representation [4], where a solution is represented by a two-dimensional array, $\pi = \{\pi_1, \pi_2, ..., \pi_f\}$. There are f rows, each of which is used to show the permutation of one factory F_l, $\pi_l = (\pi_{l,1}, \pi_{l,2}, ..., \pi_{l,\eta_l})$, $l = 1, 2, ..., f$, where η_l is the total number of jobs assigned to factory F_l. For example, there are two factories and five jobs, and it is assumed that job 1, 2 and 5 are placed in factory F_1, and job 3 and 4 are assigned to factory F_2, as a result, the solution can be represented by $\pi = \{\pi_1, \pi_2\}$, where $\pi_1 = (1,2,5)$ and $\pi_2 = (3,4)$.

```
Procedure Job_Insertion ( )
  set the initial solution π as the reference solution π_ref
  for i := 1 to f do
    for j := 1 to n_i do // where n_i is the number of
      Job ← π_ref (i, j) // the jobs processed in factory i
      π ← delete Job inside π
      π ← replace Job to the best position among all factories
    endfor
  endfor
```

Figure 3. The pseudo codes of the Job_Insertion ( )

```
Procedure Job_Exchange ( )
  find factory F_max with the largest makespan
  for i := 1 to n_max do // where n_max is the number of factory F_max
    for j := 1 to f do
      if j ≠ F_max do
        exchange job J_i with each job in factory j and find the best job pairs
      endif
    endfor
  endfor
  exchange the best job pairs
```

Figure 4. The pseudo codes of the Job_Exchange ( )

3.2. Population initialization

It is clear that the quality and the diversity are two important factors for an initial population. To balance the quality and the diversity, we create the individuals in different ways. At first, the NEH2 and the VND(a), both of which are first proposed in [2], are carried out to produce two highly-qualified individuals respectively, while the rest of the individuals are all created according to the makespans of all the factories. In the method, we obtain a random permutation of the jobs at first, and then we reassign the jobs one by one to the factory with the minimized makespan. The pseudo codes of this initialization method are presented in figure 1.

```
procedure crossover
  set the best individual in the population as parent 1
  set a randomly-selected individual in the population as parent 2
  //where parent 1 and parent 2 are different
  for i := 1 to f do // where f is the number of the factories
    select the random cutting points of each factory for both parent 1 and parent 2
    save the right sides of parent 1 and parent 2 to R2 and R1 respectively
    check parent 1 and R1, and save the remaining jobs to L1 in the order of the appearance
    check parent 2 and R2, and save the remaining jobs to L2 in the order of the appearance
    obtain child 1 by reinserting the jobs in L1 to the best slot of R1
    obtain child 2 by reinserting the jobs in L2 to the best slot of R2
  endfor
```

Figure 5. The pseudo codes of the crossover operator.

```
Procedure Localsearch
  flag ← ture
  π ← perform Job_Exchange ( ) on π
  while flag do
    flag ← false
    π' ← perform Job_Insertion ( ) on π
    if C_max(π') < C_max(π) then
      flag ← true
      π ← π'
    endif
  endwhile
```

Figure 6. The pseudo codes of the proposed local search method.

3.3. Design of the operators

In the HGA, there are three operators, the selection operator, the mutation operator and the crossover operator respectively, all of them should be designed carefully since the performance of the algorithm are effected significantly by them. In the following part, the details about the operators are presented. With the selection mechanism, the individuals are ranked according to the fitness and selected for the next iteration. In the paper, the fitness of individuals is equal to $C_{\max}$, which can show the differences among the individuals.

With the mutation operator, the individual should be changed slightly to escape from the local optimization with some good permutation remained. In the mutation operator proposed in the paper, k jobs are selected randomly, where k is a random number which is not more than the half of the total number of jobs. And then the k jobs are reinserted into the best position in all the factories. In this way, on the one hand, it is easy to escape from the local optimization, on the other hand, the new individual can remain the relative positions of the most unchangeable jobs. The pseudo codes of the mutation operator are presented in figure 2.

The crossover operator is applied to two selected individuals for obtaining two new individuals with better fitness. Our crossover operator, for both parents, selects cutting points randomly for all the factories, and the sets of jobs on the right sides of child 1 and child 2 are denoted by $R1$ and $R2$ respectively which are obtained by exchanging the sets of jobs on the right sides of parent 1 and parent 2. The remaining jobs in each sequence of both children are kept in the order of their appearance of their parents, and then they are reinserted to the best position of the right sequence of the corresponding child one by one. In the paper, the two selected individuals are the best one and a randomly-selected one which is different from the best one. The pseudo codes of the crossover operator are presented in figure 5.

```
Procedure HGA
  set the parameters: Psize, P_crossover, P_mutation
  while the stopping criterion is not satisfied do
    evaluate the fitness of each individual
    apply the local search method on the best individual and a randomly-selected individual
    update the best solution
    apply the selection operator
    make a randomly-selected individual replaced by the best solution
    select another individual randomly that is different from the new best individual
    produce a random number r
    if r < P_crossover then
      apply crossover operator
    endif
    produce a random number r'
    if r' < P_mutation then
      apply mutation operator
    endif
  endwhile
```

Figure 7. The general idea of the HGA.

3.4. Local search method

The local search has been proved an effective method for the PFSP in the related literature. To further improve the solution we have obtained, we present two new local search methods, Job_Insertion ( ) and Job_Exchange ( ), and combine them to the mentioned GA operators. In the Job_Insertion ( ), the reference solution is proposed, at the begin of the Job_Insertion ( ), we set the initial solution as the reference solution, and test the jobs in the order of the reference solution one by one to find the best position and place them to it. The pseudo codes of the Job_Insertion ( ) are presented in figure 3. In the Job_Exchange ( ), at first, we should find the factory $F_{\max}$ with the maximum makespan among the factories, and then exchange the jobs in factory $F_{\max}$ one by one with the jobs in the other factories to find the best job pairs (one job from factory $F_{\max}$, and the other one from one of the other factories), and exchange the positions between them. The pseudo codes of the Job_Exchange ( ) are presented in figure 4. In the method of the proposed local search, the Job_Exchange ( ) operator is used to the individual at first. And then the Job_Insertion ( ) is applied to the new obtained individual repeatedly until no improvement can be obtained. Figure 6 presents the pseudo codes of the proposed local search method.

3.5. An overview of the HGA

The general idea of the HGA is presented in figure 7. At first, the parameters, $Psize$, $P_{crossover}$, $P_{mutation}$, should be set, where $Psize$ is the size of the population, $P_{crossover}$ and $P_{mutation}$ are the crossover rate and the mutation rate respectively. Secondly, the fitness of each individual should be calculated which is used to rank the individuals. Thirdly, the local search method is applied to the best individual and a randomly-selected individual, then the best solution should be updated which is used to replace a randomly-selected individual different from the best one. Fourthly, besides the new best individual, another individual should be selected randomly that is for the following crossover operation and mutation operation. At last, we create two random numbers r and r' for the crossover operation and the mutation operation respectively. While r is less than $P_{crossover}$, the crossover operator should be employed to the best individual and the randomly-selected individual, and if r' is less than $P_{mutation}$, both the individuals are operated by the mutation operator.

4. Computational Experiment

The 135 problem instances used to test the performance of the algorithms are listed as follows. The combinations of $n \times m \times f$ are: $\{100, 200, 300, 400, 500\} \times \{5, 8, 10\} \times \{2, 3, 4\}$, and the setup times are generated by $\text{Setuptime} = (1 + rand(\ )\%99) \times \text{Factor} / 100$, where Factor is selected from $\{25, 50, 100\}$ randomly, and $rand(\ )$ is a random number. All of the algorithms are coded in C++ and compiled with Visual Studio 2013, and the algorithms share most important functions in the code. All the experiments in this paper are performed on windows 10 machine with 1 processor, i3-8100, and 8GB of RAM memory.

Table 1. RPI values for $t = 20 \cdot m \cdot n$ Milliseconds

Instance Size	$f=2$		$f=3$		$f=4$	
	HGA(1)	HGA	HGA(1)	HGA	HGA(1)	HGA
100×5	15.31	**2.50**	14.22	**2.42**	15.10	**2.57**
100×8	14.84	**1.88**	16.03	**2.5**	16.15	**2.71**
100×10	15.21	**2.57**	15.50	**2.09**	14.70	**2.64**
200×5	11.65	**1.86**	12.99	**1.42**	12.21	**1.98**
200×8	12.46	**1.35**	12.78	**1.19**	14.50	**2.10**
200×10	12.57	**2.36**	12.01	**1.81**	13.32	**2.44**
300×5	11.26	**1.24**	12.01	**1.38**	13.52	**2.31**
300×8	10.51	**1.49**	11.26	**2.03**	12.97	**1.39**
300×10	10.54	**1.20**	11.76	**1.55**	9.57	**1.80**
400×5	10.86	**2.37**	9.51	**1.24**	11.04	**1.64**
400×8	9.57	**1.70**	11.54	**1.57**	11.27	**1.26**
400×10	9.59	**2.25**	10.43	**1.58**	9.87	**1.36**
500×5	9.73	**4.04**	10.25	**2.42**	10.70	**3.12**
500×8	9.39	**3.72**	9.06	**2.70**	10.24	**3.96**
500×10	9.77	**2.94**	9.79	**2.84**	10.07	**2.79**
mean	11.55	**2.23**	11.94	**1.92**	12.35	**2.27**

In this paper, we apply the termination criterion with a predefined CPU time, $t = C \cdot m \cdot n$, the algorithm has elapsed, where C is a predefined value. The 3 parameters, $Psize$, $P_{crossover}$, $P_{mutation}$, are fixed as follows: $Psize$=40, $P_{crossover}$=0.1, $P_{mutation}$=0.1, and all of them are obtained by the Orthogonal experimental design with an orthogonal array $L_{16}(4^3)$ within $10 \cdot m \cdot n$ milliseconds.

We compare the presented HGA with the HGA(1) proposed in [12]. In the paper, both the HGA(1) and the HGA are tested to solve the 135 instances mentioned in this section with 5 replications, and they are also terminated with the same CPU time where the value of C has been tested at 3 levels, 10, 20 and 30 respectively in the paper. The HGA shows better performance among the experiments with the three levels. Due to the limited space of the paper, we just list the RPI values of C=20 in table 1. From table 1, we can see that the means of the RPI generated by the presented HGA are much smaller than those of the HGA(1) nevertheless the different factories involved.

5. Conclusion

In the paper, we address the SDST/DPFSP with minimization of the makespan. We propose a HGA for such a problem, and the results of the numerical experiments show that the developed algorithm has a better performance than the existing HGA(1). In the future, we will further improve the HGA by developing more effective local search methods and the operators, and more related algorithms should be compared together.

Acknowledgments

This research is partially supported by the National Science Foundation of China 51575212 and 61174187.

References

[1] Quan-Ke Pan, Liang Gao, Ling Wang, 2019, *Applied Mathematical Modelling*, Volume 73, Pages 327-348
[2] B. Naderi, Ruben Ruiz, 2010, *Computers & Operations Research*, Volume 37, Issue 4, 754-768
[3] Quan-Ke Pan, Liang Gao, Ling Wang, Jing Liang, Xin-Yu Li, 2019, *Expert Systems with Applications*, Volume 124, 309-324
[4] Ruben Ruiz, Quan-Ke Pan, Bahman Naderi, 2019, *Omega*, Volume 83, 213-222
[5] Quan-Ke Pan, Liang Gao, Li Xin-Yu, Framinan M. Jose, 2019, *Applied Soft Computing,*, Volume 81, 105492
[6] Hafewa Bargaoui, Olfa Belkahla Driss, Khaled Ghedira, 2017, *Computers & Industrial Engineering*, Volume 111, 239-250
[7] J. Wang, L. Wang and J. Shen, "*A hybrid discrete cuckoo search for distributed permutation flowshop scheduling problem*," 2016 IEEE Congress on Evolutionary Computation (CEC), Vancouver, BC, 2016, 2240-2246.
[8] Fernandez-Viagas V, Framinan JM, 2015, *International Journal Of Production Research*, Vol. 53, No. 4, 1111–112m
[9] Yan Li, Zhigang Chen, Yan Li, "*The distributed permutation flowshop scheduling problem: A genetic algorithm approach,*" 2015, ICMII 2015, Atlantis Press, 2352-538X
[10] J. Wang, L. Wang and J. Shen, "*A hybrid discrete cuckoo search for distributed permutation flowshop scheduling problem*," 2016 IEEE Congress on Evolutionary Computation (CEC), Vancouver, BC, 2016, 2240-2246.
[11] Lin Shih-Wei, Ying Kuo-Ching, Huang Chien-Yi, 2013, *International Journal of Production Research*, Volume 51, Issue 16, 5029-5038
[12] Gao Jian, Chen Rong, Deng Wu, 2013, *International Journal of Production Research*, Volume 51, Issue 3, 641-651
[13] Mohammad Mirabi, 2010, *The International Journal of Advanced Manufacturing Technology*, Volume 55, Issue 1-4, 317-326

[14] Roger Z. Rios-Mercado, Jonathan F. Bard, 1998, *Computers & Operations Research*, Volume 25, Issue 5, 351-366
[15] Roger Z Rios-Mercado, Jonathan F Bard, 2003, *Journal of Combinatorial Optimization*, Volume 7, Issue 3, 291-318
[16] P. Kaweegitbundit, 2011, *Advanced Materials Research*, Volume 339, 332-335
[17] Vanchipura R, Sridharan R, 2013, *The International Journal of Advanced Manufacturing Technology*, Volume 67, Issue 5-8, 1337-1353
[18] Rajesh Vanchipura, R. Sridharan, A. Subash Babu, 2014, *Journal of Manufacturing Systems*, Volume 33, Issue 1, 65-75
[19] Wang Y, Dong X, Chen P, Lin Y, 2014, *Advances in Intelligent Systems and Computing*, vol 277, 329-338
[20] Imene Benkalai, Djamal Rebaine, Caroline Gagné, Pierre Baptiste, 2016, *IFAC-PapersOnLine*, Volume 49, Issue 12, 408-413
[21] A. Sioud and C. Gagne, "*An MBO algorithm for a flow shop problem with sequence-dependent setup times*," 2016 12th World Congress on Intelligent Control and Automation (WCICA), Guilin, 2016, pp. 2471-2474.

Research on Multi-uav Cooperative Tracking Target Based on PSO Predictive Control Algorithm

Wang Xudong[1,*], **Zhou Wei**[1] **and Wang Xiaowei**[1]

[1]No.9, Taihu Street, Taihu Town, Tongzhou District, Beijing

[*]Wanghx96@163.com

Abstract. In the multiple unmanned aerial vehicle (multi-uav) cooperative tracking target control system, the control system is extremely complex due to the nonlinear control model and constraints. In this paper, multi-uav cooperative tracking target motion modeling is adopted,optimization performance indicators is established, and multi-objective rolling optimization is transformed to single-objective optimization. Finally, particle swarm optimization (PSO) prediction control algorithm is adopted for optimization solution, realizing real-time control of multi-uav cooperative tracking target, and its feasibility is verified by simulation.

1. Introduction

Multi-uav cooperative tracking target control not only considers the relative motion between uavs and targets, but also considers the relative motion between uavs. Therefore, while improving the flight control accuracy, the balance between the two control targets should also be taken into account. As a result, in the control system, not only the control model is nonlinear, but also the constraint is nonlinear[1].To solve this problem, this paper conducts multi-uav cooperative tracking target motion modeling, establishes optimization performance index, converts multi-objective rolling optimization to single-objective optimization, and finally proposes PSO prediction control algorithm for optimization solution, realizing real-time control of multi-uav cooperative tracking target.

2. Multi-uav cooperative tracking target flight motion modeling

With the ground coordinate system as the reference coordinate system,the target tracking of multi-uav is usually at the same height, so they are projected onto the two-dimensional plane for research[2], the specific motion analysis diagram is shown in figure 1.

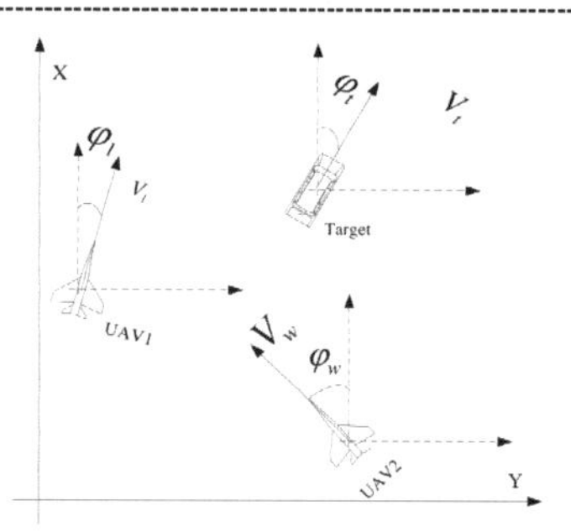

Figure 1. Schematic diagram of cooperative tracking target movement

In the figure, v_t is the target velocity, and the included Angle with the X-axis of the ground coordinate system is φ_t ; v_l is the velocity of UAV1, and the included Angle with the X-axis of the ground coordinate system is φ_l ; v_w is the velocity of UAV2, and the included Angle with the X-axis of the ground coordinate system is φ_w ; ω is the yaw angular velocity.

Suppose UAV1 coordinate is (x_l , y_l), UAV2 coordinate is (x_w , y_w), and target coordinate is (x_t , y_t), and the UAV1 motion model is as follows.

$$\dot{x}_l = v_l \cos(\varphi_l) \tag{1}$$

$$\dot{y}_l = v_l \sin(\varphi_l) \tag{2}$$

The UAV2 motion model is as follows.

$$\dot{x}_w = v_w \cos(\varphi_w) \tag{3}$$

$$\dot{y}_w = v_w \sin(\varphi_w) \tag{4}$$

$$\dot{\varphi}_w = \omega \tag{5}$$

The target motion model is as follows.

$$\dot{x}_t = v_t \cos(\varphi_t) \tag{6}$$

$$\dot{y}_t = v_t \sin(\varphi_t) \tag{7}$$

3. Discrete model and predictive control analysis of multi-uav cooperative tracking target

Discretization of the motion equation of UAV1.

$$x_l(k+1) = x_l(k) + v_l(k)\cos(\varphi_l(k))\Delta T \tag{8}$$

$$y_l(k+1) = y_l(k) + v_l(k)\sin(\varphi_l(k))\Delta T \tag{9}$$

Discretization of the motion equation of UAV2.

$$x_w(k+1) = x_w(k) + v_w(k)\cos(\varphi_w(k))\Delta T \tag{10}$$

$$y_w(k+1) = y_w(k) + v_w(k)\sin(\varphi_w(k))\Delta T \tag{11}$$

$$\varphi_w(k+1) = \varphi_w(k) + \omega_w \Delta T \tag{12}$$

$$r_2(k+1) = \sqrt{(x_w(k+1) - x_t(k+1))^2 + (y_w(k+1) - y_t(k+1))^2} \tag{13}$$

$$\theta\ (k+1) = \arctan\frac{y_l(k+1) - y_t(k+1)}{x_l(k+1) - x_t(k+1)} - \arctan\frac{y_w(k+1) - y_t(k+1)}{x_w(k+1) - x_t(k+1)} \tag{14}$$

Discretization of the motion equation of the target.

$$x_t(k+1) = x_t(k) + v_t(k)\cos(\varphi_t(k)) \tag{15}$$

$$x_t(k+1) = x_t(k) + v_t(k)\cos(\varphi_t(k)) \tag{16}$$

Will type (8), (9), (10), (11), (15), (16) into (13), (14)

$$\begin{aligned} r_2(k+1) &= \sqrt{\begin{matrix}(x_w(k) + v_w(k)\cos(\varphi_w(k))\Delta T - x_t(k+1))^2 - \\ (y_w(k) + v_w(k)\sin(\varphi_w(k))\Delta T - y_t(k+1))^2\end{matrix}} \\ &= \sqrt{\begin{matrix}(x_w(k) + v_w(k)\cos(\varphi_w(k)))\Delta T - (x_t(k) + v_t(k)\cos(\varphi_t(k))))^2 - \\ (y_w(k) + v_w(k)\sin(\varphi_w(k)))\Delta T - (y_t(k) + v_t(k)\sin(\varphi_t(k))))^2\end{matrix}} \end{aligned} \tag{17}$$

$$\begin{aligned} \theta(k+1) = {} & \arctan\frac{y_l(k) + v_l(k)\sin(\varphi_l(k))\Delta T - (y_t(k) + v_t(k)\sin(\varphi_t(k)))}{x_l(k) + v_l(k)\cos(\varphi_l(k))\Delta T - (x_t(k) + v_t(k)\cos(\varphi_t(k)))} \\ & - \arctan\frac{y_w(k) + v_w(k)\sin(\varphi_w(k))\Delta T - (y_t(k) + v_t(k)\sin(\varphi_t(k)))}{x_w(k) + v_w(k)\cos(\varphi_w(k))\Delta T - (x_t(k) + v_t(k)\cos(\varphi_t(k)))} \end{aligned} \tag{18}$$

Where ω is the angular velocity of yaw, r_1 is the distance between UAV1 and the target, r_2 is the distance between UAV2 and the target, and θ is the included Angle between UAV1 and UAV2 and the target.

In the rolling optimization process, the sampling period is relatively short. Assuming that the speed and yaw Angle of UAV1 in the sampling period remain unchanged, that is, the wind speed remains unchanged[3], then:

$$v_t(k+i) = v_t(k+i-1) = v_t(k) \tag{19}$$

$$\varphi_t(k+i) = \varphi_t(k+i-1) = \varphi_t(k) \tag{20}$$

You can get:

$$\begin{aligned} r_2(k+2) &= \sqrt{\begin{array}{l}(x_w(k+1)+v_w(k+1)\cos(\varphi_w(k+1))\Delta T-(x_t(k+1)+v_t(k+1)\cos(\varphi_t(k+1))))^2-\\ (y_w(k+1)+v_w(k+1)\sin(\varphi_w(k+1)))\Delta T-(y_t(k+1)+v_t(k+1)\sin(\varphi_t(k+1))))^2\end{array}} \\ &= \sqrt{\begin{array}{l}(x_w(k)+v_w(k)\cos(\varphi_w(k))\Delta T+v_w(k+1)\cos(\varphi_w(k)+\omega_w(k)\Delta T)\Delta T)-\\ (x_t(k)+v_t(k)\cos(\varphi_t(k))\Delta T+v_t(k)\cos(\varphi_t(k)))\Delta T)^2-\\ (y_w(k)+v_w(k)\sin(\varphi_w(k))\Delta T+v_w(k+1)\sin(\varphi_w(k)+\omega_w(k)\Delta T)\Delta T\\ -(y_t(k)+v_t(k)\sin(\varphi_t(k))\Delta T+v_t(k)\sin(\varphi_t(k))\Delta T))^2\end{array}} \end{aligned} \tag{21}$$

$$\begin{aligned} \theta(k+2) &= \arctan\frac{y_l(k+1)+v_l(k+1)\sin(\varphi_l(k+1))\Delta T-(y_t(k+1)+v_t(k+1)\sin(\varphi_t(k+1))\Delta T)}{x_l(k+1)+v_l(k+1)\cos(\varphi_l(k+1))\Delta T-(x_t(k+1)+v_t(k+1)\cos(\varphi_t(k+1)\Delta T)} \\ &\quad -\arctan\frac{y_w(k+1)+v_w(k+1)\sin(\varphi_w(k+1))\Delta T-(y_t(k+1)+v_t(k+1)\sin(\varphi_t(k+1))\Delta T)}{x_w(k+1)+v_w(k+1)\cos(\varphi_w(k+1))\Delta T-(x_t(k+1)+v_t(k+1)\cos(\varphi_t(k+1))\Delta T)} \\ &= \arctan\frac{y_l(k+1)+v_l(k)\sin(\varphi_l(k))\Delta T-(y_t(k+1)+v_t(k)\sin(\varphi_t(k))\Delta T)}{x_l(k+1)+v_l(k)\cos(\varphi_l(k))\Delta T-(x_t(k+1)+v_t(k)\cos(\varphi_t(k)\Delta T)} \\ &\quad -\arctan\frac{y_w(k+1)+v_w(k+1)\sin(\varphi_w(k)+\omega_w(k)\Delta T)\Delta T-(y_t(k+1)+v_t(k)\sin(\varphi(k))\Delta T)}{x_w(k+1)+v_w(k+1)\cos(\varphi_w(k)+\omega_w(k)\Delta T)\Delta T-(x_t(k+1)+v_t(k)\cos(\varphi_t(k))\Delta T)} \\ &= \arctan\frac{y_l(k)+2v_l(k)\sin(\varphi_l(k))\Delta T-(y_t(k)+2v_t(k)\sin(\varphi_t(k))\Delta T)}{x_l(k)+2v_l(k)\cos(\varphi_l(k))\Delta T-(x_t(k)+2v_t(k)\cos(\varphi_t(k))\Delta T)} \\ &\quad -\arctan\frac{y_w(k)+v_w(k)\sin(\varphi_w(k))\Delta T+v_w(k+1)\sin(\varphi_w(k)+\omega_w(k)\Delta T)\Delta T-(y_t(k)+2v_t(k)\sin(\varphi_t(k))\Delta T)}{x_w(k)+v_w(k)\cos(\varphi_w(k))\Delta T+v_w(k+1)\cos(\varphi_w(k)+\omega_w(k)\Delta T)\Delta T-(x_t(k)+2v_t(k)\cos(\varphi_t(k))\Delta T)} \end{aligned} \tag{22}$$

In the control time domain N, the predicted output is:

$$r_2(k+3)\text{, } r_2(k+4)\text{, } r_2(k+5)\text{, } \ldots\ r_2(k+N-1)\text{, } r_2(k+N) \tag{23}$$

$$\theta(k+3)\text{, } \theta(k+4)\text{, } \theta(k+5)\text{, } \ldots\ \theta(k+N-1)\text{, } \theta(k+N) \tag{24}$$

As you can see,the predicted output sequence is a function based on future control variables $v_w(k), v_w(k+1), \mathrm{L}\ v_w(k+N-1)$ and $\omega_w(k), \omega_w(k+1), \mathrm{L}\ \omega_w(k+N-1)$, which is obviously a multi-input multi-output nonlinear control problem[4].The values of $v_w(k), v_w(k+1), \mathrm{L}\ v_w(k+N-1)$ and $\omega_w(k), \omega_w(k+1), \mathrm{L}\ \omega_w(k+N-1)$ can be solved by the optimization algorithm, which can be used as the output of the control quantity at the next moment to perform the rolling optimization solution successively. The specific idea is shown in figure 2.

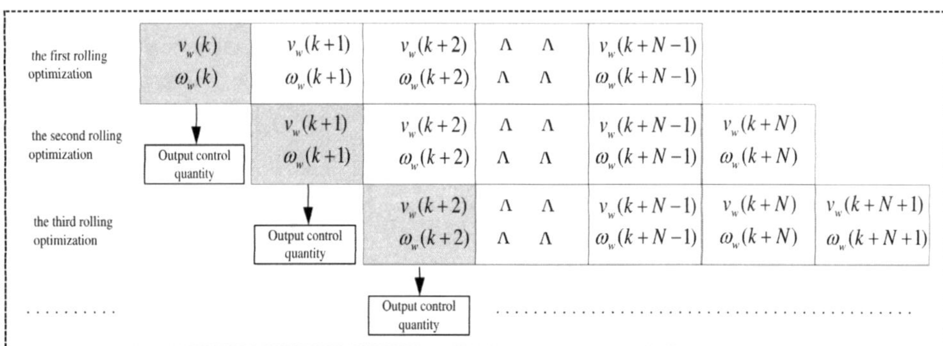

Figure 2. Rolling optimization process diagram

4. Determination of predictive performance indicators

For the problem of multi-uav cooperative tracking target and prediction control, its rolling optimization goal is to find appropriate $v_w(k), v_w(k+1), \cdots v_w(k+N-1)$ and $\omega_w(k), \omega_w(k+1), \cdots \omega_w(k+N-1)$, so that the prediction output in the whole optimization time domain is as close as possible to the reference trajectory.In this paper, the target predetermined value is used as the reference trajectory, and the error square of the target predetermined value and the predicted output in the time domain is used as the prediction performance index.

Assuming that θ_c is the included angle between the two UAV and the target, and r_c is the distance between UAV2 and the target, the following performance indicators can be obtained:

$$J_1 = \sum_{i=1}^{N} (r_2^{\,2}(k+i) - r_c^{\,2})^2$$
$$J_2 = \sum_{i=1}^{N} (\tan(\theta_2(k+i) - \tan\theta_c)^2 \tag{25}$$

J_1 and J_2 are functions based on future control variables $v_w(k), v_w(k+1), \cdots v_w(k+N-1)$ and $\omega_w(k), \omega_w(k+1), \cdots \omega_w(k+N-1)$. Optimization constraints of control variables are introduced as follows:

$$\begin{cases} v_w(k+i-1) - \Delta v < v_w(k+i) < v_w(k+i-1) + \Delta v & i \in 0,1,2\cdots N-1 \\ \omega_{\min} < \omega_w(k+i) < \omega_{\max} & i \in 0,1,2\cdots N-1 \\ v_{\min} < v_w(k+i) < v_{\max} & i \in 0,1,2\cdots N-1 \end{cases} \tag{26}$$

This problem becomes a multi-objective nonlinear optimization problem with constraints.

5. The transformation from multi-objective optimization to single-objective optimization

In the process of multi-uav cooperative target tracking, it is not necessary to accurately control the relative distance between the uav and the target to a constant value,as long as the output fluctuates within a small range around the set value, the set value target is considered to have been achieved[5].Firstly, the performance index J_1 of relative distance is optimized by a single objective to make UAV2 reach the radius required for tracking target. Then, J_1 is disturbed by a small target by setting the radius to find out the feasible solution space. Then, the optimized value of J_2 is found in the feasible solution to achieve hierarchical multi-objective optimization.

Specific ideas are as follows:

If $r_2(k) > r_c + r_{error}$ or $r_2(k) < r_c - r_{error}$, then:

$$\min(J_1) = \min(\sum_{i=1}^{N} (r_2^{\,2}(k+i) - r_c^{\,2})^2)$$

St..

$$\begin{cases} v_w(k+i-1) - \Delta v < v(k+i) < v_w(k+i-1) + \Delta v & i \in 0,1,2\cdots N-1 \\ \omega_{\min} < \omega_w(k+i) < \omega_{\max} & i \in 0,1,2\cdots N-1 \\ v_{\min} < v_w(k+i) < v_{\max} & i \in 0,1,2\cdots N-1 \end{cases} \tag{27}$$

If $r_c - r_{error} =< r_2(k) =< r_c + r_{error}$, add a slight disturbance error, and the value is positive r_{error}, then:

$$\sum_{i=1}^{N} (r_2^{\,2}(k+i) - r_c^{\,2})^2) < r_{error}^{\,2} \times N$$

St..

IOP Conf. Series: Materials Science and Engineering **646** (2019) 012038 doi:10.1088/1757-899X/646/1/012038

$$\begin{cases} v_w(k+i-1)-\Delta v < v(k+i) < v_w(k+i-1)+\Delta v & i \in 0,1,2\text{L } N-1 \\ \omega_{\min} < \omega_w(k+i) < \omega_{\max} & i \in 0,1,2\text{L } N-1 \\ v_{\min} < v_w(k+i) < v_{\max} & i \in 0,1,2\text{L } N-1 \end{cases} \quad (28)$$

Solve a set M of (v,ω), and get:

$$\min(J_2) = \min(\sum_{i=1}^{N}(\tan(\theta_2(k+i) - \tan\theta_c)^2)$$

St..

$$(v,\omega) \in M \quad (29)$$

Unify the above two questions to get:

$$\min(J_2) = \min(\sum_{i=1}^{N}(\tan(\theta_2(k+i) - \tan\theta_c)^2)$$

St..

$$\begin{cases} v_w(k+i-1)-\Delta v < v(k+i) < v_w(k+i-1)+\Delta v & i \in 0,1,2\text{L } N-1 \\ \omega_{\min} < \omega_w(k+i) < \omega_{\max} & i \in 0,1,2\text{L } N-1 \\ v_{\min} < v_w(k+i) < v_{\max} & i \in 0,1,2\text{L } N-1 \\ \sum_{i=1}^{N}(r_2^{\,2}(k+i) - r_c^{\,2})^2) < r_{error}^{\;\;2} \times N & i \in 0,1,2\text{L } N-1 \end{cases} \quad (30)$$

Then the above problems become two nonlinear multivariable constrained single objective optimiza-tion problems.

6. Nonlinear rolling optimization based on PSO

According to the basic principle of pso, after proper constraint processing and local convergence processing, the nonlinear rolling optimization process based on pso is obtained as follows.

6.1. Particle swarm initialization and the determination of search termination conditions

The particle population size is set to 25, the initial position of the particle is set to 0, and the particle velocity component is represented by the control variables $v_w(k), v_w(k+1),\text{L } v_w(k+N-1)$ and $\omega_w(k), \omega_w(k+1),\text{L } \omega_w(k+N-1)$. The initial value of the particle is divided into the initial value setting of the optimization variable of the first sampling point and the initial value setting of the optimization variable of the non-first sampling point.The initial value of the velocity control variable $v_w(k), v_w(k+1),\text{L } v_w(k+N-1)$ of the first sampling point particle swarm optimization algorithm is set as $v_w(k-1)$, and the initial value of the angular velocity control variable $\omega_w(k), \omega_w(k+1),\text{L } \omega_w(k+N-1)$ is set as 0.The control quantity value optimized by the former sampling point is taken as the initial value of the latter sampling point, and so on (as shown in figure 3).

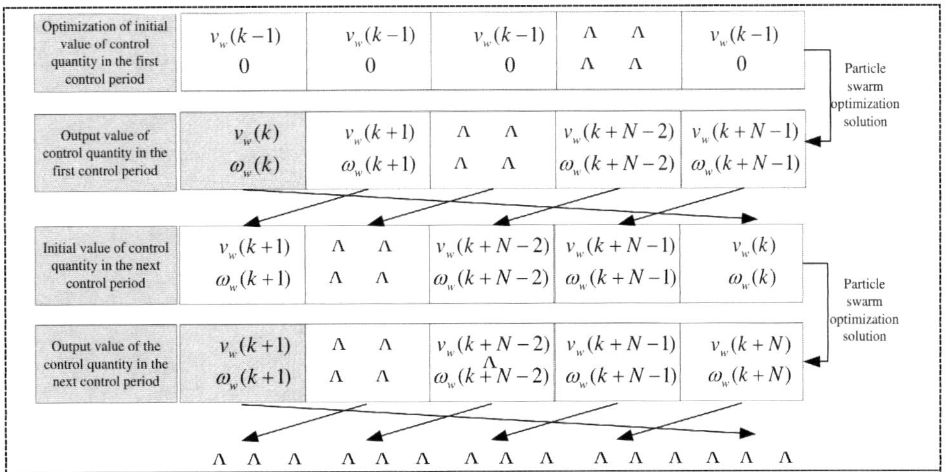

Figure 3. Initial value setting for particle swarm optimization

The maximum number of iterations is taken as the termination condition of the convergence of PSO, the larger the number of iterations, the higher the control accuracy, and the larger the calculation amount, the specific value is determined according to the hardware of the controller.

6.2. Fitness function and particle swarm variation calculation

Optimization performance index J can be used as the fitness of particle swarm, but it is easy to be locally optimal. Genetic algorithm is adopted for local optimal processing, and the variation calculation formula is as follows.

$$child_1(x_i) = p_i parent_1(x_i) + (1 - p_i) parent_2(x_i)$$
$$child_2(x_i) = p_i parent_2(x_i) + (1 - p_i) parent_1(x_i) \tag{31}$$

$$child_1(V_i) = \frac{parent_1(V_i) + parent_2(V_i)}{|parent_1(V_i) + parent_2(V_i)|} \cdot |parent_1(V_i)|$$

$$child_2(V_i) = \frac{parent_1(V_i) + parent_2(V_i)}{|parent_1(V_i) + parent_2(V_i)|} \cdot |parent_2(V_i)| \tag{32}$$

p_i is a random number between [0,1] (empirical value is about 0.2), child means child particle, parent means parent particle[6].

6.3. Particle swarm selection, movement and evaluation

In the calculation of feasible solution and infeasible solution method, the calculation of constraint violation degree is very important.First, a formula of constraint violation degree must be given.

Suppose M constraints are: $g_i(x) <= 0 \qquad i \in M$

Substitute the value of particle velocity into $g_i(x)$, if $g_i(x) > 0$, the violation degree of constraint increases by 1, otherwise it is 0, the constraint violation value *converse* is as follows.

$$converse = \sum_{i=1}^{M} k_i \quad if\ g_i(x) > 0 \quad k_i = 1 \ \ else \ \ k_i = 0 \tag{33}$$

For particle selection strategy, the comparison criteria are:

- When both particles are feasible, their objective function value is compared, and the particle with small objective function value is better.
- When both particles are infeasible, if their constraint violation degree is less than the given threshold, the particle with small objective function value is optimal. If it is larger than the threshold, the particle with small constraint violation degree is preferred.

- When particle X is not feasible and particle Y is feasible, if the violation degree of partial constraints of particle X is less than the given threshold, the particle with small objective function value is optimal. Otherwise the Y particle is optimal.

In order to reserve a large number of unfeasible solution particles at the early stage of evolution and a small number of unfeasible solution particles at the later stage of evolution, the following threshold change formula is proposed.

$$\varepsilon_{converse} = (1 - \frac{N_{now}}{N_{\max}}) \times converse_{\max} \tag{34}$$

Where N_{now} is the current iteration number, $N_{\max}$ is the maximum iteration number, and $converse_{\max}$ is the maximum possible violation value.

6.4. Output of control quantity

The optimization results are obtained through the above steps, and the control quantity is output according to the optimization results.

7. The simulation results

Suppose the tracking scenario is:the target first moves at the angular velocity of $-\pi/100\ rad/s$, after 100s, it turns to the X axis for straight motion, after 200s, it turns to the angle of $\pi/3$ with the Y axis for straight motion, after 300s, it moves in the Y axis for straight motion, there is angular velocity noise of $\pm 0.001 rad/s$ when the target moves in a straight direction, and the target position has estimation error.The Initial conditions for cooperative tracking target is shown in table 1.

Table 1. Initial conditions for cooperative tracking target

Initial position and angle of the target	Initial velocity and angular velocity of the target	Target position estimation error	Initial position and angle of UAV1	Initial position and angle of UAV2
$(0,0,\pi/2)$	(20,0)	±5	$(-2000,-1800,\pi/4)$	$(-2000,-1900,\pi/4)$
Initial velocity and angular velocity of UAV1	**Initial velocity and angular velocity of UAV2**	**Min and max cruising speed of UAV1**	**Min and max cruising speed of UAV2**	**Target tracking distance and target Angle**
(40,0)	(40,0)	(35,45)	(25,45)	$(1000,\pi/4)$

Prediction time domain N=5, uav angular velocity change rate is limited to (-0.1,0.1),Matlab Simulink toolbox was used for simulation, and the simulation results are shown in figure 4-8.

IOP Conf. Series: Materials Science and Engineering **646** (2019) 012038 doi:10.1088/1757-899X/646/1/012038

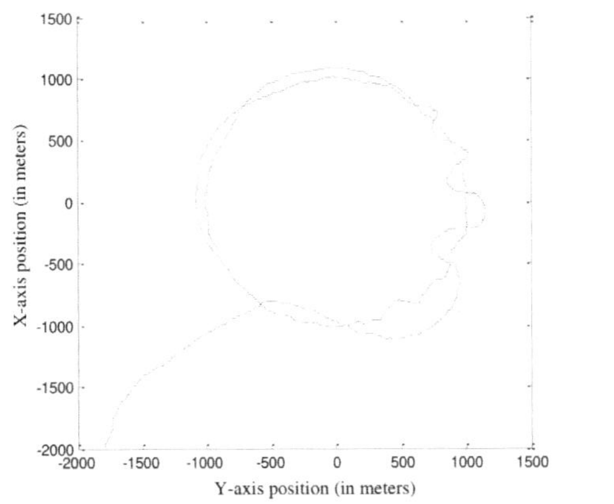

Figure 4. Position of UAV1 relative to target.

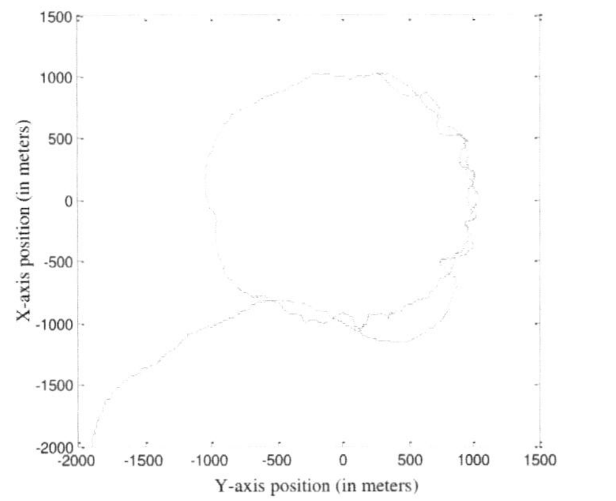

Figure 5. Position of UAV2 relative to target.

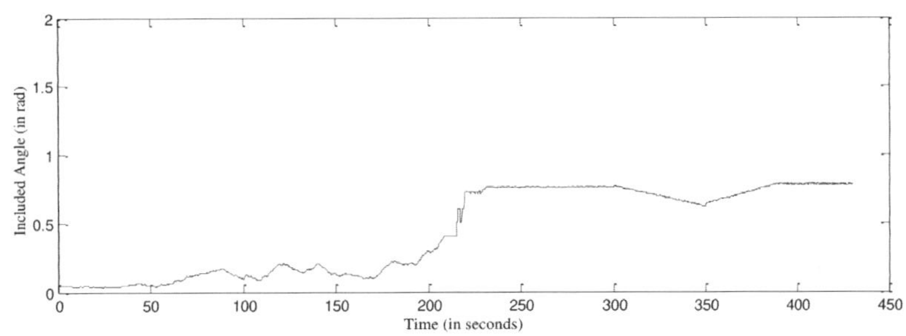

Figure 6. The Angle between UAV2 and UAV1 and the target

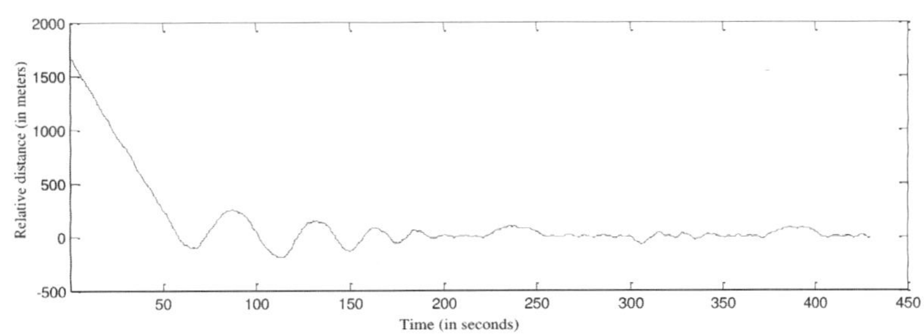

Figure 7. The error between the relative distance between UAV1 and the target and the preset distance

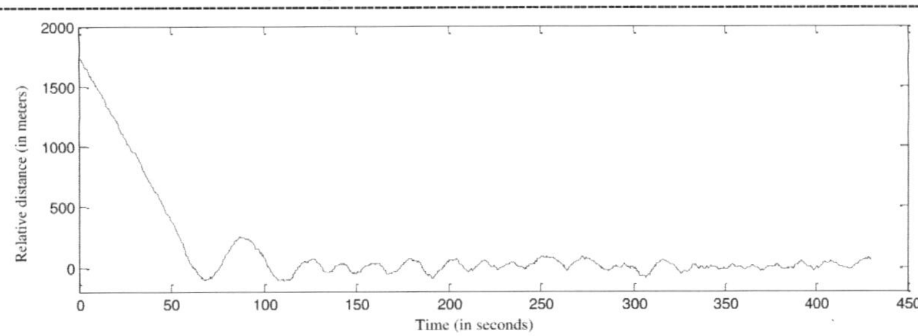

IOP Conf. Series: Materials Science and Engineering **646** (2019) 012038 doi:10.1088/1757-899X/646/1/012038

Figure 8. The error between the relative distance between UAV2 and the target and the preset distance

Through simulation, it can be seen that when the target moves at a certain angular speed or suddenly turns, the tracking distance error of the UAV1 and the UAV2 is larger than that of the target when it moves in a straight line, and the included angle error between the two drones and the target is also larger.However, as long as the final target can move in a straight line and at a uniform speed, the error can finally be within the allowable range through the adjustment of the controller, and target tracking can be well completed. The angle between the following uav and the piloting uav will eventually remain at $\pi/4$ .Moreover, the time of a rolling optimization of uav predictive control by the simulation program in ordinary PC is less than 0.2 seconds. When running in high-speed dedicated chip, the speed will only be faster, which can obviously meet the real-time requirements of control. Therefore, pso based predictive control algorithm for target tracking control can not only meet the control requirements, but also meet the real-time requirements.

References

[1] D Shim H Kim and S Sastry 2015 *Decentralized reflective model predictive control of multiple flying robots in dynamic environment* Conference on Decision and Control U.S.A p543~549

[2] Regina N Zanzi M 2009 *2D tracking and over-flight of a target by means of a nonlinear guidance law for UAV* Aerospace conference p1~11

[3] ZhuSenqiang WangDanwei ChenQijun 2014 *Standoff tracking control of moving target in unknown wind* CDC p763~765

[4] Ramachandra Sattigeri 2016 *Adaptive Estimation and Control with Application to Vision-based Autonomous Formation Flight* Georgia Institute of Technology.

[5] Richard Wise 2016 *UAV Control and Guidance for Autonomous Cooperative Tracking of a Moving Target* A research proposal submitted in partial fulfilllment of the requirements for the degree of Doctor of Philosophy University of Washington

[6] Ludington B Reimann J Vachtsevanos G Barlas I 2006 *Target Tracking with Unmanned Aerial Vehicles: From Single to Swarm Vehicle Autonomy and Intelligence Control and Automation* 14th Mediterranean Conference Mediterranean p1~6

Empirical Research on Connotation and Elements of Small and Micro Entrepreneurial Ability of Students in Vocational Colleges

Hua Hefeng[1,a] and Bao Yanli[2,b]

[1]Wuxi City College of Vocational Technology, NO.12,Qianou Road, Wuxi, China
[2]Business School, Hohai University, NO.12, Fochengxi Road, Nanjing, China

[a]hfhua0802@126.com; [b]sarah1826@163.com

Abstract: Higher vocational college students' entrepreneurial ability is based on the characteristics of higher vocational students and that of vocational colleges. It has similarities in entrepreneurial ability, but also has their own characteristics. Different from emphasis on scientific and technological entrepreneurship, higher vocational college students focus on the fields such as service and sales, and the small and micro entrepreneurial ability of higher vocational students emphasizes the use of skills and opportunity development. This paper makes an empirical study on the small and micro entrepreneurial ability of higher vocational students, and finds that their entrepreneurial ability is composed of skills practice ability, opportunity development ability, entrepreneurial operation ability and failure handling ability. The small and micro entrepreneurship of higher vocational students is an important part of college students' entrepreneurship. The development of their ability will improve the employment competitiveness of higher vocational students and the success rate of entrepreneurship.

1. Introduction

In the era of mass entrepreneurship and innovation, the upsurge of college students' entrepreneurship is emerging. According to the survey of Mycos on the graduates' entrepreneurship from 2012 to 2017, the proportion of graduates' entrepreneurship has increased steadily in the past five years, from 2.0% in 2012 to 2.9% in 2017. College students' entrepreneurship is widely accepted by the society, and the environment of college students' entrepreneurship is improving year by year. The research of college students' entrepreneurship involves many aspects, such as factors affecting entrepreneurship, entrepreneurial willingness, entrepreneurial ability, entrepreneurial performance, entrepreneurial environment construction, and entrepreneurial support service system. The professional background of college students' entrepreneurship involves economics, engineering, art and other majors. Although there are many studies on college students' entrepreneurship, the research on entrepreneurship of higher vocational college students is not deep enough. The reason is that the entrepreneurship of college students involves more technological and innovative entrepreneurship, while the entrepreneurship of higher vocational college students mostly focuses on small and micro entrepreneurship such as skills and services, and there are not enough focus of social attention and support on it. On the other hand, the existing entrepreneurship research in higher vocational colleges is more based on the concept and category of undergraduate students' entrepreneurship, lacking analysis on entrepreneurial characteristics and creation of higher vocational students. Difference between entrepreneurship of higher vocational students and that of undergraduates cannot be clearly revealed, which makes the topic of entrepreneurship of higher vocational students lack legitimacy. Therefore, the concept and scope of higher vocational students' entrepreneurship needs to be clarified before its unique problems being discussed. The research on small and micro entrepreneurship can maximize the advantages of practical skills of vocational students and find out the knowledge shortcomings and capacity bottlenecks of vocational students' entrepreneurship, and make up for the weaknesses in order to promote the cultivation of vocational students' entrepreneurial ability and improve the success rate of these students.

2. Literature review

2.1 Entrepreneurial Ability of College Students

The cultivation of college students' entrepreneurial ability is a key issue in college entrepreneurship education. The entrepreneurial ability of college students determines the overall level of entrepreneurship and the success rate of entrepreneurship [1]. College students usually experiences opportunity mining, organizational operation, relationship coordination and so on in the process of starting a business. The smooth operation of these processes depends on the use of their entrepreneurial ability. Therefore, the connotation of college students' entrepreneurial ability has been widely studied.

The perspective of entrepreneurship mainly includes perspective of trait, opportunity and management. The perspective of trait focuses on the personality, perseverance, ability and knowledge of entrepreneurs; the perspective of opportunity focuses on the ability of entrepreneurs to identify and develop opportunities; and the perspective of management focuses on the ability of entrepreneurs to manage enterprises [2]. The entrepreneurship itself includes opportunities, strategies, networks, leadership and other related abilities [3]. College students' entrepreneurial ability is based on the perspective of common entrepreneurial ability, but on this basis, it also has the characteristics of students' entrepreneurship, emphasizing the integration of specialty and entrepreneurship, the learning of entrepreneurship methods, and the bearing of failure.

According to relevant literature, there are several types of entrepreneurial ability of college students. Firstly, professional competence plus general competence. Gao Guijuan and Su Yang divide college students' entrepreneurial ability into professional ability, methodological ability and social ability through empirical research [4]. This classification not only takes into account the general entrepreneurial ability, but also takes into account the particularity of students' major. Secondly, students' entrepreneurial ability are based on process of starting a business. Yang Daojian etc. divide entrepreneurial ability of college students into opportunity discovery ability, strategic decision-making ability, resource integration ability, organizational management ability, innovation and creativity ability and frustration tolerance ability through case studies [5]. This classification combines entrepreneurial ability and the process of starting a business, which shows the practicality and complexity of startup process. Thirdly, based on the competency of entrepreneurship, Wang Hui and Zhang Huihua combine case studies and empirical studies, and divide entrepreneurial ability of college students into opportunity grasp, relationship competency, innovation creativity, practical learning and resource integration [6]. As the core competence for starting businesses, entrepreneurial competence is the premier indicator to judge whether students' businesses can succeed or not in the future. Fourthly, entrepreneurial ability of students with different majors. For example, Zhang Wenqiang points out that entrepreneurial ability of college students majoring in finance, politics and law includes professional knowledge, professional skills, innovative ability and professional ability [7]. Through empirical research, Gao Shuyu and Zou Xiaodong found that the entrepreneurial ability of engineering science and technology talents includes technological innovation ability, market-grasping ability, team leadership ability and self-management ability [8]. Since the knowledge background and technology level of each specialty are different, the evolution mode of entrepreneurship path is also different, which results in different entrepreneurship abilities. Fifth, entrepreneurial ability from the perspective of failure. Wang Fei puts forward the ability of psychological rehabilitation of entrepreneurship, that is, the ability of college entrepreneurs to recover quickly from failures [9]. Scholars also discussed the repeated willingness of college students after their failure to start a business [10]. Entrepreneurial failure is common in students' entrepreneurship. Coping with entrepreneurial failure well is an important part of students' entrepreneurship learning. The above five perspectives reflect the entrepreneurial ability of college students from various aspects. To summarize, we can find that the entrepreneurial ability of college students not only includes the ability of opportunity identification, resource integration and management execution, but also includes the professional ability, practical learning ability, frustration resistance and other elements with distinct characteristics of college students' entrepreneurship.

2.2 Entrepreneurial Ability of Higher Vocational College Students.
The research on the entrepreneurial ability of students in higher vocational colleges mainly focuses on its connotation and constituent elements. Firstly, basic ability plus special ability. Hong Xiao and Ding Wei divide the entrepreneurial ability of higher vocational students into basic ability, special ability and development ability [11]. Basic ability involves goal management, active control and opportunity mining; special ability includes professional, social and management competence; Development ability includes innovation competence, strategic competence and resource integration competence. Secondly, based on the competency of entrepreneurship, Lu Xiaoli divided the entrepreneurship of higher vocational students into interpersonal communication ability, opportunity recognition ability, innovation creativity, resource integration ability and practical learning ability through questionnaire survey [12]. Finally, entrepreneurial ability from the perspective of entrepreneurship. For example, Chen Shihui proposes that we should cultivate entrepreneur talents in Higher Vocational Colleges under the guidance of entrepreneurship, emphasizing dedication, creative thinking, cooperation and leadership [13].

The entrepreneurial ability of higher vocational students basically follows the research ideas and methods of college students' entrepreneurial ability. But due to insufficient research, the research results rarely show the unique requirements and constraints of higher vocational students' entrepreneurial ability. In general, students' entrepreneurial ability cannot reflect the process of starting small and micro businesses, and the influencing factors cannot be revealed. Therefore, characteristics of vocational college students' small and micro entrepreneurial ability needs to be deeply analyzed, and the connotation of it should be clarified more accurately to reflect the content of small and micro entrepreneurial ability.

3. Empirical Study
On the basis of theoretical exploration of students' small and micro entrepreneurial ability in higher vocational colleges, empirical research is needed to support and verify the composition of entrepreneurial ability. Based on the existing research scales of entrepreneurship and students' entrepreneurship, this study designs the scales combined with field interviews and expert opinions of relevant enterprises, and uses the form of Lickett 5-level scale to determine the ranks, from "disagreement" to "consent", expressed in 1-5 respectively(see Table 1). Subsequently, empirical research was carried out through questionnaire, data collection and analysis. Questionnaires are mainly distributed in vocational colleges, science and technology entrepreneurship parks, college students' entrepreneurship incubators and senior students in Wuxi, Suzhou and Nanjing in China. A total of 310 questionnaires were distributed and 262 were done and returned. 61 invalid questionnaires were excluded and 201 valid questionnaires were obtained. The validity rate of the questionnaires was 76.7%.

Table 1. Questionnaire on Small And Micro entrepreneurial Ability

Dimension	Item	Journal Title
Skills Practical Ability	I can continue to learn in Entrepreneurship(Learning)	Gao Guijuan, Su Yang[4]; Wang Hui, Zhang Huihua[6]; Zhang Wenqiang [7]
	I have enough knowledge of entrepreneurship(Knowledge)	
	When I have difficulties in starting a business, I can stick to it(Perseverance)	
	I have an open mind and can maintain the flexibility of entrepreneurship(Flexibility)	
Opportunity Development Ability	I'm looking for ways to identify entrepreneurial opportunities(Opportunity Identification)	Yang Daojian et al. [5]; Lu Xiaoli [12]
	I will look for ways to assess entrepreneurial opportunities.(Opportunity Assessment)	
	I will integrate and utilize entrepreneurial opportunities for Entrepreneurship(Opportunities Utilization)	
Entrepreneurial Operating Ability	I have the ability of organization and management.(Organization)	Chen Shihui[13]; Hong Xiao, Ding Wei[11]
	I will establish and maintain good business relations through various channels.(Relation)	
	I pay attention to the long-term development of start-up enterprises and have strategic awareness.(Strategy)	
Failure Handling Ability	I can learn from the failure of entrepreneurship(Failure Learning)	Wang Fei[9]; Jia Tianming, Lei Lianghai[10]
	When I fail to start a business, I will start a second business.(Second Start-up)	
	I can handle the emotions of business failure.(Failure Emotion)	

3.1 Exploratory Factor Analysis

In order to test the structural validity of small and micro entrepreneurial ability of higher vocational students, the dimension structure of entrepreneurial ability is determined by exploratory factor analysis, and then the factor structure model of entrepreneurial ability and the appropriateness of the collected data are determined by confirmatory factor analysis. Therefore, 62 questionnaires were randomly selected from 201 valid questionnaires for exploratory factor analysis. Before factor analysis, the correlation between indicators is tested. The results of KMO is 0.776, higher than 0.7, and Bartlett sphericity test value is 471.018, corresponding probability P value is 0.000, lower than the significance level of 0.001, indicating that it is suitable for factor analysis. Principal component method is used for Factor extraction to extract the factors whose eigenvalue is greater than 1, and the maximum variance method is used for rotation. Through SPSS23 software analysis, a total of four factors were extracted, and the cumulative explanatory variation of these four factors is 75.85%. The factor load of each item is greater than 0.5. The results are shown in Table 2. Later, the reliability of the scale is tested, and the Cronbach's Alpha coefficient is calculated by calculating the overall correlation coefficient (CITC) of four variables (see Table 3), all the CITC coefficients are greater than 0.35 and the clonal Cronbach's Alpha coefficients is greater than 0.7, which shows that the items have good internal consistency and the scale has good reliability.

3.2 Confirmative Factor Analysis

After exploratory factor analysis, the remaining 139 questionnaires from 201 valid questionnaires are analyzed by confirmatory factor analysis to determine whether the structural model of entrepreneurial ability fits the actual data. Firstly, reliability analysis is carried out, and the reliability is judged by calculating CITC and Cronbach's Alpha. The results are shown in Table 4 that the reliability of the scale is good. AMOS23 software is used to carry out confirmatory factor analysis. The results are shown in Figure 1. The standardized load of items in the dimensions measured is greater than 0.5, and it is highly significant in statistics, indicating that each item has a high convergence validity. The fitness index of the model and data is shown in Table 5. CMIN/DF is 1.525, GFI, TLI, CFI, NFI and IFI are all higher than 0.9. AGFI is very close to 0.9, RMSEA is 0.062 and less than the standard 0.1. It is shown that the observed data and structure model are well fitted. The division and measurement of this study are effective.

4. Result and Enlightenment

4.1 Research Result

Table 2. Analysis Results of Exploratory Factors

	Component			
	1	2	3	4
Learning	**.742**	.113	.319	.259
Knowledge	**.764**	.270	.043	.315
Perseverance	**.795**	.112	.096	.273
Flexibility	**.889**	.082	.015	.092
Opportunity Identification	-.015	.077	**.852**	-.007
Opportunity Assessment	.183	.071	**.840**	.097
Opportunities Utilization	.130	-.001	**.840**	.171
Organization	.413	.198	.163	**.719**
Relation	.257	.176	.115	**.806**
Strategy	.175	.132	.049	**.838**
Failure Learning	.188	**.807**	.072	.246
Second Start-up	.139	**.870**	.044	.162
Failure Emotion	.091	**.875**	.046	.052

Table 3. Reliability Test of Exploratory Factor Analysis

	Corrected Item-Total Correlation	**Cronbach's Alpha if Item Deleted**	**Cronbach's Alpha**
Learning	.707	.85	
Knowledge	.748	.834	.875
Perseverance	.732	.841	
Flexibility	.745	.836	
Opportunity Identification	.642	.785	
Opportunity assessment	.683	.744	.819
Opportunities Utilization	.698	.726	
Organization	.706	.745	
Relation	.694	.755	.829
Srategy	.66	.788	
Failure Learning	.706	.798	
Second Start-up	.76	.752	.851
Failure Emotion	.707	.783	

Through literature research on students' entrepreneurial ability and higher vocational students' entrepreneurial ability, combined with field interviews, the scale of entrepreneurial ability for small and micro businesses of higher vocational students is designed.

The scale divides it into four dimensions: skills practice ability, opportunity development ability, entrepreneurial operation ability and failure handling ability. Using large-scale questionnaire survey, exploratory factor analysis and confirmatory factor analysis were carried out on the questionnaire data.

Table 4. Reliability Test of Confirmatory Factor Analysis

	Corrected Item-Total Correlation	Cronbach's Alpha if Item Deleted	Cronbach's Alpha
Learning	.725	.839	
Knowledge	.764	.824	.874
Perseverance	.701	.848	
Flexibility	.731	.839	
Opportunity Identification	.668	.825	
Opportunity assessment	.743	.761	.847
Opportunities Utilization	.732	.765	
Organization	.699	.774	
Relation	.715	.763	.842
Srategy	.701	.784	
Failure Learning	.692	.812	
Second Start-up	.773	.727	.849
Failure Emotion	.690	.814	

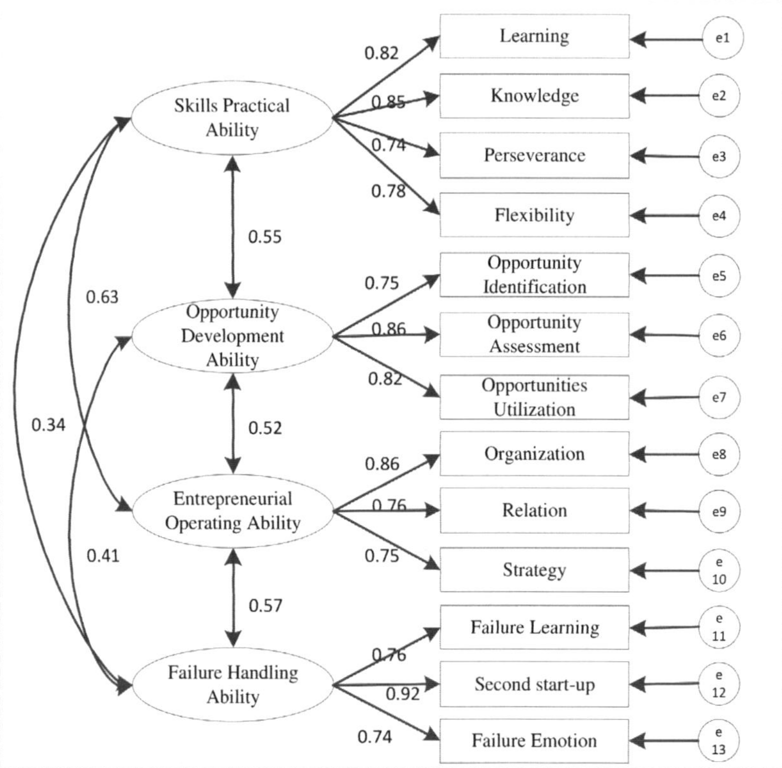

Figure 1. Measurement Results of Confirmatory Factor Analysis

Table 5. Fitting Index of Confirmative Factor Analysis

Item	CMIN/DF	GFI	AGFI	TLI	CFI	RMSEA	NFI	IFI
Number	1.525	.906	.855	.957	.967	.062	0.912	.968

The results show that the scale has passed the reliability test, and has good content validity and structural validity. This research deeply excavates the connotation of students starting small and micro businesses, which will help to analysis the key influencing factors and follow-up performance of students' small and micro entrepreneurship in the future.

4.2 Research Enlightenment

4.2.1 Skills Practical Ability Needs to be Deepened. Higher vocational colleges lay emphasis on skill training courses, and the practical operation ability is much stronger. In the process of starting businesses, higher vocational students make use of their own advantages, relying on their skills and strengths, which is the extension and expansion of vocational students' skill learning. Small and micro entrepreneurship activities can integrate students' knowledge, encourage students to continue learning, and cultivate students' perseverance in doing things. The twists and turns of starting businesses can also encourage students to reflect on themselves and improve their practical skills, flexibly adjust the process, and calmly face the uncertainty.

4.2.2 Opportunity Development Ability Need to be Strengthened. In more than 1,000 higher vocational colleges in China, majors focus mainly on technology and commerce (Zhang Zhixiang, 2012). The entrepreneurial opportunities of higher vocational students lie in skills and service-oriented small and micro businesses. Through analysis of their own knowledge and skills, higher vocational students can divide the feasible areas to start business and find their own advantageous opportunities. However, due to the diversity and complexity of entrepreneurship opportunities, students tend to be blind and conformist when evaluating and utilizing these opportunities. Therefore, colleges' entrepreneurship guidance agencies and social entrepreneurship centers are needed to guide the students and screen the opportunities.

4.2.3 Entrepreneurial Operation Ability Needs to be Improved. Students have to experience gradual process to master operation ability of start-up enterprises, and they need to hone their skills in practice. In order to make up for students' deficiencies in entrepreneurial operation ability, higher vocational colleges can adopt the way of gradual incubation to promote the growth of students' enterprises. For example, through the process of college incubator - social Incubator - social accelerator, students have room for gradual growth. At the same time, the deficiencies in the relationship of start-ups can be integrated and exploited through the relationship network of colleges and alumni resources, so as to help start-ups to quickly establish useful social network relationships. Defects in strategic management of start-up enterprises can be solved by introducing entrepreneurship mentors or consultants to guide enterprises to establish the direction of sustainable development. The improvement of organizational management of start-up enterprises requires that entrepreneurs gradually explore and grow in the real operation, and when necessary, external forces can be brought in to assist managers.

4.2.4Failure Handling Ability Needs Attention. Higher vocational college students have a high failure rate in entrepreneurship, and their frustration tolerance is weak. Therefore, more frustration tolerance education and psychological counseling should be carried out when students start their own businesses. The course of entrepreneurship can emphasize its risk in the teaching, so that students can be prepared for the hardship and failure in the early stage. The simulation training of entrepreneurship can extend to the whole process of entrepreneurship, adding the situation of despair and failure, so that students can experience the whole process from rise to decline, and enhance students' intuitive understanding of failure.

5. Conclusion

Higher vocational students' entrepreneurship is an important part of college students' entrepreneurship, which mainly focuses on service and retail and other small businesses. Through empirical research, it is found that vocational students' small and micro entrepreneurial ability includes skills practice ability, opportunity development ability, entrepreneurial operation ability and failure handling ability, among which skills practice ability is the advantage of vocational students. The entrepreneurial ability of higher vocational students is the core factor of their success in startups. And it is also the quality skills that colleges focus on cultivating. By studying the connotation and composition of the entrepreneurial ability of higher vocational students, it is convenient for colleges to cultivate and improve their entrepreneurial weaknesses.

Acknowledgements

This paper is supported by Youth Fund Project of Humanities and Social Sciences of the Ministry of Education of China (14YJC880021), Talent Supporting Project of Wuxi City College of Vocational Technology.

Reference

[1] Yang Xiaohui 2010 *Higher Education in China* **18** 42-44.
[2] Yin Miaomiao, Cai Li 2012 *Foreign Economy and Management* **12** 1-11,19.
[3] Yin Miaomiao, Fei Yupeng 2013 *Foreign Economy and Management* **10** 22-30.
[4] Gao Guijuan, Su Yang 2013 *Development and Evaluation of Higher Education* **3** 27-35.
[5] Yang Daojian, Zhao Xicang, Chen Wenjuan, Zhu Yongyue 2014 *Scientific and Technological Progress and Policy* **20** 151-156.
[6] Wang Hui, Zhang Huihua 2012 *Journal of National College of Educational Administration* **2** 81-86.
[7] hang Wenqiang 2012 *Henan Social Sciences* **4** 71-74.
[8] Gao Shuyu, Zou Xiaodong, Chen Hanchong 2013 *Research on Higher Engineering Education* **4** 28-34.
[9] Wang Fei 2017 *Journal of Higher Educational Management* **6** 60-66.
[10] Jia Tianming, Lei Lianghai 2016 *Shanghai Economic Research* **2** 56-63.
[11] Hong Xiao, Ding Wei 2013 *Education and Vocation* **6** 101-102.
[12] Lu Xiaoli 2015 *Research on Higher Engineering Education* **3** 152-156.
[13] Chen Shihui 2015 *Education and Vocation* **12** 45-47.

Research Status and Development Trend of Bioheuristic Computing

Chai zhijun* **and Li zhao**

Dept. of Graduate management, Naval Aviation University, Yantai, China

*275131218@qq.com

Abstract: Bio-inspired computation aims to study the biology function, characteristic and echanism of the various levels of nature from biological individual, population, colony until ecosystem, and set up a relevant model and computing method, so as to serve the scientific research and engineering application of human society. A few future directions and research challenges are presented, such as parallel bio-in-spired computation, bio-inspired computation with reasoning and knowledge, bio-inspired dynamics computation, bio-inspired computation based on quorumsensing,artificialbrain,evolutionaryhardware,bigdata,swarmrobot,virtualbiolo-gical,cloudcomputing,etc.

1. Introduction

In the field of biological system research, various kinds of biological heuristic algorithms emerge one after another, with diverse forms, different ideas, modeling and analysis tools of their own characteristics. At the same time, since all kinds of biological intelligent computing modes are derived from "nature" and simulate the biological intelligent behavior rules in nature, they are always designed to solve certain kinds of computing and optimization problems, and these computing modes are relatively uniform[1][2].

Biological inspired computing research in nature is the purpose of the different levels of biological individual, population, community and even ecosystem functions, characteristics and mechanism, establish a corresponding simulation model and calculation method, to serve the scientific research and engineering application of human society. It has been proved that bioheuristic computing has good performance in solving problems difficult to be solved by traditional methods such as large-scale computation and NP difficulty. In addition, the biological inspired computing has potential parallelism, distribution and refactoring, can be easily applied to engineering calculation under the network environment, management, control, and division of labor.

2. Research status of bioinspired computing

With people's exploration of the biological world, new research results have been emerging in the field of bioinspired computing since the 1990s, showing vigorous vitality and bright academic research prospects. Bioheuristic computing is an interdisciplinary subject between lifescience and computational science. The theories of genetics, biological immunology, nervous system and endocrinology in life science, artificial intelligence in information science, adaptive theory, optimal control and cognitive science and basic mathematics all provide important theoretical support for the development of bioinspired computing, shown as figure1.[2-3]

2.1 Research status of bioinspired computing

In 1994, the first Computational Intelligence conference titled "Computational Intelligence: Imitating the life", the theme of the seminar got scholars' attention. Among them, bioinspired computing attracts many scholars to devote themselves to the research and development of this field because of its unique charm.

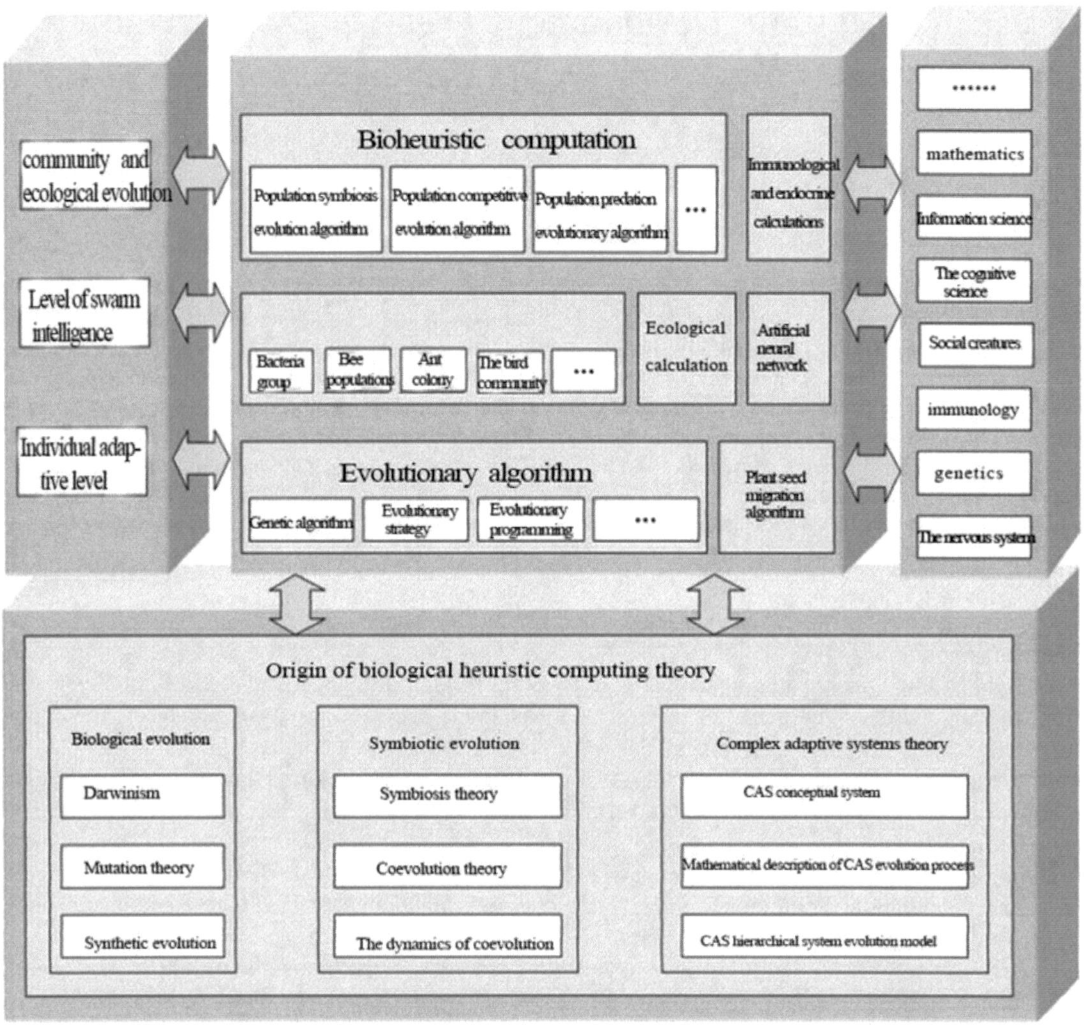

Figure 1. Bio-inspired computation theoretical origin and supporting.

2.2 Existing bio-inspired computing algorithms

As people understanding of society, especially in the past decade, biological heuristic calculation have been continuously developed, and derived a lot of new calculation method. The institute of computational technology, Chinese academy of sciences, tongji university, a university of electronic science and technology of China, Peking University, western jiaotong university and other research institutions have conducted research on bio-inspired computing, and published some monographs, such as simulated evolutionary computing of computational intelligence. Table 1 lists the proposed and widely studied biological heuristic algorithms and their heuristics. [4]

Table 1. Bio-inspired computation algorithm.

EVOCATOR	ALGORITHM NAME	BASIC IDEA	PRESENTER AND SUBMISSION TIME
molecular	DNA computing	Biomolecular structures are simulated and computations are performed using molecular biotechnology	Adleman L M,US(1994)
cells	membrane algorithm	A computational model that simulates the multilevel structure of cells	Nishida T Y,Japan(2005)
immune	artificial immune system	Mimics how the natural immune system works	Dasgupta D,US (1998)
Neural network	back propagation hopfield	A multilayer feedforward network with error inverse propagation algorithm. Considering the transmission delay of output and input in time, the neural is clarified network and dynamics.	Bryson A E and Ho Y C,US (1969) Hopfield J J,US (1982)
Biological evolution	evolutionary programming evolutionary strategies	It simulates biological evolution from a holistic perspective, emphasizing species evolution The idea of biological evolution through natural mutation and natural selection	Fogel L J,US (1963) Schwefel H P, Germany (1965)
Ants	ant colony optimization	Ant pheromone foraging behavior	Dorigo M,Maniezzo V Italy (1991)
bird	particle swarm optimization swallow swarm optimization cuckoo search	Flock flight and foraging behavior Swallows work in different roles to find food Cuckoo parasitic breeding mechanism and levy flight search mode	Kennedy and Eberhart,US (1995) Neshat M,Sepidnam G and Sargolzaei M, Iran (2013) Yang X Sand Deb S,UK(2009)
fish	artificial fish swarm algorithm	Fish feed and swim in groups	Li xiao-lei,,China (2002)
bacteria	bacterial foraging optimization bacterial colony chemotaxis	Bacterial chemotactic foraging behavior Bacterial swarm pharmacodynamics	Passino K M,US (2002) Müller S D，Marchetto J Swedish (2002)
fireflies	glowworm swarm optimization	Fireflies use their glow to attract mates or feed	Krishnanand and Ghose D, India (2005)
bees	artificial bee colony	The act of bees collecting honey	Karaboga D, Turkey (2005)
human	self-organizing migrating algorithm	Group self-organizing migration behavior in social environment	Zelinka I and Lampinen J, Czech (2000)
plant	invasive weed optimization	Simulate the process of weed invasion	Mehrabian A R and Lucas C, Iran (2006)

2.3. Research status of bioinspired computing

Existing biological heuristic algorithms have different objects of inspiration, but they all simulate the process of simple individuals cooperating to solve complex problems.Therefore, the internal structure of the algorithm has something in common: namely, within a certain geographical range, there are multiple individuals with simple capabilities, most of whom are isomorphic in structure and function;There is no central control in the population, and the cooperation among individuals is distributed.Individuals interact and collaborate by following simple rules.Moreover, the calculation mode of the above algorithm is relatively uniform, which is based on the adaptive behavior of evolutionary unit, and achieves the final goal through the iterative search method of "generation + test" feature.[5]

3. Prospect of application of biological heuristic computing

Since its inception, bioheuristic computing has been widely used in many fields of natural science and engineering science, and has shown great advantages and potential.Historically, the peak of bioinspired computing research was around 2010.At this stage, a variety of bio-inspired computational methods appeared, and the research content was gradually broadened and the application was gradually oriented to reality.But since 2015, research seems to have fallen into a slump.The reason is that the practical application of the algorithm has not been better used.But with big data, cloud computing and the urgent need to explore the complex world in the future, biologically-inspired computing with intelligent emergence mechanism will show great vitality in the following fields in the future.

3.1 Human Brain

The brain is one of the most complex system of nature, research and development of the human brain has always been a prospective research direction is of great significance. In 2014, the European commission announced the launch of the human brain project. In 2010, China launched a program on artificial brain research and related major guidance documents in the outline of medium - and long-term scientific and technological development. In 2013, the brain activity mapping project was launched in the United States. Countries such as Germany and Japan are also deploying brain science courses. Hugo DE garis, a computer professor at Utah state university, is the world's first inventor of an artificial brain. His CBM is a kind of brain manufacturing machine, using genetic algorithm to design the neural network, then the neural network modeling of electronic. In 2011, researchers at the university of southern California used carbon nanotubes to create a circuit that mimics the function of the brain's synapses. In 2012, scientists at the university of Waterloo in Canada designed a large brain called Spaun, made up of 2.5 million simulated neurons that perform eight different types of tasks. In 2012, IBM simulated a computing architecture with 530 billion neurons and 137 trillion synapses. Artificial brain manufacturing requires two levels of technical support: intelligent hardware technology that simulates brain nerves and intelligent computing technology that simulates brain thinking. Biological belongs to the behavior of complex systems and features can provide inspiration for the design of an artificial brain and evolution, adaptive, self-learning and self - grouping, highly parallel and interactive environment, can rearrange a new route, the study knowledge, discover new tasks and solve problems. [6]

3.2 Evolution hardware

Compared with the traditional fixed structure of the hardware, the hardware evolution (evolvable hardware, EHW) depends on the biological heuristic calculation, its hardware structure can be realized with the change of environment automatic hardware structure and function of refactoring. At present, the basic hardware architecture of the computer is still von neumann. Turing's Turing machine model laid the foundation for the logical working mode of modern computers. These two principles have been with computers for nearly 100 years. In contrast, evolutionary hardware is clearly a new technological baby with broader and better prospects. The computing, based on the biological heuristic evolutionary hardware research tasks include: evolutionary computation coding scheme, special application coding scheme, evolutionary computation algorithm parameters adaptive adjusting mechanism, the evolutionary computation fitness evaluation mechanism, the evolutionary computation fault-tolerant mechanism and parallel evolutionary computation, etc.

3.3 Big data

Big data is another revolution after mobile Internet, Internet of things and cloud computing. At present, there is a growing social forecasting, decision-making and business practices are based on the bjective analysis of large data. Effective analysis and processing of large data, for example, would bring huge economic value, promote social progress. The core value of big data is data analysis. Under the background of big data, data size and the difficulty of handling need analysis of the problems are extremely complex, with large scale and high dimension, strong constraints, strong dynamic and multi-objective characteristics. Faced with these characteristics, traditional data analysis methods are no longer applicable. At the same time, biological heuristic computation characteristic of the big data revolution brought great opportunities for biological heuristic calculations. At present, the international conference on large data, such as the IEEE conference on computing and big data services, ACM international conference on data management and domestic computer association of China's most influential and largest of the big data technology conference will be data analysis method, on the basis of included in this study, the biological inspired computing problem, pointed out that in big data environment, biological inspired computing can be used to complete the following data analysis: a large-scale data analysis, high-dimensional data analysis, trend analysis, data analysis, strong constraint optimization of abnormal data category analysis, multi-scale analysis, extract feature selection, uncertainty of data analysis, etc.

3.4 Swarm robo

In the complex and changeable working environment, some single-task robots are unable to complete, while group robot systems can complete complex tasks that are difficult or impossible to complete by single robot systems with faster, more reliable, lower cost and better performance.Swarm robot's applications provide a wide range of applications and challenges for bioinspired computing.Future wars are increasingly fierce, and as military operations or other tasks require, single-platform uavs urgently need to develop toward multi-platform "clusters."Cluster cooperative behavior and control method are the important basis of uavcluster.Currently, many countries around the world are using bioinspired computing to implement unmanned aerial vehicle (uav) swarm warfare, such as swarm warfare or Wolf warfare.In addition to cooperative uav operations, swarm robots can also complete environmental monitoring, hazardous environment operations, underwater and space exploration, hazard location, resource exploration and disaster search.Based on bioheuristic computation, the research tasks of collaborative work of swarm robots include: aggregation control, connectivity control, formation control, distribution control, path planning, multi-target tracking, multi-task division and multi-task coordination.[7]

3.5 Cloud computing

Cloud computing is an emerging computing mode developed by distributed computing, parallel computing and network technology, and is the core technology of the next generation network computing platform.The term "cloud" refers to a large pool of computer resources distributed on a network that are largely idle through virtualization.Through cloud computing, computing nodes can automatically access the resource pool, and all kinds of resource acquisition and services can fully realize dynamic autonomy without human participation.As cloud computing has massive, heterogeneous, distributed and diversified resources, it will involve many complex computing tasks in terms of resource optimization and data management. The implementation of these tasks needs to rely on bio-inspired computing.Based on biological inspired computing, realize the cloud computing environment complex computing tasks, task scheduling optimization, like a cloud cloud resources (including digital resources, platform, resources and performance, etc) scheduling optimization, network performance optimization, the huge amounts of data analysis and forecast data management tasks such as research work includes the following: efficient real-time algorithm, dynamic multi-objective algorithm, adaptive dynamic algorithm, dealing with large-scale algorithm, parallel and distributed algorithms and efficient processing a np-hard problem and mass data of multivariate clustering algorithms, etc.[8]

4. Conclusion

Cloud computing is an emerging computing mode developed by distributed computing, parallel computing and network technology, which is the core technology of the next-generation network computing platform. The term "cloud" refers to the distribution of a large number of computer in the network resources, these resources through virtualization basically is in the idle state. Through cloud computing, computing nodes can automatically access the resource pool, and various resource acquisition and services can fully realize dynamic autonomy without anyone participating. Cloud computing is massive, heterogeneous, distributed and diverse resources, the resource optimization and data management will involve many complex computing tasks. The realization of these tasks depends on bionic computing. Based on bio-inspired computing, realize complex computing tasks in cloud computing environment, task scheduling optimization, such as cloud resources (including digital resources, platforms, resources and performance, etc.) scheduling optimization, network performance optimization, a large number of data analysis and prediction data management tasks, such as research work includes the following: efficient real-time algorithm, dynamic multi-objective algorithm, adaptive dynamic algorithm, processing large-scale algorithm, parallel and distributed algorithm, and efficient np problem processing and multi-dimensional clustering algorithm of massive data

References

[1] Holland J H1995How adaptation builds complexity(USA: Addlson-Wesley Publishing Company) p72-77.
[2] Holland J H1992 Adaptation in natural and artificial systemsAn introductory analysis with application to biology, control and artificial intelligence(Cambridge, MA: MIT Press)p43-65.
[3] Parpinelli R S2011 New inspirations in swarm intelligenceA survey(Amsterdam:International Journal of Bio-Inspired Computation)p1-15.
[4] Xing B, Gao W J2013 Innovative computational intelligenceA rough guide to 134 clever algorithms(Berlin, Germany: Springer).
[5] Goh C K,TanKC 2009 A competitive-cooperative coevolutionary paradigm for dynamic multiobjective optimization(USA:IEEE Transactions on Evolutionary Computation)p103-127.
[6] Oca M A M D, GarridoL, Aguirre J L2005Effects of inter-agent communication in ant-based clustering algorithmsA case study on communication policies in swarm systemsLNCS 3789: Advances in Artificial Intelligence(Berlin, Germany: Springer)p254-263.
[7] Garnier S,GautraisJ,TheraulazG 2007 The biological principles of swarm intelligenceSwarm Intelligence1(1): p3 -31.
[8] Zhang J,Zhan Z H,LinY,et al2011 Evolutionary computation meets machine learningA survey(USA: IEEE Computational Intelligence Magazine)6(4):p 68-75.

IOP Conf. Series: Materials Science and Engineering **646** (2019) 012041 doi:10.1088/1757-899X/646/1/012041

Study on Low Characteristic Formulation Optimization of Solid Propellant Based on Improved Particle Swarm Optimization

Jiu-Ling Zhao*

Rocket Force University of Engineering, Xi'an, PR, China

*zhaojiuling111@163.com

Abstract. The combustion product of solid propellant produces strong plume flow field downstream of nozzle outlet, which improves plume characteristics, reduces plume characteristic signals, and improves stealth characteristics and survivability of missile. In this paper, an intelligent algorithm is proposed to optimize the formulation of solid propellant with low characteristic signal. An improved multi-objective particle swarm optimization algorithm is used to solve the formulation optimization model, and several feasible formulations for realizing low characteristic signal are obtained. On this basis, considering the energy performance, a formula screening method is given to calculate various characteristic signals and energy properties of the feasible formulation. The comprehensive measurement can not only take into account the optimization target requirements of propellant energy performance and plume characteristic signals, solve the conflict problem of reducing multiple characteristic signals, but also improve the efficiency of formulation design, and provide a reference for formulation designers.

1. Introduction

With the wide application of solid propellant in all kinds of strategic and tactical missile weapons, the diversified requirements of future operational tasks put forward more complex requirements for propellant performance. During the working process of rocket motor, the combustion products of solid propellant produce strong plume flow field downstream of nozzle exit, accompanied by a variety of complex physical and chemical phenomena, resulting in a variety of characteristic signals. Plume is a complex flow field, and the formation process is affected by many factors. Therefore, the plume characteristics are improved and the plume characteristic signals are reduced.[1] [2]Designing propellant formulation by experiencing and experimenting or adjusting oxygen balance of formulation, or using less or no smoke-producing ingredients, or replacing all or part of smoke-producing ingredients with new materials, or adding inhibitors, or adjusting the composition ratio of formulation components, often requires many tests and repeated tests, which is costly and inefficient. At present, it is only used to reduce smoke. The single characteristic signal is difficult to solve the problem of reducing the conflict of multiple characteristic signals. In addition, most of the theoretical modeling methods such as test and thermodynamic calculation are only preliminarily applied to the analysis and prediction of plume characteristic signals [3].

In this paper, an intelligent algorithm is proposed to optimize the formulation of solid propellant with low characteristic signal, which is a new research direction in the field of Engineering application.

The comprehensive measurement of energy performance can not only take into account the optimization objective requirements of propellant energy performance and plume characteristic signals, solve the problem of reducing the conflict of multiple characteristic signals, but also improve the efficiency of formulation design, and provide a reference for formulation designers.

2. Multi-objective particle swarm optimization algorithm

To overcome the shortcomings of particle swarm optimization (PSO) which can not be directly applied to solving multi-objective multi-constrained problems and is easy to fall into local optimum, this paper proposes an improved multi-objective particle swarm optimization (MPSO) algorithm.[3][4][5]Taking the minimization problem as an example, the specific steps to improve the algorithm are as follows:

(1) Initialize algorithm parameters.

The parameters of the algorithm include population particle number N , upper limit of external file particle number W , acceleration coefficient c_1 and c_2 , inertia weight ω , maximum number of iterations M_{itera} , particle mutation probability P_m , penalty coefficient $\boldsymbol{d}_{ip} = [d_{1p}, d_{2p}, \text{L}, d_{ip}, \text{L}, d_{np}]$, threshold δ of average convergence degree of particles in external file, judgment of condition interval algebra Δt and threshold τ , and random method is used to generate particles $\boldsymbol{x}$ and their velocities $\boldsymbol{v}$, which are limited to given boundary constraints when generating particles. In the range of [$\boldsymbol{L}$ $\boldsymbol{U}$], e. g. (1), the maximum $\boldsymbol{v}_{\min}$ and minimum $\boldsymbol{v}_{\max}$ velocities of particles, e. g. (2) are set.

$$\begin{aligned} \boldsymbol{x} &= \boldsymbol{L} + \text{rand} \cdot (\boldsymbol{U} - \boldsymbol{L}) \\ \boldsymbol{v} &= \frac{\text{rand}}{2} \cdot (\boldsymbol{U} - \boldsymbol{L}) \end{aligned} \tag{1}$$

Among them, $\text{rand} \in [0\ 1]$.

$$\begin{aligned} \boldsymbol{v}_{\min} &= \frac{\boldsymbol{L} - \boldsymbol{U}}{2} \\ \boldsymbol{v}_{\max} &= \frac{\boldsymbol{U} - \boldsymbol{L}}{2} \end{aligned} \tag{2}$$

(2) Computing the fitness of population particles. The penalty function method is used to take the inequality constraints (except boundary constraints) as penalty term, and fitness function is added. The substitution method is used to replace one decision variable in the equality constraints in the optimization model with other decision variables, and the fitness function is substituted. The maximization objective function in the optimization model is transformed into the minimization objective function to calculate the fitness of particles. The corresponding value is shown in equation (3).

$$eval_i(\boldsymbol{x}') = f_i(\boldsymbol{x}') + \boldsymbol{d}_{ip} \cdot \max(g_p(\boldsymbol{x}') - a_p, 0) \text{ or } \boldsymbol{d}_{ip} \cdot \min(g_p(\boldsymbol{x}') - a_p, 0) \tag{3}$$

Among them, $\boldsymbol{x}'$ represents the vector of decision variables after dealing with equality constraints $h_q(\boldsymbol{x}) = b_q$ by the substitution method, the fitness function $eval_i(\boldsymbol{x}')$ with the first objective $\boldsymbol{x}'$, and the mathematical model $f_i(\boldsymbol{x}')$ with the first objective $\boldsymbol{x}'$. At that time $g_p(\boldsymbol{x}) \le a_p$, the penalty item was $\boldsymbol{d}_{ip} \cdot \max(g_p(\boldsymbol{x}') - a_p, 0)$, at that time $g_p(\boldsymbol{x}) \ge a_p$, the penalty item was $\boldsymbol{d}_{ip} \cdot \min(g_p(\boldsymbol{x}') - a_p, 0)$, $\boldsymbol{d}_{ip} = [d_{1p}, d_{2p}, \text{L}, d_{ip}, \text{L}, d_{np}]$ indicating the coefficient of the penalty item.

(3) Establish external archives to store non-dominant solutions. When a new particle is generated, it is compared with the non-dominant relationship of the particle in the external file. If the new particle is dominated by more than two particles in the external file, it is discarded and the particle is regenerated for comparison again. If the new particle is the non-dominant solution of the particle in the current external file or is dominated by only one particle in the external file, it is stored in the

external file. At the same time, discarding the particles dominated by the particle in the external archives can properly improve the diversity of particles in the external archives; if the number of particles C_W in the external archives exceeds its upper limit W after adding new particles, discarding the particles with the highest density in the external archives and maintaining the uniform distribution of the non-dominated frontier. Finally, calculating the relationship between the new particles and the remaining particles in the external archives. Crowding distance [97], e.g. (4).

$$D(\boldsymbol{x},\boldsymbol{x}_w)=\sqrt{\sum_{i=1}^{n}(eval_i(\boldsymbol{x}')-eval_i(\boldsymbol{x}'_w))^2} \tag{4}$$

Among them, $\boldsymbol{x}$ represents the new particle and $\boldsymbol{x}_w$ represents the particle w in the external file.

$$M(\boldsymbol{x}_w)=\frac{1}{\min(D(\boldsymbol{x}_w,\boldsymbol{x}_1),D(\boldsymbol{x}_w,\boldsymbol{x}_2),\cdots,D(\boldsymbol{x}_w,\boldsymbol{x}_{w-1}),D(\boldsymbol{x}_w,\boldsymbol{x}_{w+1}),\cdots D(\boldsymbol{x}_w,\boldsymbol{x}_{C_W}))} \tag{5}$$

Among them, $M(\boldsymbol{x}_w)$ represents the density of the particle w.

(4) Choose the best location for the whole and the individual [96]. Based on the non-dominant relationship and spatial distance of the particles in the external archives, the global best position $\boldsymbol{g}_b$ of the population is selected. Based on the non-dominant relationship between the current position of the particles in the population and the historical best position $\boldsymbol{p}_b$ and the utility function value of the fitness value, the best position of the individuals in the population is selected.

When choosing the best position of the population, we first calculate the spatial distance $\frac{C_W}{3}$ of the particles in the external archives, such as formula (6). Then we rank the particles in the external archives according to the spatial distance from large to small, select the first particle, and use roulette to select the best position of the population. That is, if the random number rand $\in[0\ 1]$ is within the probability of the spatial distance of a particle, we regard the particle as the best population in the whole. Location.

$$\begin{aligned}
D_1&=\min(D(\boldsymbol{x}_w,\boldsymbol{x}_1),D(\boldsymbol{x}_w,\boldsymbol{x}_2),\cdots,D(\boldsymbol{x}_w,\boldsymbol{x}_{w-1}),D(\boldsymbol{x}_w,\boldsymbol{x}_{w+1}),\cdots D(\boldsymbol{x}_w,\boldsymbol{x}_{C_W}))\\
D_2&=\min(\{D(\boldsymbol{x}_w,\boldsymbol{x}_1),D(\boldsymbol{x}_w,\boldsymbol{x}_2),\cdots,D(\boldsymbol{x}_w,\boldsymbol{x}_{w-1}),D(\boldsymbol{x}_w,\boldsymbol{x}_{w+1}),\cdots D(\boldsymbol{x}_w,\boldsymbol{x}_{C_W})\}-\{D_1\})\\
D_w&=(D_1+D_2)/2
\end{aligned} \tag{6}$$

Among them, D_1 represents the minimum crowding distance between the particle w and other particles in the external file, D_2 represents the second minimum crowding distance between the particle w and other particles in the external file, and the spatial distance D_w between the particle w in the external file. When choosing the best position of an individual, if the current position updated by a particle in the population is in a dominant position with a certain position in its historical best position, it will be regarded as the best position of the individual; if the current position updated by a particle is not dominated by its historical best position, the position with the minimum utility function value of fitness value will be regarded as the particle. The best position of an individual, such as equation (7), where the initial particle or step (7) is the same as its historical best position in the first iteration of the particle obtained by mutation operation.

$$U_{value}=\sum_{i=1}^{n}eval_i(\boldsymbol{x}') \tag{7}$$

Among them, U_{value} is the utility function of the fitness value of the particle $\boldsymbol{x}'$.

(5) Update the velocity and position of particles $\boldsymbol{x}$. Particles are updated according to the velocity and displacement formulas, such as formula (3.1), to adjust the particles beyond the boundaries $[\boldsymbol{L}\ \ \boldsymbol{U}]$ of decision variables, maximum $\boldsymbol{v}_{\max}$ or minimum $\boldsymbol{v}_{\min}$ velocities;

When the particle's velocity v_j exceeds the maximum $v_{\max}^j$ or minimum $v_{\min}^j$ velocity, then $v_j{=}v_{\max}^j$ or $v_j{=}v_{\min}^j$, when the particle exceeds the boundary constraint range $[\boldsymbol{L}\ \ \boldsymbol{U}]$, then $x_j = L_j$ or $x_j = U_j$, and $v_j = -v_j$.

where

$\boldsymbol{v} = [v_1, v_2, \cdots, v_j, \cdots, v_D]$, $\boldsymbol{v}_{\max} = [v_{\max}^1, v_{\max}^2, \text{L}\ , v_{\max}^j, \text{L}\ , v_{\max}^D]$, $\boldsymbol{v}_{\min} = [v_{\min}^1, v_{\min}^2, \text{L}\ , v_{\min}^j, \text{L}\ , v_{\min}^D]$.

$j = 1, 2, \text{L}\ , \text{D}$, D is the number of decision variables.

(6) Update the fitness value of the particle and the non-dominant solution in the external file. According to step (2), the fitness value of the newly generated particles in the population is calculated. According to step (3), the external files are maintained and the current non-dominant solution is updated.

(7) Maintaining population diversity. The average convergence degree C of particles in external files is calculated as feedback information of the convergence degree of the algorithm. If it is less than the threshold value δ, the mutation operation is carried out by randomly selecting the particles based on the external files, and then the fitness value of the mutated particles in the population is calculated according to step (2), otherwise step (8) is carried out directly.

The specific process of particle mutation operation is as follows: Firstly, the average convergence degree C of particles in external files is calculated as feedback information of algorithm convergence, such as formula (8). If it is less than the threshold value δ, particle $\frac{C_W}{20}$ in the external file is selected randomly. If the random number $\text{rand} \in [0\ 1]$ is less than the given probability P_m, then particle $\frac{C_W}{20}$ is selected again for mutation operation, such as formula (9). Then a mutated particle $\frac{C_W}{30}$ and an unmutated particle $\frac{C_W}{30}$ are used to replace the particle $\frac{C_W}{20}$ with the highest density in the population. The mutated particle $\frac{C_W}{60}$ can keep the diversity of the population appropriately and unchanged. Particles with different operations can provide more information for searching non-dominant frontiers and improve the efficiency of the algorithm.

$$C = \sum_{j=1}^{C_W} \sum_{i=1}^{n} (eval_i(\boldsymbol{x}'_j) - eval_i^{\min})^2 \tag{8}$$

Among them,

$$eavl_i^{\min}{=}\min(eavl_i(\boldsymbol{x}'_1), eavl_i(\boldsymbol{x}'_2), \text{L}\ , eavl_i(\boldsymbol{x}'_{C_W}))$$

(8) If the population reaches the maximum number of iterations M_{itera} or the difference ΔC between the average convergence degree Δt of particles in the adjacent two generations of external archives is less than the threshold value τ, then the particles in the external archives are output, i.e. the non-dominant solution is obtained, otherwise the search is continued in step 5.

3. Engineering application

The formulation optimization model for reducing the radiation intensity and primary smoke resistance of solid propellant in 2 μm-5 μm infrared band is established. The calculation and analysis are carried out as an example, e.g. (9). Among the HMX/Al/AP/HTPB propellants, the lower boundary values L_{Al}, L_{AP}, L_{HMX}, L_{HTPB}, L_{fer} of aluminium powder, AP, HMX, HTPB, ferrocene mass fraction are 4%, 60%, 0, 10%, respectively, while the upper boundary values U_{Al}, U_{AP}, U_{HMX}, U_{HTPB}, U_{fer} are 18%, 75%, 1.7%, 20%, 11%. The values of aluminium powder, ammonia perchlorate, ferrocene, HMX, HTPB mass fraction. $I(T_E)_{max}$ and $AGARDP_{max}$ are determined by the optimum target requirements of different types of missile weapons. The theoretical specific impulse of the agent HMX/Al/AP/HTPB ranges from 262s to 264s. In this case, the lower limit $I_{S\min}$ of theoretical specific impulse is 262s and the lower limit A_L of oxygen coefficient is 0.55.

$$\begin{cases} \min \ (AGARDP, I(T_E)) \\ \text{s.t.} \\ x_{HMX} + x_{Al} + x_{AP} + x_{HTPB} + x_{fer} = 1 \\ 0.89 < x_{HMX} + x_{Al} + x_{AP} + x_{fer} < 0.90 \\ AGARDP \le 1 \\ I(T_E) \le 70000 \\ I_S \ge 262 \\ 0.9 \le A_{oc} \le 1 \\ 0 \le x_{HMX} \le 20\% \\ 4\% \le x_{Al} \le 18\% \\ 60\% \le x_{AP} \le 75\% \\ 10\% \le x_{HTPB} \le 11\% \\ 0 \le x_{fer} \le 1.7\% \end{cases} \tag{9}$$

For each formulation example, the combustion temperature and nozzle exit temperature of propellant need to be constantly re-determined, and on this basis, the equilibrium constant and total enthalpy and entropy of combustion products need to be constantly updated. The maximum number M of iterations of the improved multi-objective particle swarm optimization algorithm is 260. An improved multi-objective particle swarm optimization (MPSO) algorithm is used to solve the formulation optimization model, and the approximate non-dominant frontier of primary smoke resistance and infrared radiation intensity is obtained, as shown in Figure 1.

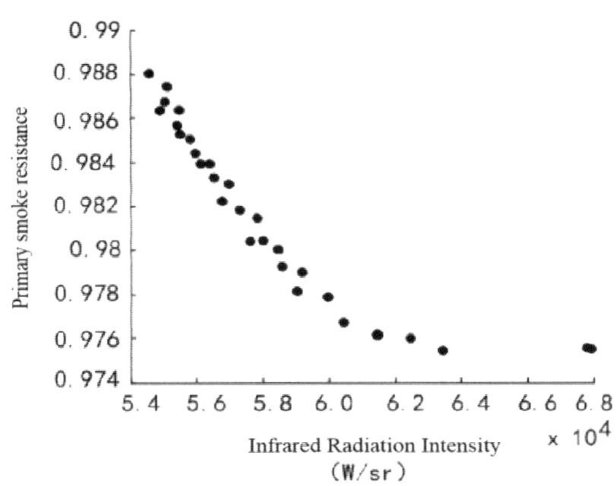

Figure 1. Approximate Non-dominated Frontier

IOP Conf. Series: Materials Science and Engineering **646** (2019) 012041 doi:10.1088/1757-899X/646/1/012041

It can be seen from Figure 1 that the reduction of primary smoke and infrared radiation intensity of solid propellant is conflicting, but the obtained approximate non-dominant front solves the problem of reducing the conflict between them. The corresponding formulations are all feasible formulations for realizing low characteristic signal, which can reduce primary smoke and infrared radiation intensity to the greatest extent on the basis of satisfying certain theoretical specific impulse. According to the different influence degree of primary smoke and infrared radiation intensity of solid propellant on missile weapon, the feasible formula $C_{\mathrm{EP}ij}$ is sorted by formula selection method according to the order from big to small, and finally the largest formula $C_{\mathrm{EP}ij}$ is screened out, which can provide reference for formula designer.

4. Conclusion

Firstly, this paper introduces the basic principle of standard algorithm, the general model of formula optimization design and the related concepts of non-dominant relationship. The equality constraints, boundary constraints and inequality constraints of decision variables are dealt with by substitution method, random generation method and penalty function method, respectively, to form fitness function, which solves the problem that particle swarm optimization can not deal with constraints directly.

Secondly, based on the internal population and the established external archives, the global and individual best positions are selected respectively, which solves the problem that the particle swarm optimization algorithm can not directly solve the multi-objective optimization problem. In addition, a particle mutation operation method based on the external archives technology is proposed, which keeps the diversity of the population, avoids the algorithm falling into the local optimum, and maximizes the number of iterations and external archives. The change of the average convergence degree of the middle particle is used as the double judgment basis for the termination of the algorithm, which improves the efficiency of the algorithm. Thus, an improved multi-objective particle swarm optimization algorithm is formed to calculate the feasible formula for realizing the low characteristic signal.

Finally, the classical example is solved by the algorithm, and how to set the maximum iteration times of the algorithm is given for the specific example. The comparison results show that the convergence and uniformity of the proposed algorithm have obvious advantages, which verifies the effectiveness of the improved algorithm.

Reference

[1] Yang Qing, Li Jie, Zuo Guo Ping. Effects of solid propellant components on the radiation characteristics of plumes. [C]. / thirty-second proceedings of the annual meeting, 2015.

[2] Feng Jinzhi, Chen Xing, Zheng Songlin. An improved multi-objective particle swarm optimization algorithm and its application [J]. computer application research, 2014, 31 (3): 675-678.

[3] Mu A Q, Cao D X, Wang X H. A modified particle swarm optimization algorithm [J]. Natural Science, 2015, 26 (2): 151-155

[4] Xu Heming. Research on multi-objective particle swarm optimization algorithm [D]. Shanghai: Shanghai Jiaotong University, 2013.

[5] Hu Wang, Gary G Y, Zhang Xin. Multi-objective particle swarm optimization algorithm based on Pareto entropy [J]. Journal of Software, 2014, 25 (5): 1025-1050.

Divide-and-Conquer CBS Design for Multi-Agent Path Finding in Large-scale Scenario

Mengxia Chen, Yiming Liu and Hejiao Huang*

Harbin Institute of Technology, Shenzhen 518055, China

*E-mail: huanghejiao@hit.edu.cn

Abstract. *Multi-Agent Path Finding* (MAPF) problem is intensively studied in theoretical computer science, robotics and so on. The key for MAPF problem is to plan a conflict-free path for each agent with different start and goal positions, and to minimize the cost of the paths. *Conflict-Based Search* (CBS) is one of the optimal algorithms, which can ensure that the optimal solution is obtained, in the case of small-scale maps and low number of agents. However, in many real-world multi-agent systems, the scale of the map is very large and the number of agents is more. In most cases, CBS is not applicable any more. Therefore, we develop Divide and Conquer CBS, called DC-CBS algorithm which divides large-scale map and the original problem into several smaller ones. For each subproblem, we use the CBS to get the optimal solution. The experimental results show that for the large-scale scenario, the DC-CBS algorithm can get the result more quickly and plan the paths for agents successfully. As for small-scale situation, the performance of DC-CBS is almost as good as that for CBS.

1. Introduction

The *Multi-Agent Path Finding* (MAPF) problem includes a undirected graph $G = (V, E)$ and a team of k agents $a_i \dots a_k$ where each agent a_i has a start position $s_i \epsilon V$ and goal position $g_i \epsilon V$. Solving MAPF optimally is NP-hard [1], as the state space grows exponentially with the number of agents. The common objective of MAPF is to find the valid path for every agent without conflicts, and minimize the total cost of the paths. MAPF problem has practical applications including robotics, traffic control, warehouse applications and assembly planning [2]. Recently, many researches have paid attention to the MAPF problem of small maps. However, the results of researches cannot perform well in large-scale maps with a great many agents. For example, the Kiva is a large-scale automatic warehouse which has hundreds of agents [3], it is a challenge to plan paths for massive agents rapidly.

In this paper, we propose a Divide-and-Conquer CBS (DC-CBS) algorithm to solve the large-scale map situation. DC-CBS algorithm is an improvement based on CBS algorithm [4] which is one of the optimal algorithms to solve the MAPF problem. CBS can get the best solution, but it runs slowly. The key of DC-CBS algorithm is to combine divide-and-conquer idea with CBS algorithm, divide large-scale map and the original problem into several smaller ones. The experimental results show that the DC-CBS algorithm can get the result more quickly than CBS, and the quality of solution is applicable although it is somewhat worse than the optimal solution.

2. Related work

Algorithms for solving MAPF can be divided into three classes: *optimal solvers*, *suboptimal solvers*, *bounded suboptimal solvers*. When the number of agents is relatively small and the aim is to find the minimum solution, optimal solvers are adopted. While suboptimal solvers are usually used when the number of agents is large.

A number of optimal solvers are based on the A* algorithm. A relevant variant is *Enhanced Partial Expansion A** (EPEA*) [5], which sorts all successors of a given state to their f value and only the successors that $f \leq C^*$ are generated. Standley presented the *independence detection* (ID) and *operation decomposition* (OD) technique [6]. OD introduces the intermediate states by considering a single agent at a time which reduces the branching factors of A* algorithm successfully. Furthermore, M* can dynamically change the branching factor according to the encountered conflicts. *Recursive M** (RM*) is an enhanced version which divides the conflicting agents into several small groups with independent conflicts. Recently, the *increasing cost tree search* (ICTS) [7] and the *conflict based search* (CBS) algorithms are one type of search-based optimal solver. CBS is a two-level search algorithm that the low-level plans the path and the high-level resolves the conflicts between the agents.

For suboptimal solvers, most of them are unbounded and not provide any guarantee on the quality of the planned path. A prominent suboptimal MAPF algorithm is the *Cooperative A**(CA*) [8] and its variants HCA* and WHCA*. CA* and its variants plan paths for every agent separately. If the path is returned successfully, it will be written a *global reservation table* and not have conflicts with the following agents. Importantly, these algorithms have a good performance in practice but not complete. Another type of suboptimal solvers is rule based that formulate specific rules for different scenarios.

The suboptimal algorithm cannot guarantee the quality of solution, but the bounded suboptimal search algorithm can accept a parameter w and ensure that the solution is less than or equal to $w \times C^*$(C* is the cost of the optimal solution). The bounded suboptimal search algorithm achieves a balance between the quality and running time of the solution. WA*[9] and others [10] are well-known bounded suboptimal algorithm. Optimal and suboptimal algorithms have their own advantages and disadvantages, and they behave differently in different scenarios. In this paper, we combine the search algorithm with the actual situation and introduce a variant of CBS.

3. Problem description

3.1. MAPF

MAPF is defined as an undirected graph $G = (V, E)$ and a set of agents labelled $a_1 a_2 \dots a_k$, each agent a_i is allocated a task with a unique start position $s_i \epsilon V$ and goal position $g_i \epsilon V$. At each time step, an agent takes an action to move to adjacent or wait in the current position. For each action *move* or *wait*, the cost is one. The *cost* of an agent is equal to the number of time steps required from the start position to goal position. For the whole problem, the total cost is the sum of the cost for all agents.

We use the term *path* only in the context of a single agent and use the *solution* to denote all paths for the given agents. A path for agent a_i is a sequence of its *move* and *wait* actions, which can be recorded as a sequence $P = v_1 v_2 \dots v_k \dots$ where $V_i \epsilon V$ and $i = 0,1,2 \dots k \dots$. If both agents a_i and a_j occupy vertex v at time point t, we record this *point conflict* as a tuple (a_i, a_j, v, t). In order to solve it, constraints (a_i, v, t) and (a_j, v, t) are added to agent a_i and a_j respectively. A *constraint* (a_i, v, t) means a_i cannot occupy vertex v at time step t.

In MAPF problem, the agents a_i and a_j are not allowed to cross the same edge (v_1, v_2) at opposite direction simultaneously. If happens, we call it *edge conflict* (a_i, a_j, v_1, v_2, t) and add *edge constraint* (a_i, v_1, v_2, t) which means agent a_i is forbidden to move from v_1 to v_2 at the time t to solve it.

3.2. The Conflict Based Search algorithm (CBS)

CBS is a two-level algorithm, including high and low levels. The high-level of CBS searches the constraint tree (CT), which is a binary tree. The pseudo-code for the high level is shown in Algorithm 1. Each node $N \epsilon CT$ contains three types of information.

(1) **a set of new constraints** for agents (*N.constraints*)

(2) **a current solution** for MAPF problem (*N.solution*)

(3) **the cost** of N.solution (*N.cost*)

The root node R of CT has no constraints (line 1). The low-level search is invoked and return one shortest path for every agent a_i (line 2). After calculating the cost of R.solution, the root node R is inserted into open table. The high-level search performs a best-first search on the CT while the nodes are ordered by their costs. Then get a node N from open table, we validate the N.solution (line 7). If there is no conflict between any two agents, the CT node N would be regarded as a goal node and the N.solution would be returned (line 9). Otherwise, the first conflict $C_n = (a_i, v_1, v_2, t)$ and the node N is declared as non-goal. To solve the conflict and guarantee getting the optimal solution, node N generates two child CT nodes that inherit the constraints from its parent and are added a new constraint respectively. For example, a new (a_i, v_1, t) constraint is added for agent a_i (line 14). Then the low-level is invoked to replan, which is responsible for planning an optimal path for a single agent that satisfies its all constraints while completely ignoring the other agents (line 16).

Algorithm 1 High-level of CBS

```
input: MAPFinstance
1: R.constraints ← ∅
2: R.solution ← find paths invoking the low − level()
3: R.cost ← SIC(R.solution)
4: insert R to Opne
5: while Open not empty do
6:     N ← best node from Open
7:     Validate the paths in N until a conflict occurs
8:     if N has no conflict then
9:         return N.solution
10:    end if
11:    C ← first conflict(a_i, a_j, v, t) in N
12:    for each agent a_i in C do
13:        A ← new node
14:        A.constraints ← N.constraints + (a_i, s, t)
15:        A.solution ← N.solution
16:        Update A.solution by invoking low−level(a_i)
17:        A.cost ← SIC(A.solution)
18:        Insert A to Open
19:    end for
20: end while
```

Algorithm 2 DC-CBS

```
input: MAPFinstance
upperProblem ← change_to_upper()
upperSolution ← find individual paths by A*
for each path p_i in upperSoluton do
    subProblem.startNode ← p_i.front()
    //set the start nodes of subProblem
    p_i.pop_front()
end for
while point exsists in upperSolution do
    for path p_i in upperSolution do
        tempNode ← SG−A*(p_i.startNode, endSet)
        subProblem.endNode ← tempNode
        //set the end nodes of subProblem
        p_i.pop_front()
        Update subSolution by invoking CBS(subProblem
        subPro.startNodes ← subPro.endNodes
    end for
end while
```

Algorithm 3 SG-A*

```
input: (startNode, endSet)
openlist.push_back(startNode)
while (!openlist.empty()) do
    curPoint ← node with lowest F value
    openlist.remove(curPoint)
    closelist.push_back(curPoint)
    if curPoint in endSet then
        openlist.clear()
        closelist.clear()
        return curPoint
    end if
    get surroundPoints of curPoint
    for point p in surroundPoints do
        if P in closelist then
            do nothing
        end if
        if P in openlist then
            update F value
        end if
        if P not in openlist then
            add p into openlist
        end if
    end for
end while
```

The time complexity of CBS is discussed as follows. In the worst case, the branching factor of CT is 2 and 2^{dp} nodes must be expanded where dp is the depth of CT. When generating a CT node, a new constraint is added. In the worst case, every agent will be constraint to avoid every other vertex except one at every time step. We use C^* denotes the total steps of all agents. Thus, the upper bound of CT nodes is $2^{|V| \times C^*}$. For each node, the low level is invoked and expands at most $|V| \times C^*$ (single-agent) states. The number of states expanded in the low-level is $(2^{|V| \times C^*} \times |V| \times C^*)$.

4. Algorithm Design

4.1. Divide and Conquer

Divide and conquer is to divide a complex problem into two or more identical subproblems, then divide these subproblems into smaller ones until the smallest subproblems can be solved directly. And the solution of the original problem is the combination of the solutions of the subproblems. In this paper, the *divide and conquer* idea is applied to CBS algorithm. By transforming the problem of large-scale map into small-scale subproblems, the large-scale problem can be solved by finding solutions for subproblems. Experimental results indicate that the solving time using this method can be well shortened.

4.2. Divide and Conquer Conflict-Based Search algorithm (DC-CBS)

The core idea of DC-CBS is to divide the actual map into blocks and abstract into smaller regional map, namely upper grid. Correspondingly, we can call actual map lower grid. Each upper point corresponds to a region in the lower grid. As shown in Figure 1(a) and Figure 1(b), the size of lower grid and upper grid is 12×12 and 3×3 respectively. The lower grid is divided into 9 regions and each one shrinks to an upper point in the upper grid. Points and regions with the same color correspond to each other. For example, the red region in Figure 1(a) corresponds to the red point in Figure 1(b) and so on.

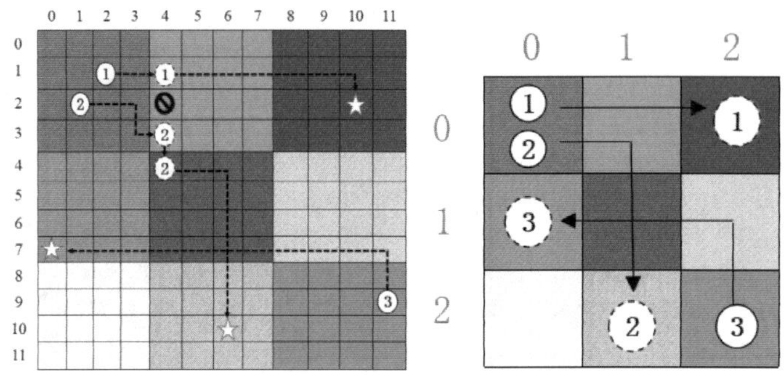

Figure 1. (a).an example of lower grid. The solid circles are the start points, while the dashed circles mean the temporary end points. The stars mean the actual end points. The dashed arrows represent the paths for agents. (b) the corresponding upper grid for (a). The solid circles are the start points and the the dashed circles are end points. The solid arrows represent the rough paths for agents

In the process of solving the MAPF, the start and goal points of the agents in the lower grid are mapped to the upper grid firstly. Then the rough paths are obtained by A* algorithm. That means, the direction of all agents to perform tasks are decided on the upper grid. Next, the two adjacent upper points from rough paths are used to search the temporary start and end points in the lower grid. Actually, a edge in the rough path corresponds to a task of lower grid. After traversing all the edges into tasks and solving all the subproblems, the paths of the whole MAPF are planned completely.

In terms of the upper grid, it is mainly used to plan the rough path of all agents and guide the direction of agents. Rough path refers to the sequence of directions that need to be changed from the start to the goal point, which is composed by a series of continuous coordinate points. For example, as shown in Figure 1(a), assuming there are three tasks, start points from (1,2), (2,1), (9,11) and end points are (2,10), (10,6), (7,0) respectively. Three tasks are projected into the upper grid, as shown in the Figure 1(b). After mapping, the rough paths can be planned as follows:

$A1$ $(0,0) \rightarrow (0,1) \rightarrow (0,2)$, $A2$ $(0,0) \rightarrow (0,1) \rightarrow (1,1) \rightarrow (2,1)$, $A3$ $(2,2) \rightarrow (1,2) \rightarrow (1,1) \rightarrow (1,0)$

Pseudo-code for DC-CBS is shown in Algorithm 2. Because an upper point corresponds to many points of the lower, there is no conflict between rough paths. We adopt A* algorithm without considering conflicts to plan the rough path for each agent (line 2).

The main function of the lower grid is to generate small MAPF subproblems and call CBS to solve them. Suppose the rough path has been planned in the upper grid, we use two adjacent upper points to plan actual path for every agent. The key of planning the actual path is how to determine the temporary end points that must be passed in the process of executing the task. We propose the Search Goal A* (SG-A*) algorithm to search the temporary endpoints and it can ensure the end point is closest one. The biggest difference between A* and SG-A* is that the latter uses a set of endpoints instead of a fixed one. The endpoint set is composed of allowable nodes in a lower region corresponding to a single upper point. The pseudo-code for SG-A* is shown in Algorithm 3. SG-A* will stop search when an allowable point p in the endpoint set. Point p will be taken as a temporary endpoint and returned (line 6 - line 9).

For example, as shown in Figure 1(a) and Figure 1(b), in the upper grid, A_1 and A_2 need to go from (0,0) to (0,1). The actual start point of A_1 is (1,2) (Figure 1(a)) and its endpoint set contains all allowable nodes corresponding to upper nodes (0,1) (Figure 1(b)). The SG-A* algorithm finds the nearest temporary endpoint (1,4) (Figure 1(a)). Next, A_1 directly plans the path to the actual end point (2,10) (Figure 1(a)) to complete the planning task, because the actual endpoint is in the adjacent region. Similarly, the actual start point of A_2 is (2,1) (Figure 1(a)). Due to (2,4) position is an obstacle and (1,4) position has been occupied by A_1, (3,4) is chosen as the current optimal position.

When the temporary endpoints of all agents are determined, they are combined into a small-scale MAPF problem and then CBS algorithm is invoked to solve it (line 14). If the original MAPF problem is not planned completely, the latest temporary endpoints of these agents will be taken as the new start points (line 15). And the next temporary endpoints will be searched again. The combination of all subsolution is the solution of the original MAPF problem.

4.3. Algorithm Analysis

The time complexity analysis of CBS is also applied for DC-CBS. In the worst case, for example, the subproblems adopt the whole map and expand all points, the time complexity of DC-CBS is $O(2^{|V| \times C^*} \times |V| \times C^*)$. But in practice, DC-CBS algorithm runs faster and takes less time to find the solution. The original problem is divided into several subproblems, in which the start point and the end point of each task are very close and it takes little time to plan the path. Hence, the running time of DC-CBS is shorter than that of CBS.

5. Experimental Results

We have done extensive experiments to compare the algorithms. Our experiments were conducted on an Inter(R) Core i5 and processor operates at 3.10GHz and has 4GB of RAM. All code for experiments was written in C++ and experiments were conducted on Visual Studio 2015.

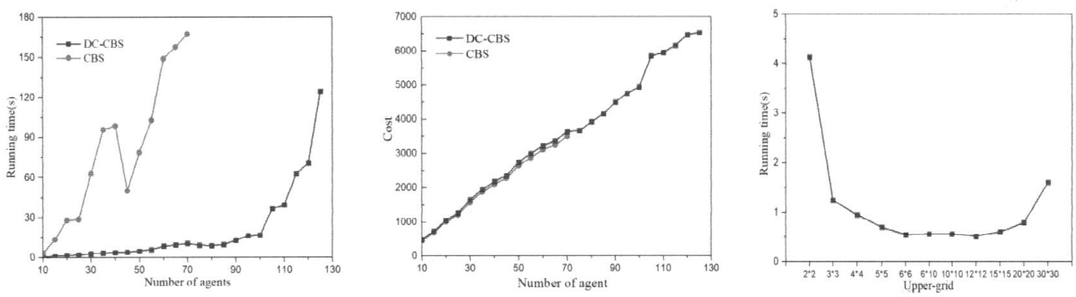

Figure 2. Comparison between CBS and DC-CBS in terms of run time (a) and cost (b). The influence of different upper grid size is shown as (c).

Firstly, we compare the DC-CBS algorithm with the original CBS algorithm. We use 80×80 4-connected grid with generating 10% obstacles randomly. At the same time, each agent is arranged a task with unique start and goal points. For DC-CBS algorithm, the size of upper grid is 5×5.

Figure 2(a) shows the running time(s) for CBS and DC-CBS. Compared with CBS algorithm (labeled as the red line), the DC-CBS algorithm (labeled as the blue line) clearly outperforms it. It should be noted that CBS algorithm cannot find a solution within four minutes while the DC-CBS algorithm could obtain the solution with half a minute when the number of agents is over 70. When the number of agents reaches 45, as shown in Figure 2(a), the running time is greatly reduced. The reason is that agent tasks are generated randomly and it is perhaps that the distance of the random start and goal points is short. All in all, DC-CBS is much better than CBS algorithm in running time. Figure 2(b) presents the relationship between the number of agents and the total cost of paths generated by CBS and DC-CBS respectively. It is well known that CBS is the optimal algorithm that can obtain the lowest cost. And it can be seen that the gap between the two algorithms is quite small when the number of agents is no more than 70. As for large-scale cases (e.g. 100 agents), DC-CBS still perform well while CBS is not available any more.

Furthermore, the size of upper grid also has a significant effect on running time. Figure 2 show the running time of 20 random tasks on a 60×60 4-connected grid with 10% obstacle rate. As shown in the figure, the size of upper grid has a little impact on the running time when the grid is between 5×5 and 15×15. Too large or too small size of upper grid would increase running time. When the size of grid is too small, such as 1×1, DC-CBS is the same as the original algorithm. And when the size of the upper grid is larger than 15×15., the running time is also increase. The main reason is that the running time spent on dividing subproblems has increased. Consequently, DC-CBS can greatly reduce the running time with the appropriate size of upper grid.

From the above experiments, it is obvious that DC-CBS takes less time to solve large-scale problem combined with moderate size of upper grid, and the quality of the solution is almost as good as the optimal solution.

6. Conclusion

We studied the MAPF problem in large-scale scenarios. In most cases, CBS cannot obtain the solution in three minutes, which is not acceptable in practical application. Therefore, we propose the DC-CBS algorithm which combines Divide-and-Conquer technique with CBS algorithm. DC-CBS divides the actual map into upper and lower grids, where a region of lower grid shrinks to a point of upper grid. Then the original problem is decomposed into several subproblems. In the upper grid, we introduce the rough path which is planned by A* without considering conflicts to guide directions for the agents. In the lower grid, SG-A* is proposed to search the temporary endpoint. It guarantees the temporary endpoint is the best one which is closest to the start point. Experimental results show that DC-CBS is more efficient to get the solution than CBS and the quality is almost as good as the optimal solution.

Acknowledgments

This work is financially supported by National Key R&D Program of China under Grant No. 2017YFB0803002 and No. 2016YFB0800804, National Natural Science Foundation of China under Grant No. 61672195 and No. 61732022.

References

[1] Jingjin Yu and Steven M. Lavalle. Structure and intractability of optimal multi-robot path planning on graphs. In *Twenty-seventh Aaai Conference on Artificial Intelligence*, 2013.

[2] Kurt Dresner and Peter Stone. A multiagent approach to autonomous intersection management. *Journal of Artificial Intelligence Research*, 31(3):591–656, 2008.

[3] Peter R. Wurman, Raffaello D&Aposandrea, and Mick Mountz. Coordinating hundreds of cooperative, autonomous vehicles in warehouses. In *National Conference on Innovative Applications of Artificial Intelligence*, 2007.

[4] Guni Sharon, Roni Stern, Ariel Felner, and Nathan R. Sturtevant. Conflict-based search for optimal multi-agent pathfinding. In *Twenty-sixth Aaai Conference on Artificial Intelligence*, pages 40–66, 2012.

[5] Ariel Felner, Meir Goldenberg, Guni Sharon, Roni Stern, Tal Beja, Nathan Sturtevant, Jonathan Schaeffer, and Robert C Holte. Partial-expansion a* with selective node generation. In *Twenty-sixth Aaai Conference on Artificial Intelligence*, 2012.

[6] Trevor Scott Standley. Finding optimal solutions to cooperative pathfinding problems. In *Twenty-Fourth AAAI Conference on Artificial Intelligence*, 2010.

[7] Guni Sharon, Roni Stern, Meir Goldenberg, and Ariel Felner. The increasing cost tree search for optimal multi-agent pathfinding. *Artificial Intelligence*, 195(195):470–495, 2013.

[8] David Silver. Cooperative pathfinding. In *Aaai Conference on Artificial Intelligence & Interactive Digital Entertainment*, 2005.

[9] Ira Pohl. Heuristic search viewed as path finding in a graph. *Artificial Intelligence*, 1(3):193–204, 1970.

[10] Jordan Tyler Thayer and Wheeler Ruml. Faster than weighted a*: An optimistic approach to bounded suboptimal search. In *ICAPS*, pages 355–362, 2008.

Chinese Named Entity Recognition for Clothing Knowledge Graph Construction

Ming Zhu[1] and De-sheng Zhen[1,*]

[1]School of Computer Science and Technology, Donghua University, Shanghai 201620, China

[*]2397646126@qq.com

Abstract. Clothing knowledge graph is a kind of vertical domain knowledge base constructed for the description of clothing knowledge in the field of textile and apparel. In this paper, based on the limitations of the clothing knowledge graph in the effect of entity extraction, the deep learning model and the statistical model are combined. A Chinese named entity recognition method based on CNN-BiLSTM-CRF is proposed. Firstly, the convolutional neural network(CNN) is used to extract the text features, and the character-level vectors with morphological features of the words are trained. Then the bi-directional long short term memory networks(LSTM) is used to learn the context features, and the vector representation of the context of each word is output. Finally, the conditional random fields(CRF) model is used for self-learning. Get the best tag sequence for the sentence. The method can automatically recognize the text, and does not rely on the artificial feature to obtain the semantic category information. Finally, the experimental data and evaluation methods are introduced. The experimental results show that the Chinese named entity recognition method based on CNN-BiLSTM-CRF is superior to other models in all indicators, indicating the effectiveness of the method.

1. Introduction

In the context of the era of big data, with the emergence of massive data and the cross-application of multiple data sources, traditional data management modes and query methods are subject to certain constraints. In recent years, with the development of knowledge graph technology, the knowledge base with semantic processing capability and open interconnection ability has created great value in many fields such as intelligent search[1], knowledge quiz[2], recommendation system[3] and so on. Knowledge graph is a structured semantic knowledge base for describing various entities or concepts and their relationships in the real world. It constitutes a huge semantic network graph. Nodes represent entities or concepts, while edges are composed of attributes or relationships[4]. With the rapid development of e-commerce, online shopping has become an important channel for consumers to purchase clothing. Compared with the traditional market, there are more varieties and types of clothing sold online. Therefore, how to organize the clothing data and make it appear in front of the user in a structured and connected way is very important. Constructing knowledge graph in the field of clothing has an important research significance and application value for the realization of clothing knowledge quiz. Although the knowledge graph has developed rapidly, there is no mature and professional knowledge graph in the clothing field. Therefore, this paper constructs the conceptual layer of clothing knowledge graph, and takes knowledge acquisition as the starting point. A Chinese named entity recognition method based on CNN-BiLSTM-CRF is proposed. The clothing entity and attribute can be automatically extracted from unstructured clothing text data. The experimental results show that the model achieves good results compared with other named entity recognition models without relying on any artificial features, and the F value can reach 91.2%.

2. Related Works

In the process of constructing the knowledge graph, it is necessary to perform physical extraction on the structured, semi-structured and unstructured text data of the clothing. For structured and semi-structured data, we only need to learn its fixed structure to extract entities and related relationships effectively, but extracting entities from unstructured text is a very difficult research task. Therefore, the rule-based approach gradually fades out the field of named entity recognition research. At present, there are many commonly used machine learning models to solve the problem of named entity recognition, such as maximum entropy model [5], hidden Markov model [6], conditional random field [7] and so on. Conditional random field is a discriminant model that does not require strict independence assumptions and overcomes the shortcomings of mark bias inherent in the maximum entropy model. Conditional random field is very suitable for sequence analysis. In recent years, it has been widely used in sequence labeling tasks, and it has outstanding performance in named entity recognition. The machine learning-based entity extraction method faces the problems of uneven data quality and low professional labeling when it is used in professional fields. Deep learning [8] is a new field in machine learning research. In recent years, deep learning has gradually been applied to the field of natural language processing. Based on the deep learning method, it can learn autonomously from the original data because it does not need to be artificially set. Therefore, it can reduce the interference of human beings on data, find deeper and more abstract features, and has become a research hotspot in recent years. Literature [9] uses a long short term memory neural network model to improve the performance of word segmentation, but this method can not obtain the semantic information behind the sentence. Literature [10] proposes to use CRF as the processing method of the output processing layer on the basis of bi-directional LSTM, which effectively improves the performance of the model.

3. Definition And Construction of the Concept Layer

The construction of the concept layer is the construction of the entire clothing knowledge graph framework. The construction of the knowledge graph concept layer is equivalent to the establishment of Ontology [11]. The most basic ontology includes concepts, concept levels, attributes, attribute value types, relationships, relationship definition fields, concept sets and relationship values, concept sets. This paper uses the protege5.2.0 tool and uses the five-tuple form to describe the clothing knowledge ontology model, defined as CMO={C,P,H^c,I,R}, where C represents all conceptual collections representing clothing, and P represents all the set of attributes, H^c represents the hierarchical structure of the ontology concept, that is, the classification structure of the concept in C, I represents the instance set, and R represents the relationship rule. First, give a hierarchical definition of the concept of the clothing field.Followed by the concept of attribute definitions and constraints. For example, in the clothing ontology, the domain of "isPartof" is the tops and bottoms, the value domain is the full body; the domain of "isFitfor" is the clothing, and the value domain is the human. These attributes can be used to relate concepts in the ontology. Such classes and classes can be associated by attributes. Then there is the definition of rules or axioms. According to the rules, some scattered and implicit knowledge can be inferred, and the original information can be supplemented and improved. Finally, between each stage and stage of the ontology life cycle, a certain reference frame is used to make technical judgments on the ontology and software environment and documents, to acquire clothing knowledge, and finally to generate a clothing knowledge base.

4. Chinese Named Entity Recognition Based on CNN-BiLSTM-CRF

In the process of constructing the clothing ontology knowledge base, a large number of clothing entities and their attributes need to be acquired. For structured and semi-structured data, lack of clothing expertise leads to incomplete clothing knowledge. This part of the knowledge can be obtained from unstructured text data such as documents. Therefore, using deep learning methods, features in unstructured text can be extracted automatically without manual participation. Therefore, this paper proposes a Chinese named entity recognition method based on CNN-BiLSTM-CRF, which can automatically extract clothing named entities from unstructured data. Based on BiLSTM-CRF Chinese named entity recognition method, the combination of deep learning model and statistical learning model can effectively extract text context information. At the same time, In order to make use of these spatial semantic information, a convolutional neural network is introduced to extract spatial semantic features. The model framework is shown in Figure 1. From the figure, the model is divided into four parts: the word vector representation part, the CNN part, the BiLSTM part and the CRF part.

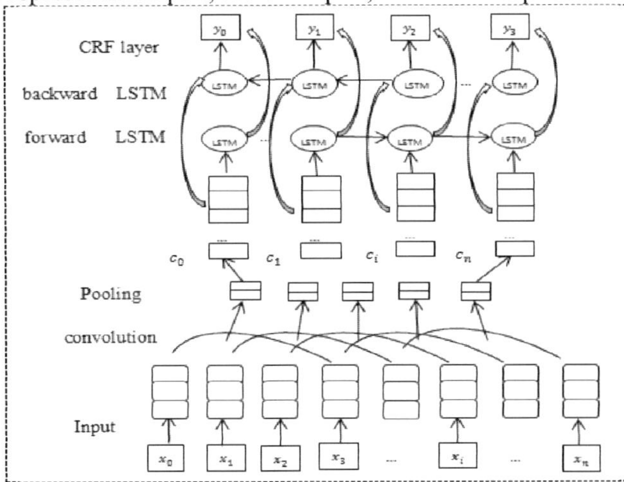

Figure 1. CNN-BiLSTM-CRF structural model

4.1. Convolutional Neural Network model Layer

This layer uses the convolutional layer to extract features from the input matrix vector. Using CNN to combine the word vector of the input word with the CNN window sliding to connect the current word with the preceding and following words, calculate the influence of the words before and after the current word, and the generated words are used to represent the feature of the word. After the convolution is completed, the word is extracted. The context information between the character and the character constitutes a new word and sentence representation feature, and the word vector is expanded in detail, and then input into the lower layer neural network.

The convolution operation uses a filter $w = R^{hk}$ to join the words with h as a window to form a new feature. For example, $x_{i:j+h-1}$ words are combined into a new feature c_i.

$$c_i = f(w \cdot x_{i:j+h-1} + b) \quad (1)$$

Where $b \in R$ represents the bias term, f() represents the nonlinear excitation function, and filter w combines each possible window word to form a new feature map $c = [c_0, c_1, \ldots, c_{n-h+1}]$. Then, we apply the maximum pooling operation on the feature map, and take the maximum value $c = \max\{c\}$ as the feature corresponding to this particular filter. This step can obtain the most important features and can adapt to the variable. The length of the sentence.

4.2. Bi-directional Long Short Term Memory Networks Model Layer

The one-way LSTM neural network model can only obtain the above information of the sentence, and the context information of the sentence after the sentence cannot be obtained. In order to compensate for the fact that the one-way LSTM cannot obtain the context information behind the sentence, this layer uses the bi-directional long short term neural network (BiLSTM) model. The BiLSTM receives the character vector as an input and obtains a hidden layer representation of the character by encoding. The hidden layer representation consists of the output of the forward LSTM and the output of the backward LSTM. The forward sequence of the forward LSTM received character vector sequence is input, $c = (c_0, c_1, c_2, \dots, c_n)$, and the forward representation sequence is obtained by forward coding $(\overrightarrow{H_1}, \overrightarrow{H_2}, \dots, \overrightarrow{H_n})$. $\overrightarrow{H_t}$ denotes the forward representation of the t-th character, and the formula is as follows:

$$\overrightarrow{H_t} = \mathrm{LSTM}([c_0, c_1, c_2, \dots, c_n]) \quad (2)$$

Where LSTM(·) represents the calculation of the LSTM model. The backward LSTM receives the reverse sequence of the sequence of character vectors as input,$c = (c_n, c_{n-1}, \dots, c_1, c_0)$, and obtains the backward representation sequence $(\overleftarrow{H_n}, \overleftarrow{H_{n-1}}, \dots, \overleftarrow{H_1})$, $\overleftarrow{H_t}$ denotes the backward representation of the t-th character, and the formula is as follows:

$$\overleftarrow{H_t} = \mathrm{LSTM}([c_n, c_{n-1}, \dots, c_1, c_0]) \quad (3)$$

Where LSTM(·) represents the calculation of the LSTM model. The forward and backward directions are combined to obtain the output H_t of the BiLSTM layer, where $H_t = [\overrightarrow{H_t} \cdot \overleftarrow{H_t}]$.

4.3. Conditional Random Fields Model Layer

The CRF layer assigns a marker to each character, and calculates the score of the entire sequence. The sequence score consists of the character marker score and the marker transition score, and the label sequence with the highest score is selected as the final prediction result. The hidden layer output of BiLSTM, $H = (H_1, H_2, \dots, H_n)$, is the input to the CRF, where N is the length of the input sequence and H_i is the ith word of the input vector. Then $Y = (y_0, y_1, y_2, \dots, y_n)$ is the output tag sequence corresponding to Z. For a given input sequence Z with a value of z, the conditional probability of taking y on the label sequence Y is $p(y|z)$. The specific formula is as follows:

$$p(y|H) = \frac{1}{S(H)} \exp\left(\sum_{t,k} \lambda_k t_k(y_{n-1}, y_n, H, n) + \sum_{n,l} u_l s_l(y_n, H, n)\right) \quad (4)$$

$$S(H) = \sum_y \exp\left(\sum_{n,k} \lambda_k t_k(y_{n-1}, y_n, H, n) + \sum_{n,l} u_l s_l(y_n, H, n)\right) \quad (5)$$

Where n=1,2,...,N, $t_k(\dots)$and $s_l(\dots)$are eigenfunctions, $\lambda_k$and u_l are the weights corresponding to $t_k(\dots)$ and $s_l(\dots)$, respectively, and S(z) is the normalization factor . In the training phase, the maximum likelihood estimation is used to calculate the optimal labeling sequence. The likelihood logarithm of the training set is $\sum_n \log p(y|z)$, and the y is finally obtained by training to obtain the highest conditional probability. The input sequence is annotated to get the best tag sequence for the sentence and the formula is as follows:

$$y = \underset{y \in Y}{\mathrm{argmax}}(y|H) \quad (6)$$

5. Experiment

5.1. Experimental Data

The experimental data comes from 12,000 articles of clothing products obtained by a clothing store in Taobao through crawling technology. Among them, 8400 pieces of comment data are randomly selected as the training set, and 3600 pieces of comment data are randomly selected as the verification set. The sequence is marked for the task, and each character in all the comment sentences has a corresponding label correspondence. This experiment uses the "BIEO" labeling system to identify whether each word in the sentence is a component of the named entity and distinguish the entity boundary. Where "B" represents the first word of the entity, "I" represents the other part of the entity, "E" represents the last word of the entity, and "O" represents the word that is not an entity. The main identification task of this experiment is to identify the named entities in the clothing ontology knowledge base in the clothing review, such as style, layout,color and other named entities.

5.2. Evaluation Methods

In this experiment, the precision (P), recall rate(R) and F value(F) were used to measure the model. Among them, the accuracy rate refers to the ratio of the number of correctly identified entities to the total number of identified entities. The recall rate refers to the ratio of the number of correctly identified entities to the total number of entities, and the F value is the harmonic mean of the accuracy and recall rate. . The specific formulas of each indicator are as follows:

$$P = \frac{n}{M} \times 100\% \tag{7}$$

$$R = \frac{n}{N} \times 100\% \tag{8}$$

$$F = \frac{2PR}{P+R} \times 100\% \tag{9}$$

Where M represents the number of entities identified, N represents the total number of entities in the test set, and the number of entities correctly identified in n generations.

5.3. Model Comparison Analysis

In order to have a more comprehensive evaluation of the performance of the model in this chapter, this experiment designed a number of models for comparative analysis. The experimental results are shown in Table 1.

Table 1. Comparison of CNN-BiLSTM-CRF and other model experiments (%)

number	model name	precision	recall	F value
1	CRF	75.68	79.83	77.70
2	BiLSTM	83.36	86.68	84.99
3	BiLSTM-CRF	88.34	91.26	89.78
4	CNN-BiLSTM	85.74	89.52	87.59
5	CNN-BiLSTM-CRF	88.95	93.56	91.20

It can be seen from the experimental results in Table 1 that although the single models of CRF and BiLSTM can complete the entity recognition tasks to a certain extent, on the whole, these models are inferior to other combined models, indicating that the single model has been Increasingly, it is not possible to meet the needs of actual named entity recognition tasks. By comparing the model 4 and the model 5, it can be seen that the effect of introducing CRF on the basis of BiLSTM is slightly better than that of introducing the convolutional neural network, which explains in a certain sense that the introduction of CRF has a recognition effect on the named entity. Promotion has a significant effect. Finally, Model 5 is 1.42% higher than the F value of Model 3, which indicates that feature extraction using convolutional neural networks can better represent text and improve model performance. In summary, CNN-BiLSTM-CRF can achieve better experimental results than the general deep learning model.

6. Discussion

In this paper, the task of identifying the entity in the clothing domain knowledge graph, the character vector is obtained by convolutional neural network to supplement the word vector, and then the CNN-BiLSTM-CRF neural network model method is constructed. The experimental results show that the model achieves the best results. The main discussions are as follows:

First of all, in the knowledge graph of the clothing field, the knowledge of the artificial feature domain has a great influence on the result of the entity extraction, which leads to the increase of the cost of the system and the decline of the generalization ability. Therefore, the constructed CNN-BiLSTM-CRF deep neural network model does not use any In the case of artificial features, it is better to obtain machine learning methods than using a large number of rich features and domain knowledge. Secondly, this paper uses CNN network to obtain character vectors representing word morphological features to supplement the deficiencies of word vectors. The model can identify the special characters more effectively, which improves the performance of the model. Finally, in order to obtain a more accurate recognition result, the output of the CNN-BiLSTM network is decoded by CRF to obtain an optimal marker sequence. The integration of CRF improves the recognition performance of clothing named entities with multiple modifiers and blurred boundaries. Therefore, in the process of entity extraction of clothing knowledge map, the CNN-BiLSTM-CRF fusion model proposed in this paper is an effective way to effectively improve the recognition performance.

7. Conclusion

This paper defines and constructs the conceptual layer of the knowledge graph of the clothing field, and studies an entity automatic extraction method based on the knowledge graph of the clothing field, aiming to build a relatively high quality knowledge base for the clothing field. Firstly, the concept layer is constructed in a top-down manner, and the whole clothing knowledge graph framework is constructed. In the process of construction, named entity recognition and extraction are performed from unstructured clothing data. And focus on how to automatically extract garment entities from unstructured text data through deep learning models. Experiments show that the model effect is better than other models, and the F value reaches 91.20%.

Building clothing knowledge graph is a complex and challenging task. There are many more areas where the clothing knowledge graph needs improvement. For example, the knowledge of the clothing field is not rich enough, and it is also necessary to seek a multi-source heterogeneous data source to expand the knowledge base. Knowledge fusion, including entity alignment and entity linking, automatically updates the established clothing knowledge graph. In general, this paper studies the clothing knowledge graph based on unstructured entity extraction, which fills the lack of knowledge graph in the domestic clothing field. For the intelligent question and answer in the field of clothing, accurate search has laid the foundation and is of great significance.

References

[1] David Amerland's. Google Semantic Search[M]. 2013.
[2] Rao Z Y, Zhang Y X. Research on Intelligent Customer Service System Based on Knowledge Map[J]. Electric Power Information & Communication Technology, 2017.
[3] Zhang J, Pourreza M, Ramachandran R, et al. Facilitating Data-Centric Recommendation in Knowledge Graph[C]// 2018 IEEE 4th International Conference on Collaboration and Internet Computing (CIC). 2018.
[4] Zeng-Lin X U , Yong-Pan S , Li-Rong H E , et al. Review on Knowledge Graph Techniques[J]. Journal of University of Electronic Science & Technology of China, 2016.
[5] Guo J, Zhao J, Zeng S. Structural reliability analysis based on analytical maximum entropy method using polynomial chaos expansion[J]. Structural & Multidisciplinary Optimization, 2018(1):1-17.
[6] Leos-Barajas V, Photopoulou T, Langrock R, et al. Analysis of animal accelerometer data using hidden Markov models [J]. Methods in Ecology and Evolution, 2017, 8 (2): 161-173.
[7] Teichmann M T T, Cipolla R. Convolutional CRFs for Semantic Segmentation[J]. 2018.
[8] Lecun Y, Bengio Y, Hinton G. Deep learning.[J]. Nature, 2015, 521(7553):436.
[9] Chen Xinchi, Qiu Xipeng, Zhu Chenxi, et al. Long short- term memory neural networks for Chinese word segmentation[J]. Conference on Empirical Methods in Natural Language Processing, 2015, 295(1):1197－1206.
[10] Dong C, Zhang J, Zong C, et al. Character Based LSTM- CRF with Radical- Level features for chinese named entity recognition[C]. International Conference on Computer Processing of Oriental Languages. Springer International Publishing. 2016:239－250.
[11] Rodríguezgarcía M Á, Hoehndorf R. Inferring ontology graph structures using OWL reasoning[J]. Bmc Bioinformatics, 2018, 19(1):7.

IOP Conf. Series: Materials Science and Engineering **646** (2019) 012044 doi:10.1088/1757-899X/646/1/012044

Document Segmentation Method Based on Style Feature Fusion

Gang Liu[1,2,a], Kai Wang[1,b], Wangyang Liu[2,c], Xu Cheng[2,d] and Tao Li[1,e]

[1]Harbin Engineering University, Harbin, Heilongjiang , China
[2]CETC big data research institute company limited, Guiyang, Guizhou,China

[a]liugang@hrbeu.edu.cn; [b]2503613988@qq.com;[c]liuwangyang@cetcbigdata.com; [d]chengxu@cetcbigdata.com;[e]389083838@qq.com;

Abstract. Style crack refers to the position where the author's identity changes in the article completed by multiple authors. This paper summarizes the current situation and theory of related fields at home and abroad, and proposes a multi-feature based document segmentation method for plagiarism detection. Seven text style features are used for style crack recognition. Through the result of feature extraction, the combination of multi-feature fusion and unsupervised machine learning algorithm is used to classify the features based on extraction, and the clustering algorithm is used to cluster the style features so as to find the location of style cracks. Experiments show that the method is effective and scientific, and achieves good results.

1. Background and significance of the project

With the advent of the data age,people's research on essays and plagiarism detection are not only based on simple string judgments, but also on the author's writing style and writing habits. The writing style can not only see an author's writing habits, but also can be used in the plagiarism detection system and user portrait technology, and also helps the author's identification, and better plagiarism detection system. Providing a new perspective, new directions, and strong support for the authors of online anonymous articles[1]. Different people will travel their own unique styles when writing, mainly in terms of words, sentences, paragraphs, rhetoric, emotions, etc. These are the writing habits that the author has inadvertently developed, so through the writing style of the article. Feature extraction, inferring the attribution of an article through style features is effective[2].

2. Related theory and work basis.

The early style research mainly used statistical methods, mainly to count the laws of vocabulary, sentences and paragraphs, and use the laws of statistics to agree on a person's style. The extraction of style features was the first to study single features. As single features could not satisfy the experimental results, multi-feature fusion also came into being. In recent years, the development of machine learning and neural networks has introduced machine learning and neural network algorithms into style extraction and author recognition, and has achieved good results.[3]

Because of the variety and difficulty of Chinese, it is obviously more difficult to extract Chinese style than that of foreign language. Chinese needs to take into account the accuracy of word segmentation system and the complex sentence structure. Although the extraction of Chinese style is more difficult than that of foreign countries, the study of style has also received widespread attention.

IOP Conf. Series: Materials Science and Engineering **646** (2019) 012044 doi:10.1088/1757-899X/646/1/012044

3. Multi-feature extraction and fusion

3.1. One-dimensional style feature

3.1.1. Word length. In terms of Chinese style, it is possible to observe the author's usage habits of two-word words, three-word words, four-word words and four-word words in terms of word use according to the length of the vocabulary after the word segmentation. Some studies have found that the frequency of use of double-word words, Zhang Ailing's frequency of use is 0.17, and Lu Xun's frequency of use is as high as 0.43. From this we can see that the length of the word can be seen in the style of the article.

In English, the number of letters is statistically counted. The range of average word lengths in Chinese is relatively small, and the average word length is between 2-4. However, the experimental results can be seen by narrowing the comparison range. Finally, the word length is taken as a parameter of the final classification.

3.1.2. Average sentence length. The average sentence length is the usage habit of the statistical author for the length of the text sentence, and the frequency of use of the long and short sentences. The length of each sentence is counted, and the average sum is averaged. The average sentence length is marked with ".", "!", and "?", and the average value of the length words in the statistical sentence is taken as the parameter of the last dimension.

3.1.3. Emotional bias. Emotional analysis has always been the analysis of emotional words to analyze emotions. Fan Xia uses neural networks for emotional analysis and verifies their effectiveness and compares their results [4]. Sailunaz uses machine learning to calculate user impact scores based on various user-based and Twitter-based parameters, and to perform sentiment analysis on Twitter [5]. This paper uses the sentiment dictionary trained in the network to analyze the article emotionally.

3.2. Multidimensional style features

3.2.1. Lexical features. The use of words can completely extract a person's literary skills, the same can be based on the richness of the use of words to judge an author's writing style. The lexical features [6] can define the length of the word, the word frequency, the proportion and the density, and define the vocabulary characteristics as shown in Table 3.1:

Table 3.1 Vocabulary Characteristics Table

Numbering	Lexical features	Numbering	Lexical features
1	Total number of words	5	Self-created words/total words
2	Number of words in two words	6	Interjection / total number of words
3	Number of words in three words	7	Number of different words / total number of words
4	Number of words in four words	8	Lexical density

3.2.2. Special punctuation. Special punctuation features statistical colon, semicolon, thousand percent, unit symbol, left and right quotation marks, left and right brackets, exclamation mark, ellipsis, dash, question mark and comma.

The special punctuation uses the simple statistical method. By using the statistics of the special punctuation above as the punctuation parameter of the final classification algorithm, it is of course necessary to delete the punctuation with the punctuation frequency of 0, so the text dimension is up to 11 dimensions.

3.2.3. Synonym. Synonym is a unique branch of Chinese style, which can extract the strength of Chinese literature. It can also extract the author's language skills and ability to control words in the use of synonyms. The usage habits of synonyms can also be seen in an author's habit of using words. This feature is based on the use of synonyms, based on the use of synonyms, to summarize the usage habits of synonyms, and then summarize the author's writing habits.

3.2.4. Function word. This paper increases the number of virtual words, and uses the custom vocabulary table as a benchmark to calculate the use of the virtual words in the virtual vocabulary. First of all, the virtual vocabulary table is produced. The source of the vocabulary table is "Now the Chinese Dictionary of Function Words". There are 840 virtual words in the vocabulary list, which are the same as the synonym table. The virtual vocabulary table is too large. The vocabulary table contains some uncommon and uncommon function words, which will affect Calculation of results. Based on the news set, the TF-IDF statistics are performed on the virtual words of the virtual vocabulary, and the TF-IDF is deleted too low. Through multiple cleanings, it was finally streamlined to 230 virtual words. The choice of 230 virtual words can be controlled first in a reasonable dimension.

4. Style crack recognition

4.1. Style crack

Style cracking refers to the position where a text style changes. In other words, an article may be completed by different authors. Therefore, it is especially important to perform author-based segmentation techniques before author identification, that is, to find out each article in this article. The corresponding part of the author's text. By writing style segmentation, the goal is to find the style crack points, where the style and style change. Style crack recognition is mainly through the technique of feature extraction and classification algorithm, and the use of sliding window, dimension reduction and other techniques. Finally find out the style crack points.

4.2. Sliding window

The sliding window is composed of five sentences as a whole for style feature recognition. Each time you slide down a sentence, the sliding window contains five sentences at a time, and each window is styled. When the style changes, each style and the last result gradually change until the style is similar. The degree is close to the same, and there is a style crack in this position.

Because the length of the paper is small, the five sentences contain less information, and there is a large possibility that there will be an accidental phenomenon. It is assumed that there is a great possibility that the position of each style crack occurs at the end of each segment. Therefore, in order to improve the accuracy rate, only the recall rate can be sacrificed. It is assumed that each style crack will necessarily occur at the end of the paragraph, that is, each paragraph in the article is assumed to have one and only one author.

4.3. Parameter weighting

Aiming at the feature redundancy problem caused by the third chapter's style feature analysis, each parameter takes different weights in the process of style crack identification, so when looking for style cracks, we need to find the weights of each parameter, and then identify style cracks by adjusting the weights of parameters.

Algorithm description: Firstly, the data set is preprocessed, the data set is extracted, and the data set is stored in the file in disorder. In the case that the weights of other feature parameters are the same, the idea of controlling variable method is used first. The control word length parameter (WLP) is calculated from 0.01 to 0.99, and the other parameters are 0.5. The highest quality of the parameter WLP is obtained under the condition that other parameters are unchanged, and the highest quality is the lowest degree of similarity between the two texts. When the other parameters are unchanged, the average sentence length parameter (ASLC) is calculated from 0.01 to 0.99, and the optimal value of the ASLC is obtained, thereby traversing all the parameters, and then the optimal result of the previous parameter is used as a reference. Continue the above method to loop until the parameter optimal value is unchanged, and obtain the parameter weight recombination. The purpose of this step is to find the validity of each parameter by the parameter weight method and delete the influence of the invalid parameter.It is found that some of the parameters weights are too small, and these characteristics have little positive impact on the results, but will affect the efficiency of the experiment, so delete this parameter.

5. Style clustering based on style features

5.1. Experimental idea

Text feature extraction is the main method of style recognition. This method classifies the style features hierarchically and extracts the features from the perspective of hierarchy. A mapping relationship between features and articles is added here. Feature extraction includes two kinds of features: one-dimensional feature and multi-dimensional feature. Parameter weight method is used to optimize feature. The results of each feature extraction are used as the input of the final k-means++ classifier. Style cracks are found through sliding windows. Style cracks are segmented by identifying style cracks.

The corpus selects news corpus, and extracts 1300 news articles from 20 people and 40 people respectively. The subjects of the corpus include news about tourism collected from People's Daily and news about sports selected from Hupu. Because there are some impurities in news collection, such as time, picture, picture introduction and photographer's name. Firstly, impurities are dealt with and the text is selected. Store the news in a CSV file and label it with the author's name. The 1300 news articles were divided into 1150 training sets and 150 test sets. The ratio of training sets to test sets was about 9:1. In order to verify the accuracy of small pages, 150 test sets are divided into 100 pages stored by length and 50 pages stored by paragraph. 50 pages are about 215 news pages.

5.2. Weight optimization of parameters

In the corpus, 100 articles of 5 authors were randomly selected to form a small sample training set for training the parameter weight method. Firstly, the training set is preprocessed, and each author's document set is put down into a txt file, and the style feature extraction stroke style feature vector is performed on each author's document set. The first step is to extract the average sentence length parameter, and then perform the word segmentation process. After the word segmentation process, the average word length, lexical features, and special punctuation marks are extracted. Next, the virtual word is extracted to perform the function word tf-idf algorithm, and the synonym is filled with the synonym vector. Calculate the emotional bias of the training set. The experimental results are shown in Table 5.1:

Table 5.1 Parameter Weight Method Results

	Word length	Average sentence length	emotion analysis	Lexical features	Special symbol	Synonym	Function word	Text similarity result
Author 1 and author 2	0.27	0.27	0.12	0.52	0.42	0.89	0.65	74.6586
Author 1 and author 3	0.18	0.32	0.13	0.62	0.21	0.82	0.88	54.7158
Author 1 and author 4	0.45	0.17	0.07	0.87	0.55	0.92	0.88	82.9728
Author 1 and author 5	0.31	0.31	0.21	0.62	0.52	0.82	0.94	62.3541
Author 2 and author 3	0.52	0.12	0.12	0.43	0.31	0.85	0.84	76.2359
Author 2 and author 4	0.64	0.08	0.14	0.76	0.66	0.76	0.76	92.8413
Author 5 and author 5	0.12	0.07	0.08	0.44	0.62	0.97	0.65	60.3281
Author 3 and author 4	0.54	0.34	0.12	0.45	0.29	0.67	0.78	87.3741
Author 3 and author 5	0.21	0.31	0.11	0.77	0.19	0.78	0.90	59.3951
Author 4and author 5	0.41	0.10	0.12	0.79	0.62	0.84	0.84	67.4821

The results of the parameter weight method can be seen that the author's four words are obviously different from other authors in the length of the word. After the experiment, it is found that the author's average length of the word is 3.2132, the three characters and the four words occupy a large proportion, while the other authors are in the 2- Between 3. The length of the sentence has little effect on the result. The lengths of the second, fourth and fifth sentences are similar, and the gaps between the other sentences are also small. The reason why sentiment analysis has the least impact on the article is that the five authors are all objective news, and the emotional shift to the supervisor is small. There are no obvious rules for the parameters of lexical features, indicating that the influence weights of multiple features are different, but it is definitely a positive influence on the recognition of style cracks. The use of special punctuation is more even, and only the special punctuation of author three is similar to everyone. The effects on synonyms and virtual words are obvious, the parameters are large, and the effect on the results is also large. From the experimental results, synonyms and function words have a greater impact on the results, but other characteristics can also reflect their writing style in special circumstances. The style similarity calculation was carried out by extracting the results from the writing style, and it was found that the authors had higher similarity of one, three and five.

5.3. Style crack recognition results and analysis

5.3.1. News set experiment. The style crack identification data set randomly extracts 20 news articles of 5 authors, splits according to paragraphs, and uses paragraphs as a part. At the beginning of the experiment, sliding window technology is used. Each time you slide down a sentence, the number of sentences per window For the five experiments, the clustering results are poor in the results of K-means++ clustering, because each change is a sentence, the magnitude of the change is small, and each change is not obvious, leading to clustering. The window results are less biased. In addition to the position where the pattern breaks at the end of the paragraph, the accuracy will pick up. In the clustering process, the result of k is uncertain, which is caused by the inaccuracy of the K-means++ algorithm. Therefore, many clustering errors occur. The test results are shown in Table 5.2. The visualization is shown in Figure 5.1.

Table 5.2 Style crack recognition results using sliding window

Author	Accuracy/%	Recall/%	F-Measure/%
Author one	66.1	87.5	75.3
Author two	61.6	81.4	70.1
Author three	58.4	75.3	65.8
Author four	68.8	71.7	70.2
Author five	56.9	80.7	66.7

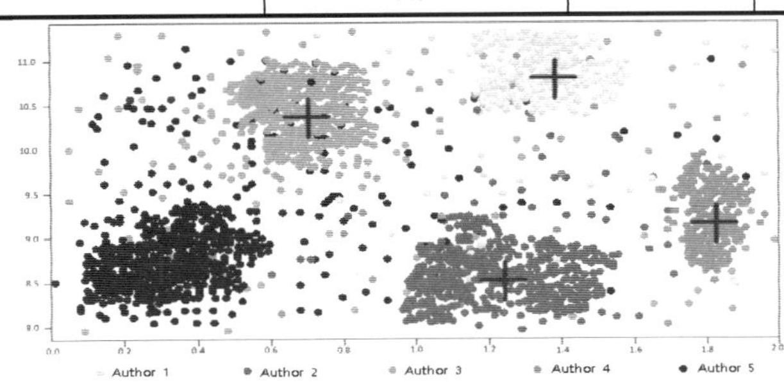

Figure 5.1 Visualization based on sliding window K-means

Finally, the slide window is discarded. The paragraph conversion symbol is used to treat each paragraph as an author. The style feature extraction is performed in units of each paragraph. The k-means clustering algorithm is performed according to the extracted style features. See Table 5.3, the visualization is shown in Figure 5.2

Table 5.3 Style crack identification based on paragraphs

Author	Accuracy/%	Recall/%	F-Measure/%
Author 1	74.3	90.5	81.6
Author 2	75.6	84.7	79.9
Author 3	66.8	85.0	74.8
Author 4	71.5	81.7	76.3
Author 5	67.9	92.7	78.4

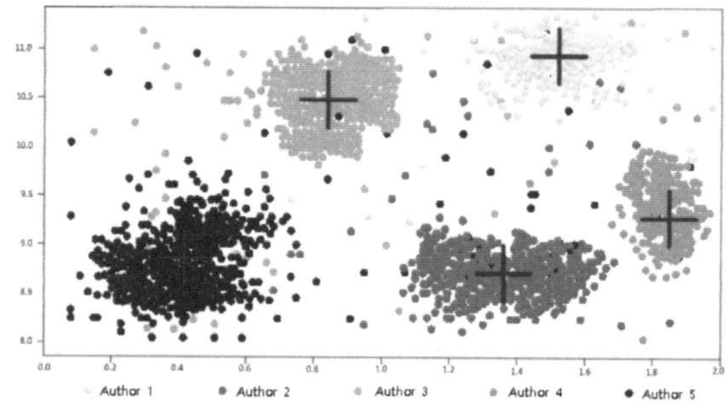

Figure 5.2 Paragraph-based K-means visualization

Although the sliding window is proposed to find all the style crack points as much as possible, since the change of the sentence is not obvious each time a sentence is changed, the style clustering effect is general. The effect of pattern crack recognition based on paragraphs is better than the use of sliding window, which has improved in accuracy and recall rate, and can also increase the evaluation value by 10 percentage points.

5.3.2. Fiction experiment. The original detection of 40 times after the Dream of Red Mansions has always been the main object of the discussion of the writers. The final experiment of segmentation crack recognition is to use the Dream of Red Mansion as the background, and each time as a whole, the style feature extraction, in which the virtual words are no longer used in this paper. The vocabulary table uses 22 vocabulary words. The style clustering based on style feature extraction is performed on 120 times. The result statistics are divided into the first 40 rounds, the middle 40 rounds, and the last 40 rounds. The k-value of the k-means algorithm is 2, the results are shown in Table 5.4, and the visualization is shown in Figure 5.3. Shown

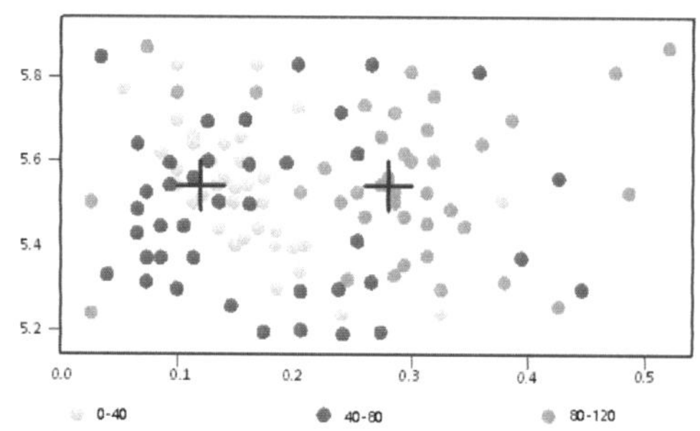

Figure 5.3 Visualization of Dream of Red Mansions Result Analysis

Table 5.4 Analysis of the results of the Dream of Red Mansions

	Xueqin Cao	Er Gao	F-Measure/%
First 40 chapters	36	4	90.0
40 chapters in the middle	33	6	82.5
latter 40 chapters	8	32	82.5

It can be seen from the results that the accuracy of the first 80 times is relatively high, and the last 40 times is relatively low. Through the experimental results of individual features, it is found that the average sentence length, sentiment analysis and virtual words have a great influence on the results. There are obvious differences in the length of the sentence after the first 80 rounds. The reason for the emotional analysis is that the first 80 times Positive, the latter 40 negative, the virtual word has the greatest impact on the experimental results, 22 virtual vocabulary has a positive impact on the experimental results.

6. Conclusion

In this paper, style crack recognition is the result of the extraction of style features, which is a combination of multi-feature fusion and unsupervised machine learning algorithms. Among them, multi-feature fusion is to extract the author's style features better, and machine learning is based on sliding windows or paragraphs. Based on the extracted features, the clustering algorithm is used to cluster the style features to find the style cracks. position. Experiments on the news corpus and the novel corpus are carried out respectively. It is concluded that the segment-based crack recognition is 10% higher than the experimental results based on the sliding window. Therefore, the experimental idea based on the sliding window still needs further improvement.

Reference

[1] Franco-Salvador M, Rosso P, Montes-y-Gómez M. A systematic study of knowledge graph analysis for cross-language plagiarism detection[J]. Information Processing & Management, 2016, 52(4): 550-570.
[2] Juola P, Mikros G K, Vinsick S. Correlations and Potential Cross-Linguistic Indicators of Writing Style[J]. Journal of Quantitative Linguistics, 2019, 26(2): 146-171.
[3] Tschuggnall M, Gerrier T, Specht G. StyleExplorer: A Toolkit for Textual Writing Style Visualization[C]//European Conference on Information Retrieval. Springer, Cham, 2019: 220-224.
[4] Fan Xia,Zhi Zhang,Study of text emotion analysis based on deep learning[J]. IEEE. 2018 13th: 2158-2297p.
[5] Sailunaz,Kashfia,Emotion and Sentiment Analysis from Twitter Text. PRISM,2018: 107533.
[6] Jin Q, Li C, Chen S, et al. Speech emotion recognition with acoustic and lexical features[C]//2015 IEEE international conference on acoustics, speech and signal processing (ICASSP). IEEE, 2015: 4749-4753.

IOP Conf. Series: Materials Science and Engineering **646** (2019) 012045 doi:10.1088/1757-899X/646/1/012045

A Novel Multi-information Fusion Algorithm Based on Population for Wind Power System

Yufeng Wang[1] and Chunyu Xu[2]

[1]Nanyang Institute of Technology, School of Software, Nanyang, Henan, China
[2]Nanyang Institute of Technology, School of Computer and Information Engineering, Nanyang, Henan, China

E-mail: wangyufeng@whu.edu.cn

Abstract. Reserve capacity optimization of power system plays an important role in large-scale Wind Power System. In this paper, a novel multi-information fusion algorithm based on population (MIFA-P) is proposed, which can balance the problem of reserve capacity optimization. MIFA-P determines the type of optimization problem by counting the maximum number of projections on the principal characteristic axis of the optimal individuals of the current population. Therefore, the corresponding class of evolutionary algorithm is chosen to solve the problem. In the process of evolution, each evolutionary algorithm achieves the organic integration through sharing population information. Finally, the validity and applicability of MIFA-P algorithm for large-scale decision variable black box optimization problem are verified by solving a practical scenario optimization problem of reserve capacity distribution in power system after wind power integration.

1. Introduction

Wind energy is a clean and renewable energy source. It plays an important role in energy saving and emission reduction and improving economic benefits of power grid. However, the scale of wind power generation is uncontrollable, because the wind power of wind farm is affected by natural climate [1]. When wind power is integrated into the power grid, it will directly lead to the uncontrollability of the entire power grid system.

In order to ensure the stability of the whole power grid system, conventional energy reserve capacity is put into use at multiple nodes of the whole power grid [2]. Therefore, how the reserve capacity of conventional energy is distributed in each node has a great influence on the stability of the whole power system. Inspired by nature, a variety of meta-heuristic algorithms have been proposed recently to deal with optimization problems, such as Evolution-based Algorithms [3], Swarm Intelligence-based Algorithms [4] and Cooperative coevolution-based Algorithms [5]. These have been successfully deal with different types of optimization problems. However, these meta-heuristic algorithms have failed to make full use of valuable information available from the individuals in previous iterations to guide their current and later search in the black-box real world optimization problems.

In this paper, we propose a novel multi-information fusion algorithm based on population, called MIFA-P. MIFA-P uses mathematical statistics to approximate classify optimization problems. Then, by sharing population information, it chooses a better algorithm to solve the corresponding type optimization problems, so as to improve the performance of the whole algorithm for solving unknown type optimization problems. In the process of evolutionary algorithm solving, the problem type is checked every k generations, and then the better algorithm is chosen to solve the problem. Because MIFA-P can learn the historical evolution information of the population in the whole process of solving, and automatically select the corresponding algorithm by calculating the projection relationship of each decision variable on the principal feature axis.

2. Probabilistic Principal Component Analysis

Probabilistic Principal Component Analysis (PPCA) model is a Bayesian treatment of Principal Component Analysis (PCA), and can be run generatively to provide samples from its distribution [6].

PPCA firstly introduces latent variables explicitly and corresponds to the principal component subspace of the original data. Then, the observed variables are sampled under the condition of this latent variable [7]. That is, the d-dimensional decision variable is added a Gaussian noise by a linear transformation of the m-dimensional implicit variable.

$$\mathbf{x} = \mathbf{Wz} + \boldsymbol{\mu} + \boldsymbol{\varepsilon} \tag{1}$$

Where $\mathbf{W}$ is a rotation matrix, the size of which is $D \times M$ dimension, $\mathbf{z}$ is the latent variable vector, $\mathbf{z} = (z_1, z_2, \cdots, z_M)$, $\boldsymbol{\mu}$ is the mean value of the current population, and $\boldsymbol{\varepsilon}$ is the additional Gauss noise vector, $\boldsymbol{\varepsilon} = (\varepsilon_1, \varepsilon_2, \cdots, \varepsilon_D)$.

$$p(\mathbf{z}) = N(\mathbf{z} \mid 0, I) \tag{2}$$

$$p(\mathbf{x} \mid \mathbf{z}) = N(\mathbf{x} \mid \mathbf{Wz} + \boldsymbol{\mu}, \sigma^2 I) \tag{3}$$

$$p(\boldsymbol{\varepsilon}) = N(\boldsymbol{\varepsilon} \mid 0, \sigma^2 I) \tag{4}$$

We can see that the linear Gaussian latent subspace model is fully determined by its parameters: $\mathbf{W}$, $\boldsymbol{\mu}$ and σ. In high-dimensional space, the calculation cost of solving model parameters directly and accurately is high. Therefore, the expectation maximization algorithm is used to estimate its parameters. Specific methods are as follows: In step **E**, the distribution parameters are initialized, the old parameters are used as fixed observation points, the distribution of hidden variables in the current algorithm is calculated, and the expected values are calculated. Then, the latent variable distribution is substituted into the logarithmic likelihood function of the complete data to calculate the maximum likelihood estimate. In M-step, the likelihood estimates calculated in E-step are maximized. Repeat steps **E** and **M** until the termination conditions are met. The logarithmic likelihood function of $\mathbf{X}$ is:

$$\ln p(\mathbf{X}, \mathbf{Z} \mid \boldsymbol{\mu}, \mathbf{W}, \sigma^2) = \sum_{n=1}^{N} \{\ln p(x_n \mid z_n) + \ln p(z_n)\} \tag{5}$$

Where $\mathbf{X}$ is the latent space corresponding to the original space, and the exact maximum likelihood solution of the mean $\boldsymbol{\mu}$ is:

$$\boldsymbol{\mu} = \overline{x} = \frac{1}{N} \sum_{n=1}^{N} x_n \tag{6}$$

In step **E**, the old parameters are used to calculate the expected values.

$$\mathrm{E}[z_n] = \mathbf{M}^{-1} \mathbf{W}^T (x_n - \overline{x}) \tag{7}$$

$$\mathrm{E}[z_n z_n^T] = \sigma^2 \mathbf{M}^{-1} + \mathrm{E}[z_n] \mathrm{E}[z_n]^T \tag{8}$$

In step **M**, the new parameter values are estimated.

$$\mathbf{W}_{\text{new}} = \left[\sum_{n=1}^{N} (x_n - \overline{x}) \mathrm{E}[z_n]^T \right] \left[\sum_{n=1}^{N} \mathrm{E}[z_n z_n^T] \right]^{-1} \tag{9}$$

$$\sigma_{\text{new}}^2 = \frac{1}{Nd}\sum_{n=1}^{N}\{\| x_n - \bar{x} \|^2 - 2\mathrm{E}[z_n]^T \mathbf{W}_{\text{new}}^T (x_n - \bar{x}) + Tr(\mathrm{E}[z_n z_n^T]\mathbf{W}_{\text{new}}^T \mathbf{W}_{\text{new}})\} \tag{10}$$

3. Proposed Approach

3.1. Basic Ideas

Each optimization algorithm can effectively solve some specified types of optimization problems. In the black box optimization problem, the hybrid algorithm can significantly improve the efficiency of solving the black box optimization problem by detecting the type of the problem and choosing the optimal algorithm with better effect in solving the specified type of optimization problem. The basic framework diagram of MIFA-P is shown in Figure 1.

MIFA-P is divided into five steps: initialization population, evaluation population, selection of dominant population, problem type detection and selection algorithm. Among them, the problem type detection step is to classify the problems by projecting historical information of each decision variable on the principal feature axis in the statistical dominant individuals. The selection algorithm step is to automatically select the better algorithm to solve the type of problem according to the information detected by the detection step of the problem type. There are three kinds of algorithms in the candidate pool: swarm intelligence-based algorithms, cooperative coevolution-based algorithms and estimation of distribution algorithm.

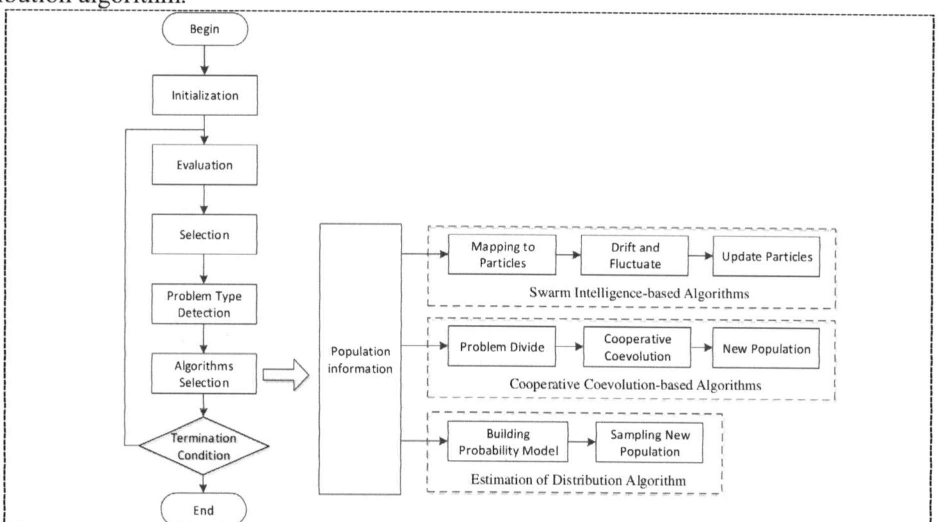

Figure 1. Basic Framework Diagram of MIFA-P

In problem type detection, MIFA-P runs 25 times independently. In the process of evolution, the number of occurrences of the optimal projection value of each dimension decision variable on the principal feature axis is recorded.

$$p_{\mathbf{v}}^{\mathbf{u}} = |\mathbf{u}|\cos\theta = \frac{|\mathbf{u}| \times |\mathbf{v}| \times \cos\theta}{|\mathbf{v}|} = \frac{\mathbf{u} \cdot \mathbf{v}}{|\mathbf{v}|} \tag{11}$$

Where θ is the angle between vectors $\mathbf{u}$ and vectors $\mathbf{v}$, $\mathbf{v}$ is the principal eigenvector, that is, the eigenvector corresponding to the maximum eigenvalue, $\mathbf{u}$ is the normalized value of the current optimal individual.

$$\phi = \frac{\sum_{u=1}^{D} \max p_{\mathbf{v}}^{\mathbf{u}}}{D!} \tag{12}$$

Where ϕ is algorithm selection coefficient. If $\phi = [0.75, 1]$, MIFA-P select cooperative coevolution-based algorithms as based algorithm. If $\phi = [0.5, 0.75]$, MIFA-P select swarm intelligence-based algorithms as based algorithm. If $\phi = [0, 0.5]$, MIFA-P select estimation of distribution algorithm.

3.2. The flowchart of MIFA-P

Algorithms 1 show the pseudocode of MIFA-P. The algorithm consists of four parts: initialization, constraint checking, problem type checking and selection. Among them, code 1 is population initialization; code 3-6 are system reliability constraints detection part punishes individuals who do not meet the constraints; code 7-13 are problem detection part, according to the type of problem, chooses PSO, DECC and EDA algorithm to generate new population; code 14-18 are population selection part, the current generation population and the previous generation population. The best individuals of the group are combined.

Algorithm 1 General Framework of an MIFA-P

Input: Population Size *N*; Mixing rate ω;

Output: Optimal Population Pop_g;

1. random initialization population
2. **While** the termination condition is not met **do**
3. Non-sequential Monte-Carlo simulation is used to test whether each individual in a new population satisfies the system reliability constraints
4. **If** Individual dissatisfaction constraint **then**
5. Punishment of individuals
6. **End If**
7. Calculate and rank all individual fitness functions satisfying system reliability constraints;
8. Problem characteristics of detection based on Pop_g^{Se}
9. Select the corresponding algorithm to cross-mutate the population and generate a new population
10. Non-sequential Monte-Carlo simulation is used to test whether each individual in a new population satisfies the system reliability constraints.
11. **If** Individual dissatisfaction constraint **then**
12. **GOTO** Step10
13. **End If**
14. **If** g > 0 then
15. $Pop_g = Top_{\omega * N}(Pop_{g-1}) \cap Top_{(1-\omega)*N}(Pop_g)$
16. **Else**
17. $Pop_g = Pop_g^{Se}$
18. **End If**
19. g=g+1;
20. **End While**

3.3. Problem Mapping

According to the requirements of Technical Specification for Power System Design (DL/T5429-2009), the maximum reserve capacity of the system is 15%~20% of the total rated load of the system. Therefore, the decision variables range from zero to 20% of the total system load. The fitness function is:

$$F(\mathbf{x}) = \sum_{i=1}^{D} x_i \tag{13}$$

Where $\mathbf{x}$ is an individual in the population, $\mathbf{x} = (x_1, x_2, \cdots, x_D)$. D is the dimension of the problem, which represents the number of nodes that can add spare capacity. Each decision variable ranges from zero to 20% of the total system load.

4. Experimental Results

4.1. Parameter setting

In order to fully verify the performance of MIFA-P algorithm in solving practical optimization problems, this section uses the IEEE-RTS79 system for reliability testing. Assuming that all the equipment in the system is in normal operation and no overload occurs, the transmission power of the transmission line takes the maximum load value of the system. In this section, the reliability evaluation module is simulated by non-sequential Monte Carlo method, and the probability sampling of wind speed and output power are determined by Weibull distribution model. The capacity of each wind turbine in the wind turbine is 1500 kW, and the forced outage rate of the wind turbine is 0.05. Among them, the starting wind speed of the fan is 3 m/s, the rated wind speed is 10.5 m/s, and the cut-out wind speed is 25 m/s, respectively.

4.2. Analysis of Standby Capacity of Power System with Fixed Wind Capacitance Connected to Grid

In order to test the effectiveness of MIFA-P algorithm in the optimization of reserve capacity distribution of power system, this section analyses the optimization of reserve capacity distribution of power system after the fixed wind capacity is connected to the grid. By adding 60 MW wind turbines and reducing the reserve capacity of 60 MW conventional units, the distribution of additional reserve capacity in power system is calculated on the premise that the two indicators of the probability of load loss and the expected value of insufficient energy are not lower than the index value of the system before the wind power is connected to the grid. The experimental results are shown in Table 1 and Figure 2.

Table 1. Standby Capacity Increment When 60 MW Wind Capacity is Incorporated（MW）

Algorithm	Generator Node										Sum
	1	2	7	13	15	16	18	21	22	23	
GA	8.42	14.88	5.57	4.06	3.03	1.99	1.48	3.92	0.65	2.83	**46.83**
PSO	8.91	14.53	6.32	4.13	2.95	1.81	1.67	4.12	0.63	2.92	**47.99**
MIFA-P	7.22	14.57	6.01	3.36	2.37	1.35	0	4.64	0.33	2.78	**42.63**

Table 1 and Figure 2 shows that when 60 MW wind turbines are integrated into the grid system, under the premise that the two indicators of load loss probability and power shortage expectation are lower than those before wind power grid connection, GA algorithm needs to increase 46.83 MW of conventional energy reserve capacity, PSO algorithm needs to increase 47.99 MW of reserve capacity, MIFA-P algorithm needs to increase 42.63 MW of reserve capacity. The capacity allocation scheme increased by MIFA-P algorithm is 4.2 MW less than that of GA algorithm, accounting for 7% of the grid-connected capacity of wind power. The capacity allocation scheme increased by MIFA-P algorithm is 5.36 MW less than that of PSO algorithm, accounting for 9% of the wind power Grid-connected capacity. When the nodes of power system continue to increase, the economic benefits will be more obvious. The convergence curves of the three contrast algorithms are shown in Figure 2. Figure 2 shows that the convergence speed of MIFA-P algorithm is faster than that of GA and PSO algorithm when the 60 MW fixed-capacity wind turbines are connected to the grid and the three algorithms of GA, PSO and MIFA-P are used to optimize the reserve capacity distribution.

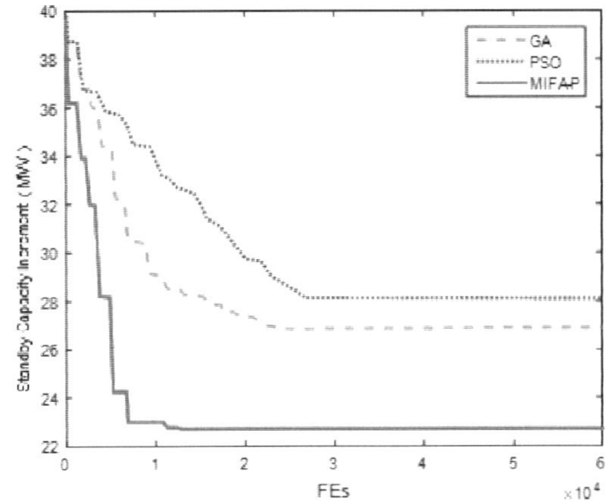

Figure 2. Standby Capacity Increment Graphs

5. Conclusions

In this paper, a new multi-agent intelligent algorithm MIFA-P based on population sharing is proposed. The algorithm judges the type of the optimization problem by counting the occurrence times of the maximum projection value of the current optimal individual on the principal feature axis. Thus, the evolutionary algorithm of corresponding classes is chosen to solve the problem. In the process of evolution, each evolutionary algorithm achieves organic fusion by sharing population information. Finally, the effectiveness and applicability of MIFA-P algorithm in solving large-scale decision variables black box optimization problems are verified by solving a practical scenario optimization problem of reserve capacity distribution after wind power grid connection.

Acknowledgments

This research was supported by the Key Research Projects of Henan Science and Technology Department (No.192102310477), the Foundation and Frontier Projects of Nanyang (No. JCQY2018012), and the Research on Education and Teaching Reform Program of NYIST (No. NIT2017JY-001 and NIT2017JY-032).

References

[1] Taylor, J. W., Mcsharry, P. E., & Buizza, R. (2009). Wind power density forecasting using ensemble predictions and time series models. IEEE Transactions on Energy Conversion, 24(3), 775-782.

[2] Li, W., Zhou, J., Xie, K., & Xiong, X. (2008). Power system risk assessment using a hybrid method of fuzzy set and monte carlo simulation. IEEE Transactions on Power Systems, 23(2), 332-343.

[3] Fogel, David B. (1997). Evolutionary algorithms in theory and practice. Complexity, 2(4), 26–27.

[4] Karaboga, Dervis, and B. Akay. (2009). A survey: algorithms simulating bee swarm intelligence. Artificial Intelligence Review, 31(1-4), 61-85.

[5] Yang Zhenyu, K. Tang, and X. Yao. (2008). Large scale evolutionary optimization using cooperative coevolution. Information Sciences, 178(15), 2985-2999.

[6] Bishop, C. M. (2006). Pattern Recognition and Machine Learning (Information Science and Statistics), 467-473.

[7] Dong, W., Wang, Y., & Zhou, M. (2018). A latent space-based estimation of distribution algorithm for large-scale global optimization. Soft Computing (3) 153-170

Crawling Chinese-Myanmar Parallel Corpus: Automatic Collection, Screening and Cleaning Corpus

Kai Xiong[1], Rui Yuan[1], Wenxue He[1,*], Yanmei Jing[1], Yansheng Wang[1], Qiqi He[2] and Huafu Li[1]

[1]School of Physics and Electronic Information, Yunnan Normal University, Kunming, 650500, China
[2]School of Information Science and Technology, Yunnan Normal University, Kunming, 650500, China

[*]E-mail: wendell31132@hotmail.com

Abstract. The collection of Chinese-Myanmar Parallel Corpus (CMPC) is the key step in the natural language processing (NLP) and training Machine Translation Engine (MTE) of Southeast Asia minority languages. As the scarcity of CMPC resources that efficient corpus collection methods are worth studying extremely. Traditional corpus collection methods include manual collection, text recognition of books and Internet crawlers, etc. Among them, the most efficient method to collect corpus is internet crawler preached by many. Traditional Internet crawler algorithm is interfere easily by a lot of spamming and advertising that lead to the time-consuming and low-precision. We propose a web crawler mechanism combines acquisition automatically technology bilingual website list, crawling corpus and cleaning corpus to obtain high quality parallel corpus. Firstly, using the hyperlinks to recursively access related corpus websites through building the website graph. Furthermore, the breadth-first, Backline and PageRank crawler framework used to build a corpus selection model based on crawling with threshold, matching link, ranking the heat of page, through this, the CMPC can be found accurately. Finally, the corpus cleaning model based on the HTML parsing to determine a set of standardized token sequences. By testing the Chinese-Myanmar reptile algorithm established in this paper, the experimental results show that our benchmarks this model exceeds previous published benchmarks. Up to now, we have obtained 1.1 million parallel corpus pairs of Chinese-Myanmar.

1. Introduction

The CMPC is a one-to-one correspondence between the Burmese and its corresponding Chinese translation. It provides indispensable training data for training the Chinese-Myanmar (CM) machine translation model [1], and is also an important basic resource for lexicography [2] and cross-language information retrieval [3-5].

The two traditional methods obtained by the Chinese-Myanmar candidate website are: Method 1: Using a search engine to collect the target URL. PTMiner [6], STRAND [7] and can obtain a large number of links through search engines and anchor texts, and then use these links as seed links which to recursive downloads. WPDE [8] adds ALT information for the picture. Such methods can automatically discover candidate sites, but the candidate resources are mixed and contain a large amount of irrelevant page site information. Method 2 defines the target source. BITS [9] pre-collects a number of related topics, and then recursively downloads. Although this method can effectively find the list of candidate websites, the number of selected web pages is limited and the selection of the website needs manual intervention, and the website cannot be automatically and quickly determined. This paper proposes a website heuristic collection method, which solves the problem that the source resources of CMPC are small and the coverage of the field is limited. The initial URL seed set used to collect CMPC crawlers must be highly relevant to the topic defined in advance. It does not have to collect all the web pages, only crawl those pages related to the theme [10-11], generally three ways to develop a crawling strategy for web pages. 1. Depth-first traversal [12], the method's crawl logic can lead to excessive depth, so that the crawled pages lack corresponding value, and will also affect the crawl hit rate and crawl efficiency. 2. Best-first traversal [13], this method does not necessarily crawl the optimal selection every time, and may even bypass the optimal page. 3. Breadth-first traversal [14], the pages crawled by this method are closer to the links in the seed set. Pages are more likely to be required and less likely to be repeated. However, the efficiency of the algorithm is reduced after downloading and filtering a large number of pages. Therefore, this paper proposes its own crawling algorithm based on breadth-first traversal combined with backlink $I_B(p)$ and PageRank $I_R(p)$.

This paper proposes a program to collect CMPC from the web, including the discovery and acquisition of CMCW, web crawlers based on Java framework, collection and cleaning corpus. The main advantages of the program in this paper (i)Previous studies mostly obtained corpus resources from a specific bilingual website (government website or conference website), and the collected resources are often small in scale and limited in scope. We first confirms the CM website list and uses it as a heuristic information to use the engine to obtain valuable bilingual hybrid websites, thus expanding the source of bilingual parallel corpus. (ii)This paper develops a crawling priority system based on the Java framework, completes the collection of CMCW from the Web, and downloads the candidate website pages. Using the threshold capture and stop operation mode, the crawler program preferentially accesses the high I(p) page, avoiding the overlap of the download pages, thereby reducing the browsing and downloading of irrelevant pages, and the crawling efficiency is improved. (iii)After obtaining the CMCW, use the regular expression processing mechanism to remove the labels in the original corpus text, standardize the text format, symbols, and obtain high-quality parallel statements. The CMPC automatic acquisition system is not only easy to construct and maintain, but also able to acquire data accurately and quickly.

2. Chinese-Myanmar parallel website acquisition

In this chapter, we present a new solution to obtain the candidate website of CMPC. First and foremost, we represent the CMPC website as the graph structure. Furthermore, we propose a site acquisition model to identify cross-text information and hyperlinks between two parallel crawl processes. By crawling the identified parallel hyperlinks, we can recursively obtain a list of CMPC websites and download candidate website pages. At the same time, we optimized the traditional parallel crawling process, increased the download rate, and avoided the overlap of corpus pages.

Due to the complexity of web page style and website maintenance mechanism, the CM website uses different naming schemes for parallel documents, which also makes it difficult to perform crawling based on URL patterns. Therefore, we propose a new parallel corpus acquisition mechanism. The mechanism use the CM parallel website list as seed information to represent each website list information as Crawl-proc. Then, using the graph alignment structure to communicate and coordinate between the website lists, which discover new parallel hyperlinks and enrich the website list of the CMPC. When downloading pages within multiple retrieval processes, the same page may be downloaded repeatedly. To coordinate the process and prevent overlap, we optimized the acquisition engine. In figure 1, we show the framework for the Chinese-Myanmar parallel website.

IOP Conf. Series: Materials Science and Engineering **646** (2019) 012046 doi:10.1088/1757-899X/646/1/012046

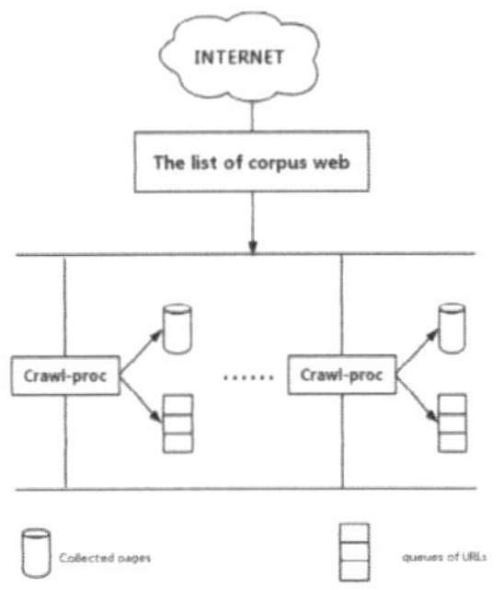

Figure 1. CMPC website acquisition framework

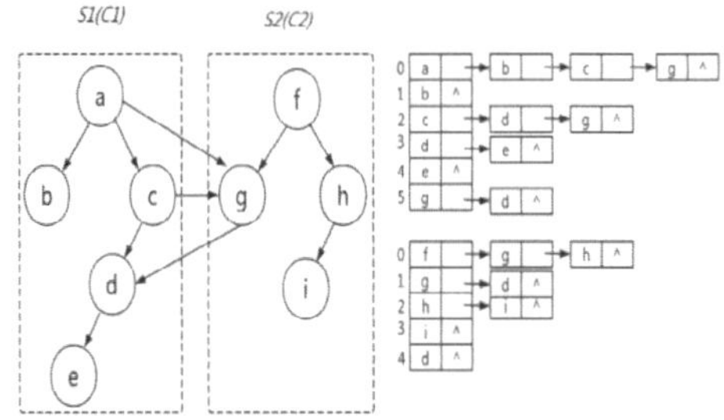

Figure 2. Communication and coordination during crawling

The graph alignment structure completes the communication and coordination between the website lists. Using the graph model, the logical structure of our corpus page is represented as a graph where each node belongs to a predefined node type. In all nodes, we assume that the Web has S_1 and S_2 partitions. At the same time, each Crawl-proc crawl from the root page a and f of each site. When Crawl-proc processes the inter-partition link, C_1 notifies page g after downloading page a (and c), and C_2 transfers the URL of page d to C_1 after downloading page h. The alignment model is shown in figure 2.

3. Crawling Chinese-Myanmar parallel corpus

In this section, we use the hypertext information from the corpus page we obtained in the previous section. A model for crawling CMPC models and corpus cleaning models was proposed. First of all, the traditional method of crawling CMPC is to use automatic language tagging. This method removes all interfaces that are not CM language, resulting in the loss and confusion of a large number of parallel corpora. In order to solve this problem, we added a CM corpus selection model based on threshold crawling, link matching, and page heat ranking. We used many sorting structure directories of web sites to extract the most relevant CM language resources. Secondly, one of us corpus cleaning model is proposed, which relies on the parsing of the underlying HTML of the page to determine a set of standardized token sequences. The first step in this process is to linearize the HTML document. Parallel corpora generates a linear sequence of three tokens through a common set of labels; the second step is to align the linearized sequence using standard dynamic programming techniques, including normalization, tokenization stemming, and stemming and lemmatization techniques.

3.1. Chinese-Myanmar corpus selection model

Due to the lack of small language resources, search engines cannot retrieve every web interface. Therefore, it is very important to develop effective crawling strategies to prioritize CM corpus pages.

First, we designed the CM corpus selection model. Among them, we put forward the page importance metrics, and judge the relevance of the page to the corpus, page priority and corpus standards. At the same time, we use threshold crawling and detection to detect page hits and visits. Page importance metrics:

- $I(p)$- Number of pages.
- $I_S(p, Q)$A measure of text similarity between web pages p and Q.
- Inverse document frequency: Inverse document frequency can be multiplied by the number of times the word appears in the document to calculate the page importance.
- Number of backlinks $I_B(p)$: Number of web URLs that the search engine points to the page.
- G-Threshold condition, any page with $I(p) \geqslant G$ is called a "hot" page.

Since the system sorts the pages after the crawl is complete, we calculate the PageRank as the page rank using a term weighting strategy that calculates the inverse document frequency for all crawl pages.

$$I_R(p) = (1-d) + d\sum_{i=0}^{n} \frac{IR_{(t_i)}}{c_i} \quad (1)$$

where $t_{[0,\dots,n]}$ is the set of pages running to time t, c_i is the number of backlinks of page t_i, and d is the probability of random access to the next page, referred to as the damping factor.

In order to keep the crawler's page vector space near the corpus list, we added an average similarity rating between the page and the list.

$$similarity = \frac{\sum_{p\in C(i)} \frac{\sum_{k\in p\cap v} w_{pk} w_{vk}}{\sqrt{\left(\sum_{k\in p} w_{pk}^2\right)\left(\sum_{k\in v} w_{vk}^2\right)}}}{|C(i)|} \quad (2)$$

where V is the vector representing the list , $w_{pk}$is the TF-IDF weight of term k in p where inverse document frequency is computed from the crawl set $C(i)$ once vital pages have been visited by each crawler.

As each crawler page set grows over time, we can map the reptile's trajectory over time and evaluate the system. In order to make our experiments easy to manage, we consider that the CM corpus page constitutes the entire Web, and only evaluate the performance in this case. Figure 3 shows a simplified control flow chart for the CM corpus selection model, which consists mainly of a parallel page selection engine and a crawler.

Second, we implement and evaluate the corpus selection model framework by combining three types of crawling-first, Backlink and PageRank crawling algorithms. Figure 4 shows our web crawler algorithm. Among them, the crawler manages three main data structures:

- The URL queue contains URLs that have been viewed and need to be accessed. Once the page is accessed, it is stored (including its URL) in the crawled page face.$I_S(p,Q)$ - A measure of text similarity between web pages p and Q.
- Links contain form pairs (u1, u2), where URL u2 is displayed as URL u1 in the visited page. Number of backlinks $I_B(p)$: Number of web URLs that the search engine points to the page.
- Three types of crawling algorithms: breadth-first ,Backlink ,and PageRank

3.2. Chinese-Myanmar corpus cleaning model

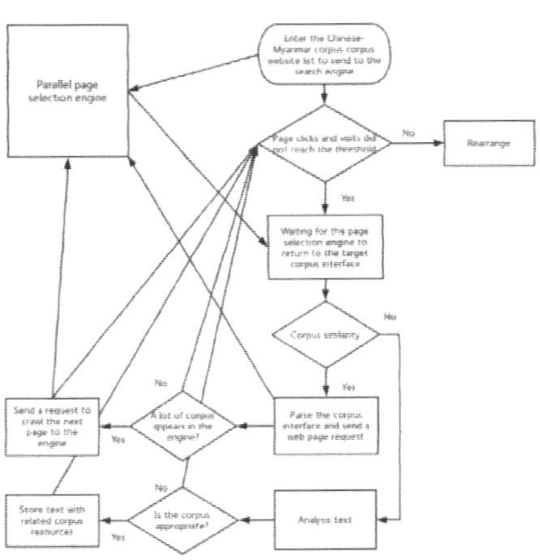

Figure 3. Simplified control flow chart of Chinese-Myanmar corpus selection model

```
Algorithm 1: Crawling algorithm
Input: Vietnamese corpus URLs
Parameter declaration
InitQueue(url_Q);
for i ← 1 to num do
    p ← starting_urls[i];
    EnQueue(url_Q,p);
while (!QueueEmpty(url_Q)) do
    url ← DeQueue(url_Q,p);
    page ← crawl_page(url);
    EnQueue(crawled_pages,(url,page));
    u ← QueueTraverse(crawled_pages);
    EnQueue(links,(url,u));
    if (u ∉ url_Q) ∧ ((u,−) ∉ crawled_pages) then
        EnQueue(url_Q,u);
        breath first:
        BFS(starting_urls[ ],num);
        backlink count:IB'(p);
        u ← QueueTraverse(url_Q);
        backlink count[u] = numberofterms(−,u)inlinks
        sort url_Q by backlink count[u]
        PageRank:IR'(p);
        solvethefollowingsetofequations :
        IR[u] = (1 − 0.9) + 0.9 Σ_i IR[v_i]/c_i
```

Figure 4. Web crawler algorithm

After crawling the CMPC, we obtained the entire HTML source code. This is what the browser needs to render the page. But most of this is useless to us. We need to use regular expressions to extract all the plain text that is visible on the crawled page surface. Simultaneously, "page extraction" and "post-processing, persistence" are the two stages of the crawler. Separating the modules can make the code structure clearer, which can be executed on separate threads and on different machines. Therefore, we define a CM corpus cleaning model responsible for the processing of the extracted results, including calculations, persistence to files, and databases. It relies on parsing the underlying HTML of the page to determine a set of standardized token sequences. In the actual CM parallel website, we found that many authors use tags to format text, which would destabilize the corpus for nonlinear structures. Therefore, we use the fast linear structure alignment algorithm to complete the parsing of HTML. Standard dynamic programming techniques align linearized sequences, including normalization, tokenization, stemming, and lemmatization techniques. The CM corpus cleaning model extracts titles and sentences from the corpus interface and displays them. Then we save them all in the database.

4. Experimental result

4.1. Web list and Targets Dataset

In order to evaluate our candidate website list acquisition algorithm, we started running random breadth-first crawling through sites such as Burma library to collect candidate website lists and target data sets. Since our dataset was downloaded by the crawler in a specific way, it may not be able to correctly represent the actual situation. Table 1 shows a few sample topics.

There are the precision is used to evaluate the web list performance. The formulations are as follows:

$$Precision = \frac{AP}{TP} \tag{3}$$

where AP denotes the number of appropriate pages, TP denotes the total number of pages, $0 \leq Precision \leq 1$

Table1. A few sample web list. The descriptions are truncated for space limitations

weblist	Description	Targets
Daily newspapers	Kyemon (The Mirror)	http://www.kicnews.org/
	newspaper	http://www.kamayutmedia.com/
	Myanma Alin (The Light of	http://burma.irrawaddy.org/
	Myanmar)	http://frontiermyanmar.net/mm
	Myawady Daily	https://www.facebook.com/ElevenMediaGroup/
	New Light of Myanmar	http://burmese.dvb.no/
	The Yadanabon	http://www.bbc.co.uk/burmese/
Digital media list	Duwun	http://www.duwun.com.mm/
	Newsday Myanmar	http://www.newsdaymyanmar.com/
Published overseas	Mizzima News Agency	http://mizzimaburmese.com/
	Mandalay Gazette	http://www.mandalaygazette.com/
	Freedom News Group	http://www.freedomnewsgroup.com/

In the next experiment, we used the crawler program designed in this article to download these lists: we tested the download efficiency of the crawler program by using 20 pages from the 2, 8, 32, and 64 process downloads in figure 5 . After that, we analyze the score of hot pages crawling at a given point ,the P_{ST} , We consider using the threshold condition G for crawling in 3, 10 and 100 in Section 3 According to the popular page definition, H is 806,200 (58%), 291,900 (21%) and 140,390 (10.1%) pages were considered popular among the 1,390,000 pages crawled. In figure 6, the horizontal axis is part of the Burmese webpage that crawls over time. At the right end of the horizontal axis, all 1,390,000 pages are accessed.

The experimental results show that as our definition of popular pages becomes more restrictive (larger G), the crawler can find popular pages faster. Even if G is large, it is always difficult to find the "last set" of popular pages. That is, on the right side of 0.8 on the horizontal axis, the reptile finds the popular interface at roughly the same rate as the random reptile.

Then we looked at all the documents to see if they were suitable for the corpus and completed the evaluation of the list of networks. The result is shown in table 2.

Table 2. Web list precision results

web list	processes	Total		Appropriate			
		Docs	Words	Docs	%	Words	%
Daily newspapers	2	57	2,321,888	47	82.46%	1,189,459	51.23%
	8	64	2,521,212	42	65.63%	1,079,394	42.81%
	32	65	2,853,710	29	44.62%	921,968	32.31%
	64	67	2,887,755	25	37.31%	827,535	28.66%
Digital media list	2	77	2,934,642	67	87.01%	1,570,224	53.51%
	8	82	2,987,674	57	69.51%	1,337,166	44.76%
	32	85	3,022,155	41	48.24%	1,031,087	34.12%
	64	85	3,089,788	37	43.53%	981,462	31.76%
Published overseas	2	28	1,606,312	22	78.57%	791,682	49.29%
	8	31	1,723,220	18	58.06%	672,612	39.03%
	32	33	1,801,123	15	45.45%	589,459	32.73%
	64	35	1,832,001	13	37.14%	523,428	28.57%

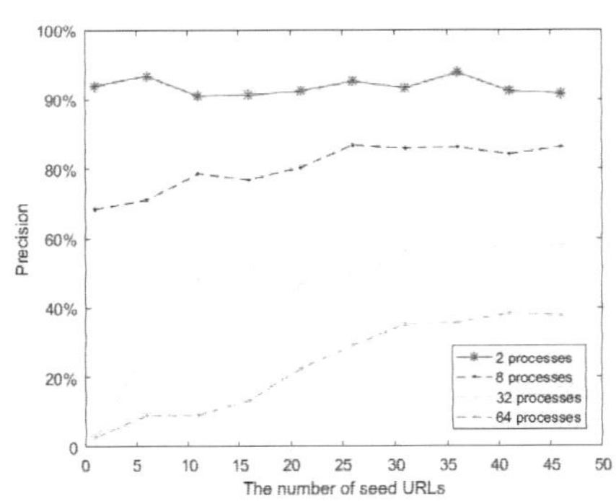

Figure 5.the precision that use web list approach.

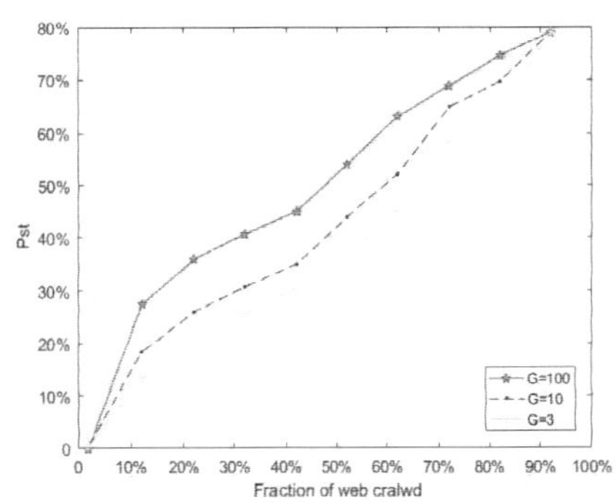

Figure 6. Fraction of web crawled.

4.2. Web crawl experiment

In this section, we evaluate the performance and efficiency of the crawling algorithm described in Section 2. There are two general-purpose experimental evaluation approaches to exhibits the performances of our approaches .The first metric is the simple recall level based on the target pages:

$$recall = \frac{|Ci \cap T|}{|T|} \tag{4}$$

where C(i) is the set of pages crawled by crawler i and T is the target set. This measure allows us to determine how well a crawler can locate a few highly relevant pages. However, there may well be many other relevant pages that are not specified in the target set. To assess a crawler's capability to locate these other relevant pages, we have to estimate the relevance of any crawled page. We do so using the lexical similarity between a page and the list to guide link selection. The second metric we measure the mean similarity between the list and the set of pages crawled. It has been defined in section 3.Both recall and similarity can be plotted against the number of pages crawled to obtain a trajectory over time that displays the behavior of the crawl. We now outline the results of a number of crawls carried out to evaluate and compare performance and efficiency for the crawling algorithms described in Section 3

The performance of the three crawlers – PageRank, Breadth-First, Backlink, is compared in figure 7 and figure 8. We ran three variations of each crawling algorithm for values of our model.

Recall and similarity yield qualitatively consistent results. As expected Breadth-First performs poorly and thus constitutes a baseline.it is no surprise that PageRank performs Optimum performance.

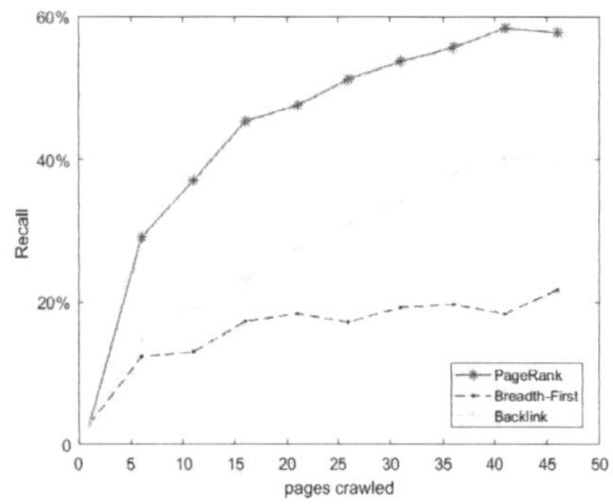

Figure 7. Average target recall by the PageRank, Breadth-First and Backlink crawlers.

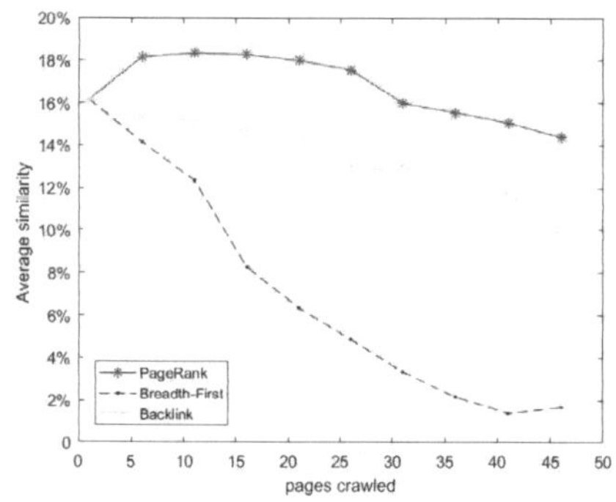

Figure 8. Average similarity by the PageRank, Breadth-First and Backlink crawlers.

5. Contribution

Aiming at the practical problems that CM parallel sentences are difficult to obtain at low cost and high efficiency, we propose a crawler framework that automatically obtains a list of bilingual corpus websites, corpus crawling and corpus cleaning. The framework mainly includes: 1 the engine and website list automatically obtains the CM corpus candidate website, 2 combines the breadth-first, Backlink and PageRank crawler algorithms to implement the CM corpus crawling, 3 parses the HTML to determine a set of standardized token sequences. Compared with other traditional methods, this method has the advantage of automatically obtaining a large number of effective CM websites and broadening the source of corpus. Compared with other traditional methods, this method has the advantage of automatically obtaining a large number of effective CM websites and broadening the source of corpus. The web crawler uses the regular expression processing mechanism to filter irrelevant pages quickly and efficiently capture high-quality CMPC. The conclusion is that the method proposed in this paper is indeed effective in the acquisition of CMPC with scarce information. The method can extract parallel sentence pairs on the CM parallel website with low cost and high efficiency, and provide the basis for Chinese and Myanmar natural language processing and machine translation engine training. In future research, we need to improve the adaptability and crawling efficiency of the framework, optimize the intelligent capture and proofreading capabilities of the method to quickly build a large-scale parallel corpus.

Acknowledgment

This work was supported in part by the research is funded by the National Natural Science Foundation of China 61841113.

References

[1] Brown P F, Cocke J, Pietra S, et al. A statistical approach to machine translation[J]. *Computational Linguistics, 1990,* 16(2):79-85.

[2] Sun Le,Jin Y B, Du Lin S, et al. Automatic Extraction of Bilingual Term Lexicon from Parallel Corpora[J].*Journal Of Chinese Information Processing* , 2000, 14(6): 33-39.

[3] Zhao xin, Deng Ming-xing Web-based Cross Language Information Retrieval Technique [J],*Computer and Modernization* ,2009,(2)： 35-38.

[4] Chen J, Nie J Y. Automatic construction of parallel English-Chinese corpus for cross-language information retrieval[C]// Applied Natural Language Processing Conference. 2000:21-28.

[5] Lu Y, Zhang X, Zheng D.Automatic English-Chinese Parallel Corpus Acquisition and Sentences Extraction[J].*Journal of Computational Information Systems*,2013,9(6):2365-2372.

[6] Nie J Y, Simard M, Isabelle P, et al. Cross-language information retrieval based on parallel texts and automatic mining of parallel texts from the Web[J]. 1999:253-269.

[7] Resnik, P.and N. A. Smith. The Web as a Parallel Corpus[J]. *Computational Linguistics*,2003,29(3):223-248.

[8] Chen J,Chau R, Yeh, C.H.: Discovering parallel text from the world wide web. In: Proceedings of he 2nd Workshop on Australasian Information Security, Data Mining and Web Intelligence, and Software Internationalization, Dunedin, New Zealand, Australian Computer Society, Inc. 2004:157-161

[9] Yi S, Shi S, Wang H, et al. Parallel Approach and Platform for Large-Scale WEB Data Extraction[C]// International Conference on Advanced Cloud & Big Data. 2013.

[10] Furche T , Gottlob G , Grasso G , et al. OXPath: A language for scalable data extraction, automation, and crawling on the deep web[J]. *VLDB Journal*, 2013, 22(1):47-72.

[11] Baroni M, Bernardini S. BootCaT: Bootstrapping Corpora and Terms from the Web[C]//LREC. 2004: 1313-1316.

[12] Bazin A. A depth-first search algorithm for computing pseudo-closed sets[J]. *Discrete Applied Mathematics,* 2018.

[13] Kishimoto A, Fukunaga A, Botea A. Parallel Best-First Search for Optimal Sequential Planning (Research Statement)[J]. 2009.

[14] Feng Z, Lin H, Zhai J, et al. An adaptive breadth-first search algorithm on integrated architectures[J]. *Journal of Supercomputing*, 2018(3–4):1-21.

IOP Conf. Series: Materials Science and Engineering **646** (2019) 012047 doi:10.1088/1757-899X/646/1/012047

Grey Level Co-occurrence Matrix (GLCM) as a Radiomics Feature for Artificial Intelligence (AI) Assisted Positron Emission Tomography (PET) Images Analysis

Mahbubunnabi Tamal

Imam Abdulrahman Bin Faisal University, PO Box 1982, Dammam 31441, Kingdom of Saudi Arabia

Email: mtamal@yahoo.com; mtamal@iau.edu.sa

Abstract. Positron Emission Tomography (PET) allows tumour microenvironment to be studied in vivo with high sensitivity and specificity. Inter- and intra-tumour morphological and phenotypic heterogeneity or pattern provided by PET images are of critical importance. The traditional practice of visual interpretation of these images are not sufficient enough to extract all the information embedded in the images. On the other hand, simultaneous development of automated and reproducible analysis methodologies makes it possible to extract large amount of quantitative features from these images which is termed as radiomics. Analysis of these radiomics feature using artificial intelligence (AI) can significantly improve individualized treatment selection and monitoring. Grey level co-occurrence matrix (GLCM), a member of texture based radiomics feature family is widely used as a biomarker of heterogeneity and can provide information of the tumour microenvironment. The GLCM can subsequently be used for artificial intelligence (AI) assisted tumour diagnosis, monitoring of progression and treatment planning as well as for monitoring response to therapeutic intervention. This aim of the study was to investigate the accuracy and robustness of PET based GLCM in varying image acquisition and analysis conditions using phantom data. It has been observed that GLCM based textural features (e.g., correlation, entropy, homogeneity, energy contrast and dissimilarity) are not only dependent on the volume but also on the quantization level. They are also dependent on signal-to-noise ratio (SNR) and image contrast. The dependencies of these features to the varying imaging conditions are also not linear and cannot always be directly related. To use these GLCM derived textural features as biomarkers for AI assisted analysis, all the information regarding the textural features should always be included along with the changes in volumes and contrast of the PET images in the training dataset.

1. Introduction

PET radiotracer uptake in tumour is often heterogeneous due to different biological characteristics of tumour cells (e.g., cell proliferation, cell death, differential metabolic activity, vascular structure etc.) and large amount of quantitative features from these images can be extracted which is termed as radiomics. Artificial intelligence (AI) assisted accurate quantification of tumour radiomic features [1] has the potential to be used as a tumour staging and prognostic biomarker [2-3]. Among a number of radiomic features describing tumour heterogeneity [4-6], textural features (homogeneity, correlation, energy, contrast, dissimilarity and entropy) – a second order heterogeneity metric extracted from quantifier based grey level co-occurrence matrices (GLCMs) [7] accounting for both spatial and

IOP Conf. Series: Materials Science and Engineering **646** (2019) 012047 doi:10.1088/1757-899X/646/1/012047

intensity information have shown to be capable of staging tumour [8] as well as to predict response [9, 10] for FDG PET images at varying levels.

GLCMs are generated using quantized or resampled intensities within a volume of interests (VOIs) [10] where intensities are resampled in an integer number of bins with the number of bins being power of 2. Textural features extracted from these GLCMs have been reported to be strongly dependent on the metabolically active volume (MATV) using simulated data [11] and confirmed on clinical data [12-17]. Intensity quantization substantially affects the texture indices and thus should be chosen carefully [12, 18]. Reducing quantization always decreases homogeneity [19] and prognostic impact of the textural features is influenced by quantization level [20]. Several groups have suggested using either quantization level 32 [12] or 64 [10,15]. Quantization level 150 or higher also has been proposed in other studies [11,13]. No statistically significant differences have been reported in an another study [10]. Three textural features - homogeneity, dissimilarity and entropy are found to be robust to delineation method and partial volume effects (PVE) [15]. A separate study suggested that smoothing and segmentation have only a small effect compared to quantization [18].

Non uniform selections of parameters and methods across studies make the choice of best textural feature based on MATV, quantization and segmentation challenging and its relationship with the tumour biological characteristics indistinguishable [12,14]. Relationship between volume and quantization has not been explicitly investigated in these studies. No systematic report is available in the literature regarding the effects of image contrast and noise on segmentation and textural features. This aim of the study was to investigate the accuracy and robustness of PET based GLCM in varying image acquisition and analysis conditions using phantom data.

2. Materials and Methods

The torso NEMA phantom containing six spheres (Figure 1) with 10, 13, 17, 22, 28 and 37 mm diameters correspond to 0.52, 1.15, 2.57, 5.58, 11.49 and 26.52 ml volume respectively was filled with ^{18}F solutions. Two different contrasts (2:1 and 4:1) between the spheres and the background were created by reducing the radioactivity in the background.

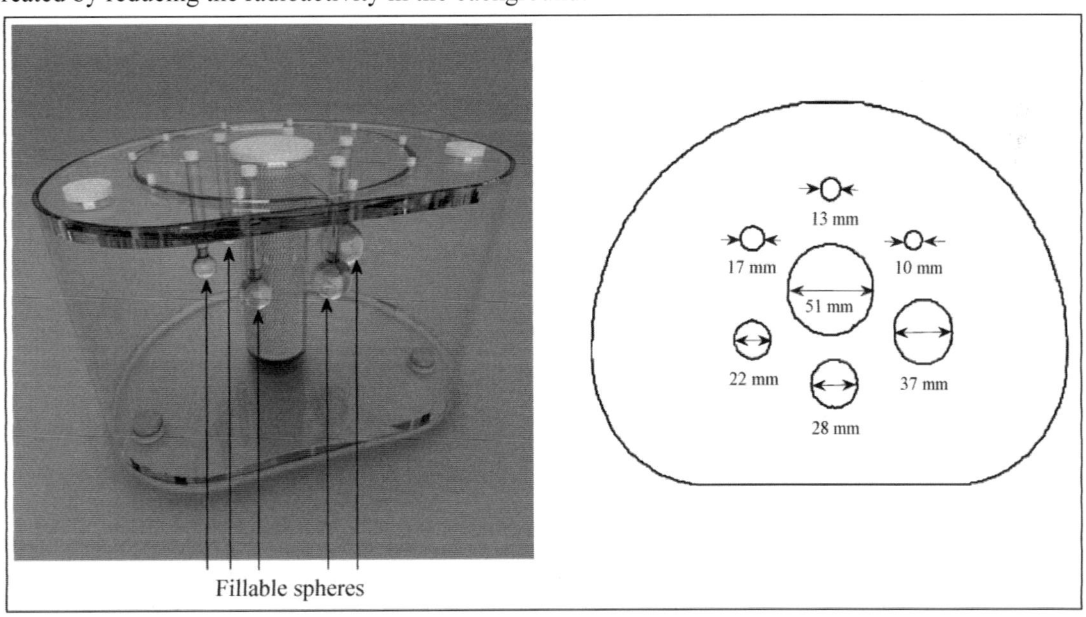

Figure 1: Torso NEMA phantom with a cold insert and six fillable spheres (left) with 10, 13, 17, 22, 28 and 37 mm diameters correspond to 0.52, 1.15, 2.57, 5.58, 11.49 and 26.52 ml volume respectively (right).

The phantom data were acquired in 3D mode on the TrueV PET-CT scanner (Siemens, USA) for 120 minutes which provides 109 image planes or slices covering a 21.6 cm axial FOV (field of view). Images were reconstructed into a 256×256×109 matrix with voxel dimensions of 2.67×2.67×2.00 mm using OSEM reconstruction algorithm with 4 iterations and 21 subsets for five different scan durations (900, 1200, 2000, 4000 and 7200 seconds corresponding to 15, 20, 33.3, 66.6 and 120 minutes respectively) to represent different levels of signal-to-noise ratio (SNR). The starting time of each static frame were shifted to reconstruct five different overlapping realizations for the first four durations. All the reconstructed images were then smoothed with a 4-mm FWHM (full width at half maximum) Gaussian filter after applying decay correction.

All the spheres were delineated using three different segmentation methods. First volume of interest (VOI_{true}) was estimated using the calculated boundaries based on the known diameter and position of each sphere. The second delineation method was a fixed threshold set to 40% (I_{40T}) of the maximum intensity (I_{max}) within the sphere giving a VOI noted as VOI_{40T} [21]. The final volume of interest (VOI_A) was estimated using an adaptive threshold based method as described by (Schaefer et al), where the threshold intensity (I_A) is given by

$$I_A = (\alpha \times I_{70}) + (\beta \times I_{bg}) \quad (1)$$

I_{70} is the mean intensity in a contour containing all voxels with a value greater than 70% of the Imax in the sphere and I_{bg} is the mean background intensity within a sphere of size 26.52 ml located away from all the spheres to avoid partial volume effect (PVE). Both the threshold based methods were applied separately on each roughly delineated VOI containing a sphere to generate the corresponding VOIs. The α and β parameters for the adaptive threshold were calculated using the mean value of optimal cutoff intensities ($I_{optimal}$). $I_{optimal}$ of each hot sphere is calculated using optimal threshold ($T_{optimal}$) and I_{max}. $T_{optimal}$ is estimated as the percentage threshold value of I_{max} which provides the best matched thresholded volume with the VOI_{true} for the uniform sphere phantom.

Quantization of intensities of each VOI was carried out by normalizing the intensities (between 0 and 1) and multiplying the normalized intensities by different quantized values, {Q= 8, 16, 32, 64, 128 and 256}. Grey level co-occurrence matrix (GLCM) was derived for each normalized and quantized VOI data. Several textural features (homogeneity, correlation, energy, contrast, dissimilarity and entropy), a second order heterogeneity measures, were then estimated from these GLCM data.

3. Results

It has been observed that all the textural features are dependent on the quantization value at varying degree. Figure 2 shows the relationships between the mean textural features of five realizations and quantization values for VOI_{true}. Homogeneity exponentially decreases with the increase of quantization levels. Separations among homogeneity for different spheres remain unchanged for different quantization levels. Correlation remains constant with quantization for all spheres from quantization level 32 onwards. However, there are clear separations among correlations for different spheres. Contrast and dissimilarity increase approximately linearly with the increase of quantization levels. Separation among the spheres increases with the increase of quantization levels. Energy decreases and entropy increases with quantization levels. For volumes less than 5.58 cm^3 both entropy and energy remain unchanged after quantization 32. However, it keeps on changing with quantization for bigger spheres and requires higher quantization level to remain unaffected. Higher quantization level would make volumes appear as heterogeneous. On the contrary, low quantization level would make them appear as homogeneous. A compromise is required while choosing appropriate quantization level. Considering all six textural features, quantization level 64 or 32 appears to provide the best compromise. Quantization level 64 has been chosen in this study to generate all the textural features unless mentioned otherwise.

IOP Conf. Series: Materials Science and Engineering **646** (2019) 012047 doi:10.1088/1757-899X/646/1/012047

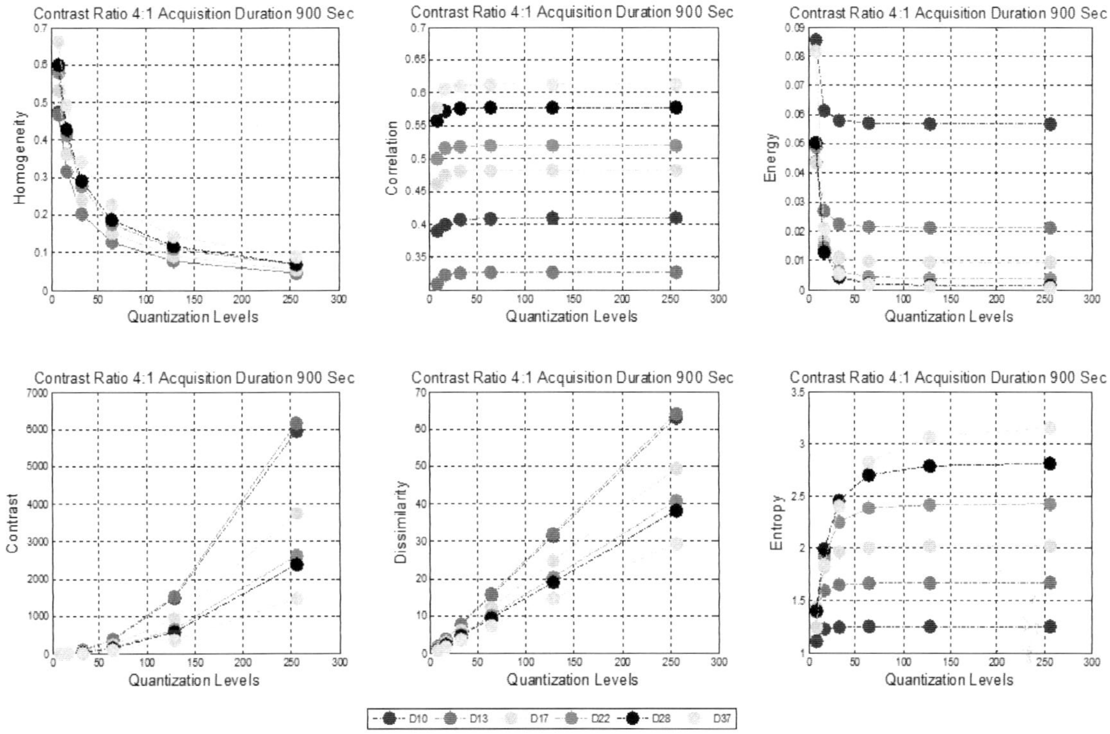

Figure 2: Mean textural features of five realizations against quantization for VOI_{true} for all six spheres. The features shown here are for 4:1 contrast and 900 seconds acquisition duration.

Dependency of textural feature on sphere volume for contrast 4:1 is shown in Figure 3. Features for the spheres located at the background also show dependency on the volumes. Homogeneity and entropy increase with volumes, whereas contrast, dissimilarity and energy decrease. There are subtle differences between the spheres and backgrounds for homogeneity, contrast and dissimilarity showing their dependency on the volume edge. Entropy and energy are robust to edge as shown by very good agreement between the sphere and background. The separation between sphere and background for correlation indicates that it is more dependent on the intensity variations. All the features reaches plateau with the increase of volume at varying rates.

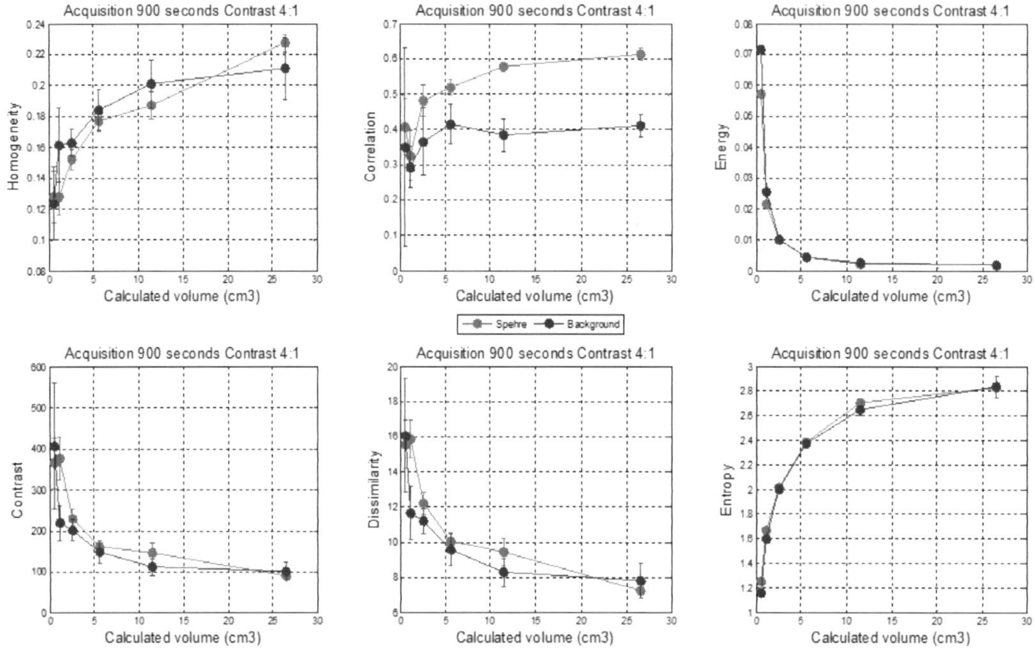

Figure 3: Textural features against volumes for both sphere and background. To analyze the texture in the background same volume as the sphere were used.

Figure 4 compares the relationships between textural features and acquisition durations for VOI_{true} and VOI_{40T} for contrast 2:1. Textural features derived using VOI_{40T} are significantly different than those of VOI_{40T} for smaller volumes. As the volume increases the differences between them reduces. The textural features also vary with the noise as the VOI_{40T} vary with the noise. With an adaptive segmentation method, all textural features become independent of noise for volume greater than 2.57 cm^3 and match closely with the features generated using VOI_{true} (Figure 5).

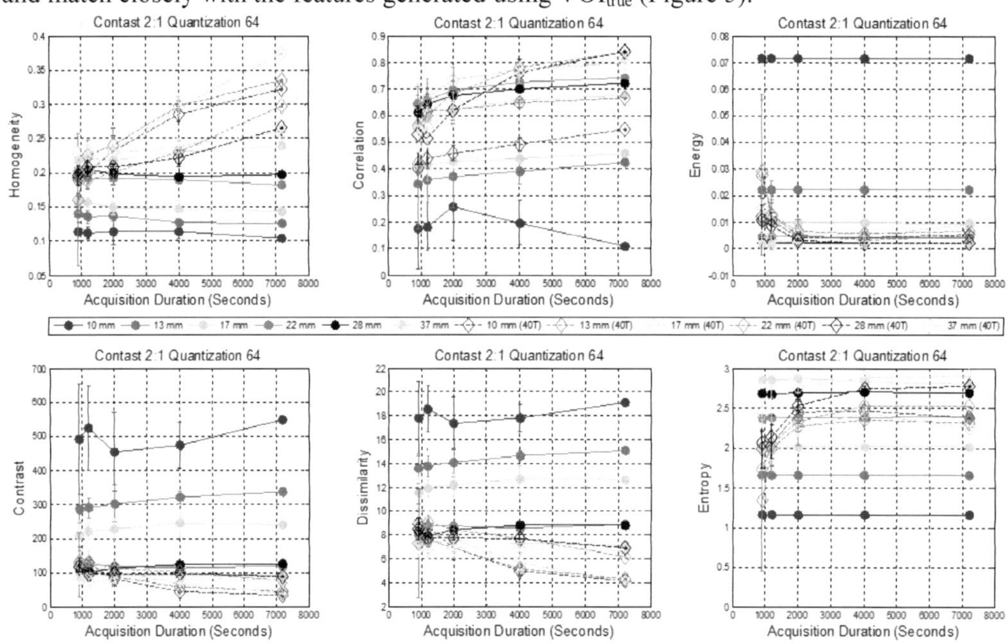

Figure 4: Comparison of textural features extracted using VOI_{40T} and VOI_{true} for different acquisition durations with contrast 2:1 and quantization level 64.

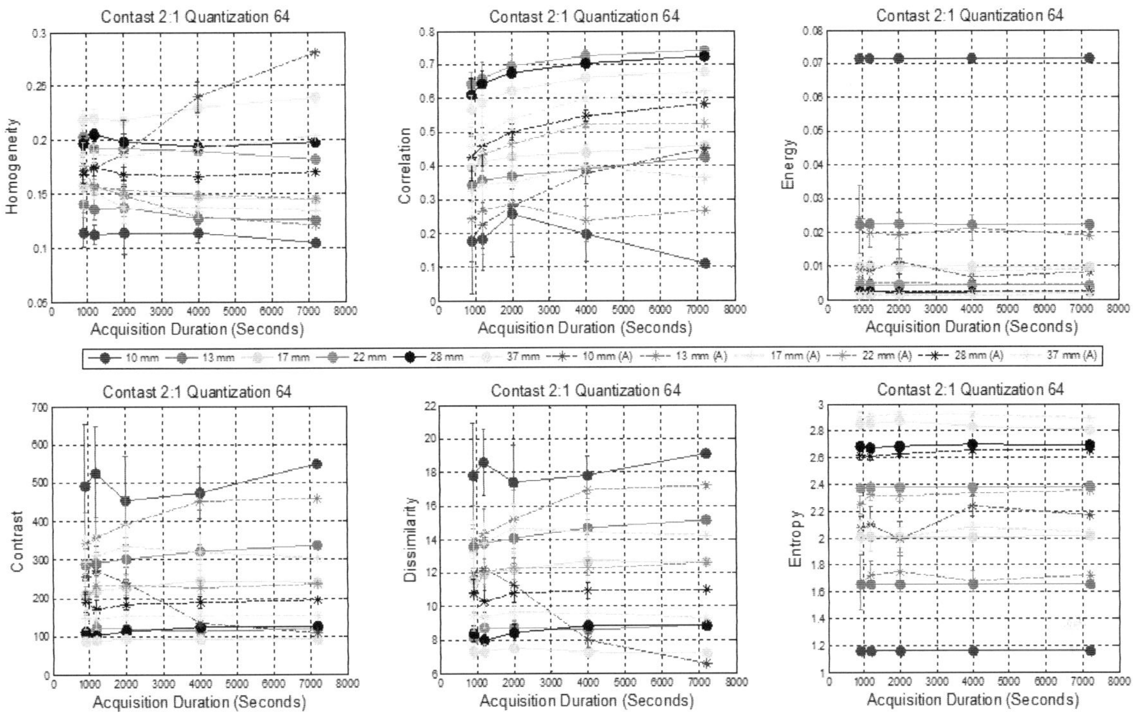

Figure 5: Comparison of textural features extracted using VOI_A and VOI_{true} for different acquisition durations with contrast 2:1 and quantization level 64.

4. Discussion

To use textural features as a tumour staging and prognostic biomarker using AI, better understanding of relationships of textural features with MATV, quantization and segmentation are very important. Investigation of spheres filled with same homogeneous activity reveals that bigger the volume wider the range of intensities making quantization sensitive to the volume of lesions. In such cases, higher quantization makes bigger homogeneous spheres appear as heterogeneous compared to the smaller ones. Lower quantization level removes the dependency on volume by forcing the intensities to be homogeneous and eliminating the heterogeneity information. Considering the characteristics of all the textural features for the homogeneous spheres over a range of volumes, it appears that quantization level 32 or 64 should be preferred and the findings are similar to the finding of previous studies [10,12].

All six textural features are dependent on volumes at varying degrees with entropy and energy being the most sensitive ones. Spheres of similar volumes placed in the background reveals that PVE effect on textural features is far smaller than the effect of volume. Dependency of entropy on MATV significantly reduces for volumes greater than 45 cm^3 for quantization 256 [13]. However, a different study suggested to use volume greater than 10 cm^3 [17] for quantization 64. Dependency of quantization on volume investigated in this study explains the reason for finding two different cut-off volumes. Investigation on heterogeneous spheres suggested that if response occurs as a result of combined changes in volume and heterogeneity, entropy and energy are only able to display changes in volumes (not heterogeneity), making them unsuitable for prognostic biomarkers of heterogeneity. High sensitivity of correlation to intensity also makes it less suitable to report changes in heterogeneity.

Two threshold based delineation methods (40% fixed and adaptive) were employed to investigate the effects of segmentation on textural features. The volumes generated using these two methods are substantially different. Since VOIs delineated using 40% threshold are different from each other, textural features generated using these VOIs are also different with the actual lesion volumes being the same. However, since VOIs generated using adaptive threshold matches with the VOI_{true}, textural features are closer to the true textural features compare to VOI_{40T}. These results suggested that texture indices are highly sensitive to the segmentation method. The results are consistent with previously published ones [12,15,22].

Volume delineated by a robust segmentation method is capable of generating textural features such as homogeneity, contrast and dissimilarity that are capable of capturing tracer uptake heterogeneity if the volume changes between scans are minimal. Since homogeneity directly related to volume, it can only be used as a feature of image heterogeneity if the changes of volume and homogeneity are in opposite directions, i.e., if the combined multiplicative changes of volumes and homogeneity are either zero or negative. On the other hand, as contrast and dissimilarity are inversely related to volume they can be used as an image heterogeneity feature if the combined multiplicative changes of volumes and homogeneity are either zero or positive. Since contrast is approximately two times more sensitive to volumes compared to dissimilarity, homogeneity and dissimilarity are the two textural features that should be used to measure heterogeneity. These two features also provide complementary heterogeneity information which can be used for cross validation.

5. Conclusion

Homogeneous regions appear heterogeneous on PET images as quantified by textural features. Textural features generated using GLCM depends on quantization and volume. Since these features differentially vary with volume, regions should be segmented using methods are that are robust to variations in contrast and noise using quantization level 64. Small scale heterogeneity phantom studies suggest that homogeneity and dissimilarity are the most suitable textural features to be used as heterogeneity measures where there are combined changes in both heterogeneity and volume due to treatment. Further investigations are required with different heterogeneous phantoms to fully understand the volume effects on these textural indices. Nonetheless, to use these textural features as prognostic biomarkers for an AI assisted system, changes in textural features between baseline and treatment scans should be utilized along with the changes in volumes to train the system.

References

[1] Choi JY. Radiomics and Deep Learning in Clinical Imaging: What Should We Do? Nuclear Medicine and Molecular Imaging. Mar 2018;52:89–90.

[2] Eary JF, O'Sullivan F, O'Sullivan J, Conrad EU. Spatial heterogeneity in sarcoma 18F-FDG uptake as a predictor of patient outcome. Journal of nuclear medicine : official publication, Society of Nuclear Medicine. Dec 2008;49(12):1973-1979.

[3] Asselin MC, O'Connor JP, Boellaard R, Thacker NA, Jackson A. Quantifying heterogeneity in human tumours using MRI and PET. European journal of cancer. Mar 2012;48(4):447-455.

[4] El Naqa I, Grigsby P, Apte A, et al. Exploring feature-based approaches in PET images for predicting cancer treatment outcomes. Pattern recognition. Jun 1 2009;42(6):1162-1171.

[5] Miwa K, Inubushi M, Wagatsuma K, et al. FDG uptake heterogeneity evaluated by fractal analysis improves the differential diagnosis of pulmonary nodules. European journal of radiology. Apr 2014;83(4):715-719.

[6] van Velden FH, Cheebsumon P, Yaqub M, et al. Evaluation of a cumulative SUV-volume histogram method for parameterizing heterogeneous intratumoural FDG uptake in non-small cell lung cancer PET studies. European journal of nuclear medicine and molecular imaging. Sep 2011;38(9):1636-1647.

[7] Haralick RM, Shanmugam K, Dinstein IH. Textural Features for Image Classification. Systems, Man and Cybernetics, IEEE Transactions on. 1973;SMC-3(6):610-621.

[8] Mu W, Chen Z, Liang Y, et al. Staging of cervical cancer based on tumor heterogeneity characterized by texture features on (18)F-FDG PET images. Physics in medicine and biology. Jul 7 2015;60(13):5123-5139.

[9] Cook GJ, Yip C, Siddique M, et al. Are pretreatment 18F-FDG PET tumor textural features in non-small cell lung cancer associated with response and survival after chemoradiotherapy? Journal of nuclear medicine : official publication, Society of Nuclear Medicine. Jan 2013;54(1):19-26.

[10] Tixier F, Le Rest CC, Hatt M, et al. Intratumor heterogeneity characterized by textural features on baseline 18F-FDG PET images predicts response to concomitant radiochemotherapy in esophageal cancer. Journal of nuclear medicine : official publication, Society of Nuclear Medicine. Mar 2011;52(3):369-378.

[11] Brooks FJ, Grigsby PW. Low-order non-spatial effects dominate second-order spatial effects in the texture quantifier analysis of 18F-FDG-PET images. PloS one. 2015;10(2):e0116574.

[12] Orlhac F, Soussan M, Maisonobe JA, Garcia CA, Vanderlinden B, Buvat I. Tumor texture analysis in 18F-FDG PET: relationships between texture parameters, histogram indices, standardized uptake values, metabolic volumes, and total lesion glycolysis. Journal of nuclear medicine : official publication, Society of Nuclear Medicine. Mar 2014;55(3):414-422.

[13] Brooks FJ, Grigsby PW. The effect of small tumor volumes on studies of intratumoral heterogeneity of tracer uptake. Journal of nuclear medicine : official publication, Society of Nuclear Medicine. Jan 2014;55(1):37-42.

[14] Brooks FJ. On some misconceptions about tumor heterogeneity quantification. European journal of nuclear medicine and molecular imaging. Aug 2013;40(8):1292-1294.

[15] Hatt M, Tixier F, Cheze Le Rest C, Pradier O, Visvikis D. Robustness of intratumour (1)(8)F-FDG PET uptake heterogeneity quantification for therapy response prediction in oesophageal carcinoma. European journal of nuclear medicine and molecular imaging. Oct 2013;40(11):1662-1671.

[16] Willaime JM, Turkheimer FE, Kenny LM, Aboagye EO. Quantification of intra-tumour cell proliferation heterogeneity using imaging descriptors of 18F fluorothymidine-positron emission tomography. Physics in medicine and biology. Jan 21 2013;58(2):187-203.

[17] Hatt M, Majdoub M, Vallieres M, et al. 18F-FDG PET uptake characterization through texture analysis: investigating the complementary nature of heterogeneity and functional tumor volume in a multi-cancer site patient cohort. Journal of nuclear medicine : official publication, Society of Nuclear Medicine. Jan 2015;56(1):38-44.

[18] Doumou G, Siddique M, Tsoumpas C, Goh V, Cook GJ. The precision of textural analysis in F-FDG-PET scans of oesophageal cancer. European radiology. May 21 2015.

[19] Brooks FJ, Grigsby PW. FDG uptake heterogeneity in FIGO IIb cervical carcinoma does not predict pelvic lymph node involvement. Radiation oncology. 2013;8:294.

[20] Cheng NM, Fang YH, Chang JT, et al. Textural features of pretreatment 18F-FDG PET/CT images: prognostic significance in patients with advanced T-stage oropharyngeal squamous cell carcinoma. Journal of nuclear medicine : official publication, Society of Nuclear Medicine. Oct 2013;54(10):1703-1709.

[21] Erdi YE, Mawlawi O, Larson SM, et al. Segmentation of lung lesion volume by adaptive positron emission tomography image thresholding. Cancer. Dec 15 1997;80(12 Suppl):2505-2509.

[22] Tixier F, Hatt M, Le Rest CC, Le Pogam A, Corcos L, Visvikis D. Reproducibility of tumor uptake heterogeneity characterization through textural feature analysis in 18F-FDG PET. Journal of nuclear medicine : official publication, Society of Nuclear Medicine. May 2012;53(5):693-700.

IOP Conf. Series: Materials Science and Engineering **646** (2019) 012048 doi:10.1088/1757-899X/646/1/012048

Automated Blood Cell Detection and Counting via Deep Learning for Microfluidic Point-of-Care Medical Devices

Tiancheng Xia[1], Richard Jiang[1], YongQing Fu[2] and Nanlin Jin[3]

[1]School of Computing & Communication, Lancaster University, Lancaster, UK, LA1 4WY
[2]Faculty of Engineering & Enviornment, Northumbria University, Newcastle, UK, NE1 8ST
[3]Computer and Information Sciences, Northumbria University, Newcastle, UK, NE1 8ST

Correspondent e-mail: r.jiang2@lancaster.ac.uk

Abstract. Automated *in-vitro* cell detection and counting have been a key theme for artificial and intelligent biological analysis such as biopsy, drug analysis and decease diagnosis. Along with the rapid development of microfluidics and lab-on-chip technologies, *in-vitro* live cell analysis has been one of the critical tasks for both research and industry communities. However, it is a great challenge to obtain and then predict the precise information of live cells from numerous microscopic videos and images. In this paper, we investigated *in-vitro* detection of white blood cells using deep neural networks, and discussed how state-of-the-art machine learning techniques could fulfil the needs of medical diagnosis. The approach we used in this study was based on Faster Region-based Convolutional Neural Networks (Faster RCNNs), and a transfer learning process was applied to apply this technique to the microscopic detection of blood cells. Our experimental results demonstrated that fast and efficient analysis of blood cells via automated microscopic imaging can achieve much better accuracy and faster speed than the conventionally applied methods, implying a promising future of this technology to be applied to the microfluidic point-of-care medical devices.

1. Introduction

Microfluidic technologies [1, 2] have recently found wide-range applications in biological and medical applications, such as lab-on-chip and point-of-care (POC) diagnostic devices, which revolutionised the personalized medicine and rapid disease diagnosis. Point-of-care testing (or bedside testing) is generally defined as medical diagnostic testing at or near the point of caor in other words, at the time and place of patient care. This contrasts with the conventional treatment, in which testing was wholly or mostly confined to the medical laboratory. In this case, the specimens are often taken away from the point of care and then hours even days will be waited for the results, during which the point of care is asked to wait before the critical information is obtained.

Such the POC diagnosis has facilitated a paradigm shift from therapeutic treatments to predictive, personalized and preventive ones [1, 2], and from the conventional diagnostic tests performed inside the clinical labs settings to near-patient ones. This will enable doctors to have timely diagnostic information to make quick decisions regarding to further diagnosis or immediate treatments. At the same time, patients are hugely benefited from the ease of usages of the POC devices, which allow them to personally monitor their own health in reliable and quantified ways, simply being at home. There is no requirement for the tested samples to be delivered to a lab and no need for the results to be transmitted physically manually or over long distance and time. Doctors, nurses, or even patients could perform the tests and immediately receives the results on the spot, thereby, saving huge amount of time with the rapid diagnosis.

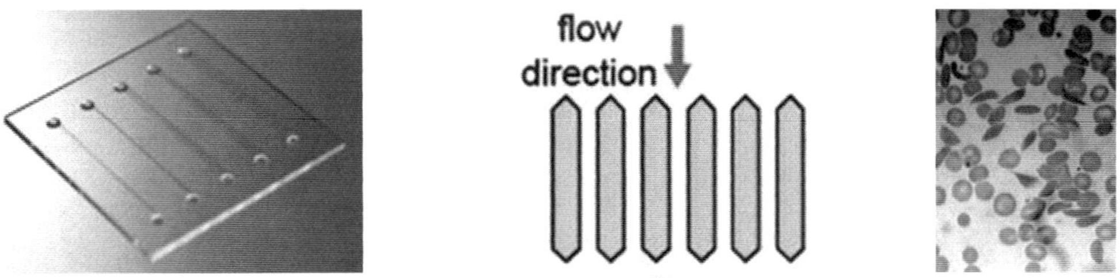

a) b) c)

Figure 1. Microfluidic point-of-care diagnostics of blood cells. a) Microfluidic device; b) Blood flow illustration; c) Microscopic inspection of blood cells.

Among various lab-on-chip based diagnosis tasks [1, 2], microfluidics based blood cell diagnostics as shown in Fig. 1 is one of critical ones. White blood cell (WBC) counting, a routine test to measure the number of white blood cells for a patient, is an important means for medical diagnosis. A low WBC count may be linked to a toxic reaction, a viral infection, or side-effect from chemotherapy, or a disease in the bone marrow which limits the body's ability to produce the normal WBCs, whereas a high WBC count might be linked to an imperative sign of an infection or leukaemia.

In hospitals, the conventional diagnostic procedures for blood cell analysis involve microscopic inspection of peripheral blood samples. For decades, this operation is performed by experienced operators with two main analysis procedures: classification and counting of the cells. However, lacking of automation and intelligent procedures has become a critical barrier for the integration of microscopic image analysis into microfluidic POC diagnostic system [3, 4]. So far, blood cell counting in the commonly used POC devices is mostly done using flow cytometry, which is based on monitoring changes of the fluorescence signals as a quantitative tool [4], however, this is far inferior and less accurate compared to cell-by-cell counting based on the microscopic images. As it is well-known, microscopic cell counting is non-invasive and free of fluorescence dyes, and it is based on the lab-on-chip devices and hence could drastically reduce the cost of the POC diagnostics.

In this work, aiming at the integration of automated microscopic image analysis into microfluidic POC device for blood cell counting, we investigated the methods of live cell detection techniques based on a recently developed artificial intelligence (AI) approach. Live-cell imaging experiments [3-6] provide the possibilities of identification, tracking, and analysis of cells from the microscopic images using a computer vision-based AI method. These experiments have been applied for wide-range applications such as biomedicine, material science and organic chemistry. Automated tracking of cell populations *in-vitro* using time-lapse microscopy images helps high-throughput spatiotemporal measurements of a wide variety of cell behaviours, including migration (translocation), mitosis (division), quiescence (inactivity) and apoptosis (death) of cells in addition to the reconstruction of cell lineages (mother–daughter relations) [7]. These capabilities are of enormous values in many areas of biomedical engineering, such as stem cell research, oncological studies, tissue engineering, drug discovery, genomics, and proteomics [7, 8].

Early live cell detection and tracking methods conventionally apply classic computer vision techniques such as segmentation algorithms, motion detection, level-set methods, image descriptors (such as SIFT/LBP/HOG). However, fully automatic cell tracking faces many challenges [6-10] including poor contrasts with high noise levels, irregular cell contours, difficulty for entry and exit of the cells, all of which make it difficult to deploy to commercial applications such as microfluidic POC devices. On the other hand, classic machine learning approaches such as subspace learning [11-12], ensemble approaches [13-14] and random forests [15] may work well on pattern analysis tasks. However, in medical diagnosis, patients and users may demand an extremely high accuracy to avoid any medical misdiagnosis that could cause the fatal results. Hence, it is extremely challenging when AI is applied for medical care, which is a bit tricky in the sense that any misclassification will take legal or ethical consequences, and often result in a huge loss to compensate the victim of the misdiagnosis.

Recently, deep neural networks have been witnessed with a huge success in many challenging pattern analysis tasks [16-18]. Particularly for image analysis, convolutional neural networks (CNN) [19-20] has been used for the detection of in-vitro stem cells and achieved a considerably high accuracy (around 90%). In this paper, targeting at exploiting this technique for microfluidic blood cell analysis, we will examine the potentials of the recently-proposed object detection methods and explore its potential for fully-automated microfluidic POC technology. In this work, we select a method called Faster Region-based CNN (Faster-RCNN) for white blood cell examination, and investigate its robustness for potential commercial usages.

2. Faster RCNNs based Object Detection

In the past five years, deep learning based object detection has been evolving into a series of different methods, including different versions called R-CNN [21], Fast R-CNN [22] and Faster R-CNN [23], in an order with the evolution of objection detection. Region proposal networks (RPNs) are critical in all these types of approaches. R-CNN and Fast R-CNN firstly estimate the possible region proposals using a method called selective search (SS) [20]. A CNN-based network is then employed to classify the object and also detect its bounding box. The main dissimilarity between these two is that R-CNNs feed the region proposals at a pixel level into CNN for detection while the Fast R-CNN based networks take the region proposals at a feature map level. In both R-CNN and Fast R-CNN, the region proposal unit (i.e. SS) and the detection network (classifier) are decoupled. Obviously, such decoupling is not a best approach. For instance, the SS may have a false negative value, and this error will then pass over to the detection network successively. Apparently, it could be better to couple both stages with each other and associate them together. In the Faster R-CNN, RPNs using SS [20] is replaced by RPN using CNN, and this CNN is shared with detection network, and hence SS is removed from its cascaded process.

2.1. Region Proposal network

In object detection using Faster RCNN, RPN is the critical component and has been proven very efficient. Its purpose is to generate multiple potential region candidates identified within a given image. RPN produces the proposals for the objects, and owns a specialized architecture itself. Here, we can further breakdown the RPN architecture into different components.

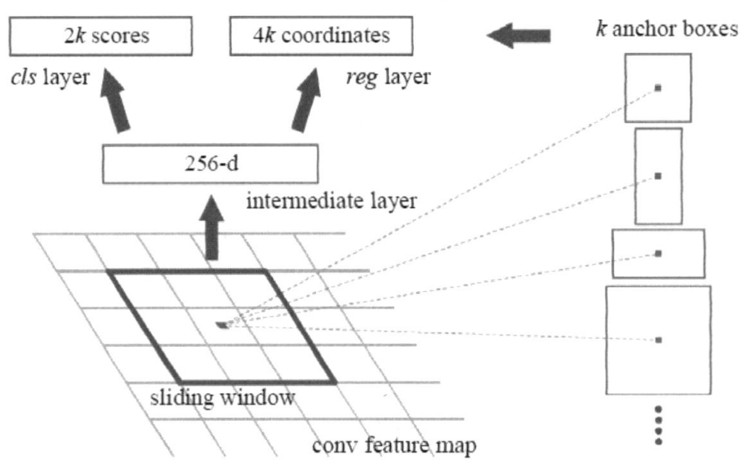

Figure 2. The typical RPN architecture [23].

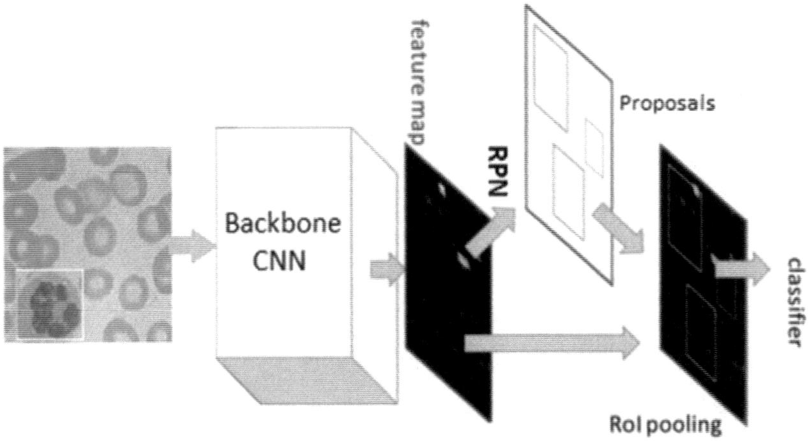

Figure 3. Overall architecture of Faster R-CNN detection network.

As shown in Fig. 2, RPN has a classifier as well as a regressor. Here, we bring in the concept of anchors, the central point of the sliding window. In the AlexNet, the dimension of the sliding window is 256-dim; in the VGG-16, it is 512-dim. The classifier calculates the probability of a proposed region having the expected object. The regressor regresses the coordinates of the region to make them fit better with the objects.

For any image, there are two important parameters: aspect-ratio and scale. Here, aspect ratio is the width of image divided by the height of image, and scale is the size of the image. If we choose 3 scales and 3 aspect-ratios, we will then have total of 9 proposals possible for each pixel. This is the way to estimate the value of k, and for this case, k=9. For the whole image, number of anchors is then counted as $W*H*k$.

This region proposal algorithm is robust to translations, with a merit of the translational invariance. It has multi-scale anchors. Instead of *"Pyramid of Filters"*, this algorithm has *"Pyramid of Anchors"*. As a result, this is more cost effective than previously proposed algorithms such as Fast RCNN. These anchors are assigned labels based on two factors: 1) The anchors with highest Intersection-over-Union (IoU) overlap with a ground truth box; 2) the anchors with IoU overlap higher than 0.7.

As it is expected, the RPN is an algorithm that needs to be trained, with regards to its Loss Function:

$$L(\{p_i\},\{t_i\}) = \frac{1}{N_{cls}}\sum_i L_{cls}(p_i,p_i^*) + \lambda\frac{1}{N_{reg}}\sum_i p_i^* L_{reg}(t_i,t_i^*) \quad (1)$$

The first term is the classification loss over two classes (object or non-object). The second term denotes how much the regression loss of bounding boxes is when there is an object (i.e. $p_i* = 1$) in that region. Therefore, RPN networks will check over the locations that contain objects. The corresponding locations and bounding boxes will then be handed over to the detection network to identify which class the object is and what bounding box that object has. While the regions could be overlapped with each other, a strategy called non-maximum suppression (NMS) is then used to reduce the number of proposals, typically from more than 6000 to N (N=300).

2.2. Detection Network

Next to the RPN, the following part in Faster R-CNN is illustrated in Fig. 3. First of all, ROI pooling is carried out. The pooled area is then passed through the CNN layers and two fully-connected branches as the class softmax and the bounding box regressor, respectively. Since the CNN backbone layers are shared in extracting the feature maps with different outputs at the end, training procedure is very different from the ones for RCNN and Fast RCNN. The training steps of Faster RCNN include,

1) Re-train (fine-tune) RPN with the pre-trained model (such as AlexNet);

Table 1. The test results on the WBC dataset

No of Test Images	False Positive	False Negative	Overall Miss Rate	Overall Accuracy
314	1	4	1.3%	98.4%

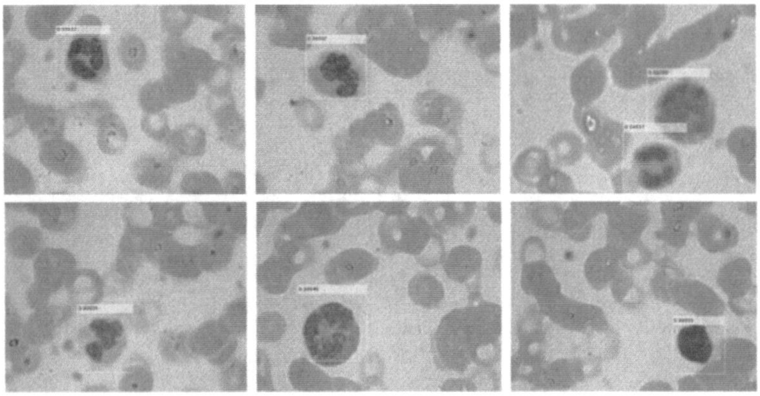

Figure 4. In-vitro white blood cell detection.

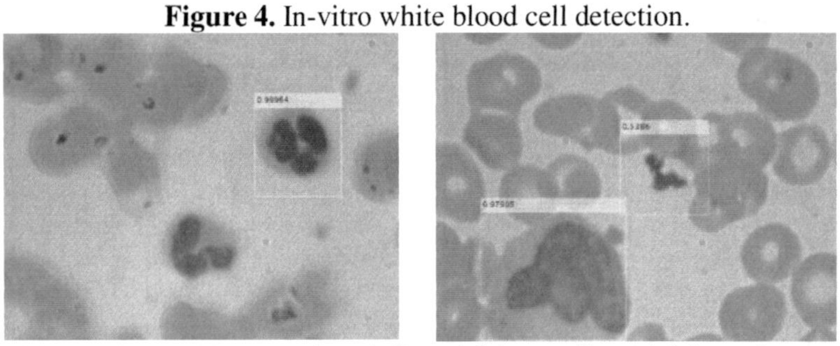

Figure 5. The false negative (left) and false posiive (right) test images

2) Re-train (fine-tune) a separate detection network with imagenet pre-trained model;
3) Use the detector network to initialize RPN training, while fixing the shared CNN layers and only tuning unique layers of RPN;
4) Keep the CNN layers fixed, fine-tune the unique layers of the detector network.

After the above steps, the object detector is ready to use on the trained classes of objects.

3. Using Faster RCNN for Blood Cell Detector

3.1. Experimental Setup

While mostly Faster RCNN detector has been trained with large datasets such as imagenet, it can be re-trained for specific tasks via a mechanism called transfer learning, as detailed in Section 2. In this work, we aim to examine the performance of Faster RCNN detector for our microfluidic blood cell analysis. Obviously, blood cells are not a class in the original imagenet classes, and thus we will need to re-train the detector so that it could be used to detect blood cells. To enable this, we used a small dataset of white blood cells for the transfer learning, freezed the backbone layers and re-trained the parameters of the FC layers for two classes, namely white blood cell and other objects. As shown in Fig. 3, an RPN is trained to quickly identify all WBC-like regions, and then a cascaded classifier is trained to determine if the detected region contains a WBC.

In our experiment, we analysed a small dataset consisting of 364 images [24], which were split into two subsets: 50 images for training and 314 images for testing. In the training setup, the learning rate was set as 1×10^{-4} for the RPN and 1×10^{-5} for the CNN. The maximum number of training epochs was set to be 15 for all the stages. The input layer of neurons was set to be 32×32×3.

3.2. Experimental Results

After our WBC detector was trained on the small train dataset, we run the test on the rest of images, e.g., 314 test images. Fig. 4 shows the detection results from our re-trained detector on the tested images, while the detected WBCs are shown in the yellow rectangle with their probability specified in the yellow labels. Table 1 lists the statistic results of the experiment using the test dataset. Among all the 314 test images, we have only four WBCs missed in the detection (namely False Negative, FN), and one detection is wrongly selected as positive (False Positive, FP). In overall, the miss rate is only 1.3% and the detection accuracy is around 98.4%. Fig. 5 shows the FN and FP detection cases, where in the left image a white blood cell was missed in the detection and in the right image a wrong region was misclassified as a white blood cell.

4. Further Discussion

Such a high detection accuracy achieved by our CNN detector implies the aoolicability to use this technique for the real-life medical purpose. However, test results on the well-selected samples may still yet to be limited. More vigorous tests are needed to establish sufficient confidence for investors to support any real business-targeted development.

Mostly, to guarantee a reliable or acceptable accuracy for medical use, the accuracy under any conditions needs to be higher than 90% at a relatively fast speed. It may still have a long way to go before the real-life applications. Just like face recognition, test accuracy on datasets can easily be boosted to 99% but the real life test in the wild can have a low accuracy less than 20%, as reported recently by London police. However, our research may throw lights on its further exploitation for microfluidic POC device design.

5. Conclusion

In this work, we investigated the fully automatic WBC counting method based on microscopy image analysis using deep learning and validated the accuracy of the CNN based method in experiments. While it is a new technique to count blood cells in microfluidic devices for the POC service, our work shows a new pathway from image analysis that can be potentially applied for POC devices, though it may still be a long way to go to enable the pragmatic use in clinic practice.

References

[1] Chin, C. D., *et al*, 2012, Commercialization of microfluidic point-of-care diagnostic devices, *Lab Chip*, 2(12), p.2118-2134.

[2] Quesada-Gonzále D and Merkoçi A, 2018, Nanomaterial-based devices for point-of-care diagnostic applications. *Chemical Society Reviews*. 47 (13), p.4697–4709.

[3] Labati R. D., *et al*, 2011, ALL-IDB: the acute lymphoblastic leukemia image database for image processing, *Proc. of the 2011 IEEE Int. Conf. Image Processing*, Brussels, Belgium, p.2045-2048.

[4] Basiji D. A., *et al*, 2007, Cellular image analysis and imaging by flow cytometry, *Clin. Lab. Med.*, 27 (3), p.653–670.

[5] Mualla F., *et al*, 2013, Automatic cell detection in bright-field microscope images using SIFT, random forests, and hierarchical clustering, *IEEE Trans. Medical Imaging*, 32 (12). p.2274-2286.

[6] Jiang R., *et al*, 2010, Live-cell tracking using SIFT features in DIC microscopic videos, *IEEE Tran. Biomed Eng.*, 57, p.2219–2228.

[7] Harder, N., et al. 'Large-scale tracking and classification for automatic analysis of cell migration and proliferation, and experimental optimization of high-throughput screens of neuroblastoma cells', *Cytometry Part A*, (2015): 524-540.

[8] Armitage, J. P., 'Regulation of the expression and positioning of chemotaxis and motor proteins in Rhodobacter sphaeroides', *Diss. University of Oxford*, 2010.

[9] Tuysuzoglu, A., 'Robust inversion and detection techniques for improved imaging performance',

Diss. Boston University, 2014.

[10] Fernyhough, E. N., 'Automated Segmentation of Structures Essential to Cell Movement', *Diss. University of Leeds*, 2016.

[11] Jiang R, et al, 2010, Multimodal Biometric Human Recognition for Perceptual Human–Computer Interaction, *IEEE Trans. Systems, Man, & Cybernetics Part C*, 40(5), p.676.

[12] Jiang R, et al, 2010, Face recognition in global harmonic subspace, *IEEE Trans. Information Forensics and Security*, 5(3), p.416-424.

[13] Jiang R, et al, 2017, Emotion recognition from scrambled facial images via many graph embedding, *Pattern Recognition*, 67, p.245-251.

[14] Jiang R, et al, 2016, Face recognition in the scrambled domain via salience-aware ensembles of many kernels, *IEEE Trans. Information Forensics and Security*, 11(8), p.1807-1817.

[15] Jiang R, et al, 2016, Privacy-protected facial biometric verification via fuzzy forest learning, *IEEE Trans. Fuzzy Systems*, 24(4), p.779-790.

[16] Jiang Z, et al, 2019, Social Behavioral Phenotyping of Drosophila with a 2D-3D Hybrid CNN Framework, IEEE Access, p.100.

[17] Storey G, et al, 2018, Integrated Deep Model for Face Detection and Landmark Localisation from 'in the wild' Images, IEEE Access, p.200.

[18] Storey G, et al, 2017, Role for 2D image generated 3D face models in the rehabilitation of facial palsy, IET Healthcare Technology Letters.

[19] Wang, Y., *et al*, 'Stem cell motion-tracking by using deep neural networks with multi-output', *Neural Computing and Applications*, 2017, pp.1-13.

[20] Zitnick C L and Dollar P, 2014, Edge boxes: Locating object proposals from edges, European Conference on Computer Vision (ECCV), p.1.

[21] Girshick R, 2014, Rich feature hierarchies for accurate object detection and semantic segmentation, *IEEE Conference on Computer Vision and Pattern Recognition*, p.1.

[22] Girshick R, 2015, Fast R-CNN, *IEEE International Conference on Computer Vision*, p.1.

[23] Ren, S., He, K., Girshick, R., Sun, J.: Faster R-CNN: towards real-time object detection with region proposal networks. In: NIPS (2015)

[24] BCCD Dataset, http://github.com/akshaylamba/all_CELL_data, accessed by Feb 2018.

A Novel Ground Plane Detection Method Using an RGB-D Sensor

Li Chen[1,*], Jun Zhou[1] and Xuepeng Chu[1]

[1]School of Electrical and Mechanical Engineering, HoHai University, Changzhou, China

[*]hhuc_cl@hhu.edu.cn

Abstract. A novel ground plane detection method is put forward based on three-dimensional (3D) point cloud data obtained using an RGB-D sensor. It consists of three stages: data pre-processing, occupancy grid map construction and ground plane segmentation. In order to obtain more accurate 3D point cloud data, the weight median filter (WM) is applied to recover the invalid depth pixel in the depth image. Different from the traditional approaches which process the 3D point cloud data directly, our algorithm transforms the 3D point cloud data into an occupancy grid map. Considering that the occupancy in the occupancy grid map and the distance from the point to the ground plane are distinct between ground points and other points, we will get a part of the 3D points that is definitely in the ground. The points selected are back-projected to the pixel coordinate system to get the result of the ground plane detection. In terms of experiment, the sensor is mounted on the mobile robot, Turtlebot2. The proposed method can detect more than 95 percent of the ground point. It produces accurate ground plane detection in different scenes.

1. Introduction

It is crucial for 3D navigation and environment perception in robotics fields. In order to perceive the 3D world, the robots are equipped with 3D sensors such as laser radars and stereo cameras. Lidars are relatively expensive and big consumers of energy. It is complex to build imaging systems with stereo cameras. Imaging systems with stereo cameras is difficult to deal with textureless regions. In recent years, Microsoft Kinect sensors (Kinect) have created many opportunities for multimedia computing. Kinect's [1] impact has extended far beyond the gaming industry. With its wide availability and low cost, a quantity of researchers and practitioners in computer science and robotics adopt it in the field of 3D modelling and human-computer interaction. Nevertheless, it doesn't work well in many cases, such as strong light exposure, transparent object, reflective or IR absorbing surfaces etc.

Several previous works approach to the ground plane detection problem using a v-disparity representation [2]. Hough transform [3], is a relatively mature algorithm that can be employed not only for the detection of lines and circles on 2D images, but also for 3D space for plane detection. Another common approach is RANSAC algorithm [4, 5, 6]. Hahnel et al. [7] randomly selected 3D points. Then they found the point set with the most points by region growth method and fit the plane parameters. Hoang Vu et al. [8] combines gradient threshold segmentation and mean height evaluation to segment ground plane.

RANSAC algorithm realizes plane extraction even the image includes other planes. Whereas, the result of detection with the RANSAC algorithm is wrong in the scene where other large planes appear.

These large planes contain more 3D points than the ground plane. Hadsell et al. [9] propose a stereo-based solution to segmentation in which plane fitting on point cloud data is also used. The disadvantage is that the solution requires one to specify a-priori the number of planes to extract, and thus is not suitable for multi-plane environments.

In this paper, we present a novel and simple algorithm to detect the ground plane using occupancy grid map. It is not new to use occupancy grid map to detect the ground plane [10], but our approach gives a different method to calculate the occupancy. The paper is organized as follows: Section 2 includes some work about data pre-processing. Then our proposed algorithm is shown in Section 3. Section 4 includes experiments of our proposed algorithm. Finally, the paper is concluded in Section 5.

2. Data pre-processing

The depth image obtained by Kinect contains noise and black holes, so it is necessary to pre-process the depth image. Color and depth images must be converted to 3D point cloud data before.

2.1 Image pre-processing

A filter algorithm which is called the weight median filter (WM) is proposed. The WM filter involves two steps, row filling (see in figure 1) and weight median filling.

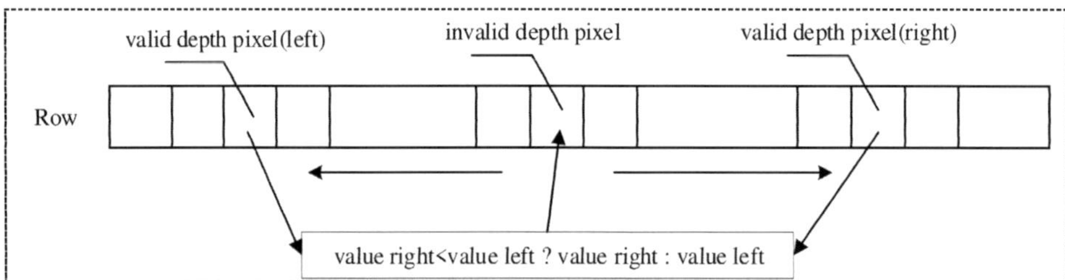

Figure 1. the process of row filling

If you find the valid depth pixel in only one direction, you will fill the invalid depth pixel with it. In weight median filling, the weight in the fixed-size window is determined by pixel distance and depth values. The weight calculation formula is as follows,

$$w = e^{\left(-\frac{((u-u_0)^2+(v-v_0)^2)}{sig\_wnd} - \frac{(depth-depth_0)^2}{sig\_dis}\right)} \quad (1)$$

where sig_wnd represents half the size of the window and sig_dis is a constant. (u_0, v_0) is the center of the window.

2.2 Coordinate System Conversion

In coordinate transformation, the Kinect is considered as the center of the world coordinate system. In (2), we assume a pinhole camera having no skew.

$$\begin{pmatrix} x_k \\ y_k \\ z_k \end{pmatrix} = \begin{pmatrix} \frac{i_k - c_x}{f_x} z_k \\ \frac{j_k - c_y}{f_y} z_k \\ z_k \end{pmatrix} \quad (2)$$

Z_k represents the value of Z-axis coordinate with the corresponding 2D pixel (i_k, j_k). $(i_k, j_k)^T$ is the pixel coordinates in the kth depth image. $(c_x, c_y)^T$ is the camera's principal point, in pixels. f_x and f_y

represent focal length. In our experiments, the parameters of RGB-camera were retrieved from the datasets or from the calibration procedure of the device.

Synchronization of RGB-camera and Depth-camera is guaranteed by the timestamp from Microsoft Software Development Kit (SDK). On the basis of Eq.2, we can obtain the 3D point cloud data of the scene.

3. The Proposed Method

A new approach to detect ground plane (see in figure 2) is proposed which includes occupancy grid map construction, ground plane fitting and ground plane segmentation.

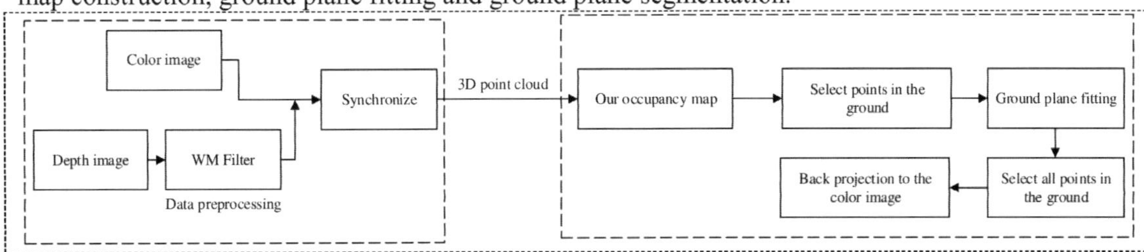

Figure 2. Overview of the proposed ground plane detection algorithm

3.1 Occupancy Grid Map Construction

It is the key step in the entire algorithm. The point in the occupancy grid map is either occupied or free. We can obtain more accurate data about obstacle in the occupancy grid map. This paper modifies the process of occupancy calculation to adapt to our algorithm. The paper updates the occupancy with binary Bayes filter.

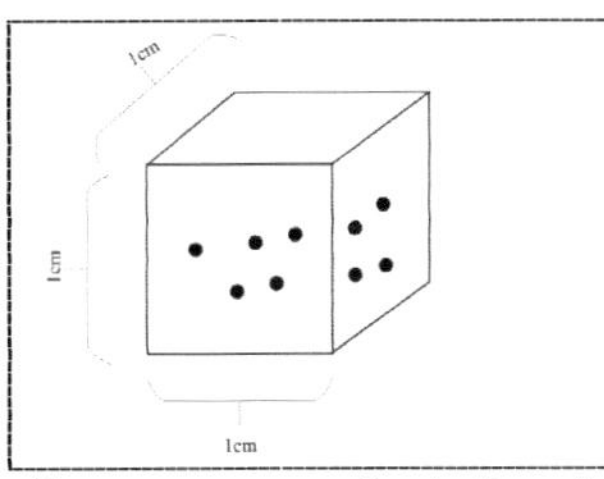

Figure 3. some points in the same grid

The classic method of occupancy calculation is an estimation of the grid occupancy in the two dimensions, time and distance. We find that some points in the ground are mapped to the same grid (see in figure 3.). If you follow the classic method of occupancy calculation, the occupancy of points in the ground will be big. This is not suitable for our algorithm. In the paper, the points at the same location with different height are considered to construct occupancy grid map. Our method to calculate the occupancy is shown in figure 4.

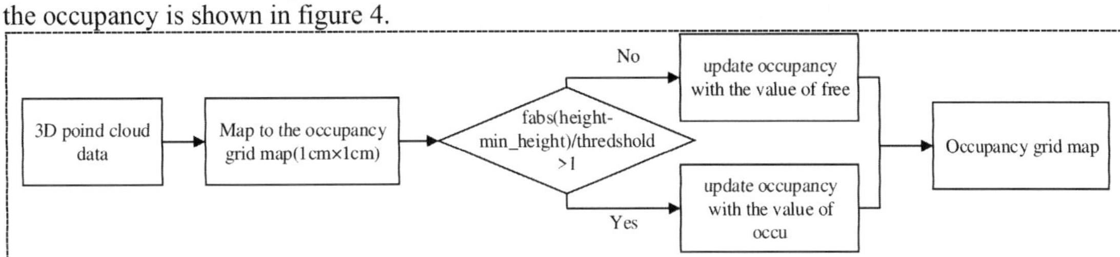

Figure 4. our method to construct the occupancy grid map

In general, the lowest height point is definitely in the ground. Each point in the 3D point cloud data is compared to this point in height. The result of the comparison will guide to update the occupancy.

3.2 Ground Plane Segmentation

If a pixel is in the ground plane, the occupancy of the pixel is below zero. Some points on the ground will be selected based on the occupancy. Then these filtered 3D point cloud data are used for ground plane fitting. This paper employs the way of Jann Poppinga [11] to obtain the normal vector of the ground plane.

Suppose we have a set of 3D points $r_i = (x_i, y_i, z_i)^T, i = 1...k$ in the ground plane and we want to get the normal vector of these data. It can be shown that this is the eigenvector problem under the assumption of the isotropic Gaussian noise. The gravity center of the given data is defined as:

$$r_G = \frac{1}{k}\sum_{i=1}^{k} r_i \tag{3}$$

Using this, the following matrix M_k is defined:

$$M_k = \sum_{i=1}^{k}(r_i - r_G)(r_i - r_G)^T \tag{4}$$

Then the normal vector of the ground plane is equal to the eigenvector n which corresponds to the smallest eigenvalue of the matrix . Suppose the resulting vector is normalized. The parameters of the plane equation are solved by the normal vector and the gravity center. On the basis of Eq.5, the distance from the point to the plane is obtained. If the distance is beyond the threshold, the points will be removed as outliers. The remaining points are back-projected to the pixel coordinate system to get the result of the ground plane detection.

$$dis = \frac{|ax_i + by_i + cz_i + d|}{\sqrt{a^2 + b^2 + c^2}} \tag{5}$$

4. Experiments

Color and depth images are converted to 3D point cloud data, since the input of the system are 3D point cloud data. The resolution of all input images is 640×480. In the following figure 5,6, we show the results of experiment on a group of depth images. They were obtained with a Kinect on the TurtleBot2. In the image pre-processing, the WM filter is proposed. The image in figure 5 is the original depth image. There are many black holes in the image, especially around the object. After the WM filter method, the number of holes in the image in figure 6 decreases significantly.

Figure 5. the original depth image

Figure 6. the depth image with WM filter

The results of the detection of the three scenes are shown in figure 7,8,9. Some points are in the ground, but the occupancy of them is beyond zero. A total of two reasons lead to the problem. The first reason is that the ground is not an ideal plane and the second reason is that we can't guarantee that the Kinect is placed in a horizontal position. There are a lot of holes in the detection image without WM filter (see in figure 9). The number of points in the ground selected by occupancy (*detect1*) is less than the number of points actually in the ground (*origin*) (see in figure 10). Nevertheless, it guarantees the accuracy of the ground plane fitting. The result of the ground plane detection is correct. The result of ground plane detection does not cover all the ground in the image. The reason is that the limitation of the sensor's range.

Figure 7. result in simple scenes

Figure 8. result in complex scenes

Figure 9. result without WM filter

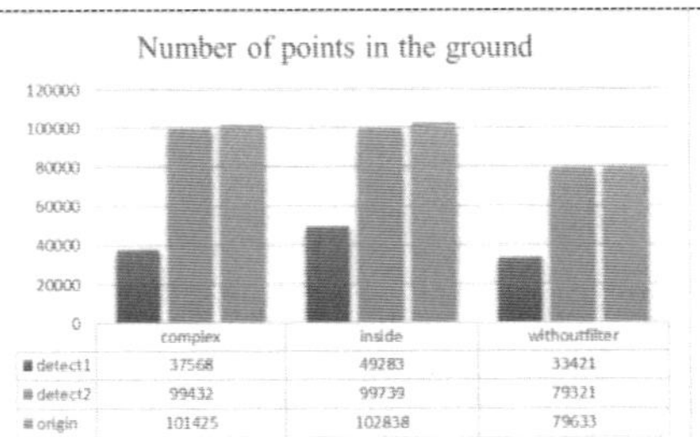

Figure 10. the number of points in the ground in three scenarios

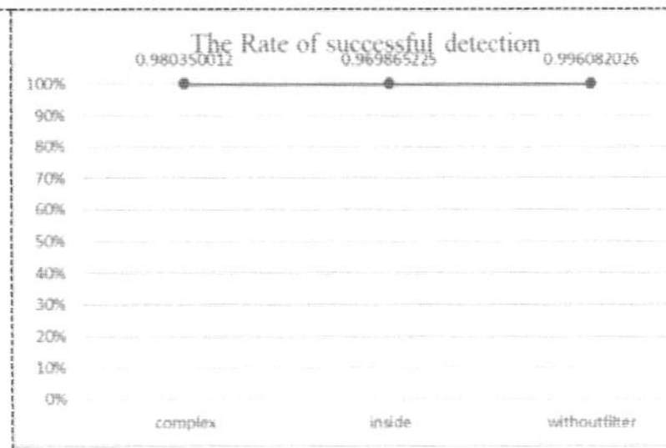

Figure 11. the rate of successful detection in three scenarios

detect1 represents the process of filtering points based on occupancy. *detect2* represents the process of filtering points based on distance. *origin* represents the original point cloud data only includes the points in the ground.

There is a significant increase in the number of points in the ground from *detect1* to *detect2*, which means that the calculation about the distance from the point to the plane plays a good role in ground plane segmentation. The Rate of successful detection describes the proportion of the ground points detected in the all ground points. The algorithm proposed can detect more than 95% of the points in the ground (see in figure 11).

Figure 12. the result of RANSAC in multi-plane environment

Figure 13. the result of RANSAC in simple scenes

Figure 14. the result of our method in multi-plane environment

Figure 15. the result of our method in simple scenes

The red points in figure 12,13 are the ground point detected by [4]. This detection algorithm is effective in the scene where there are no other large planes. Nevertheless, the result of detection is incorrect in the scene where a large plane appears. It is not suitable for multi-plane environments. The green points in figure 14,15 are the ground points detected by the proposed method. The result of ground detection is correct in the scene where a large plane appears. In terms of data pre-processing, [4] adopted Voxelization in Point Cloud Library (PCL [12]), which lead to poor real-time performance.

5. Conclusion

We put forward a new ground plane detection algorithm, which consists of three stages: data pre-processing, occupancy grid map construction and ground plane segmentation. The algorithm of ground plane detection is computationally efficient, since the 3D point cloud data are reduced to two-dimensional occupancy grid map for processing. Moreover, the effect of the weight median filter is quite obvious. The algorithm can detect more than 95 percent of the ground point and the result of the ground plane detection is accurate. We also find that the results of the algorithm are correct in different scenes, even if the sensor has a change in pitch and roll angles. These contributions come from the method to calculate the occupancy we introduced above. The algorithm proposed in this paper is not only computational efficiency but also robust.

Acknowledgments

This work was supported by the Postgraduate Research & Practice Innovation Program of Jiangsu Province no.KYCX18_0540 under grant agreement no.2018B733X14.

References

[1] Chen W, Yue H, Wu X, et al.: 'Real-time obstacle detection for legged robots using the Kinect sensor', Advanced Robotics, 2014, 28(20), pp. 1375-1387

[2] Harakeh, Ali , D. Asmar , and E. Shammas . "Ground Segmentation and Occupancy Grid Generation Using Probability Fields." International Conference on Intelligent Robots & Systems 2015.

[3] Vera E, Lucio D, Fernandes L A F, et al.: 'Hough Transform for Real-Time Plane Detection in Depth Images', Pattern Recognition Letters, 2018

[4] Ramy Ashraf Zeineldin, Nawal El-Fishawy: 'Fast and accurate ground plane detection for the visually impaired from 3D organized point clouds Conference'. 2016 SAI Computing Conference (SAI), London, United Kingdom, July, 2016, pp. 373-379

[5] R?Fer, Thomas , et al.: 'Real-Time Plane Segmentation Using RGB-D Cameras', in 'Computer Science' (2012), pp. 306-317

[6] Hong Liu, Yongqing Jin, Chenyang Zhao.: 'Real-time trust region ground plane segmentation for monocular mobile robots'. 2017 IEEE International Conference on Robotics and Biomimetics (ROBIO), Macao, China, December, 2017, pp. 952-958

[7] H?Hnel D, Burgard W, Thrun S.: 'Learning compact 3D models of indoor and outdoor environments with a mobile robot', Robotics and Autonomous Systems, 2003, 44(1), pp. 15-27

[8] Vu, Hoang , et al. "A Ground Segmentation Method Based on Gradient Fields for 3D Point Clouds.", 2018.

[9] Vijayanagar, K.R., Loghman, M., Joohee Kim.: 'Refinement of depth maps generated by low-cost depth sensors'. SoC Design Conference (ISOCC), Jeju Island, South Korea, November, 2012

[10] C. Braillon, C. Pradalier, K. Usher, J. Crowley, and C. Laugier,"Occupancy grids from stereo and optical flow data," in ISER, 2006.

[11] Jann Poppinga, Narunas Vaskevicius, et al.: 'Fast Plane Detection and Polygonalization in Noisy 3D Range Images'. IEEE/RSJ International Conference on Intelligent Robots and Systems, Nice, France, Sept, 2008, pp. 22-26

[12] Hadsell R, Sermanet P, Scoffier M, et al.: 'Learning Long-Range Vision for Autonomous Off-Road Driving'. Journal of Field Robotics, 2009, pp. 120-144

IOP Conf. Series: Materials Science and Engineering **646** (2019) 012050 doi:10.1088/1757-899X/646/1/012050

Unsupervised Learning of Depth and Visual Odometry using Photometric Calibration

Tianhong Yu

Beijing Institute of Technology

tedyth@163.com

Abstract. Depth and visual odometry estimation are two essential parts in SLAM systems. Compared with traditional algorithms, supervised learning methods have shown promising results in single view depth estimation and visual odometry estimation. However, they require large amounts of labeled data. Recently, some unsupervised approaches to estimate depth and odometry via minimizing photometric error draw great attention. In this paper, we present a novel approach to learn depth and odometry via unsupervised learning. Our method ameliorates the original photometric loss to enhance the robustness to illumination change in real scenarios. In addition, we propose a new structure of Pose-net and Explainability-net to achieve rotation-sensitive odometry results and more accurate explainability masks. The experimental results have demonstrated that our approach achieves better performance than existing unsupervised methods in both depth and odometry results.

1. Introduction

In recent years, SLAM (Simultaneous Localization And Mapping) algorithms have been widely used in automatic-driving cars or other autonomous robots. Visual odometry, as a part of SLAM, learns camera poses with only video frames as input. On the other hand, depth prediction is also an essential part in SLAM system. An accurate depth map is needed when creating a large 3D point cloud.

Since these two parts are indispensable, several works [1][2][3][4] estimate depth using the supervised method. However, it requires large amounts of labeled data, which is difficult to obtain.

Unsupervised method [5][6][7][8][9] are proposed subsequently. One of those unsupervised methods, proposed by [9], build a depth network and a pose network. They utilize photometric warp error as the training loss to jointly train the above two networks. On the basis of [9], a number of works [10] [11] [12] [13] [14] are proposed and achieve better results. However, there is still a problem in [9], that is, the photometric loss is easily affected by light changes.

Inspired by that, in this paper, we propose a framework (figure 1) which jointly estimates depth and camera poses based on [9]. We make the following contributions. 1)We apply photometric calibration to each frame. Specifically, we propose a novel loss based on the brightness of each video frames, which solves the problem of the illumination change in the real environment. 2)Two groups of separate convolutional layers are proposed in Pose-net in order to balance the weights between translation and rotation loss. 3)We apply a convolutional layer and a deconvolutional layer in our explainability network, which selectively enhances useful features and suppresses less useful ones. 4)The experimental results demonstrate that we significantly improve the performance over the existed methods in both depth estimation and odometry estimation.

2. Our Approach

In this section, we introduce our network which learns depth and camera poses from consecutive video frames in an unsupervised method.

2.1. Network structure

We mainly adopt the architecture of [9]. Our work consists of three networks, Depth-net, Pose-net and Explainability-net.

Depth-net is an encoder-decoder network similar to [15]. It takes a single image as input and output four scales of dense depth maps. The structure of Pose-net and Explainability-net is in figure 2. As we can see in figure 2, Pose-net is a simple convolution network like VGGNet in [16]. We input consecutive camera frames (3 frames or 5 frames) and output 6-DoF camera poses of each frame.

At the same time, we take full advantage of the first few layers in Pose-net and add some convolution and deconvolution layers behind them to get our Explainability-net. By means of the deconvolution layers, we obtain explainable masks in four scales. These masks will be used in computing the training loss, which will be introduced in the subsequent chapter.

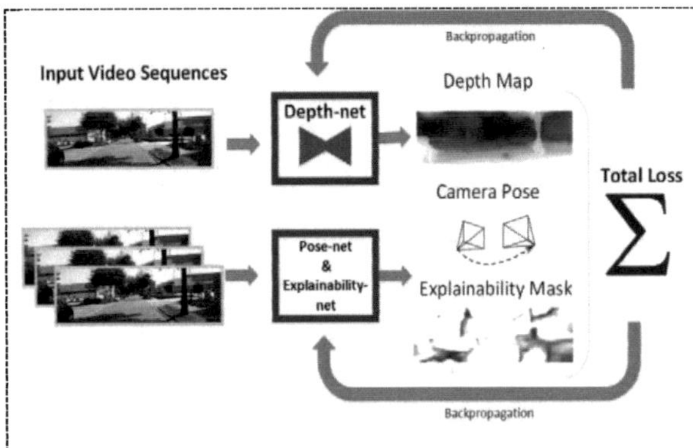

Figure 1. Overview of our framework.

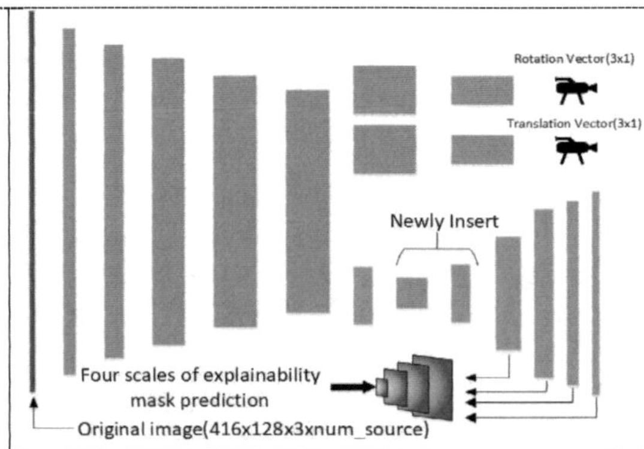

Figure 2. Structure of Pose-net and Explainability-net.

2.2. A Brightness-based loss

In [9], we assume that the surface is Lambertian so that the photo-consistency error is meaningful. However, the brightness of the scene varies while the camera moves in reality. This will cause a problem that the photometric error is not accurate. Inspired by the idea of making a photometric calibration in [11], we propose a photometric error based on the brightness of the scene which is robust to the change of brightness.

Our problem can be formalized as follows: Our network input is consecutive image sequence(take 3 as an example) I_{t-1}, I_t, I_{t+1}, including the source view and target view. Our goal is to obtain the depth D_t of each frame and camera poses T_{t-1}, T_t, T_{t+1}, representing the camera ego-motion. Then, according to [9], the pixel loss can be formulated as follows:

$$L_{pixel} = \sum_s \sum_p |I_t(p) - I_s(p)| \tag{1}$$

where p represents pixel coordinates in the I image. $I_t(p)$ is the target view image. $I_s(p)$ is the warped image from the target view to the source view. The formula about warping is as follows:

$$p_s \sim K T_{t\to s} D_t(p_t) K^{-1} p_t \tag{2}$$

where p_t represents pixel coordinates in the target image. $D_t(p_t)$ denotes the depth of p_t which is obtained by the Depth-net. K denotes the intrinsic matrix of the camera. T is the transformation from the target view to the source view obtained by the Pose-net.

However, this loss function suffers when there are significant illumination changes in the video. To solve this problem, we propose a new pixel loss instead of the original one. We use $B(p)$ to denote the brightness of one pixel in one video frame I . $B(p)$ can be represented as:

$$B(p) = 0.6 * r(p) + 0.3 * g(p) + 0.1 * b(p) \tag{3}$$

where $r(p), g(p), b(p)$ respectively represents the pixel value in red, green, blue channels of the image I .

Then, the pixel loss can be formulated as follows:

$$X(p) = I(p) - B(p) \tag{4}$$

$$L_{pixel} = \sum_{s}\sum_{p} |X_t(p) - X_s(p)| \tag{5}$$

We use $X(p)$ to describe pixel p in the image I without the influence of brightness. Therefore, we can use the difference between $X(p)$ to replace the original pixel loss.

According to the method in [9] 's work, in order to improve the performance against occlusion and moving objects problem, we add a predicted mask obtained by the network in figure 2 on the pixel loss, our final loss consists of pixel loss, smooth loss and explainability loss. It can be formulated as follows:

$$L_{final} = \sum_{l} L^l_{pixel} + \lambda_s L^l_{smooth} + \lambda_e \sum_{s} L_{reg}(E^l_s) \tag{6}$$

where l represents different image scales, s represents source images, and λ_s and λ_e are respectively the weights of smooth loss and the explainability loss. You can find the introduction of smooth loss and explainability loss in [9] 's work.

We use this final loss to jointly train the Depth-net and Pose-net and achieve a better performance we'll see in the next part.

2.3. Rotation-sensitive Pose-net

In most cases, the camera follows the car's movement which consists of translation and rotation. However, translation is more common than rotation in reality. As a result, this will cause a problem that translation is easy to fit during training time while the rotation has not been trained completely yet. In other words, the original network is not sensitive to rotation.

The above analysis motivates us to investigate the problem of the asynchrony of translation and rotation. We change the structure of the network in [9] 's work. We regard rotation and translation as two irrelevant results respectively. Specifically, we apply two groups of two separate convolutional layers before the last camera pose vectors, which can be seen in figure 2. In this way, there are no relationships between our Pose-net's results (rotation and translation), since they are trained separately. Thus we can get a more accurate rotation result throughout our rotation-sensitive network. Because of the more accurate rotation result, depth results are improved subsequently, which we'll find out in Section 3.

2.4. Improved Explainability-net

It is widely known that getting good performance via a deep convolutional network is a complicated task. In recent works, one way to solve this problem is to improve the ability of distinguishing the features.

IOP Conf. Series: Materials Science and Engineering **646** (2019) 012050 doi:10.1088/1757-899X/646/1/012050

In our work, the Explainability-net shares the first five feature encoding layers with Pose-net. After these five convolutional layers, we add a convolutional layer to get distinctive features. It is then followed by four deconvolutional layers, which results in four scales of explainable masks. We simply apply a convolutional layer after the feature map and append a deconvolutional layer subsequently to recover the scale of the feature map. In other words, we suppress our network model in the spatial dimension at first to distinguish the more important channel and enhance the ability of feature expression. Then we recover to the original spatial dimension so that the feature maps carry more important information.

3. Experimental Results

In this section, we will show experimental results for evaluating the performance of our proposed framework. Our network is implemented with the Tensorflow framework. We use GTX 1080Ti to train and test our network. We achieve real-time performance during test time.

3.1. Training configurations

Dataset: We follow [9] 's work and choose KITTI dataset to train and test our network.

Training parameters: We choose the Adam optimizer [17] as the optimizer of our framework. We utilize a small learning rate of 0.0001, $\beta_1 = 0.9$ and $\beta_2 = 0.999$. we set the batch size for training to 4 in order to improve the training speed.

3.2. Depth Estimation

Here, we compare our depth estimation results with other various methods to evaluate the performance of our pose estimation network. We fix the length of image sequences to be 3 frames, and treat the central frame as the target view and the ±1 frames as the source views. We choose the split provided by [2], which has been mentioned before.

The detailed depth estimation results are listed in table 1. Some comparison results are even supervised method. Part of results are not only trained in one dataset KITTI, but also trained on two datasets, KITTI and Cityscapes [18]. All the methods are tested in KITTI dataset. The table reports separate results for a depth cap of 50m and 80m. As we can see from the table, our method outperforms most of the methods mentioned in the table. Our results are even better than some supervised methods or methods trained on both KITTI and Cityscapes [18] dataset.

Table 1. Depth estimation results on the KITTI dataset. For training dataset, K = KITTI, and CS = Cityscapes [18]. Cap indicates the largest value in depth estimation. Currently, we usually adopt 50m and 80m as our evaluation criteria. We can find our results achieve an obvious improvement beyond other methods.

Method	Dataset	Supervision	Cap	Error Metric				Accuracy Metric		
				Abs-Rel	Sq-Rel	RMSE	Log-RMSE	$\delta<1.25$	$\delta<1.25^2$	$\delta<1.25^3$
Eigen et al.[2] Coarse	K	Depth	80m	0.214	1.605	6.563	0.292	0.673	0.884	0.957
Eigen et al.[2] Fine	K	Depth	80m	0.203	1.548	6.307	0.282	0.702	0.890	0.958
Liu et al.[4]	K	Depth	80m	0.202	1.614	6.523	0.275	0.678	0.895	0.965
Godard et al.[6]	K	Stereo	80m	0.148	1.344	5.927	0.247	0.803	0.922	0.964
Godard et al.[6]	CS+K	Stereo	80m	0.124	1.076	5.311	0.219	0.847	0.942	0.973
Zhou et al.[9]	K	-	80m	0.208	1.768	6.856	0.283	0.678	0.885	0.957

Zhou et al.[9]	CS+K	-	80m	0.198	1.836	6.565	0.275	0.718	0.901	0.960
Zhan et al.[12]	K	Stereo	80m	0.135	1.132	5.585	0.229	0.820	0.933	0.971
Wang et al.[10]	K	-	80m	0.151	1.257	5.583	0.228	0.810	0.936	0.974
Mahjourian et al.[14]	K	-	80m	0.163	1.240	6.220	0.250	0.762	0.916	0.968
Ours	**K**	**-**	**80m**	**0.149**	**1.168**	**5.643**	**0.233**	**0.801**	**0.938**	**0.970**
Garg et al.[5]	K	Stereo	50m	0.169	1.080	5.104	0.273	0.740	0.904	0.962
Zhou et al.[9]	K	-	50m	0.201	1.391	5.181	0.264	0.696	0.900	0.966
Zhou et al.[9]	CS+K	-	50m	0.190	1.436	4.975	0.258	0.735	0.915	0.968
Zhan et al.[12]	K	Stereo	50m	0.128	0.815	4.204	0.216	0.835	0.941	0.975
Mahjourian et al.[14]	K	-	50m	0.155	0.927	4.549	0.231	0.781	0.931	0.975
Ours	**K**	**-**	**50m**	**0.146**	**0.989**	**4.339**	**0.221**	**0.815**	**0.943**	**0.973**

3.3. Odometry Estimation

As more source views express more temporal information and perform better in odometry estimation experiments, we fix the length of image sequences to be 5 frames, and treat the central frame as the target view and the ±2 frames as the source views. We choose the official KITTI Odometry set as the training set, which has been mentioned before.

We show our odometry estimation results in table 2. ORB-SLAM is a classical monocular SLAM system proposed by [19]. ORB-SLAM(full) is the complete version of ORB-SLAM, including loop closure and relocalization. ORB-SLAM(short) denotes the system runs on 5-frame snippets (same as our input setting). For the sake of scale consistency, we optimize the scaling factor in advance. From the data in the table, we can discover that our method significantly outperforms our baseline [9]. And our method's performance is very close to the performance of ORB-SLAM and [14]'s method.

Table 2. Odometry estimation results on the KITTI Odometry set. We adopt ATE (Absolute Trajectory Error) proposed in [19] as our evaluation metric. Lower is better. Our method significantly outperforms our baseline [9]and is very close to the performance of ORB-SLAM and [14]'s method.

Method	Sequence 9	Sequence 10
ORB-SLAM[19](full)	0.014±0.008	0.012±0.011
ORB-SLAM[19](short)	0.064±0.141	0.064±0.130
Zhou et al.[9]	0.021±0.017	0.020±0.015
Mahjourian et al.[14]	0.013±0.010	0.012±0.011
Ours	**0.014±0.008**	**0.013±0.010**

4. Conclusion

We present a novel framework to learn depth and odometry using the unsupervised method. Our contribution is to ameliorate the original photometric loss to enhance the robustness to illumination change and propose a new structure to achieve better results. The proposed method only needs monocular video streams for training, and can provide depth and odometry results during test time.

The results of depth estimation in experiments demonstrate that our algorithm can even outperform some supervised methods and original unsupervised method. Our results of odometry estimation can keep up with traditional SLAM algorithm in some area and can be employed in real time.

Finally, it is possible to employ our framework into a real SLAM system with loop detection. This can significantly reduce drift and improve performance.

References

[1] P. Agrawal, J. Carreira, and J. Malik, "Learning to see by moving," in IEEE International Conference on Computer Vision, 2015, pp. 37–45
[2] D. Eigen, C. Puhrsch, and R. Fergus, "Depth map prediction from a single image using a multi-scale deep network," in International Conference on Neural Information Processing Systems, 2014, pp. 2366–2374
[3] F. Liu, C. Shen, and G. Lin, "Deep convolutional neural fields for depth estimation from a single image," pp. 5162–5170, 2014
[4] F. Liu, C. Shen, G. Lin, and I. Reid, "Learning depth from single monocular images using deep convolutional neural fields," IEEE Transactions on Pattern Analysis and Machine Intelligence, vol. 38, no. 10, pp. 2024–2039, 2016
[5] R. Garg, K. B. G. Vijay, G. Carneiro, and I. Reid, "Unsupervised cnn for single view depth estimation: Geometry to the rescue," pp. 740–756, 2016
[6] C. Godard, O. M. Aodha, and G. J. Brostow, "Unsupervised monocular depth estimation with left-right consistency," in Computer Vision and Pattern Recognition, 2017, pp. 6602–6611
[7] Y. Kuznietsov, J. Stckler, and B. Leibe, "Semi-supervised deep learning for monocular depth map prediction," pp. 2215–2223, 2017
[8] M. Ye, E. Johns, A. Handa, L. Zhang, P. Pratt, and G. Z. Yang, "Selfsupervised siamese learning on stereo image pairs for depth estimation in robotic surgery," 2017
[9] T. Zhou, M. Brown, N. Snavely, and D. G. Lowe, "Unsupervised learning of depth and ego-motion from video," pp. 6612–6619, 2017
[10] C. Wang, J. M. Buenaposada, R. Zhu, and S. Lucey, "Learning depth from monocular videos using direct methods," 2018
[11] J. Engel, V. Koltun, and D. Cremers, "Direct sparse odometry." IEEE Transactions on Pattern Analysis and Machine Intelligence, vol. PP, no. 99, pp. 1–1, 2016
[12] H. Zhan, R. Garg, C. S. Weerasekera, K. Li, H. Agarwal, and I. Reid, "Unsupervised learning of monocular depth estimation and visual odometry with deep feature reconstruction," 2018
[13] R. Li, S. Wang, Z. Long, and D. Gu, "Undeepvo: Monocular visual odometry through unsupervised deep learning," 2017
[14] R. Mahjourian, M. Wicke, and A. Angelova, "Unsupervised learning of depth and ego-motion from monocular video using 3d geometric constraints," 2018
[15] N. Mayer, E. Ilg, P. Husser, P. Fischer, D. Cremers, A. Dosovitskiy, and T. Brox, "A large dataset to train convolutional networks for disparity, optical flow, and scene flow estimation," in Computer Vision and Pattern Recognition, 2016, pp. 4040–4048
[16] K. Simonyan and A. Zisserman, "Very deep convolutional networks for large-scale image recognition," in ICLR, 2015, Conference Proceedings
[17] D. P. Kingma and J. Ba, "Adam: A method for stochastic optimization," Computer Science, 2014
[18] M. Cordts, M. Omran, S. Ramos, T. Rehfeld, M. Enzweiler, R. Benenson, U. Franke, S. Roth, and B. Schiele, "The cityscapes dataset for semantic urban scene understanding," 2016
[19] R. Mur-Artal and J. D. Tardos, "Orb-slam2: An open-source slam system for monocular, stereo, and rgb-d cameras," IEEE Transactions on Robotics, vol. PP, no. 99, pp. 1–8, 2016

Pose-aware Person Re-Identification with Spatial-temporal Attention

Qi Zhu

Beijing Institute of Technology

qzhu.working@gmail.com

Abstract. The Person re-ID task objective is to search for a specified target pedestrian image in a large gallery. With the development of deep learning technology, the accuracy of personnel re-identification has been greatly improved. However, previous methods lacked sufficient attention to local features when focusing on global features, causing background and occlusion objects to affect the foreground. The framework we proposed specifically designed this problem to optimize local features through human pose key-points and self-attention mechanisms. And the effectiveness of the method is demonstrated in multiple datasets.

Index Terms—Re-Identification, Pose, Attention, Spatial-temporal

1. Introduction

Person re-identification has been viewed as one of the critical problems of the general recognition task. It aims to identify a person of interest from a large scale gallery image database which captured from distinctively cameras in large time. By given a probe image of the person, Person re-identification algorithm will estimate the visual similarities between the probe image and gallery image database. Then the gallery images can be ranked descended according to the similarities as re-identification results. Person re-identification is a very challenging task. Between cameras, the illumination conditions, background clutters, occlusion, visible human body parts, and perceived posture of the person could be completely different. Even within same camera, as the person moves and covered by items, the features can vary dramatically. All of the factors generate a huge intra-class variation and become the limitation of traditional re-identification approaches.

Recently, following the development of deep learning [1, 2, 3] and better classification models [3, 4], person re-identification obtain more robust end-to-end representative features through deep learning networks. Compared with shallow learning, these deep features effectively improve the performance of person re-identification in the commonly used datasets [5, 6, 7].

Therefore, the deep feature cares more about global information. It may ignore some of the local features which include semantic information. [8] designs a simple splitting method to acquire different local features to aid the global features. [9] proposed a new approach that processes the image as different slices, then re-arrange all these pieces as new images. In the latest researches, some methods use pose information to guide the splitting of persons in the images and obtain better results.

In this paper, we present a new human re-identification framework that includes a new branch to explore feature information within the image. In this new branch, we extract image partitions with semantic information by extracting key-points of the human pose.

The spatial-temporal information extracted according to the key-point information can adequately capture the local features and supplement the difficult-to-identify parts in the global feature analysis.

The main contributions of this work can be summarized as the following three-folds:

- We propose a novel coarse-to-fine framework to learn precise pedestrian features in person re-identification task. It mainly depends on pose estimation and self-attention mechanism.
- The proposed spatial-temporal framework obtained accurate aligned parts information, and consider the different human parts based on self-attention network without extra training.
- Our approach achieves the state-of-the-art performance on various large-scale person Re-ID dataset such as Market-1501 [5], and demonstrates its effectiveness and generalization ability.

2. Related Work

2.1. Recurrent Neural Network

Recurrent Neural Network (RNN) is the deep neural network that considers temporal information as the input. It needs to memorize past information and captures the contextual dependency of the sequential inputs. Long Short Term Memory (LSTM) [1] networks are proposed to apply in different tasks. In [10], Pyramidal LSTM was proposed to solve the person re-identification.

2.2. Pose Detection

Pose estimation is a traditional research [11] in computer vision, and nowadays lots of deep learning methods [12] are proposed to solve this question. Recently, [13, 14] proposed some methods to combine the multi-scale features and learning mechanisms.

2.3. Self-Attention

The self-attention mechanism is soft attention [15] that uses attention scores to measure all salient features. The main calculation of the self-attention network is matrix multiplication without iteration, meaning that it can be easily accelerated. Recently, more and more tasks have successfully applied the self-attention mechanism and achieved excellent performance.

2.4. Person Re-identification

In recent years, many state-of-the-art Re-ID results are relying on features embedding deep learning. Due to its superior performance. Deep learning based methods have been demonstrating the re-id community in the past several years. In the two early Works [8, 16], two images are feed into the siamese model at the same time to determine if there is the same person appeared in both of the images. After these works, approaches were improved in different Ways by [17, 18, 19]. After that, deep learning methods become a popular option in re-ID.

3. The Proposed Re-ID Method

We introduce a new deep framework for the person re-identification task. As shown in Fig. 1, our framework has global and local branches. And the local branch contains three main parts such as pose estimation, representation learning and self-attention mechanism. We will give more details of each part in the following subsections.

3.1. Framework

We can get the overall structure information of the framework for our proposed deep Re-ID from Fig 1. For one training sample, the input images are sent into two branches.

The global branch is intended to extract features with global image information. In this branch, we feed the images into Resnet-50 structure and obtain global features F_{global} from the 512-dimensional fully connected layer results at the end of this branch.

In the local branch, the image is transmitted into a pre-trained pose estimation network to extract key-points of the human body. We use the obtained human body key-points to locate and separate the

IOP Conf. Series: Materials Science and Engineering **646** (2019) 012051 doi:10.1088/1757-899X/646/1/012051

original image to get the sub-image we need. Then we use Residual blocks from Resnet to extract the sub-feature group. After that, with the sub-feature group as input, we can get self-correlation features from the self-attention network. Finally, through the LSTM structure, we will take the spatial-temporal self-attention features of the sub-images.

After getting all the results of the two branches, we concatenate the global features, sub-features, and self-attention features as the final feature results.

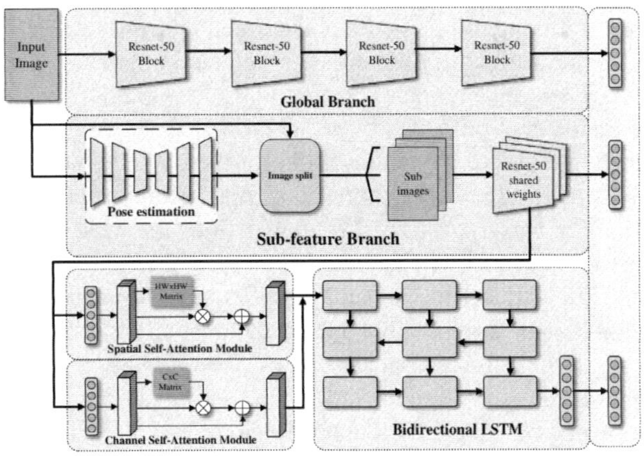

Figure 1. Overview of our framework. We take the query image as input. Our algorithm consists of two branches, a global branch and a sub-feature branch, resulting in global features, sub-features, and self-attention features.

3.2. Pose estimation

In the pose estimation task, given a RGB image I containing a human, we aim to estimate a set of joint coordinates $Y \in Y_{2D}$. These joints J composed as the human skeleton $Y_{2D} \in R^{J\times 2}$, where the J is the number of human joint. The two coordinates are pixel coordinates in I.

In our approach, this paper adopts an off-the-shelf model named Stacked Hourglass Networks [14]. This method produces 16 key-points for each Re-ID image prediction. These points divide the image into three parts, head-shoulder, upper body and lower body. According to the selected key-points, we can roughly divide the original image into three parts, in which redundant key-points ensure that we don't lose the required components while dealing with human images in different positions.

Suppose the image I_{H*W} with size $H * W$, for all the anchor points in each part, $P_{1..N} = (x_1, y_1), (x_2, y_2), \cdots, (x_N, y_N)$, We crop the regions with Eq. i.e.

$$R^{s,u,d} = \left[y_c - \frac{w}{2}, y_c + \frac{w}{2}\right] \tag{1}$$

$$y_c = \frac{y_{min} + y_{max}}{2} \tag{2}$$

$$w = \max(\max(y_1, y_2, \cdots, y_N) - \min(y_1, y_2, \cdots, y_N), \epsilon) \tag{3}$$

The ϵ is minimum dicing height to prevent inaccurate key-points locating. On the general 128×64 pixel size dataset, ϵ takes 32 as the minimum value.

After obtaining the sub-image from input image, we unify the image size to 64×64 pixels, so that the convolutional neural network group can easily extract the sub-features. Since the texture information of the image needs to be obtained here, we adopt the design of the stacked residual blocks in Resnet. Since the texture features at different locations have similarities, the network parameters between different stacked residual blocks can be shared. Reducing the quantity of parameters can help the network to converge stably. Here we obtain the local sub-feature group named F_{sub}.

3.3. Self-attention Mechanism

Attention mechanism is always used to weight the feature maps, it highlights the key-points of human and alleviates the weights of backgrounds. In self-attention module, we introduce the dual self-attention network structure improved on [20]. Here the sub-features are fed into two similar modules. In the spatial self-attention module, the feature F_{sub} is reduced by a convolutional layer to obtain the feature F'_{sub} of the $H * W * C$ dimension. F'_{sub} is reshaped and multiplied with a matrix of dimension $H * W$ to get the spatial domain attention weight, and broadcast back to the input size. Then F'_{sub} is multiplied with the weight to obtain the attention feature. Structure of the channel self-attention Module is similar to the spatial one's, only weight matrix is changed to $C * C$.

3.4. Spatial-Temporal Features

Based on the pose detection method, we obtain different parts of the human in the images. The global representation features of the image focus more on global information such as the appearance, color and pose. In such cases, only some parts are different such as the head, upper body or legs and shoes. For solving these kinds of situations, it has been proposed many methods based on part-alignment representation structure. The part-based local representation of human is an effective complementary to the global information. Then, through integrating the global and LSTM features, we finally get a spatial-temporal discriminative features for our person re-identification.

In our Re-ID framework, we choose bidirectional LSTM as the concrete implementation. The Bidirectional LSTM network we designed include three layers. The first and the third layer network are forward, and the second is reverse. Bidirectional LSTM inputs 512-dimensional sub-features group generated by the self-attention model, and outputs 512 dimension features F'_{att} which contains spatial-temporal information from self-attention and bidirectional LSTM. After that, F'_{att} concatenate with F_{global} and F'_{sub} as the final feature F_{final} of the input image.

4. Experimental Results

To evaluate the effectiveness of our approach on person re-identification performance, we conduct several groups of experiments on CUHK03-NP, Market-1501, and DukeMTMC-reID

4.1. Baseline

When choosing the baseline, we compared the identification model structure with the verification model structure. Overall, the two can achieve similar performance. The verification method incorporates hard-mining acceleration training, but even this identification model structure still has an advantage in convergence speed and total training time. This may be related to the former using all the label information in training, while the latter is only related to the weak label information. [8].

So in this paper, we use the identification model structure as the form of the baseline. The baseline model uses the Residual-50 structure. After the original fully connected layer, another fully connected layer of the category of the training set is connected for the softmax loss calculation. In the baseline training process, we started training from the ImageNet pre-trained network with the initial learning rate set to 0.1 for a total of 60 epochs.

4.2. Implementation Details

In our experiments, the image data uniformly reshape to $256 * 128$ pixels in preprocessing for ease of use. And we simply use flipping and random-cutting to augment data.

For the global branch, we use Resnet-50 network as the backbone pre-trained on ImageNet, replace the max pooling layer with average pooling layer, and training with the classification loss. In the local branch, we predict 14 key-points for each pedestrian image. The sub-images are extracted by the shared stacked residual block to obtain 128-dimensional F_{sub}. Next, the F_{sub} send to the dual self-attention network to extract the attention information and output 512-dimensional F'_{sub}. The features then flow into the 3-layer Bidirectional LSTM network to obtain 512-dimensional F'_{att}.

All experiments performed on Pytorch. In training, we set batch size to 32, max epoch 60, pre-training part learning rate 0.01, and the remaining part learning rate 0.1. Training is optimized by

SGD optimizer, which weight_decay set to $5 * 10^{-4}$ and momentumset to 0.9. All training and verification perform on one GTX 1080Ti graphics card.

4.3. Evaluation with Baseline
We validated our algorithm performance on the Market-1501, DukeMTMC-reID, and CUHK03-NP datasets with default training-testing separation scheme using indicators **Rank-1** and **MAP**. In each training and testing experiment, we adopt the default hyper parameters without specific adjustments. Results is shown in Table 1.

Table 1. Performance compared with baseline.

Market-1501	Rank 1	MAP	CUHK03-NP(manual labeled)	Rank 1	MAP
baseline	0.879	0.700	baseline	0.477	0.437
Ours	**0.909**	**0.750**	**Ours**	**0.489**	**0.452**
CUHK03-NP(auto labeled)	Rank 1	MAP	DukeMTMC-reID	Rank 1	MAP
baseline	0.419	0.388	baseline	0.761	0.582
Ours	**0.440**	**0.404**	**Ours**	**0.773**	**0.604**

4.4. Comparison with Related Works
We compare our method with several state-of-the-art methods on Market-1501 dataset. As shown in Table 2., our method achieves competitive results with the state of the art.

Table 2. Comparison with state-of-the-art methods on the Market-1501 dataset.

Method	Rank 1	MAP	Method	Rank 1	MAP
BoW+kissme	0.444	0. 208	PAN	0.827	0.634
MR CNN	0.456	0.266	Transfer	0.837	0.655
Gated SCNN	0.659	0.396	DML	0.877	0.688
SOMAnet	0.739	0.479	SVDNet+REDA	0.871	0.711
ReRank	0.771	0.636	baseline	0.879	0.700
Verif-Identif	0.795	0.599	**Ours**	**0.909**	**0.750**

5. Conclusions
In this paper, we propose a new network framework for the pedestrian re-identification task. Expand from the global branch, we introduce a new branch based on the self-correlation mechanism and LSTM network structure to extract local features. This network can be implemented with limit hardware resources and trained stably. For verification, we performed experiments on Matrket-1501, CUHK03-NP and DukeMTMC-reID, and achieved visible performance improvements in pedestrian re-identification tasks.

In the future, we will focus on the video field and improve the performance and speed by improving the structure.

References
[1] Hochreiter S, Schmidhuber J. Long short-term memory[J]. Neural computation, 1997, 9(8): 1735-1780.

[2] Srivastava N, Hinton G, Krizhevsky A, et al. Dropout: a simple way to prevent neural networks from overfitting[J]. *The Journal of Machine Learning Research*, 2014, 15(1): 1929-1958.

[3] Iandola F, Moskewicz M, Karayev S, et al. Densenet: Implementing efficient convnet descriptor pyramids[J]. *arXiv preprint* arXiv:1404.1869, 2014.

[4] Szegedy C, Ioffe S, Vanhoucke V, et al. Inception-v4, inception-resnet and the impact of residual connections on learning[C]//*Thirty-First AAAI Conference on Artificial Intelligence*. 2017.

[5] Zheng L, Shen L, Tian L, et al. Scalable person re-identification: A benchmark[C]//*Proceedings of the IEEE International Conference on Computer Vision*. 2015: 1116-1124.

[6] Li W, Zhao R, Xiao T, et al. Deepreid: Deep filter pairing neural network for person re-identification[C]//*Proceedings of the IEEE Conference on Computer Vision and Pattern Recognition*. 2014: 152-159.

[7] Ristani E, Solera F, Zou R, et al. Performance measures and a data set for multi-target, multi-camera tracking[C]//*European Conference on Computer Vision. Springer, Cham*, 2016: 17-35.

[8] Zheng L, Huang Y, Lu H, et al. Pose invariant embedding for deep person re-identification[J]. *arXiv preprint* arXiv:1701.07732, 2017.

[9] Su C, Li J, Zhang S, et al. Pose-driven deep convolutional model for person re-identification[C]//*Proceedings of the IEEE International Conference on Computer Vision*. 2017: 3960-3969.

[10] Zhang X, Luo H, Fan X, et al. Alignedreid: Surpassing human-level performance in person re-identification[J]. *arXiv preprint* arXiv:1711.08184, 2017.

[11] Cheng D S, Cristani M. Person re-identification by articulated appearance matching[M]//*Person Re-Identification. Springer, London*, 2014: 139-160.

[12] Toshev A, Szegedy C. Deeppose: Human pose estimation via deep neural networks[C]//*Proceedings of the IEEE conference on computer vision and pattern recognition*. 2014: 1653-1660.

[13] Tompson J J, Jain A, LeCun Y, et al. Joint training of a convolutional network and a graphical model for human pose estimation[C]//*Advances in neural information processing systems*. 2014: 1799-1807.

[14] Newell A, Yang K, Deng J. Stacked hourglass networks for human pose estimation[C]//*European Conference on Computer Vision. Springer, Cham*, 2016: 483-499.

[15] Lin Z, Feng M, Santos C N, et al. A structured self-attentive sentence embedding[J]. *arXiv preprint* arXiv:1703.03130, 2017.

[16] Byeon W, Breuel T M, Raue F, et al. Scene labeling with lstm recurrent neural networks[C]//*Proceedings of the IEEE Conference on Computer Vision and Pattern Recognition*. 2015: 3547-3555.

[17] Ustinova E, Ganin Y, Lempitsky V. Multi-region bilinear convolutional neural networks for person re-identification[C]//*2017 14th IEEE International Conference on Advanced Video and Signal Based Surveillance (AVSS). IEEE*, 2017: 1-6.

[18] Varior R R, Shuai B, Lu J, et al. A siamese long short-term memory architecture for human re-identification[C]//*European Conference on Computer Vision. Springer, Cham*, 2016: 135-153.

[19] Zheng L, Yang Y, Hauptmann A G. Person re-identification: Past, present and future[J]. *arXiv preprint* arXiv:1610.02984, 2016.

[20] Fu J, Liu J, Tian H, et al. Dual attention network for scene segmentation[J]. *arXiv preprint* arXiv:1809.02983, 2018.

IOP Conf. Series: Materials Science and Engineering **646** (2019) 012052 doi:10.1088/1757-899X/646/1/012052

Fusion of Deep Feature and Hand-Crafted Features for Terrain Recognition

Xue Li[a], Xin Ma[b] and Peng Song

School of Control Science and Engineering, Shandong University, Jinan, China

[a]lx_sdu@163.com; [b]maxin@sdu.edu.cn

Abstract. Terrain recognition is one of the key problems of mobile robots. It can help the robots understand the surrounding environment. With terrain prediction, the robots could realize autonomous navigation and path planning. This paper focuses on image feature selection for terrain recognition with visions. For terrain recognition tasks, feature is used to represent image information. Traditional visual features can be targeted to express the low-level information like color or texture. The deep feature is extracted by self-learning of neural network, containing richer semantic information than low-level features. There is a complementary relationship between the two. The efficiency and accuracy of terrain recognition is remarkably raised by the fusion of two above features. In the course of algorithm, combination of the off-line training model and on-line recognition model is used to identify the terrain type of the sample, which is to ensure the real-time performance. The corresponding terrain dataset--SDUterrain is established. The algorithm achieves 96% or higher classification accuracy in the experiments based on the SDUterrain Dataset, which is much higher than the single feature classification algorithm.

1. Introduction

Terrain recognition is an important research in the theory of autonomous navigation of mobile robot. Its ability to perceive the environment largely determines its ability to move autonomously in an unknown environment [1]. Visual sensors contain far more information than others and not affected by themselves. All the while, terrain recognition based on machine vision is a research hotspot of robot.

Image classification is a typical problem in machine learning. Feature is the key point. The quality of features largely determines the accuracy of classification. The complexity and variability of the field environment bring considerable challenges to terrain identification. Features which performs well in one environment is not suitable in the others. In this case, the selection of terrain features is particularly important.

Traditional machine learning relies on hand-crafted features heavily such as colors, textures, edges or shapes. Manduchi [2] proposed a color-based classification system, which recognizes the detected objects according to the preset terrain types. But color features is sensitive to illumination. Local binary patterns [3, 4] extracts the texture features from the grayscale image, which can eliminates the influence of illumination to some extent. Oliva [5] used spatial envelop feature to represent the content of images, and used Bayesian classifier to accomplish terrain recognition. Filitchkin and Byl [6] proposed the terrain classification method of SURF [7] combined with SVM [8], and apply it to a real quadruped robot-littledog, as the basis of gait planning and adjustment.

However, hand-crafted features only work in certain environment. This means that the generalization is insufficient in the unknown environment. The changing environment requires high descriptive and distinguishable features to represent the terrain image. It is not possible to do so solely on the hand-crafted features. In recent years, with the development of deep learning, it has become a new trend to complete terrain classification.

Hinton [9] first proposed the concept of deep network in 2006 and developed explosively in 2012 [10]. Deep learning has abandoned artificial design features. It can automatically learn more abstract feature at a higher level and describe the inner structure of abundant data. Deep learning shows its advantages in the terrain recognition in field environment. Liu [11] first introduced deep learning into scene terrain recognition, and proposed a terrain classification method based on deep sparse filtering network. Zhang [12] learned from [11], combining the deep convolutional neural network with the near distance learning strategy to improve the accuracy of terrain classification and make it more robust to the change of field environment. Xue [13] discusses the significance of imaging angle in terrain recognition, and proposes a differential angle imaging network, which uses both the original image and the differential image input into the convolutional neural network. And in [14], deep encoding pooling network for terrain recognition is proposed, which integrates random texture information and local spatial information in terrain images, outperforming the most advanced methods.

Moreover, the deep features in middle layers of neural network has also appeal to scholars. Donahue [15] extracted the deep features for terrain recognition in dynamic scene and aerial scene classification. Cimpoi [16, 17] improved the pooling layer and extracted the FV-CNN features. Results show that the deep features can integrate multi-scale information, and describe arbitrary shape and size of the region. It is suitable for material recognition and classification.

Inspired by [15]-[17], this paper proposed a terrain recognition method based on feature fusion. We extracted manual features, like color, LBP, and deep feature from trained convolutional neural network to represent terrain images. Finally inputting the fusion features into SVM to complete the terrain recognition.

2. Algorithm of terrain recognition

The proposed method in this paper is to solve the classification problem of several single flat terrain types with machine learning. The algorithm innovates in feature selection. We fused the hand-crafted features with deep features to improve the semantic information of the terrain features.

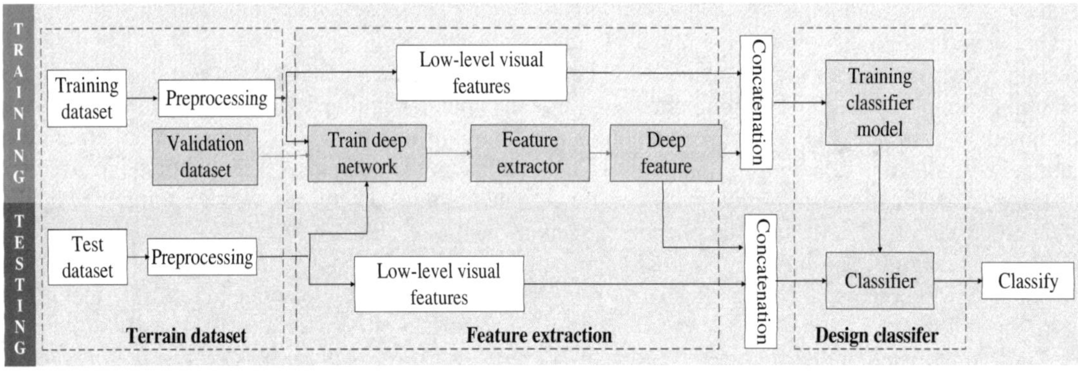

Figure 1. Terrain recognition algorithm based feature fusion.

There are two reasons for this. As mentioned earlier, traditional visual features are used to express low-level information such as colors, textures, etc. It magnifies some visual characteristics of the image which is easy to classify, but simultaneously leads to the problem that the features are not fully expressed. Only when the image is similar to the training image has excellent performance. It is difficult to distinguish all terrain types. The deep feature is extracted from the trained neural network. Its semantic information is more comprehensive than the low-level features, and the generalization is stronger. Not that the highly aggregated features is ideal. It isn't. Theoretically, deep features are highly abstract features obtained from multi-layer convolution. The performance of deep features is proportional to the size of dataset and the layers of network. On the premise of limited dataset size, high classification accuracy can only be improved by increasing layers of the network. By doing so, it must lead to the loss of some low-level information. The low-level features of the neural network are usually the basic visual information such as the color and the edge of the image, which are commonly used in traditional machine learning. Considering the above analysis, there is a semantic complementary relationship between the two. After feature fusion, more perfect feature expression can be obtained. The flow of terrain recognition algorithm in this paper is shown in figure 1.

2.1. Hand-crafted features

RGB. There are color differences between terrain types. Color features obtain the color histogram vector by counting the proportion of each color component in the whole image. By counting the RGB values of all pixels in a terrain image, the 768-dimensional color histogram features can be obtained.

LBP. Terrain images also have distinct texture properties. LBP is a global statistical texture descriptor with low computational complexity and high speed. This feature is obtained from grayscale images, which can avoid the effect of illumination changing.

CEDD. CEDD [18] combines the color and texture information of the image. It can be seen as the fusion feature of RGB and LBP. Compared with the fusion feature, the dimension is lower, and the classification accuracy is similar.

We tested the above visual features in SDUterrain dataset, which will be introduced later in this paper.

Results are showed in table 1.

Table 1. Results of composite features.

	SVM(%)	Dimension	Time(s)
CEDD	76.53	144	28.78
RGB-LBP	79.16	1024	41.02

2.2. Deep feature

The highly distinguishable features can significantly improve the accuracy of classification. Visual features such as color and texture obtain high accuracy only if the training data is similar to the test data. Images captured in the actual scene is much more complex than images in the dataset. We hope to extract more comprehensive features than hand-crafted features. As we all know, deep learning takes the lead in the field of image classification. It can be applied in two ways. One is to directly provide end-to-end classification results. Another is used to extract deep features from the middle layers of the deep learning network. Inspired by this, we extract deep features for terrain recognition.

The advantage of deep learning is that it can realize the self-learning of large-scale data, and it can get rid of the disadvantage of human intervention in traditional visual feature extraction. The structure of deep network makes it possible to extract features in layers. The deep feature of full connected layer is generally the preferred method. However, the spatial structure information of the terrain image is lost because the high dimensional feature is mapped to one directly in the full connected layer. The pooling layer is another way to extract deep feature. As shown in table 2, results of deep features from pooling layer is better than full connected layer. Therefore, the deep feature of max pooling layer is used to express the terrain image in this paper. We choose ResNet-34 [19] to train the feature extractor.

Table 2. Results of deep features.

	SVM(%)	Dimension
Full connected layers	54.5	512
Max pooling layers	94.6	512

2.3. Feature fusion of deep feature and low-level features
Lots of experiments have proved that the features extracted by trained neural network are indeed contains more semantic information than traditional visual features. Generally speaking, the larger the dataset, the better the network performs. Besides, from the premise of limited datasets, we can increase the layers of neural network to improve the performance of network [20]. Another drawback of having too many network layer is the loss of lower layer information. These portions usually are color, texture and edges as mention above. Considering this problem, we will concatenate the deep feature with the hand-crafted feature to express the terrain image. The experiment results prove that the classification ability of the feature is improved to some extent.

3. SDUterrain dataset
Dataset is the most important part of image classification. There are few published dataset in direction of terrain recognition and cannot be directly applied in this paper. We constituted a terrain images dataset captured in actual environment. Sample images are collected by Canon IXUS 175 consumer camera and Logitech Pro C920 Webcam hardware. In the field environment, the surface of terrain types is often influenced by illumination, which will frequently lead to the color or texture change. The above factors (illumination, weather, season, angle, etc.) are considered as much as possible in the process of image collection. Typical terrain in the SDUterrain is divided into six classes: sand, mud, asphalt, grass, gravel and mulch that includes 13,200 images. Samples are shown in figure 2.

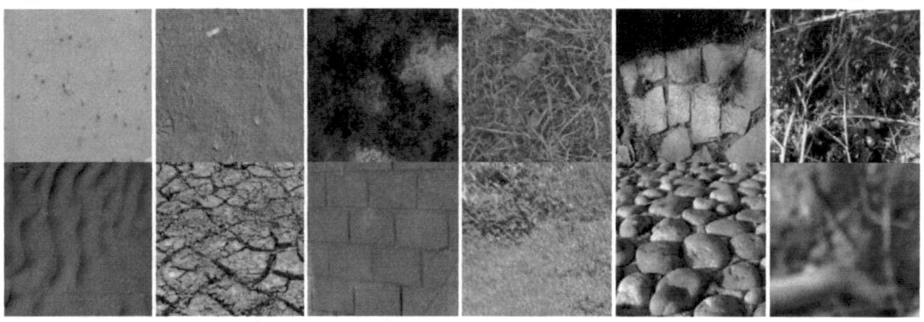
sand mud asphalt grass gravel mulch

Figure 2. Samples of six types terrain in SDUterrain dataset.

4. Experiments
We evaluate the performance of our terrain algorithm on the SDUterrain dataset.

4.1. Dividing of the SDUterrain dataset
We divide the training dataset, the validation dataset, and the test dataset in proportion with 9:1:1. Images are collected under different environmental conditions. The number of images is not exactly the same under each condition, which should be extracted at random according to the proportion to avoid vary in quality. The training dataset is used to train the deep network parameters, and the validation dataset is used to verify the performance of the training network to avoid over-fitting.

4.2. Extracting of terrain features
We first extract the traditional visual features like color and texture features. Color histogram features are extracted from RGB space, each component is 256 dimensions, a total of 768 dimensions. LBP features are extracted from grayscale maps, which lost the color information. It eliminates the effect of illumination. The feature dimension is 256. CEDD features belong to color and texture composite features, feature dimension is 144, compared with the fusion feature of color histogram and LBP, it has lower dimension and higher computation speed. It is also an ideal visual feature in terrain recognition.

Then we extract the deep features. The residual network of 34 layers is used to train as a feature extractor. We replace the last fully connected layer with the max pooling layer. The dimension of the deep feature is 512.

Last we fused the above feature in concatenation, as shown in figure 3. In this experiment, we choose two kinds of fusion feature, which are deep feature concatenated RGB color histogram feature and LBP texture feature, another is deep feature concatenated CEDD feature.

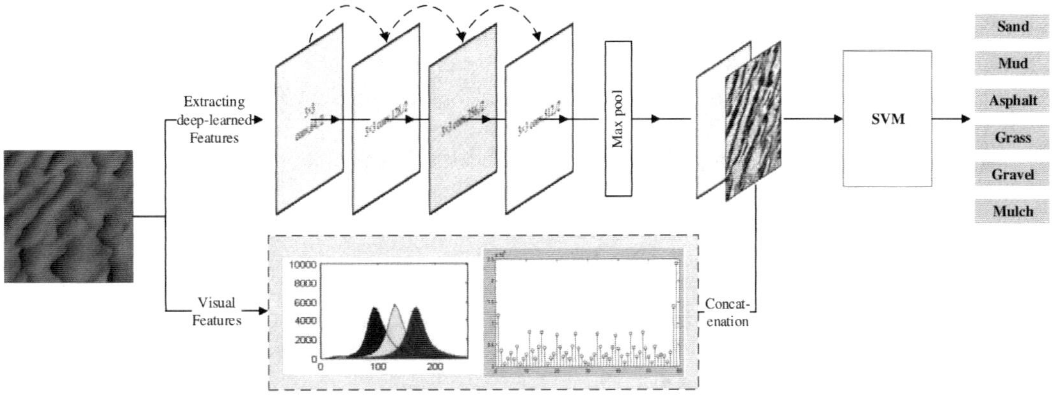

Figure 3. Low-level features and deep features extracted in terrain recognition algorithm.

4.3. Classification

Terrain features extracted from above process are inputted into the SVM for classification. Result is represented by confusion matrix. It is a kind of visualization matrix which expresses the performance of algorithm. It is usually used for supervised learning. The horizontal coordinate represents the true value of classification, and the vertical coordinate represents the expected value. The diagonal elements represent the classification accuracy of each type of terrain, and the average classification accuracy can be obtained by counting the diagonal elements. Figure 4(a) gives the average accuracy of 96.33% of the first fusion feature, and figure 4(b) gives the average accuracy of 97.00% of the other.

	Sand	Mud	Asphalt	Grass	Gravel	Mulch
Sand	**0.93**	0.02	0.04	0.00	0.01	0.01
Mud	0.03	**0.94**	0.02	0.00	0.00	0.00
Asphalt	0.02	0.00	**0.96**	0.00	0.00	0.00
Grass	0.01	0.00	0.00	**0.99**	0.01	0.00
Gravel	0.00	0.00	0.00	0.01	**0.98**	0.01
Mulch	0.00	0.01	0.00	0.00	0.01	**0.97**

(a)Deep + RGB + LBP mean accuracy: 0.9633

	Sand	Mud	Asphalt	Grass	Gravel	Mulch
Sand	**0.93**	0.02	0.04	0.00	0.01	0.00
Mud	0.03	**0.96**	0.01	0.00	0.00	0.01
Asphalt	0.01	0.01	**0.97**	0.00	0.00	0.00
Grass	0.01	0.00	0.00	**0.99**	0.01	0.00
Gravel	0.01	0.00	0.01	0.01	**0.97**	0.01
Mulch	0.00	0.01	0.00	0.00	0.01	**0.97**

(b)Deep + CEDD mean accuracy: 0.9700

Figure 4. Results based different features of the terrain recognition algorithm in SDUterrain dataset.

5. Conclusion

The terrain recognition algorithm in this paper combined the advantages of deep features and hand-crafted features. On the premise of limited dataset, we improve the performance of deep network by increasing the number of layers, and fuse the hand-crafted features to compensate the loss of low-level information. Compared with the method of traditional machine learning and deep learning, terrain recognition algorithm in this paper can further improve the classification accuracy.

Acknowledgments

This work was supported in part by the Chinese National Key Research and Development Plan (2018YFB1305803), Chinese National Natural Science Foundation (61673245), Chinese National Programs for High Technology Research and Development (2015AA042307).

References

[1] Wang T H, Sung D G, Lin P C. Terrain classification, navigation, and gait selection in a leg-wheel transformable robot by using environmental RGBD information[C]. In Proceedings of the IEEE Conference on International Automatic Control Conference. 2018: 1-1.
[2] Manduchi R, Castano A, Talukder A, et al. Obstacle detection and terrain classification for autonomous off-road navigation[J]. Autonomous Robots. 2005, 18(1): 81-102.
[3] Ojala T, Pietikäinen M, Mäenpää T. Multiresolution gray-scale and rotation invariant texture classification with local binary patterns[J]. IEEE Transactions on Pattern Analysis & Machine Intelligence. 2002, 24(7): 971-987.
[4] Mäenpää T, Pietikäinen M. Texture analysis with local binary patterns[M]. Handbook of Pattern Recognition and Computer Vision. 2005: 197-216.
[5] Oliva A, Torralba A. Modeling the shape of the scene: A holistic representation of the spatial envelope[J]. International Journal of Computer Vision. 2001, 42(3): 145-175.
[6] Filitchkin P, Byl K. Feature-based terrain classification for littledog[C]. In Proceedings of the IEEE/RSJ International Conference on Intelligent Robots and Systems. 2012: 1387-1392.
[7] Lowe D G. Distinctive image features from scale-invariant keypoints[J]. International Journal of Computer Vision. 2004, 60(2): 91-110.
[8] Cortes C, Vapnik V. Support-vector networks[J]. Machine Learning. 1995, 20(3): 273-297.
[9] Hinton G E, Osindero S, Teh Y W. A fast learning algorithm for deep belief nets[J]. Neural Computation. 2006, 18(7): 1527-1554.
[10] Krizhevsky A, Sutskever I, Hinton G E. Imagenet classification with deep convolutional neural networks[C]. Advances in Neural Information Processing Systems. 2012: 1097-1105.
[11] Liu H, Min Q, Sun C, et al. Terrain classification with Polarimetric SAR based on Deep Sparse Filtering Network[C]. In Proceedings of the IEEE Conference on Geoscience and Remote Sensing Symposium. 2016: 64-67.
[12] Zhang W, Chen Q, Zhang W, et al. Long-range terrain perception using convolutional neural networks[J]. Neurocomputing. 2018, 275: 781-787.
[13] Xue J, Zhang H, Dana K, et al. Differential angular imaging for material recognition[C]. In Proceedings of the IEEE Conference on Computer Vision and Pattern Recognition. 2017:6940-6949.
[14] Xue J, Zhang H, Dana K. Deep texture manifold for ground terrain recognition[C]. In Proceedings of the IEEE Conference on Computer Vision and Pattern Recognition. 2018: 558-567.
[15] Donahue J, Jia Y, Vinyals O, et al. Decaf: A deep convolutional activation feature for generic visual recognition[C]. In Proceedings of the International Conference on Machine Learning. 2014: 647-655.
[16] Cimpoi M, Maji S, Vedaldi A. Deep filter banks for texture recognition and segmentation[C]. In Proceedings of the IEEE Conference on Computer Vision and Pattern Recognition. 2015: 3828-3836.
[17] Cimpoi M, Maji S, Kokkinos I, et al. Describing textures in the wild[C]. In Proceedings of the IEEE Conference on Computer Vision and Pattern Recognition. 2014: 3606-3613.
[18] Zou Y, Chen W, Xie L, et al. Comparison of different approaches to visual terrain classification for outdoor mobile robots[J]. Pattern Recognition Letters. 2014, 38(38):54-62.
[19] He K, Zhang X, Ren S, et al. Deep residual learning for image recognition[C]. In Proceedings of the IEEE Conference on Computer Vision and Pattern Recognition. 2016: 770-778.
[20] Szegedy C, Liu W, Jia Y, et al. Going deeper with convolutions[C]. In Proceedings of the IEEE Conference on Computer Vision and Pattern Recognition. 2015: 1-9.

End to End Person Re-Identification Based on Attention Mechanism

Yang Li[1,2,*], Huahu Xu[1,3], and Minjie Bian[1,3]

[1] School of Computer Engineering and Science, Shanghai University, Shanghai 200444, China
[2] School of Information Technology, Shanghai Jianqiao University, Shanghai 201306, China
[3] Shanghai Shangda Hairun Information System Co., Ltd, Shanghai 200444, China

[*]E-mail: liyang162@shu.edu.cn

Abstract. This paper treats person re-identification (re-id) as a sequential model, guide person re-id with person detection, combines recurrent neural network (RNN) with attention mechanism, and proposed an end to end person re-id method for surveillance scenarios. The feature of target person is firstly extracted using ResNet, and then the feature is added to the long short-term memory (LSTM) network to guide the attention model for the region of interest in the surveillance image. Finally, combined with the information observed from the image multiple times, the most similar candidate person in the image is deduced, and the feature distance is calculated and ranked for person re-id. This paper strengthens the relationship between person detection and person re-id, and reduces the error between models. Because the number of candidate person matched with target person is reduced, this method can process person re-id task with less calculation and time. This paper also verified the effectiveness of the proposed method by experiments comparing a variety of person detection and re-id methods on several person re-id datasets.

1. Introduction

Person re-identification (re-id) aims at matching the target person from a series of person images through the appearance of persons. It can be used in many fields, such as video surveillance, suspect tracking, multi-camera person tracking, and person activity analysis. Due to the complex changes of pose, viewpoint, light, occlusion, resolution, background clutter and so on, person re-id is challenging and has attracted wide attention in recent years, and has become a rapidly developing research field [1].

Current person re-id algorithms mainly study how to match the target person image with the candidate person images which has been cropped and extracted from the complete scene image. However, in practical application scenarios, the candidate person images are usually not cropped from the complete scene images, so it is necessary to identify the target person from the complete scene images. Although person detection and person re-id are closely integrated tasks in many scenarios, but they are usually studied as two separate tasks, and the method of combining the two is not yet popular. In order to identify specific target person in surveillance scenarios, a natural method is to detect all candidate persons from the images, then match the search targets one by one, and finally select the matching targets.

In this paper, a new method of person detection and re-id based on attention mechanism is proposed, which extracts the feature of the target person as the initial memory information and pass to the person detection and re-id network to guide the process of retrieve the target person in the image. Compared with the previous methods, this paper combines the detection and re-id process more closely as an end to end solution. The introduction of attention based memory information makes the detection process more purposeful and reduces the number of re-id calculations, thus reducing the amount of calculation in the whole process.

2. Related Work

Pedestrian detection. Hand-crafted features are widely used in early years, the HOG descriptor [2] is based on local image differences, and successfully represents the special head-shoulder shape of pedestrians. A deformable part model (DPM) [3] is proposed to handle deformations and still uses HOG as basic features. More recently, the integral channel feature (ICF) detectors [4,5,6] become popular, as they achieve remarkable improvements while running fast. In recent years, convolutional neural networks (CNN) are used in pedestrian detection and make good progress [7,8,9,10]. Some works use the R-CNN architecture, which relies on ICF for proposal generation [7,8]. Aiming for an end-to-end procedure, Faster R-CNN[11] is adopted and it achieves top results by applying proper adaptations [12,13].

Person re-id. Early person re-id mainly use manually designing discriminative features [14,15], using salient regions [16] and learning distance metrics [17,18]. Most of these CNN-based methods can be divided into two categories. The first category uses the siamese model with image pairs [19,20,21,22] or triplets [23,24] as inputs. The main idea of these works is to minimize the feature distance between the same persons and maximize the feature distance between different persons. The second category regards person re-id task as a classification problem [25,26,27]. The main drawback of classification models is that they require more training data. Xiao et al. [25] propose to combine multiple datasets for training and improve feature learning by domain guided dropout. Zheng et al. [26,27] point out that classification models are able to reach higher accuracy than siamese model, even without careful sample choosing. Recently, attention mechanism [21,28,29,30] has been adopted to learn better features for person re-id.

Detection and re-id. Xu et al. [31] jointly model pedestrian commonness and uniqueness, and calculate the similarity between query and each sliding window in a rude manner. Zheng et al. [27] studied the effect of detection on re-id by combining detection with re-id using different methods. Other works propose to solve the problem in an end-to-end fashion by employing the Faster R-CNN detector [11] for pedestrian detection and share the base network between detection and re-id [1]. In [32], feature discriminative power is increased by introducing center loss during training. Liu et al. [33] improve the localization policy of Faster R-CNN by recursively shrink the search region from the whole image till achieving precise location of the target person. Xiao et al. [1] proposed a new pedestrian retrieval framework, in which detection and re-id were processed in a unified CNN. They proposed a new loss function Online Instance Matching (OIM) to train the network. In addition, they also collected and labelled a large-scale pedestrian image data set for person re-id tasks.

In this paper, the problem is treated as a sequential model. Using the information of detection to guide re-id, the recurrent neural network (RNN) and attention mechanism are combined to carry out person detection and re-id, which can reduce the computational load needed to be processed in a single cycle.

3. Methodology

3.1. Framework of Attention Based Algorithm

Traditional target detection algorithms usually deal with the whole image's visual information indiscriminately. But human beings usually will not and can't process all the information of an image in detail, just as when we are viewing a painting, although we can observe the whole painting, our attention is usually focused on a small part of the painting at a certain time period, that is, our attention to the whole picture information is not balanced. Based on this idea, this paper uses the attention mechanism [34] to carry out person detection and re-id, which can focus on the selection of interested parts from the whole image for processing. Compared with traditional methods, this method can reduce the computational load in the whole detection and re-id process.

In this paper, the problem of person detection and re-id is regarded as a sequential model problem, which is processed by combining attention model with RNN and using a memory unit containing target person information. Suppose that we can only observe an image through an observation agent, and the observation agent can only observe a part of the image at each time point. After inputting the target person to be identified, the feature is extracted as the initial memory information of the target person. The initial memory is used to help the attention model select the next observation region. The final search result can be obtained by combining the information obtained from multiple observations. Just as in reading an article, we understand the current text according to the text information we have read before, rather than dividing the information independently, that is, people's thinking is continuous. We regard the whole person detection and re-id process as a continuous process, and the RNN is widely used to deal with sequence problems. This paper uses Long Short-Term Memory (LSTM) [35] network for the task.

This section introduces the overall framework of the algorithm. The whole person detection and re-id process consists of two key parts: memory unit and neural net-work. ResNet50 [36] Part1 is used to extract the features of the target person as memory information and passed to the person re-id network. The person detection and re-id network used in this paper is a LSTM recurrent neural network using attention model. At each step of the recurrent process, only a part of the whole image can be observed. The network is guided by memory information for the observation region of the whole image. This algorithm transforms the traditional two-step process into a simple process, which reduces the number of times of calculating and ranking the feature distance of recognition, and thus reduces the amount of calculation and processing time. The block diagram of the algorithm is shown in Figure 1.

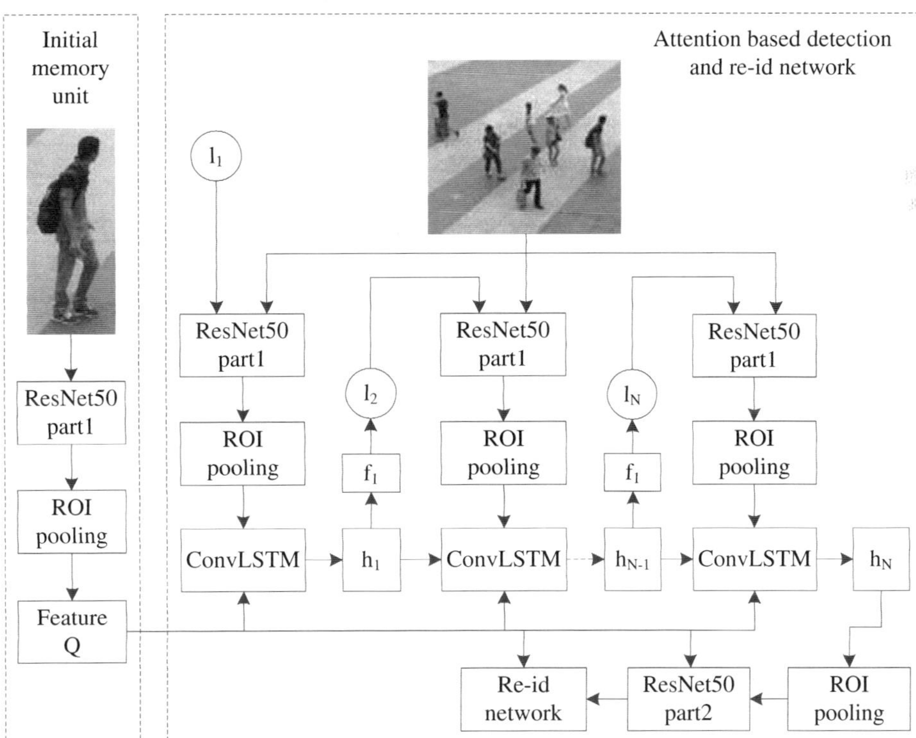

Figure 1. Block diagram of end to end attention based person re-identification algorithm.

Figure 1 shows the flow of the whole algorithm. After input the image of a target per-son, the feature extraction is carried out through the ResNet50 Part1, and the feature expression Q of the target person is obtained and passed to the detection and re-id network using attention model. The detection and re-id network in this paper is based on Convolutional LSTM (ConvLSTM) [37]. The network can only observe the image through an observation agent with limited observation range, where l_i represents the coordinates of the observation position of step i. The image and the current observation coordinate l_i constitute the observation region of step i in the cycle. The feature of current observation region is also extracted by the ResNet50 Part1 and passed to the ConvLSTM unit, the ConvLSTM unit combined with historical state h_{i-1} to generate the next state h_i. H_i determines the next observation coordinate l_{i+1} through a positioning function f_l. The final feature vector h_N is obtained after a certain number of cycles, After feature extraction through ResNet50 part2, h_N and the target person feature Q are passed to the re-id network for cosine feature distance calculation and ranking for person re-identification.

3.2. Initial Memory Unit

In the traditional process for person re-id based on surveillance scene, all candidate pedestrian images are firstly detected and cropped from surveillance images by pedestrian detection, and then all the candidate images are sent to the re-id network for feature distance calculation and ranking with the comparison with the target person.

In such a two-stage process, the feature of the target person that need to be identified is not fully utilized in the pedestrian detection stage, and is used only when com-pared with the candidate images in the re-id stage. Therefore, all pedestrians detected from the surveillance images will be used as candidate pedestrians to participate in feature distance calculation and ranking. If the detection process can be guided by the features of the target person to be retrieved, the number of person re-id processes can be reduced if the object which is similar to the target person features can be found in the image at the detection stage. Firstly, we extract the features of the target person, which can be called initial memory. We put the initial memory into the detection and re-id network, and use the initial memory to guide the next region of attention of the observation agent. In order to achieve such a process, we design an initial memory unit and use the ResNet50 to extract the feature expression of the target person. In each step of the observation process of the recurrent neural network, the feature expression is passed to the LSTM unit to help the detection and re-id network select the next region to be observed.

In this paper, two networks are used. The RNN with convolutional LSTM unit is used as the core network for detection and re-id. The ResNet50 is used to extract person image features. The ResNet-50 is divided into two parts: lower layers are used to extract the features for pedestrian detection, while the higher layers are used to extract the features for person re-id.

3.3. Recurrent Attention Model

There are two ways to process visual information, bottom-up data-driven process and top-down task-driven process. At present, most of the visual attention models are based on bottom-up, namely data-driven visual attention model. Bottom-up data-driven models usually use the underlying visual features of an image, such as colour, brightness and direction, to generate saliency maps of multi-dimensional features of an image.

Although data-driven visual attention is widely used at present, when it is applied to object detection, the effect may not be satisfactory due to the interference of the underlying visual features of the environment itself. So top-down task-driven visual model is more effective. Task-driven visual attention model is a model with prior knowledge, more targeted. Task-driven visual attention model pre-establishes visual expectations, separates the desired objects from the original image, and thus completes the extraction of regions of interest in the image. Task-driven models generally pay attention to different objects in three aspects: scenario priori information, task requirements and object features. For example, in a complex scene to find a car, and the car has four wheels, so after adding wheel features in the model, the model can quickly eliminate other interference information in the detect process, and improve the efficiency of the detection. The processing method in this paper is similar to task-driven visual attention model. Adding feature information of target person to detection and re-identification tasks is similar to generating visual expectations.

In order to add the feature information of the target person to the process, we can regard the person detection and re-identification task as a series of decision-making processes of observation agents with specific tasks and interaction with visual environment. The observation agent can only observe a part of the whole image at each time. The observation agent can extract information from a smaller region and decide how to control its perception range, such as changing the location of the observation region. Observation agents can also influence the real state of the environment by performing some actions. Since the observation agent can only observe some parts of the image at one time, it needs to learn to integrate information from different times to determine what actions to perform and where to observe at the next time. After each decision, the observation agent receives a reward value. Its goal is to obtain the maximum reward value. When the model is applied to the task of person detection and re-id, the action is detect and re-identify the target person, and the reward signal reflects whether the action is correct or not.

RNN is used in this paper to deal with above recurrent attention model, and LSTM is used to improve the traditional RNN to solve the problem of gradient disappearance and gradient explosion. In order to make more effective use of the information of the target person, the feature of the target person are added to the modified ConvLSTM unit as an independent input information. ConvLSTM is a processing structure specially designed for spatiotemporal sequence. The main disadvantage of traditional LSTM is that it uses full connection when encoding input to internal state and state transition, without considering the processing of spatial information. ConvLSTM uses convolution to make the transition between different states, which can make better use of spatial location information.

In this paper, the feature information of target person is added to ConvLSTM. On the basis of ConvLSTM unit, the algorithm adds the feature of target person extracted by Resnet-50 part1 to the input gate, forgetting gate, output gate and input information, which is used to guide the internal state of ConvLSTM to change toward the region relevant to target person, and discard the information irrelevant to target person. The feature information of the target person is independent of the processing flow of the recurrent steps, that is, the feature of the target person is a globally invariant value in the whole process of detection and re-id. In this paper, the feature information of the target person is added in the observation process, which makes the whole process of person detection and re-id more purposeful.

3.4. Training Strategy

The purpose of training is to learn parameters to maximize the reward value of the observation agent when interacting with the environment, which is a reinforcement learning strategy. The observation agent acquires the sampling of the interaction sequence through the current strategy and then changes the parameters of the agent, which increases the probability of selecting the action with high cumulative reward value and decreases the probability of obtaining low reward action.

In some situations, we know which actions are right. For example, in an object detection task, the observation agent's action in the final step must output the predictive label of the object. The object label is known in the training image dataset. After learning the optimization strategy, the observation agent can output the correct label of the training image at the end of the observation sequence. This method is a kind of supervised learning. The object detection task can be accomplished by maximizing the conditional probability of predicting the correct label of the training image.

In the person detection and re-id method described in this paper, since the environment we observe is a static image frame, we only need to transfer the h_N state to the re-id network at the last layer of the observation state for re-id matching. By using the above reinforcement learning strategy, that is, feedback the corresponding reward value through the correctness of the re-id result output by the final re-id network, to train the whole detection and re-id network.

3.5. Efficiency Analysis of the Algorithms

Compared with the method of directly connecting person detection and person re-identification in series, this algorithm can effectively reduce the amount of calculation in the whole process.

Person re-identification requires the calculation of feature extraction, feature distance and ranking between the target person image and the images of candidate persons. Therefore, when the candidate set is large, the time of re-id is long. In the traditional method of direct serialize person detection and person re-id, because the information of the target person is not fully used, all the persons detected by the detector in the surveillance video image will participate in the process of person re-id, so the calculation of the whole process is relatively large. In the real world, a person will only appear in a place at a time, so in a surveillance image, the target person will appear at most once. This method uses the features of the target person to guide the process of person detection. In a surveillance image, only one candidate person who is most similar to the target person is inferred for person re-id, thus reducing the number of persons participating in the process of person re-id, and reducing the amount of calculation and time required for the whole process.

For example, in a re-id task in which the number of images in a surveillance scene is n, the average number of persons in each image is assumed to be m. In the process of traditional two-stage person detection and re-id methods, the persons appearing in each surveillance image are detected firstly, and then, as the comparison candidates of person re-id, the feature distance between the candidates and the target persons is calculated and sorted, it needs to do such calculation with m*n candidate persons. The method in this paper only infers one candidate person in each surveillance image, that is, it only needs to calculate and rank the feature distance with n candidate persons after detection, which reduces the time of the whole process.

4. Experiments

In this section, the settings and datasets used in our experiments are firstly introduced, and then the algorithm is tested on CUHK-SYSU[1] and PRW[27] datasets, and compared with traditional methods.

4.1. Experimental Setting and Datasets

Our system is implemented with Pytorch. The ResNet50 is used for target person feature extraction, and is initialized with ImageNet pre-trained model. Conv1 to conv4_3 are used as ResNet50 part1 to extract features at detection stage, conv4_4 to conv5_3 are used as ResNet50 part2 to extract features for person re-id. In each step of the observation process of the RNN, the extracted feature is passed to the LSTM unit to help the detection and re-id network select the next region to be observed. The network is trained using SGD with a batch size of 16. The initial learning rate is 0.001, decayed by a factor of 0.1 at 60K and 80K iterations respectively and kept unchanged until the model converges at 100K iterations.

CUHK-SYSU. CUHK-SYSU is a large-scale person search database consisted of street/urban scene images captured by a hand-held camera or selected from movie snapshots. It contains 18,184 images and 96,143 pedestrian bounding boxes. There are a total of 8,432 labelled identities, and the rest of the pedestrians are served as negative samples for identification.

PRW. The PRW dataset is extracted from video frames captured with six cameras in a university campus. There are 11,816 frames annotated with 34,304 bounding boxes. Among all the pedestrians, 932 identities are tagged and the rest of them are marked as unknown persons similar to CUHK-SYSU.

4.2. Performance Comparison on CUHK-SYSU and PRW

This section compares our method with the traditional person detection methods combined with person re-id methods on CUHK-SYS and PRW datasets. In traditional methods, DPM [3], ACF [38] and Faster R-CNN [11] are selected as detectors to generate candidate regions, and DSIFT [14], LOMO [15] and IDE [27] are used as person re-id methods. In addition, it is also compared with OIM [1]. Table 1 shows the performance comparison between our algorithm and above algorithms on the CUHK-SYSU and PRW datasets.

From Table 1, it can be seen that the accuracy of our algorithm is better than most other methods. Overall, the accuracy performance is consistent between CUHK-SYSU and PRW. The accuracy of each algorithm on PRW is lower than that on CUHK-SYSU due to the poor image quality of PRW compared with CUHK-SYSU. The accuracy of our algorithm on CUHK-SYSU and PRW is slightly lower than that of OIM, but our method reduces the calculation of the whole process, at the expense of a small amount of re-id accuracy, in order to achieve faster re-id speed.

Table 1. Accuracy comparison of various methods on CUHK-SYSU and PRW.

Method	CUHK-SYSU		PRW	
	mAP(%)	Rank-1(%)	mAP(%)	Rank-1(%)
DPM+DSIFT	46.2	42.3	13.2	13.5
DPM+LOMO	59.1	67.2	15.4	49.3
DPM+IDE	61.2	67.9	22.6	50.1
ACF+DSIFT	34.1	39.8	10.7	26.5
ACF+LOMO	57.4	65.3	12.5	32.7
ACF+IDE	58.3	65.4	19.5	45.2
Faster R-CNN+DSIFT	49.6	55.3	21.8	49.7
Faster R-CNN +LOMO	69.5	76.2	23.1	51.1
Faster R-CNN +IDE	69.3	76.5	23.6	51.3
OIM	76.1	79.2	27.6	54.1
Ours	75.9	78.9	27.5	53.9

4.3. Runtime Comparison

By introducing the feature information of the target person into the process of person detection and re-id, our algorithm reduces the number of candidate person images participating in the process of calculating and ranking the feature distance, thus reducing the amount of calculation and running time of the whole process. Table 2 shows the comparison of running time between OIM and our method, which directly combines person detection and re-id on CUHK-SYSU.

Table 2. Runtime comparison on CUHK-SYS

Number of test images	OIM	Ours
100	193.2s	165.4s

From Table 2, it can be seen that the processing time of our method is improved by about 14.4% compared with OIM. Each image in CUHK-SYSU average contains about 4 to 5 pedestrians. According to the analysis in section 3.5, person re-id is a 1:N matching process. N represents the number of pedestrians in candidate set. For methods like OIM, the more pedestrians in a surveillance image, the more pedestrians will be generated in the candidate set. However, for your method, the number of pedestrians generated in the candidate set is independent of the number of pedestrians in the image, and only one candidate will be generated in each image. Therefore, the more pedestrians in each image, the speed improvement is bigger of our method compared with other algorithms. Of course, as the number of pedestrians in each image increases, the interference factors increase, and the re-id accuracy decreases slightly.

5. Conclusion

This paper designs an end to end person re-id method based on attention mechanism for surveillance image. The method regards person re-id in surveillance scene as a sequential model problem. Using LSTM network, the original two-stage process of person detection and person re-id is transformed into a unified process of detection and re-id. In this paper, we introduce an initial memory unit containing features extracted using ResNet50 for target person and combine it with person detection and re-id network to detect and re-id the target person. Compared with the two-stage process of person detection

IOP Conf. Series: Materials Science and Engineering **646** (2019) 012053 doi:10.1088/1757-899X/646/1/012053

before re-id, the method in this paper is more purposeful in the whole process. In this paper, the attention model is combined with the initial memory unit to guide the observation region of the network, which reduces the number of candidate pedestrians for person re-id, thus reducing the computational load and improve the re-id efficiency.

References

[1] Xiao, T., Li, S., Wang, B., Lin, L., Wang, X.: Joint detection and identification feature learning for person search. In: CVPR. (2017).

[2] Dalal, N., Triggs, B.: Histograms of oriented gradients for human detection. In: CVPR. (2005).

[3] Felzenszwalb, P.F., Girshick, R.B., McAllester, D., Ramanan, D.: Object detection with discriminatively trained part-based models. *IEEE Trans. PAMI* **32(9)** 1627–1645 (2010).

[4] Dollar, P., Tu, Z., Perona, P., Belongie, S.: Integral channel features. In: BMVC. (2009).

[5] Zhang, S., Bauckhage, C., Cremers, A.B.: Informed Haar-like features improve pedestrian detection. In: CVPR. (2014).

[6] Zhang, S., Benenson, R., Schiele, B.: Filtered channel features for pedestrian detection. In: CVPR. (2015).

[7] Zhang, S., Benenson, R., Omran, M., Hosang, J., Schiele, B.: How far are we from solving pedestrian detection? In: CVPR. (2016).

[8] Zhang, S., Benenson, R., Omran, M., Hosang, J., Schiele, B.: Towards reaching human performance in pedestrian detection. *IEEE Trans. PAMI* **40(4)** 973-98 (2018).

[9] Ouyang, W., Wang, X.: Joint deep learning for pedestrian detection. In: ICCV. (2013).

[10] Ouyang, W., Wang, X.: A discriminative deep model for pedestrian detection with occlusion handling. In: CVPR. (2012).

[11] Ren, S., He, K., Girshick, R., Sun, J.: Faster R-CNN: Towards real-time object detection with region proposal networks. *IEEE Trans. PAMI* **39(6)** 1137-1149 (2017).

[12] Zhang, S., Benenson, R., Schiele, B.: Citypersons: A diverse dataset for pedestrian detection. In: CVPR. (2017).

[13] Zhang, S., Yang, J., Schiele, B.: Occluded pedestrian detection through guided attention in CNNs. In: CVPR. (2018).

[14] Zhao, R., Ouyang,W.,Wang, X.: Unsupervised salience learning for person re-identification. In: CVPR. (2013).

[15] Liao, S., Hu, Y., Zhu, X., Li, S.Z.: Person re-identification by local maximal occurrence representation and metric learning. In: CVPR. (2015).

[16] Zhao, R., Oyang,W.,Wang, X.: Person re-identification by saliency learning. *IEEE Trans. PAMI* **39(2)** 356–370 (2017)

[17] Paisitkriangkrai, S., Shen, C., Van Den Hengel, A.: Learning to rank in person re-identification with metric ensembles. In: CVPR. (2015).

[18] Zhang, L., Xiang, T., Gong, S.: Learning a discriminative null space for person re-identification. In: CVPR. (2016).

[19] Ahmed, E., Jones, M., Marks, T.K.: An improved deep learning architecture for person re-identification. In: CVPR. (2015).

[20] Varior, R.R., Shuai, B., Lu, J., Xu, D., Wang, G.: A siamese long short-term memory architecture for human re-identification. In: ECCV. (2016).

[21] Liu, H., Feng, J., Qi, M., Jiang, J., Yan, S.: End-to-end comparative attention networks for person re-identification. *IEEE Transactions on Image Processing* **26(7)** (2017).

[22] Xu, J., Zhao, R., Zhu, F., Wang, H., Ouyang, W.: Attention-aware compositional network for person re-identification. In: CVPR. (2018).

[23] Ding, S., Lin, L., Wang, G., Chao, H.: Deep feature learning with relative distance comparison for person re-identification. *Pattern Recognition* **48(10)** (2015).

[24] Cheng, D., Gong, Y., Zhou, S.,Wang, J., Zheng, N.: Person re-identification by multi-channel parts-based CNN with improved triplet loss function. In: CVPR. (2016).

[25] Xiao, T., Li, H., Ouyang, W., Wang, X.: Learning deep feature representations with domain guided dropout for person re-identification. In: CVPR. (2016).
[26] Zheng, L., Bie, Z., Sun, Y.,Wang, J., Su, C.,Wang, S., Tian, Q.: MARS: A video bench-mark for large-scale person re-identification. In: ECCV. (2016).
[27] Zheng, L., Zhang, H., Sun, S., Chandraker, M., Yang, Y., Tian, Q.: Person re-identification in the wild. In: CVPR. (2017).
[28] Liu, X., Zhao, H., Tian, M., Sheng, L., Shao, J., Yi, S., Yan, J., Wang, X.: Hydraplus-net: Attentive deep features for pedestrian analysis. In: ICCV. (2017).
[29] Li, W., Zhu, X., Gong, S.: Harmonious attention network for person re-identification. arXiv preprint arXiv:1802.08122 (2018).
[30] Li, D., Chen, X., Zhang, Z., Huang, K.: Learning deep context-aware features over body and latent parts for person re-identification. In: CVPR. (2017).
[31] Xu, Y., Ma, B., Huang, R., Lin, L.: Person search in a scene by jointly modeling people commonness and person uniqueness. In: ACM'MM. (2014).
[32] Xiao, J., Xie, Y., Tillo, T., Huang, K., Wei, Y., Feng, J.: Ian: The individual aggregation network for person search. arXiv preprint arXiv:1705.05552 (2017).
[33] Liu, H., Feng, J., Jie, Z., Jayashree, K., Zhao, B., Qi, M., Jiang, J., Yan, S.: Neural person search machines. In: ICCV. (2017).
[34] Wang F, Jiang M, Qian C, et al.: Residual attention network for image classification. In: CVPR. (2017).
[35] Hochreiter S, Schmidhuber J.: Long short-term memory. *Neural Computation* **9(8)** 1735-1780 (2013).
[36] He, K., Zhang, X., Ren, S., Sun, J.: Deep residual learning for image recognition. In: CVPR. (2016).
[37] Shi X, Chen Z, Wang H, et al.: Convolutional LSTM network: A machine learning approach for precipitation nowcasting. In: CVPR. (2015).
[38] Dollár P, Appel R, Belongie S, et al.: Fast feature pyramids for object detection. *IEEE Trans. PAMI* **36(8)** 1532-1545 (2014).

IOP Conf. Series: Materials Science and Engineering **646** (2019) 012054 doi:10.1088/1757-899X/646/1/012054

Research on Facial Expression Recognition Based on Voting Model

Yang Fei* **and Guo Jiao**

Hubei University of Technology

*E-mail address: yishanyishui@163.com

Abstract. In order to improve the recognition rate of real-time classification of facial expressions,we proposed a method of facial expression recognition based on voting mechanism. Firstly, different neural network models are constructed to learn facial features. Then, the extracted features are fed into the classifier to obtain the posterior probability of various features. Finally, through the voting mechanism, the optimal decision-making level fusion is achieved to complete the facial expression classification.Experiments show that the average recognition rate of fer2013,CK+ and JAFFE database is 74.58%, 100% and 100% respectively. Compared with other recognition methods, experiental data show that this method has superior performance, improves the recognition rate and robustness of the algorithm, and ensures the universality of the algorithm.

1. Introduction

Facial expression is a common way to convey emotional information in social activities. The human face can be broken down into thousands of different expressions, but the mainstream approach divides the basic facial expressions into seven categories: happy, sad, fear, disgust, surprise, anger and neutral. Facial expressions contain a variety of information, including personal emotional status, personality characteristics and cognitive activities[1]. Moreover, many factors such as personal emotional state, psychological state and physical state are closely related to the information hidden in facial expressions. Mehrabian, a psychologist, suggests that only 7% of human communication is verbal, compared with 55% for facial expressions, which account for the most information. It can be seen that the study of facial expressions can help us get a lot of valuable and diverse information.

With the development of computer vision deep learning technology, great progress has been made in feature extraction of facial expression images[2]. At present, the main feature extraction methods are divided into two categories: feature extraction based on manual method and feature extraction based on machine learning. The main methods of manual feature extraction are LBP(Local Binary Pattern), Gabor and Hog[3]. The manual feature extraction method has good effects for specific data sets. However, the manual extraction method is not universal. When facing different data sets or data sets with a large number of pictures, its accuracy will decline. Feature extraction based on machine learning has greater generality, but it is often not accurate and takes a long time to calculate, which causes a great waste of computing resources and time[4]. When dealing with the data of large samples, the deeper network can achieve a better fitting effect. When dealing with the data of small samples, the shallow network often has a higher precision, while the deep network has a lower precision due to the lack of sufficient sample quantity support.

This paper proposes a voting mechanism model based on neural network, which integrates deep neural network, VGG19 and ResNet18, and retains the advantages of different neural networks for different sample sets, so as to further improve the original network model on the basis of retaining the original accuracy, and achieve good experimental results.

2. Related work

Traditional facial expression recognition methods generally adopt manual method to extract facial expression features and use traditional machine learning methods to determine facial expressions. In literature [5], Local Binary Pattern (LBP) was used to extract expression features to reduce the impact of facial illumination imbalance on expression recognition, and Support Vector Machine was used for expression determination. Zhao et al. [6] further enhanced the robustness of expression feature extraction by fusing LBP on three orthogonal planes, reduced the influence of facial posture and light and other factors, and determined expressions by combining k-nearest neighbors (KNN) and Hidden Markov Models (HMM).Based on the theory of non-negative matrix factorization (NMF), Zhi et al. [7] proposed the graph-preserving non-negative matrix factorization (GSNMF) method to highlight facial features and improve the recognition rate of facial expression features, and the final expression determination was also realized by KNN. In traditional expression recognition research, the efficiency of expression recognition is not considered and the accuracy of expression recognition is low due to the limitations of traditional machine learning classification methods.

As Convolutional Neural Network (CNN) shows excellent performance in computer vision fields such as image segmentation and image classification, more and more researches tend to use Convolutional Neural Network to extract facial features and improve the robustness of facial expression judgment. Kim et al. [8] determined the static expression type by integrating the results of multiple CNN in an exponential weighted decision fusion. Li et al. [9] proposed a new CNN method to maintain depth and locality, aiming to enhance discrimination between expression categories by maintaining local tightness and maximizing the gap between classes. Kample et al. [10], after analyzing algorithm differences and performance effects in multiple literatures, improved the accuracy of expression recognition by constructing cascading CNN. Although the above research improves the recognition accuracy to some extent, the complex network structure and connection mode make the training process quite tedious and difficult to achieve real-time recognition. In order to improve the recognition efficiency, Arriaga et al. [11] greatly simplified the network structure by combining the residual module and the deep separable convolution layer, so that the constructed CNN could realize the real-time recognition effect, but the model could only reach the benchmark accuracy. To sum up, the current research on facial expression recognition cannot take into account the recognition accuracy and efficiency.

3. Framework for the proposed facial expression recognition system

3.1. VGG19 and Resnet18

3.1.1 Model design. Deep convolutional neural network is used to integrate facial expression feature extraction and expression classification into an end-to-end network. VGG19 and Resnet18 were used to identify and classify facial expressions.

VGG network is a kind of classical image classification network[12]. Because it can extract image features, it is also applied to the loss function in the style transfer network. The other is used for Edge Detection of HED (Holistically - Nested Edge Detection) network[13] is developed based on VGG network. The name VGG is derived from the Visual Geometry Group at Oxford University, where the author works.

The Table 1 shows the structure of VGG19 image classification network:

IOP Conf. Series: Materials Science and Engineering **646** (2019) 012054 doi:10.1088/1757-899X/646/1/012054

Table 1.VGG19 network

ConvNet Configuration					
A	A-LRN	B	C	D	E
11 weight layers	11 weight layers	13 weight layers	16 weight layers	16 weight layers	19 weight layers
input (224 × 224 RGB image)					
conv3-64	conv3-64 **LRN**	conv3-64 **conv3-64**	conv3-64 conv3-64	conv3-64 conv3-64	conv3-64 conv3-64
maxpool					
conv3-128	conv3-128	conv3-128 **conv3-128**	conv3-128 conv3-128	conv3-128 conv3-128	conv3-128 conv3-128
maxpool					
conv3-256 conv3-256	conv3-256 conv3-256	conv3-256 conv3-256	conv3-256 conv3-256 **conv1-256**	conv3-256 conv3-256 **conv3-256**	conv3-256 conv3-256 conv3-256 **conv3-256**
maxpool					
conv3-512 conv3-512	conv3-512 conv3-512	conv3-512 conv3-512	conv3-512 conv3-512 **conv1-512**	conv3-512 conv3-512 **conv3-512**	conv3-512 conv3-512 conv3-512 **conv3-512**
maxpool					
conv3-512 conv3-512	conv3-512 conv3-512	conv3-512 conv3-512	conv3-512 conv3-512 **conv1-512**	conv3-512 conv3-512 **conv3-512**	conv3-512 conv3-512 conv3-512 **conv3-512**
maxpool					
FC-4096					
FC-4096					
FC-1000					
soft-max					

VGG19 advantages:

The structure of VGGNet is very simple. The entire network USES the same size convolution kernel size (3x3) and maximum pooling size (2x2).A combination of small filters (3x3) convolution layers is better than a large filter (5x5 or 7x7) convolution layer.It is proved that the performance can be improved by deepening the network structure.

ResNet network[14] is based on VGG19 network, which is modified, and residual element is added through short-circuit mechanism. The change is mainly reflected in ResNet directly using the convolution of stride=2 for down-sampling, and replacing the full connection layer with the global average pool layer. An important design principle of ResNet is: when the size of feature map decreases by half, the number of feature maps increases by twice, which keeps the complexity of network layer.

The Table 2 shows the structure of resnet-18 image classification network:

Table 2. ResNet18 network

Layer name	Output size	18-layer	34-layer	50-layer	101-layer	152-layer
conv1	112×112	7×7.64, stride 2				
conv3_x	56×56	3×3 max pool, stride 2				
		$\begin{bmatrix}3\times3, 64\\3\times3, 64\end{bmatrix}\times2$	$\begin{bmatrix}3\times3, 64\\3\times3, 64\end{bmatrix}\times3$	$\begin{bmatrix}1\times1, 64\\3\times3, 64\\1\times1, 256\end{bmatrix}\times3$	$\begin{bmatrix}1\times1, 64\\3\times3, 64\\1\times1, 256\end{bmatrix}\times3$	$\begin{bmatrix}1\times1, 64\\3\times3, 64\\1\times1, 256\end{bmatrix}\times3$
conv4_x	28×28	$\begin{bmatrix}3\times3, 128\\3\times3, 128\end{bmatrix}\times2$	$\begin{bmatrix}3\times3, 128\\3\times3, 128\end{bmatrix}\times4$	$\begin{bmatrix}1\times1, 128\\3\times3, 128\\1\times1, 512\end{bmatrix}\times4$	$\begin{bmatrix}1\times1, 128\\3\times3, 128\\1\times1, 512\end{bmatrix}\times4$	$\begin{bmatrix}1\times1, 128\\3\times3, 128\\1\times1, 512\end{bmatrix}\times8$
conv5_x	14×14	$\begin{bmatrix}3\times3, 256\\3\times3, 256\end{bmatrix}\times2$	$\begin{bmatrix}3\times3, 256\\3\times3, 256\end{bmatrix}\times6$	$\begin{bmatrix}1\times1, 256\\3\times3, 256\\1\times1, 1024\end{bmatrix}\times6$	$\begin{bmatrix}1\times1, 256\\3\times3, 256\\1\times1, 1024\end{bmatrix}\times23$	$\begin{bmatrix}1\times1, 256\\3\times3, 256\\1\times1, 1024\end{bmatrix}\times36$
conv3_x	7×7	$\begin{bmatrix}3\times3.512\\3\times3.512\end{bmatrix}\times2$	$\begin{bmatrix}3\times3.512\\3\times3.512\end{bmatrix}\times3$	$\begin{bmatrix}1\times1, 512\\3\times3, 512\\1\times1, 2048\end{bmatrix}\times3$	$\begin{bmatrix}1\times1, 512\\3\times3, 512\\1\times1, 2048\end{bmatrix}\times3$	$\begin{bmatrix}1\times1, 512\\3\times3, 512\\1\times1, 2048\end{bmatrix}\times3$
	1×1	Average pool, 1000-d fc, softmax				
Flops		1.8×10^9	3.6×10^9	3.8×10^9	7.6×10^9	11.3×10^9

ResNet - 18 advantages:

Compared with the traditional convolutional neural network (VGG), ResNet has the advantages of lower complexity, deeper parameter reduction, simple optimization and deeper classification accuracy.

3.1.2 Loss function design. In the design, we explore two calculation methods of loss function, the first is cross entropy loss function. The model obtained the output probability of each category after the full connection layer, but at this time, the probability was not normalized. We normalized the probability to 1 through a softmax layer, which was easier for data processing. The calculation formula of cross entropy loss function is as follows:

$$J(\theta) = -\frac{1}{m}\sum_{i=1}^{m}[y^i \log(h_\theta(x^i)) + (1-y^i)\log(1-h_\theta(x^i))] \tag{1}$$

In the softmax regression, we solved the multi-classification problem by normalizing the probability size, and class index y could take k different values (instead of 2).

The second kind of loss function is the linear SVM (Support Vector Machine) classification loss, which is abstract as the hinge loss function, and the call of hinge loss is derived from the graph of the loss function, which is a broken line.

$$J(i) = \max(0, 1 - h_\theta(x^i)) \tag{2}$$

If properly classified, the loss is 0, otherwise the loss is

$$1 - h_\theta(x^i) \tag{3}$$

SVM and Softmax classifiers are the two most commonly used classifiers. The difference is that the SVM outputs each classification score and then selects the class with the highest score. Different from SVM, Softmax classifier is a logistic classifier that is faced with the induction of multiple classifications. Its normalized classification probability is more intuitive, and its probability sum is 1.

3.2 Deep Neural Networks

DNN is deep neural network[15]. The difference between RNN (Recurrent Neural Network) circular neural network and CNN convolutional neural network is that DNN refers to the fully connected neuron structure, which does not include convolution unit or temporal correlation. This paper adopts a two-layer neural network, as shown in the Figure 1 below.

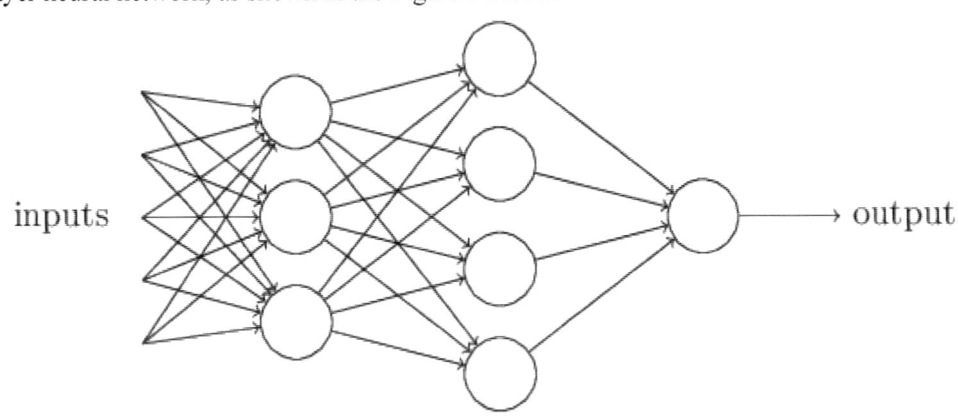

Figure 1. DNN network

3.3 Voting model

We introduce a voting mechanism model(as Figure 2) to integrate the three networks. When the same photo is input into different models, it is possible to get the same or different classification results. We select three different models mentioned above and choose the classifier which can achieve the highest accuracy. When the classification results of the three models are identical, the classification results are output. When the classification results of two models are different from the other, the classification results of most models are output. When the classification results of the three models are different, the classification results of the model with the highest accuracy are output when a single model is tested. This model retains the advantages of the original model and makes the model universal on different data sets.

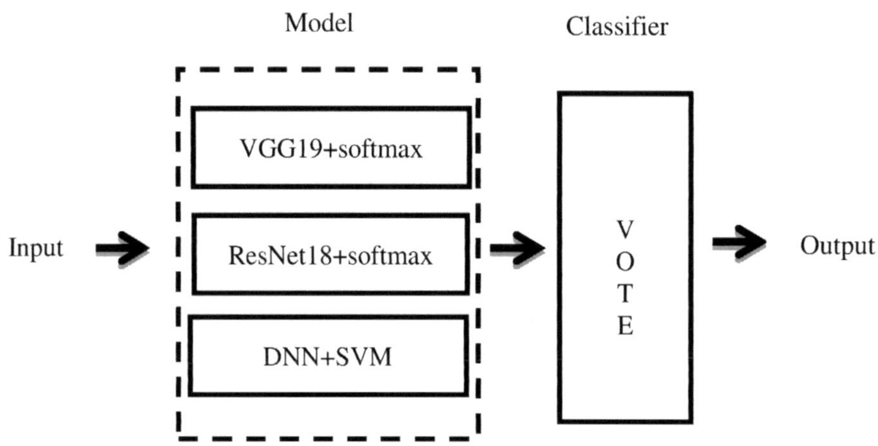

Figure 2. voting model

4. Experiment

4.1 Database

FER2013 data set consists of 28,709 training graphs, 3589 public test graphs and 3589 private test graphs. Each image is a grayscale image with 48 by 48 pixels. FER2013 has seven emojis in its database: angry, disgusted, fearful, happy, sad, surprised and neutral. This database is the data of Kaggle competition in 2013. Since it is mostly downloaded from web crawlers, there are some errors. The artificial accuracy of this database is about 65%.

The CK+ database was released in 2010 as an extension of the cohn-kanade Dataset. This database includes 123 subjects, 593 image sequences, and the last Frame of each image sequence has the label of action units, while 327 of the 593 image sequences contain the label of emotion. This database is obtained under laboratory conditions and is rigorous and reliable. CK+ is a database of comparative standards in facial expression recognition, and many articles use this data for testing.

The JAFFE data set had 213 images. Ten Japanese female students were asked to make seven facial expressions. Seven emotions include: Angry, Disgust, Fear, Happy, Sad, Surprise, Neutral. (anger, disgust, fear, happiness, sadness, surprise, neutral)

4.2 The experimental results

In order to prevent the network from over-fitting too fast, some artificial image transformations, such as rollover, rotation and cutting, can be done. This is called data enhancement. Another benefit of data manipulation is to expand the amount of data in the database, making the trained network more robust. In this experiment, at the training stage, we randomly cut images of 44*44, and randomly mirror the images, and then send them to the training. In the test phase, this paper takes an integrated approach to reduce outliers. We cut and mirror the images in the upper left, lower left, upper right, lower right and center, which makes the database 10 times larger, and then send these 10 images into the model. Then the probability is averaged, and the maximum output is the corresponding expression, which effectively reduces classification errors.

We first carried out experiments on a single model. To ensure the effectiveness of the experiment, we used 10 times of cross-validation in the experiment. The experimental results are as follows Table 3:

Table 3. experimental results

Model	classifier	Fer2013	CK+	JAFFE
VGG19	SVM	70.772	93.285	91.265
VGG19	softmax	**72.112**	94.646	92.735
ResNet18	SVM	70.549	93.343	89.285
ResNet18	softmax	71.19	94.04	90.707
Deep NN	SVM	53	**100**	**100**
Deep NN	softmax	25	22	**100**
VOTE	VOTE	**74.58**	**100**	**100**

4.3 Discussion

The depth models of VGG and Resnet18 can achieve good classification results on the FER database. And VGG's method is better than Resnet18's. Dropout can effectively reduce overfitting and improve accuracy. The Dropout method is equivalent to randomly deactivating some connections during training and then supplementing them during testing, which is equivalent to integrating several good models to make comprehensive predictions.The 10-fold cutting scheme can further reduce the error rate of identification. For the training phase, random cutting increases the data volume, which is equivalent to directly expanding the data set and slowing down the effect of overfitting. For the prediction stage, 10 times of data can predict the result at the same time, which is equivalent to the integration operation to reduce the network misjudgment.

The classification method of Softmax is better than that of SVM. Softmax takes all the categories into account, and the result of the classification, is the result of the classification in all K classes. However, the SVM method only USES binary classification each time, so it can only learn whether it is the sample, but not necessarily the correct sample, which further increases the difficulty of classification

In the effect of single model, we reached a very high level of VGG19+dropout+10crop+softmax. In the Public test set, it reaches 71.496%, and in the Private test set, it reaches 73.112%, which is also the state-of-the-art level under a single model.Deep network works better for large data sets, while shallow network works better for small data sets.The voting mechanism can retain the advantages of each model and integrate them, increasing the robustness and applicability of the model.Deep network works better for large data sets, while shallow network works better for small data sets.

5. Conclusion

This paper presents a multiple model based on voting mechanism of facial expression recognition method, the neural network model is built with expression classification experiment, respectively by using the voting mechanism to obtain all kinds of model of the time, realize the model of decision fusion, through the principle of the minority is subordinate to the majority, making multiple classifier accuracy maximization. Experimental results in FER2013, CK+ and JAFFE databases show that the method in this paper not only highlights the differences between models, but also optimizes the classification recognition rate and robustness by integrating the advantages of multiple models and optimizing the weight of each feature through voting mechanism. According to the experimental data, the recognition rate of this algorithm is improved compared with other algorithms, and the recognition universality is guaranteed. However, the time performance of the algorithm in this paper is not as good as that of the single-feature construction method because it needs multithreading to conduct three sub-experiments at the same time. In addition, in the process of expression feature fusion, in order to highlight different features and reduce the uncertainty of decision results, the next step is to introduce parallel processing technology and improve the performance of recognition methods.

References

[1] Gu, W. , Xiang, C. , Venkatesh, Y. V. , Huang, D. , & Lin, H. . (2012). Facial expression recognition using radial encoding of local gabor features and classifier synthesis. Pattern Recognition, 45(1), 80-91.
[2] Carcagnì, Pierluigi, Del Coco, M. , Leo, M. , & Distante, C. . (2015). Facial expression recognition and histograms of oriented gradients: a comprehensive study. SpringerPlus, 4(1), 645.
[3] Chao, W. L. , Ding, J. J. , & Liu, J. Z. . (2015). Facial expression recognition based on improved local binary pattern and class-regularized locality preserving projection. Signal Processing,117, 1-10.
[4] Sajjad, M. . (2017). Facial appearance and texture feature-based robust facial expression recognition framework for sentiment knowledge discovery. Cluster Computing, 21(5), 1-19.
[5] Shan C, Gong S, McOwan P W. Facial expression recognition based on local binary patterns: A comprehensive study[J]. Image and vision Computing, 2009, 27(6): 803-816.
[6] Zhao G, Pietikainen M. Dynamic texture recognition using local binary patterns with an application to facial expressions[J]. IEEE Transactions on Pattern Analysis & Machine Intelligence, 2007 (6): 915-928.
[7] Zhi R, Flierl M, Ruan Q, et al. Graph-preserving sparse nonnegative matrix factorization with application to facial expression recognition[J]. IEEE Transactions on Systems, Man, and Cybernetics, Part B (Cybernetics), 2011, 41(1): 38-52.
[8] Kim B K, Lee H, Roh J, et al. Hierarchical committee of deep cnns with exponentially-weighted decision fusion for static facial expression recognition[C]//Proceedings of the 2015 ACM on International Conference on Multimodal Interaction. ACM, 2015: 427-434.
[9] Li S, Deng W, Du J P. Reliable crowdsourcing and deep locality-preserving learning for expression recognition in the wild[C]//Proceedings of the IEEE Conference on Computer Vision and Pattern Recognition. 2017: 2852- 2861.
[10] Pramerdorfer C, Kampel M. Facial expression recognition using convolutional neural networks: state of the art[J]. arXiv preprint arXiv:1612.02903, 2016.
[11] Arriaga O, Valdenegro-Toro M, Plöger P. Real-time convolutional neural networks for emotion and gender classification[J]. arXiv preprint arXiv:1710.07557, 2017.
[12] Goodfellow I J, Erhan D, Carrier P L, et al. Challenges in representation learning: A report on three machine learning contests[C]//International Conference on Neural Information Processing. Springer, Berlin, Heidelberg, 2013: 117-124.
[13] Bulat A, Tzimiropoulos G. How far are we from solving the 2d & 3d face alignment problem(and a dataset of 230,000 3d facial landmarks)[C]//Proceedings of the IEEE International Conference on Computer Vision. 2017: 1021-1030.
[14] Ronneberger O, Fischer P, Brox T. U-net: Convolutional networks for biomedical image segmentation[C]// International Conference on Medical image computing and computer-assisted intervention. Springer, Cham, 2015: 234-241.
[15] Long J, Shelhamer E, Darrell T. Fully convolutional networks for semantic Segmentation[C]//Proceedings of the IEEE conference on computer vision and pattern recognition. 2015: 3431-3440.

Research on Intrusion Detection and Target Recognition System Based on Deep Learning

Xianwei Hu[1,3]*, Tie Li[1], Zongzhi Wu[2]and Xuan Gao[4]

[1] School of Civil and Resource Engineering, University of Science and Technology Beijing, 100083, Beijing, China
[2] State Administration of Work Safety, 100713, Beijing, China
[3] Research Institute of China National Offshore Oil Corporation Ltd., 100034, Beijing, China
[4] China National Offshore Oil Corporation Research Institute, 100028, Beijing, China

*xianweihu@outlook.com

Abstract. Intrusion target detection and recognition are of great significance to security protection of oil and gas fields. An intrusion detection system is built with the integration of infrared image acquisition module, infrared image processing module, moving target detection module and recognition module. Traditional target recognition algorithm highly relies on manual design feature extraction algorithm, which requires designer to have adequate prior knowledge, and cannot avoid the influence of subjective factors of people. Intrusion detection and target recognition system are proposed based on deep learning, which uses neural network algorithm. Deep learning model is built through feature extraction and training of acquired images of intrusion objects, and thus subsequent invasion objects are detected and recognized. Intrusion detection is achieved through simulation of human brain, which boasts of more intelligent recognition process and more accurate recognition results compared with traditional recognition method. According to applications in real scenario, the system proposed has better detection and recognition results and great practical value.

1. Introduction

Oil and gas fields have two basic attributes of value and loophole, which makes intrusion risk inevitable at oil and gas fields. The key for intrusion risk prevention at oil and gas fields is adoption of proper preventive measures to keep the risk for intrusion of oil and gas fields in a range as low as reasonably practicable (ALARP). Real-time detection and surveillance of perimeters of oil and gas fields can be realized by image acquisition, video surveillance and other technical approaches. However, it's critical to use computer for automatic recognition of intrusion targets, which can reduce manpower used for unnecessary naked-eye surveillance, monitor and recognize intrusion targets, and make alarms and prompts in time.

Traditional target recognition algorithm carries out target recognition by extracting 2D or 3D features of targets, and matching them with feature model base [1-3]. In recent years, deep learning(DL) as a relatively effective approach in the field of target recognition has been extensively used in image processing, speech recognition, object classification and other issues, but there's still no mature application plan for its use in oil and gas field. This paper has proposed an application framework for oil and gas field intrusion detection and target recognition system based on DL. It conducts feature extraction and feature training of different types of intrusion objects by DL network, and determines number of layers of network structure as well as optimal value of weight through multiple trainings, which allows it to realize fast automatic recognition of strange objects illegally approaching oil and gas fields, provides important information for monitoring and early warning system of oil and gas fields, and thus guarantees the safety of oil and gas fields in a more effective manner.

DL is both a hot research area in the field of machine learning, and an important technology for realization of artificial intelligence. Prof. Geoffrey Hinton of University of Toronto proposed the idea of DL in Science [4] for the first time in 2006, he pointed out that multi-layer artificial neural network model can better simulate human brain and conduct feature learning, and thus improve data classification capability [5]. DL refers in general to multiple hidden layers, which can simulate information processing mechanism as well as problem solving theories and methods of human brain [6,7]. It refers in particular to cascaded complex structure and artificial neural network algorithm for nonlinear processing unit to create data model [8]. DL originated from research on artificial neural network [9], which forms more abstract high-level representation through combination of low-level features to represent the class or feature of attributes, and thus find distributed representation of data. The machine learning model of multiple hidden layers can be built through DL, which can learn features of important targets based on mass training data, and eventually realize efficient classification or forecast [10].

DL is considerably studied and applied in the field of artificial neural network. There are very few researches on application of DL theories and methods in intrusion detection of Oil and Gas Field. This paper is an attempt to apply DL in intrusion detection, and build the application framework of intrusion detection and target recognition system in Oil and Gas Field.

2. Deep Learning Model

Deep Learning, as indicated in Figure 1, normally has over 2 layers of nodes, which is different from shallow neural network. It focuses on feature learning, and it has effectively overcome the commonly seen 'gradient diffusion' phenomenon of multilayer structure [11].

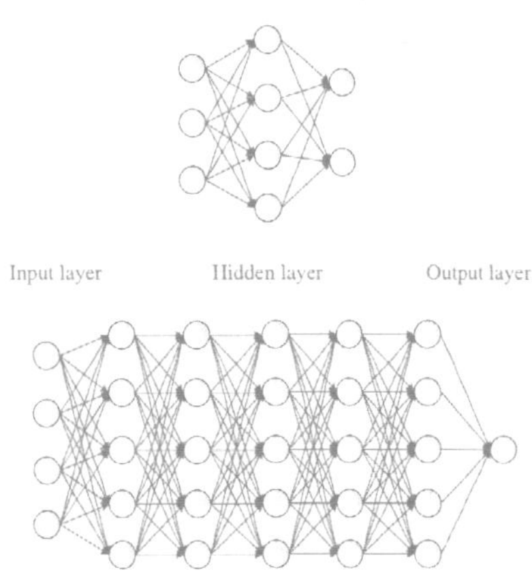

Figure 1. Neural Network and Deep Learning

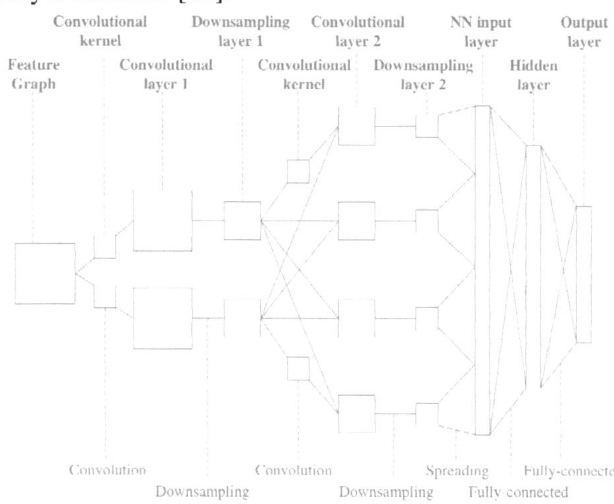

Figure 2. CNN structure

DL algorithm can automatically complete feature extraction by machine, and acquire weight of each neuron in neural network through backpropagation [12]. Convolutional neural network (CNN) [13] is one of the most extensively used network model in DL. CNN is a multilayer supervised learning neural network, which basic structure includes feature extractor and classifier. Feature extractor normally consists of several 2D plane layers of convolutional layer and pooling layer, and each plane consists of several independent neurons. Feature set of images can be acquired during convolution process, and aggregation of features of images at different positions can be conducted during pooling process to downsize data and increase speed of algorithm. Feature extractor is normally connected with classifier consists of multilayer perceptrons at its back to realize the function of integration of class discriminating feature information. Diagram of CNN structure [14] is indicated in Figure 2.

2.1 Convolutional Layer
Convolve corresponding area of convolution kernel (convolution filter) and input image in convolutional layer to get a value, calculate convolution value by constantly moving convolution kernel on the image to complete the convolution operation of the whole image, and thus output feature map of convolutional layer through a nonlinear activation function [15].

2.2 Pooling layer
In pooling layers, principle of local correlations of images is used for subsampling of output feature map of convolutional layers, which realizes data dimension reduction while keeping image feature information. Therefore, pooling layer is also called subsampling layer. Pooling layer is also known as average pooling or max-pooling based on different statistical modes [16].

2.3 Output layer
The last layer of CNN is output layer, which's also called fully connected layer. It plays the role of 'classifier' that integrates class discriminating feature information. Image features acquired after convolution or pooling operation are multidimensional features, which need to be converted to one-dimensional feature to calculate gradient. Fully connected layer can actually map tensor of any number of dimensions to tensor of designated dimensions [16]. In other words, its main role's to map features acquired after convolution and pooling operation to sample space [17], and to convert multidimensional features to one-dimensional feature.

3. Application

3.1. Design of intrusion detection system
Intrusion detection system includes infrared image acquisition module, infrared image processing module, moving target detection module and moving target recognition module, and system structure diagram is indicated in Figure 3.

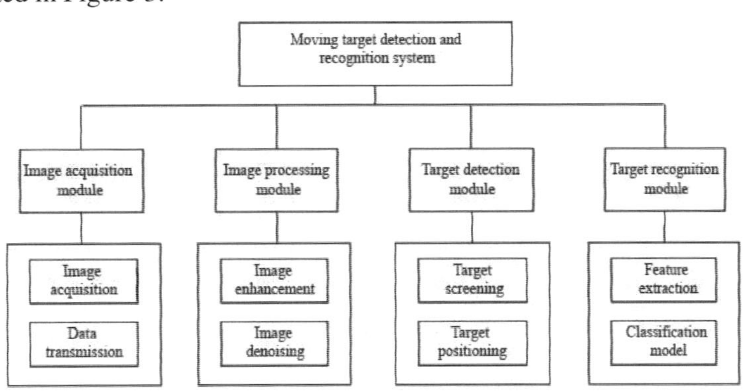

Figure 3. System Structure

Infrared image acquisition module mainly completes infrared image acquisition and conversion, it uses infrared thermal imager to acquire infrared images through 'measuring' the heat object radiates outwards, and send them to management center for signal processing through optical fiber. Infrared image enhancement module is used to conduct image enhancement and denoising processing of video images acquired to increase contrast, improve signal-to-noise ratio and resolution, strengthen detailed features of images, enhance image definition, and satisfy subsequent analysis and processing requirements of computer. Moving target detection system uses target detection algorithm to complete acquisition and positioning of targets to be recognized. Moving target recognition system trains corresponding classification model through extraction of image features and feature data, which allow it to test and recognize target objects after detecting new targets.

As the step for classification and recognition of moving targets detected, moving target recognition module is the last step of intrusion detection system, whose recognition accuracy is critical to intrusion assessment. With the influence of dynamic environments, such as camera vibration, lighting influence and background coverage, the detection and recognition of moving target is of great challenge [18].

3.2. Intrusion target recognition based on deep learning

When using traditional method for target recognition, it requires extracting features of data to be detected in the first place, and then conduct training on features extracted to acquire classification model. Traditional pattern recognition systems or machine learning systems include BP neural network, template matching method, DS evidence theory and support vector machine [8]. Those target recognition methods highly depend on selection of feature extraction algorithm. Only when the proper feature algorithm is selected, features can better describe the actual object, and target recognition boasts of higher accuracy. It requires highly specialized knowledge and a great number of engineering technologies to design feature extractor and classifier. They are not only lack of expression ability of complicated functions and classification ability of complicated issues, but also easily affected by subjective factors of people.

Target recognition method based on DL doesn't need manual design of excellent feature extractor, and features of excellent images can be acquired through self-learning process of mass data. Simulation of the mechanism of human brain for data interpretation is the key advantage of DL, and gives 'vitality' to it. DL contains multiple hidden layers, which form more abstract high-level representation through multilayer nonlinear mapping and combination of low-level features to represent the class or feature of attributes, and thus find distributed representation of data.

In short, after carrying out denoising, normalization and other preprocessing, intrusion target recognition system will convert images to be processed to data suite for analysis. It will extract features of infrared images of intrusion targets detected, extract features that can reflect nature of targets, and then acquire corresponding classification model through training on image features extracted. After that, it will carry out test on well-trained DL model, and eventually judge whether there's an intrusion and give alarms automatically. Framework structure [8] is as indicated in Figure 4.

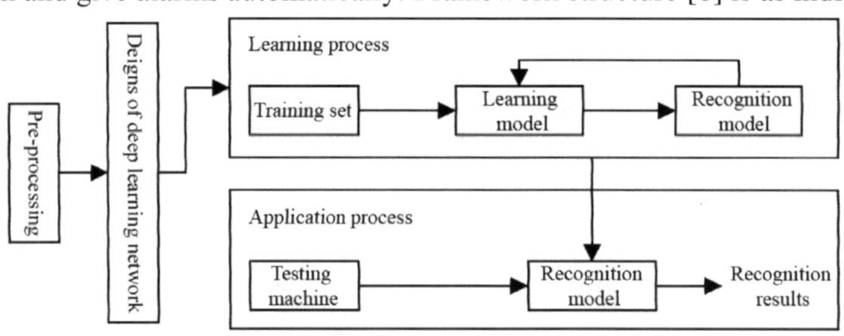

Figure 4. Intrusion target recognition framework based on DL

3.3. Structural design of intrusion target recognition network

Main intrusion risks of applicable scenarios of this article's research objects are personnel [19] and vehicles. Birds and other smaller targets pose no threat can be removed and ignored by related image processing methods, which can increase speed of algorithm.

Network structure design is a difficulty and key point for application of DL, number of input neurons, number of output neurons, number of hidden layers, number of neurons in hidden layers and learning model are needed to be selected and determined. This article uses CNN algorithm, and divides a great number of images to be recognized into training set and test set. Firstly, it extracts

IOP Conf. Series: Materials Science and Engineering **646** (2019) 012055 doi:10.1088/1757-899X/646/1/012055

features from images of training set, such as intersection points of straight lines, contour lines, peripheral points and other distinguishing features that can represent their similarity and discrimination. Secondly, it conducts feature matching through establishing corresponding relations between feature points extracted from different images. Thirdly, it conducts training and adjustment of DL model with unknown parameters to define unknown parameters in the network structure, and acquire final recognition model, such as building of personnel and vehicle target models which include size and speed of targets. When there's a new detection target, the system will run smart comparison and matching of the target extracted and database, and recognize and classify the target. When there's a moving target in surveillance area which size meets model parameters, the system will extract and track such target. The system also got the ability to filter out interferences, which allows it to eliminate inference factors in surveillance scenario, and make accurate judgments on different activities in video surveillance. The entire moving target recognition process is as indicated in Figure 5.

Target detection is carried out on the infrared images acquired through intrusion detection system, and target recognition is conducted through DL algorithm. Pedestrian and vehicle intrusions are separately detected and recognized in Figure 6 and Figure 7. The real scenarios of pedestrian and vehicle intrusion are captured by PTZ camera, see Figure 8 and Figure 9.

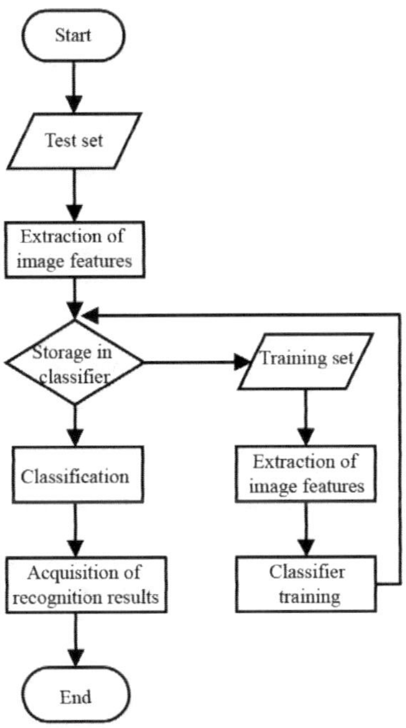

Figure 5. Moving target recognition process based on DL

Figure 6. Pedestrian intrusion–infrared image capture

Figure 7. Vehicle intrusion–infrared image capture

Figure 8. Personnel intrusion–PTZ image capture

Figure 9. Vehicle intrusion–PTZ image capture

4. Conclusions

The intrusion detection system based on DL is proposed and the application framework of intrusion detection and target recognition system in Oil and Gas Field are built. The conclusions are listed as bellow.

- The intrusion detection system in Oil and Gas Field is optimized designed, consisting of image collection, image processing, moving target detection and recognition.

The application framework proposed in Oil and Gas Field is of more accuracy compared with traditional target recognition algorithm, which can be verified both theoretically and practically. (1) Feature Extraction in traditional algorithm is low in efficiency and accuracy, while the DL in the framework proposed can work out accurate image feature through automatic configuration. (2)

Through practical application of the proposed framework, the intrusion in Oil and Gas Field is efficiently recognized and classified into categories of pedestrian and vehicle.

Acknowledgement
This research is supported by National Key Technology Research & Development Program of China (Project number: 2017YFC0804900, Subject number:2017YFC0804901).

References

[1] Ozaki N 2002 Visually enhanced CCTV digital surveillance utilizing Intranet and Internet *ISA Transactions* vol. 41 pp. 303–315.
[2] Apatcha S, Somkuan K, Preecha Y P 2010 Live sport monitoring using remote camera system *Procedia Social and Behavioral Sciences* vol. 2 pp. 49–52.
[3] Teutsch M, Muller T, Huber M, etc 2014 Low Resolution Person Detection with a Moving Thermal Infrared Camera by Hot Spot Classification *CVPR workshop* pp. 209-216.
[4] Hinton G E, Salakhutdinov R R 2006 Reducing the dimensionality of data with neural network *Science* vol. 313 no. 5786 pp. 504－507.
[5] Krizhevsky A, Sutskever I, Hinton G E 2012 Image Net classification with deep convolutional neural networks *International Conference on Neural Information Processing Systems. Curran Associates Inc.* pp. 1097-1105.
[6] Hinton G E. 2011 A better way to learn features: technical perspective *Communications of the ACM* Vol. 54 Iss. 10 pp. 94.
[7] Dean J 2019 Deep Learning for Solving Important Problems *WWW '19: The World Wide Web Conference, ACM*
[8] Cheng J Y 2018 Application Framework of Deep Learning in Radar Communication Object Recognition *Modern Radar* vol. 40 no. 8 pp. 55-59 (in Chinese).
[9] Hoffmann L 2019 Reaching new heights with artificial neural networks *Communications of the ACM* Vol. 62 Iss. 6.
[10] Zhang J J, Liu K, Khalid F, etc. 2019 Building Robust Machine Learning Systems: Current Progress, Research Challenges, and Opportunities *DAC '19: Proceedings of the 56th Annual Design Automation Conference 2019, ACM.*
[11] Yu X X, Liu Z, Geng Z Y et al 2018 A UAV Target Recognition Method for No Flying Zone Based on Deep Learning *Journal of Changchun University of Science and Technology (Natural Science Edition)* vol. 41, iss. 3, pp. 95-101 (in Chinese).
[12] Yoshinaga S, Shimada A, Nagahara H, etc 2014 Object detection based on spatiotemporal background models *Computer Vision and Image Understanding.*
[13] Lee H, Grosse R, Ranganath R et al 2009 Convolutional deep belief networks for scalable unsupervised learning of hierarchical representations *Proceedings of the 26th Annual International Conference on Machine Learning. Montreal: ACM Press* pp. 609-616.
[14] Liu Y Q, Zhan F Y, Jiang X W et al 2017 MATLAB Computer Vision and Deep Learning Practices *Publishing House of Electronics Industry* (in Chinese).
[15] Chen L F, Wu H, Cui X L et al 2018 Convolution neural network SAR image target recognition based on transfer learning *Chinese Space Science and Technology* vol. 38, iss. 6 pp. 45-51 (in Chinese).
[16] Zhou Q 2018 A method of ship moving target tracking and trajectory acquisition based on multi-features *North China University of Technology* pp. 11-12. (in Chinese).
[17] Bai Y F 2018 Research on detection and recognition of moving objects in intelligent video surveillance *Taiyuan University of Technology* pp. 14-17 (in Chinese).
[18] Kim J B, Kim H J 2002 Efficient region-based motion segmentation for a video monitoring system *Pattern Recognition Letters*
[19] Walk S, Majer N, Schindler K, and Schiele B 2010 New Features and Insights for Pedestrian Detection *In Conference on Computer Vision and Pattern Recognition* pp. 2.

Visual System for Tracking Specific Human Body Extraction of 3D Skeletal Points Based on Monocular

Baopeng Xu and Shuying Zhao*

Northeastern University, China

*568450666@qq.com

Abstract. In this paper, we introduce the Extracting specific human 3D skeleton point system based on monocular tracking. The system mainly consists of two parts. The first part is the detection and tracking of specific human body. This article uses simple online and real time tracking with a deep association metric (DEEP SORT)[1] algorithm, which is simple but effective, and meets system requirements in terms of efficiency and real-time. The second part is to extract the 3D bone points for the specific target of the tracking. We refer to Xingyi Zhou's research work[2] in this area. Utilizing the correlation between 2D pose and depth estimation subtasks, the training is end-to-end, and the algorithm introduces 3D geometric constraints to normalize 3D pose prediction, which is effective without ground truth value depth labels. In this paper, the two methods are combined by improvement, and the Extracting specific human 3D bone point system based on monocular tracking is designed. It can realize the tracking of 3D skeletal points of specific targets. The system has high practical value in human-computer interaction, virtual reality and motion recognition.

1. Introduction

The problem of human pose estimation has been extensively studied in computer vision. It has many important applications in human-computer interaction, virtual reality and motion recognition. The existing research work is divided into two categories: 2D pose estimation and 3D pose estimation. Due to the availability of large-scale 2D annotated human poses and the emergence of deep neural networks, 2D human pose estimation problems have recently achieved great success[1][3]. State-of-the-art technology enables accurate predictions in a variety of settings. The extraction of 2D bone points has made great progress, but the correct rate of understanding and analyzing human-behavior when extracting 2D bone point information is still not very high, which prompts us to have three-dimensional space for human bone points. The research of information and the practical application are more appropriate. In [4], it is also confirmed that the use of human body 3D skeletal point information is much higher than the correct rate of using 2D skeletal point information. In terms of human-computer interaction, the three-dimensional skeleton point information of the human body can be accurately obtained, so that the robot can directly respond to human instructions. In this paper, we extract the 3D skeletal points based on the information of the two skeletal points, and label the detected people to distinguish them to achieve the analysis of specific individual situations in the actual application process. It also meets the needs of human-computer interaction. In the previous computer vision to obtain three-dimensional information, they generally need binocular camera, multi-camera camera, or through the method of structure from motion (SFM)[5]. In our system, we use the deep neural network to obtain the 3D bone points of the human body through the monocular camera. The requirements for visual equipment are simpler.

2. 3D Skeleton point extraction module

For our mission, our proposed network architecture as shown in figure 1.

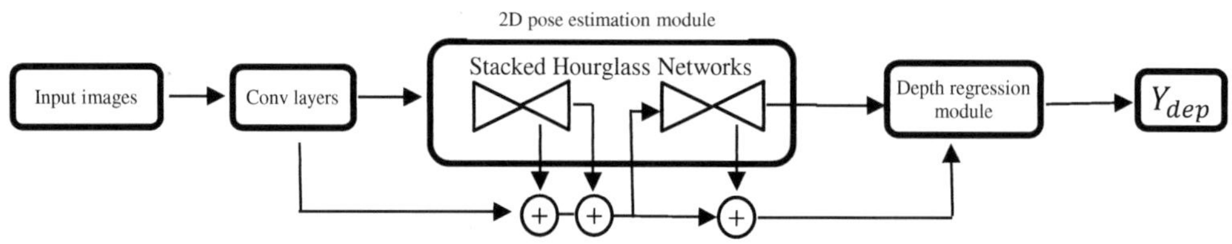

figure 1. Proposed network architecture

Here is a shallow stacking hourglass[1] model with stack=2. The deep regression module contains 4 consecutive residual and regression modules that can be considered as half hourglass.

The network is trained in a two-part datasets, the first part of images in the wild with only 2D ground truth, the second part of images in the lab with 3D ground truth.

2.1. 2D Skeleton point Estimation Module

Here we use the stacked hourglass [2] model as our system's 2D bone point extraction module, the network output is J low-resolution heat map[3], J is the number of joints. Each heat map represents the probability distribution of a joint. The peak position on these heat maps is the coordinate position of the 2D joint point we want to predict. Because these heat maps are also easily integrated with other deep layer feature maps, they serve as input to deep regression module.

To train this module, the loss function is:

$$L_{2D}(\hat{Y}_{HM},\hat{Y}_{2D}) = \sum_h^H \sum_w^W (\hat{Y}_{HM}^{(h,w)} - G(Y_{2D})^{(h,w)})^2 \quad (1)$$

The loss measures the L^2 distance between the predicted heat-maps Y_{HM} and the heat-maps $\mathrm{G}(\mathrm{Y}_{2D})$ rendered from the ground truth Y_{2D} through a Gaussian kernel [3].

2.2. Depth Regression Module

If only the position of the 2D joint point is used as the input of the depth prediction, and then the training is performed according to the corresponding 3D annotation in the datasets, but if you do this, the result is not certain. because, in general, a single 2D skeleton has multiple 3D interpretations, so as above Here, we combine the J heat maps and other deep layer feature maps output by the 2D bone point extraction module by upsampling resize as input to the 3D depth regression module. J represents the number of joints. These features, which extract semantic information at multiple levels for 2D pose estimation, provide additional cues for 3D pose recovery.

According to Zhou et al[2] research, a new loss caused by geometric constraints[6] is proposed for weakly labeled data sets. In the absence of a ground truth depth label, this geometric constraint is used as an effective regularization of depth prediction. It is based on the fact that ratios between bone lengths remain relative fixed in a human skeleton (e.g., upper/lower arms have a fixed length ratio, left/right shoulder bones share the same length).

Specifically, let R_i be a set of involved bones in a skeleton group i , Here we consider four sets of bones: R_{arm}={left/right lower/upper arms}, R_{leg}={left/right lower/upper legs} , $R_{shoulder}$={left/right shoulder bones},R_{hip}={left/right hip bones}.

We let L_e represent the length of the bone e, and $\overline{L_e}$ the normal bone length. This value is set to the average of the dataset Human3.6M used for training. The ratio of each bone $\frac{L_e}{\overline{L_e}}$ in each group R_i should be maintained. So we set the loss function of the geometric constraint of the bone to the sum of variance among $\{\frac{L_e}{\overline{L_e}}\}$ $e \in R_i$:

$$L_{geo}(\hat{Y}_{dep}|Y_{2D}) = \sum_i \frac{1}{|R_i|} \sum_{e \in R_i} (\frac{L_e}{\overline{L_e}} - \overline{r_i})^2 \quad \text{Where } \overline{r_i} = \frac{1}{|R_i|} \sum_{e \in R_i} \frac{L_e}{\overline{L_e}} \tag{2}$$

The length of the bone is a function of the position of the joint, which in turn is a function of the predicted depth. Therefore, L_{geo} is continuous and differentiable with respect to Y_{dep}. This allows us to continuously optimize the geometric loss. The loss of the depth regression module is:

$$L_{dep}(\hat{Y}_{dep}|I, Y_{2D}) = \begin{cases} \lambda_{reg} \|Y_{dep} - \hat{Y}_{dep}\|^2, if \ I \in \gamma_{3D} \\ \lambda_{geo} L_{geo}(\hat{Y}_{dep}|Y_{2D}), if I \in \gamma_{2D} \end{cases} \tag{3}$$

2.3. Training
We combine equations (1), (2), and (3) as the loss function of training:

$$L(\hat{Y}_{HM}, \hat{Y}_{dep}|I) = L_{2D}(\hat{Y}_{HM}, Y_{2D}) + L_{dep}(\hat{Y}_{dep}|I, Y_{2D}) \tag{4}$$

Since the network consists of two modules and there is a highly nonlinear nature that the geometric constraints cause loss, we divide the training process into three steps. In the first phase, we use the 2D dataset to train the 2D bone point extraction module. The second phase uses the 2D and 3D datasets to initialize the 3D bone point extraction module and fine tune the 2D bone point extraction module. At this step we do not enable geometric constraints. The third phase activates the geometric constraints.

2.4. Datasets
MPII-training. MPII dataset[7] is a large field dataset, image data is obtained from online video, and artificially annotated 16 2D joint points in the image. The data set contains 25,000 training pictures and nearly 3,000 verification pictures, and the human body has a border annotation in the image.

Human-3.6M. Human 3.6M datase[5] is widely used in 3D human pose estimation. This data set contains 3.6 million images captured by the motion capture system in an indoor environment. We reduce redundancy by interval sampling, according to the standard scheme uses five subjects (S1, S5, S6, S7, S8) as training set and (S9, S11) as our test[8] .

2.5. An exam Result display
An exam Result display as shown in figure 2.

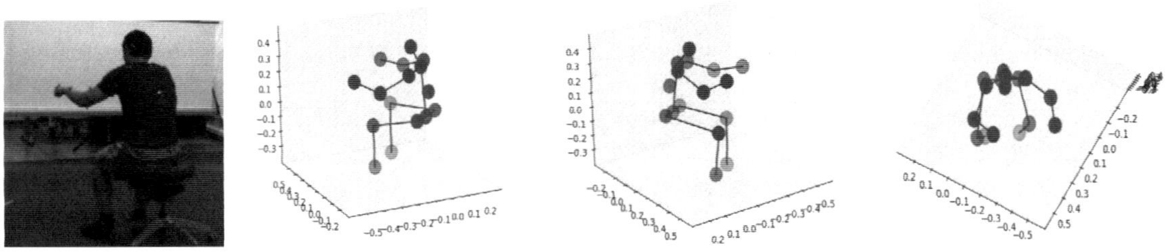

figure 2. *3D skeleton point extraction results*

3. Tracking Module

3.1. Track Handling and State Estimation
For our tracking module, we draw on the DEEP SORT algorithm[9], which uses a standard Kalman filter with a constant velocity motion and a linear observation model. The target we need to track contains the center position of the bounding box (u; v), the aspect ratio γ , height h and their

corresponding velocity in image coordinates. Among them, we use the boundary coordinates (u, v, γ, h) as the direct observation of the object state.

For a target a we want to track, we first set a threshold Amax and then calculate the number of frames since the last successful measurement of the association a. This counter is incremented during the Kalman filter prediction, when the target of the tracking is associated with the predicted amount. Reset to 0. But when the value of the counter is greater than the threshold Amax, means our mission target has left the scene we observed. Then remove from the tracking target collection. For every detection target that cannot be associated with an existing tracking target, we set these targets as tentative targets in the first three frames, and delete those targets if they are not successfully associated with the predicted values in these three frames. We need to explain that the traditional way to resolve the correlation between predicted Kalman state and newly arrived measurements is to use the Hungarian algorithm to solve the allocation problem. We use a combination of two appropriate indicators to integrate motion and appearance information and establish judgment conditions.

In terms of motion information, we can use the Mahalanobis distance between the predicted Kalman state and the new arrival amount to determine whether or not to correlate.

$$D^m(i,j) = (d_i - y_i)^T {S_i}^{-1}(d_i - y_i) \quad (5)$$

Mahalanobis distance away from the standard deviation of the average track position estimate how to consider the state detected by measurement uncertainties. We need to set a threshold for the Mahalanobis distance to determine whether it is related. This threshold is calculated by calculating the inverse $\gamma 2$ distribution when the confidence interval is 95%. For our four-dimensional measurement space, the corresponding Mahalanobis threshold is 9.4877. If the association between the i-th track and the j-th detection is acceptable, the evaluation is 1.

$$b_{i,j}^{(1)} = 1 \ if \ D^m(i,j) \le t^{(1)} \quad (6)$$

When the motion state of the target is relatively stable, the combination of Kalman prediction and Mahalanobis distance is a good correlation measure, but the prediction state derived from the Kalman filter frame is only a rough estimate of the target, when occlusion occurs or this metric becomes very inaccurate when the movement is unstable. So we introduce the appearance descriptor r_i and measure the minimum cosine distance between the i-th track and the j-th detection. This requires us to save a gallery $R_k = \{r_k^{(i)}\}_{k=1}^{L_k}$ of the last Lk = 100 associated appearance descriptors for each track k.

$$D^{cos}(i,j) = \min\left\{1 - r_j^T r_k^{(i)} \middle| r_k^{(i)} \in R_i\right\} \quad (7)$$

We also set a threshold for the cosine distance to determine if it is related.

$$b_{i,j}^{(2)} = 1 \, if \ D^{cos}(i,j) \le t^{(2)} \quad (8)$$

Appearance descriptors are derived from convolutional neural networks, In simple terms, using a convolutional neural network to extract the multidimensional feature vector of the target. so the appropriate threshold is obtained from the training set. We will introduce the structure of specific network in the following sections.

The Mahalanobis distance indicator determines the possible object position information based on motion, which is very effective for short-term prediction. The cosine distance is an indicator that takes into account the appearance information. If the motion state is unstable, the appearance information is useful for recovering the identity after occlusion. We use the weighted combination of these two metrics.

$$C_{i,j} = \lambda D^m(i,j) + (1 - \lambda) D^{cos}(i,j) \quad (9)$$

The algorithm takes into account that when the target we are tracking is occluded for a period of time, the Kalman filter adds more uncertainty to the target prediction, especially when the two targets that are being tracked compete for the same detection, the Mahalanobis distance indicator will How to add more uncertainty, so a matching cascade was introduced to solve this problem.

As input we provide the set of track T and detection D indices as well as the maximum age Amax. Then calculate the association cost to select the set that can be associated, then we iterate the age

Amax to solve the linear assignment problem in the age increment. Next, the tracking target set Tn associated with the detection in the last n frames is selected, next we solve the linear assignment between the tracking target and the mismatch detection in the set Tn, and finally update the matched and unmatched detection sets and return.

3.2. Deep Appearance Descriptor

According to the above, the appearance descriptor of the algorithm is obtained by a neural network trained by[10] a large pedestrian recognition data set containing 1.1 million images of 1261 pedestrians. The advantage of using deep neural networks is that they can be accelerated with GPU, which is very suitable for online tracking.

The neural network applied in the algorithm first accesses two convolutional layers and one maximum pooling layer, followed by six residual blocks, and maps the extracted 128-dimensional global features to the dense layer. It fits the variables we need to calculate the cosine distance.

3.3. Result display

Result display as shown in figure 3.

Figure 3. Tracking module tracking results

4. 3D Skeleton Point Extraction Modul and Tracking Module Fusion

Our main goal is to take advantage of the capabilities of the two modules above to achieve our task of tracking specific targets and extracting their 3D bone points. For our follow-up work, it provides the basis for the action recognition, human-computer interaction and other work of specific targets.

The system uses YOLOV3[8]to detect the human body. But there are two issues that need to be faced in the process of convergence. One is that the target extracted by our tracking module returns in the form of a bounding box, but the border we get sometimes cannot completely contain the target. We have statistics for all test sets, and the tracking module the result of the operation, manual labeling measurement, and the curve fitting, the initial solution is to expand the length and width of the border to the original 1.126 according to the original aspect ratio of the extracted bounding box.

The second problem is that the general shape of the human body is rectangular, and the picture input by our 3D extraction module is fixed 256 × 256 size. This can cause serious deformation of the human body and cause huge errors. The solution we take here is to scale or expand a long side of the bounding box to 256 pixels according to the original bounding box aspect ratio, and to fill or expand the short side with less than 256 pixels to complement the white pixels.

We combine the two modules described in this article to build our visual system for tracking specific human body extraction of 3D skeletal points based on monocular. The overall work flow chart of the system as shown in figure 4

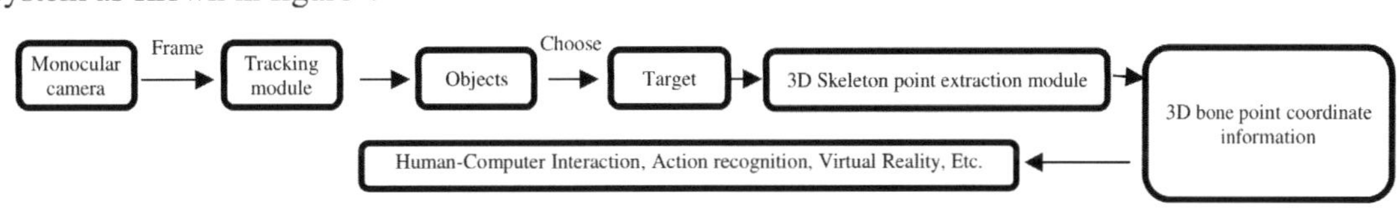

Figure 4. The overall work flow chart of the system

5. Experiment And Discussion

5.1. Single target tracking fusion experiment

Single target tracking fusion experiment as shown in figure 5

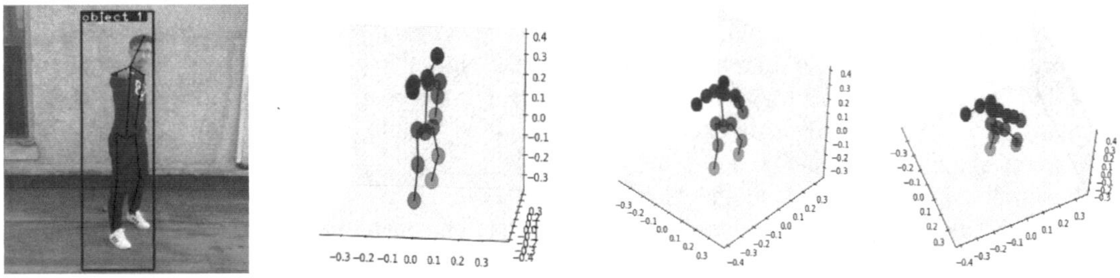

(a)Input (b) Front view (c) Side view (d) Top view

Figure 5. Single target tracking extraction skeleton point

5.2. Multi-target tracking fusion experiment

Multi-target tracking fusion experiment as shown in figure 6.

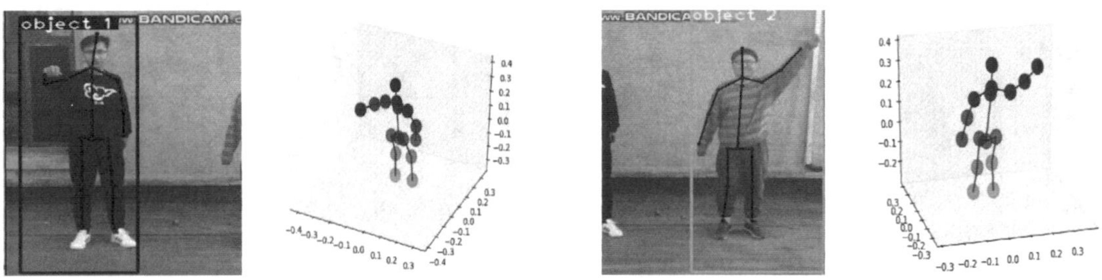

(a) Object1 (b) Object2

Figure 6. Track target with Object1 and Object2 and extract bone points

5.3. Discussion

At present, there are still many improvements in our system. First of all, the network of our system is not end-to-end. The next work can use the extraction module to extract the network of apparent information and the extraction features used by 3d skeleton points. The network is combined to strive for end-to-end training. Currently we are using single frame prediction, and we have discarded important time information. The next work can consider how to use the information in time. Another point is that our system needs to be configured to run on our robots. We can consider deploying a large amount of computing to the cloud processor while the system is running. This is also the trend of the development of robots.

Acknowledgment

Thanks to Professor Zhao Yuying for his careful guidance during the research process, and also to Xin Wu, Zhichhao Si for their help in the data collection and statistics section.

References

[1] A. Bewley, Z. Ge, L. Ott, F. Ramos, and B. Upcroft. 2016. Simple online and realtime tracking. In 2016 IEEE Int. Conf. on Image Proces. 3464–3468

[2] X. Zhou, Q. Huang, X. Sun, X. Xue, and Y. Wei. 2017. Towards 3D human pose estimation in the wild: A weakly-supervised approach. In IEEE Conference on Computer Vision and Pattern Recognition (CVPR).

[3] Shih-En Wei, Varun Ramakrishna, Takeo Kanade, Yaser Sheikh. 2016. Convolutional Pose Machines. In IEEE Conference on Computer Vision and Pattern Recognition (CVPR). 4724-4732.

[4] Sijie Yan, Yuanjun Xiong, Dahua Lin. 2018. Spatial Temporal Graph Convolutional Networks for Skeleton-Based Action Recognition. Thirty-Second AAAI Conference on Artificial Intelligence.

[5] A. Irschara, C. Zach, J.-M. Frahm, and H. Bischof. 2009. From structure-from-motion point clouds to fast location recognition. In IEEE Conference on Computer Vision and Pattern Recognition (CVPR).

[6] Akhter, I., Black, M.J. 2015. Pose-conditioned joint angle limits for 3D human pose reconstruction. In: IEEE Conference on Computer Vision and Pattern Recognition, CVPR, pp 1446–1455.

[7] M. Andriluka, L. Pishchulin, P. Gehler, and B. Schiele. 2014. 2d human pose estimation: New benchmark and state of the art nalysis. In IEEE Conference on Computer Vision and Pattern Recognition (CVPR).

[8] Li S., Chan A.B. 2015. 3D Human Pose Estimation from Monocular Images with Deep Convolutional Neural Network. In: Cremers D., Reid I., Saito H., Yang MH. (eds) Computer Vision -- ACCV 2014. ACCV 2014. Lecture Notes in Computer Science, vol 9004. Springer, Cham.

[9] N. Wojke, A. Bewley, and D. Paulus. 2017. Simple Online and Realtime Tracking with a Deep Association Metric. CoRR, abs/1703.07402.

[10] Jonathan Tompson, Ross Goroshin, Arjun Jain, Yann LeCun, Christoph Bregler. 2015. Efficient Object Localization Using Convolutional Networks. The IEEE Conference on Computer Vision and Pattern Recognition (CVPR), 648-656.

IOP Conf. Series: Materials Science and Engineering **646** (2019) 012057 doi:10.1088/1757-899X/646/1/012057

Research on Action Recognition Method Based on Weighted DTW Algorithm

Zhang Zhen[1,*] and Zhang Ya[1]

College of Mechatronic Engineering , North University of China , Taiyuan 030051, China

[*]E-mail: zbzz2012@163.com

Abstract: Action recognition technology is an important part of artificial intelligence. In order to improve the rate of action recognition, this paper designs an action recognition method based on Weighted DTW algorithm. Firstly, the Kinect2.0 is used to obtain the three-dimensional data of the human joint points. Then the quaternion method is used to define the action sequences. The weight of joint is calculated according to the participation in different types of action. Then the improved DTW algorithm is used to design the action recognition experiment. The experimental results show that the action recognition algorithm designed in this paper has better recognition rate and timeliness than traditional DTW algorithm and F-DTW algorithm.

1. Introduction

With the rapid development of the artificial intelligence industry, action recognition technology has received more and more attention[1]. Human action recognition technology can be widely used in many fields including public place situation recognition, medical and motion monitoring, and new human-computer interaction[2]. However, the human body has the features of high mobility, huge individual differences and disorder[3]. In the process of action recognition, the camera is affected by light and shadow changes, visual angle, and picture clarity. This series of random factors makes the study of human action recognition very complicated[4].

Depth sensors have developed rapidly in recent years. And the problem that the precision is insufficient when we use two-dimensional images to study action recognition is gradually solved. Kinect2.0 is a relatively mature depth sensor in recent years. Compared with the previous generation, Kinect2.0 adopts the TOF ranging principle, which is almost immune to illumination changes and textures when acquiring external action information[5]. This undoubtedly provides great help for the study of human action recognition.

The algorithm research of action recognition is of great significance for the development of human body gesture recognition. Samsu Sempena[6] et al. used a dynamic time warping method to compare video input to a defined action list. This method requires a lot of calculations when analyzing complex actions performed by multiple people ,which will reduce the efficiency of recognition. T. Vajda[7] proposed a motion recognition method based on fast dynamic time warping and feedforward neural network. This approach speeds up the identification process by introducing constraints that shorten the length of the time series. However, when the test sequence differs greatly from the action template, its recognition accuracy will decrease.

In this paper, we propose a weighted calculation method based on the participation of joint points in different action sequences, and realize the recognition of action sequences using the Weighted DTW algorithm in the experiment. Experimental results show that the algorithm has good recognition accuracy and efficiency.

2. Human action description

2.1. Feature vector establishment

In the process of realizing human action recognition, it is necessary to establish an action model as a template for identifying objects. The human skeleton obtained by the Kinect 2.0 sensor contains the pose information of the joint point[8]. As figure 1 and figure 2 show, the whole structure of skeleton continues from SpineBase as the root joint to head and extremities. In the process of identification, it is not enough to set up the action model by simply obtaining the joint point data. Converting the corresponding joint point data into action features that can be recognized is the basis for implementing action recognition.

Considering that the activity of spine changes minimally when making different movements, we use SpineBase as the reference point to establish the action feature, and calculate the joint direction data from the data flow acquired by the sensor. We define a continuous sequence of action by creating multiple joint-direction data features included in multiple frames as a set of feature vectors.

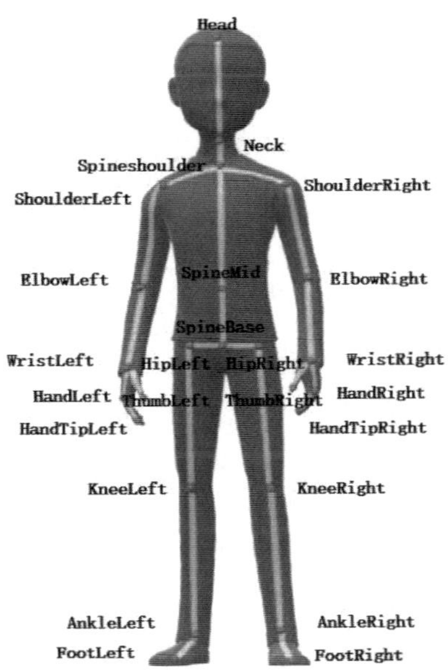

Figure 1. Skeleton map recognized by Kinect2.0.

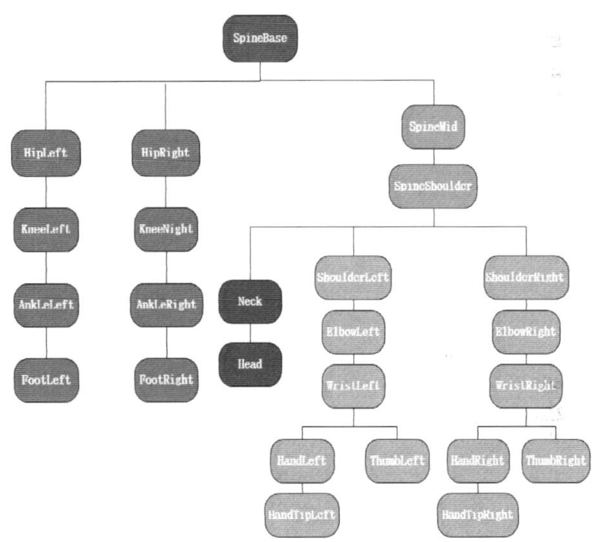

Figure 2. The joint hierarchical structure

2.2. Quaternion representation

A quaternion is a complex number with three imaginary parts, which is expressed as follows:

$$q = xi + yj + zk + w \tag{1}$$

We choose quaternions to describe the joint pose state. On the one hand, compared to Euler angles, the quaternion can avoid the gimbal lock phenomenon when the degree of freedom is insufficient caused by coincidence between two relatively rotating plane. On the other hand, quoting a quaternion can reduce the amount of calculation and the space occupied by the storage of calculation[9].When using kinect2.0 to get the joint data stream, we can call the IBody::GetJointOrientations function to get the quaternion part which contains the actions information as shown in figure 3.

We can calculate the distance between quaternions from the acquired data. Figure 4 is a schematic diagram of calculating the quaternion distance. q_1, q_2 are two unit quaternions. S_1 rotates to S_2 by $q_2q_1^{-1}$.If $q_2q_1^{-1} = [w, x, y, z]$, then define the distance between q_1 and q_2 as:

$$d(q_1, q_2) = \arccos w \tag{2}$$

We can define the method of action recognition by using $d(q_1, q_2)$ which indicates the direction and distance difference of q_1 to q_2.

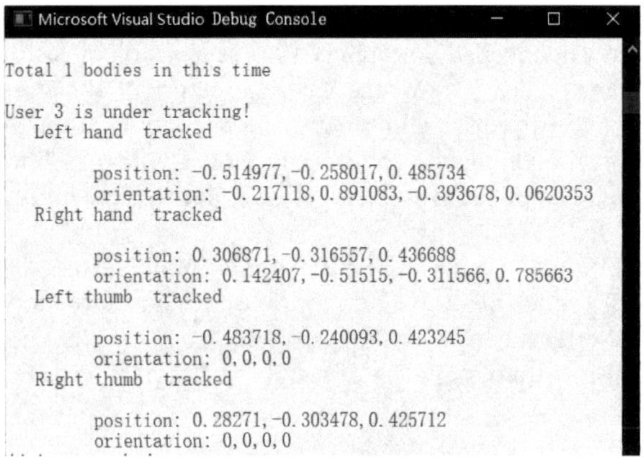

Figure 3. Quaternions data acquisition.

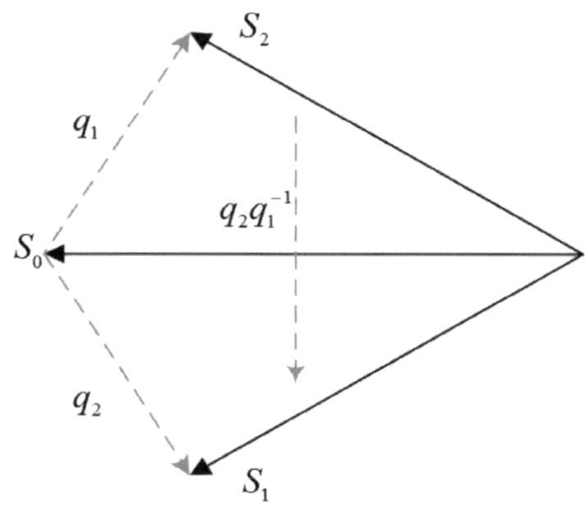

Figure 4. Quaternions distance calculation.

3. Research on action recognition algorithm

Considering that we have used data changes of multiple joint direction to define the action sequence, to implement action recognition is the process of identifying and matching for a sequence of consecutive actions that have been defined. The action process is time-ordered. In the process of action recognition, the sequence of actions of the recognition target often differs from the sequence of action templates over time due to the individual differences and the randomness of the recognition object. Therefore, it is necessary to use the corresponding algorithm to achieve target matching in the research of action recognition.

3.1. Standard Dynamic time warping

The Dynamic time warping algorithm can be effectively applied to the identification match between two action sequences. The main idea of DTW is to extend and compress two time series, and obtain the minimum warping path between the two sequences in dynamic matching[10].

Define two sequences of actions, where F is the template sequence and T is the test sequence, with lengths m and n:

$$\mathrm{F} = (f_1, f_{2,} \ldots f_i \ldots f_m) \tag{3}$$

$$T = (t_1, t_{2,} \ldots t_j \ldots t_n) \tag{4}$$

f_i and t_j are sets of joint direction data related to the i th frame and the j th frame of the action sequences F and T, respectively. Usually m≠n, therefore, we use the dynamic programming method to create an $m \times n$ matrix grid. The elements (i, j) in the matrix, which can be expressed in Euclidean Metric , represent the similarity between the eigenvalue f_i of the i th frame in the template sequence and the eigenvalue t_j of the j th frame in the test sequence.

$$d(f_i, t_j) = (f_i - t_j)^2 \tag{5}$$

We use the DTW method to adjust a curve with the highest similarity from w_1 to w_k in the square, which is the warping path as shown in figure 5. Define the warping path as W, and the element g in W is $w_g = (i, j)_g$, then:

$$W = w_1, w_2, w_3 \ldots w_g \ldots w_k \quad (6)$$

$$\max(m, n) \le k \le m + n - 1 \quad (7)$$

W satisfies the minimum matching distance between time series F and T:

$$DTW(F, T) = \min \left\{ K^{-1} \left(\sum_{k=1}^{K} w_k \right)^{1/2} \right. \quad (8)$$

K is the compensation coefficient for the planning path of different lengths.

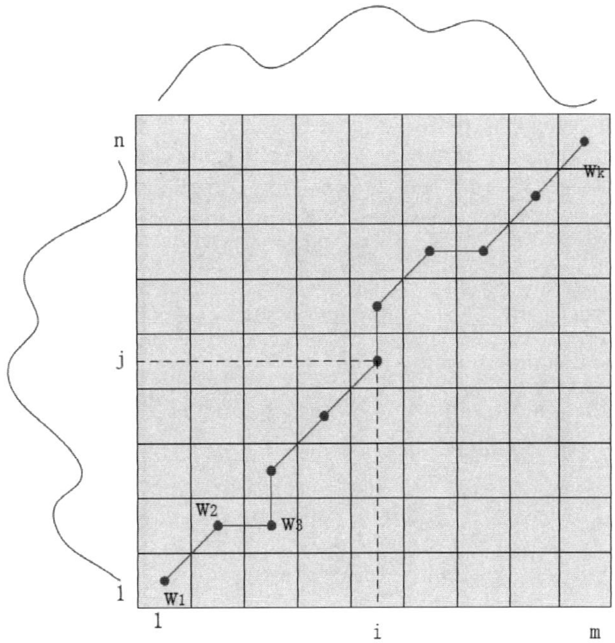

Figure 5. The best planning path by DTW

Meanwhile, We needs to meet three constraints as below.

3.1.1. Boundary conditions. The order should be specified from A to B regardless of the length of the sequence of actions, which means the planning path is from the bottom left to the top right.

3.1.2. Continuous conditions. It is forbidden to cross a point in the sequence to match. The next point in the path $w_g = (i_g, j_g)$ need to satisfy $i_g - i_{g-1} \le 1$ and $j_g - j_{g-1} \le 1$ if $w_{g-1} = (i_{g-1}, j_{g-1})$. In this way we can ensure that each of the match is aligned with an adjacent point, and all the points in the sequence participate in path regularization.

3.1.3. Monotonic conditions. The sequence of actions should be arranged in order. So the path needs to be planned according to the monotony of time.

Restricted by the three conditions[11], the warping path has only three directions to plan in the square[12], and the next grid point is limited to $(i+1, j)$, $(i, j+1)$ and $(i+1, j+1)$. From the above constraints, a cumulative distance can be defined to represent the similarity between the two sequences.

$$L(i, j) = d(f_i, t_j) + \min\{\gamma(i-1, j), \gamma(i, j-1), \gamma(i-1, j-1)\} \quad (9)$$

3.2. F-DTW algorithm

F-DTW restricts and abstracts the traditional DTW algorithm. First, it abstracts the data from the original sequence when warping the path, which is called coarse-grained. Then, in the process of coarse-grained warping path, it executes the DTW method to plan the path again, which is called fine-grained[13]. This algorithm reduces the search space, so the algorithm recognition complexity can be reduced and the system response time can be improved in sequence identification. However, when the time sequence is coarsely granulated and then expanded to a fine-grained path, the calculation accuracy of the DTW distance will be reduced.

3.3. Weighted DTW algorithm

The traditional DTW algorithm needs to calculate and match each feature vector of each frame in the action sequence in the process of warping path[14]. When calculating the total matching distance[15], the pose change of each joint point needs to be considered in the sequence matching. However, in the process of action recognition, the participation of each joint is different, and the amount of motion variation of the same joint in different action sequences is also different. In order to reduce the complexity of sequence recognition and improve the timeliness of action recognition, this paper assigns weights according to the participation degree of joint in different actions, and use the weighted DTW algorithm to realize identification matching between the template sequence and the test sequence.

We set the weight coefficient ω_k according to the degree of participation of the joint i in the action sequence. Define the function $Tru(f_i,k)$, $\beta=\left|d\left(f_{i-1}^k,f_i^k\right)\right|$, where f_i^k represents the quaternion representation of the direction data of joint k which is contained in the i th frame of the sequence f . The function satisfies the following relationship:

$$Tru(f_i,k)=\begin{cases}1, \ \beta\neq 0\\ 0, \beta=0\end{cases} \tag{10}$$

Then the weight coefficient ω_k can be expressed as:

$$\omega_k=\frac{1}{N-1}\sum_{i=1}^{N}Tru(f_i,k) \tag{11}$$

We can acquire the weighted cost function from the improved DTW algorithm:

$$imp(f_i,d_j)=\sum_{k=1}^{K}\omega_k\left|d\left(f_i^k,t_j^k\right)\right| \tag{12}$$

Test different action templates to get a gray-scale value matrix which represents the value of weight coefficient. From the gray-scale value in the square, it can be intuitively seen that the weight coefficient of joints is different in different action templates as shown in figure 6.

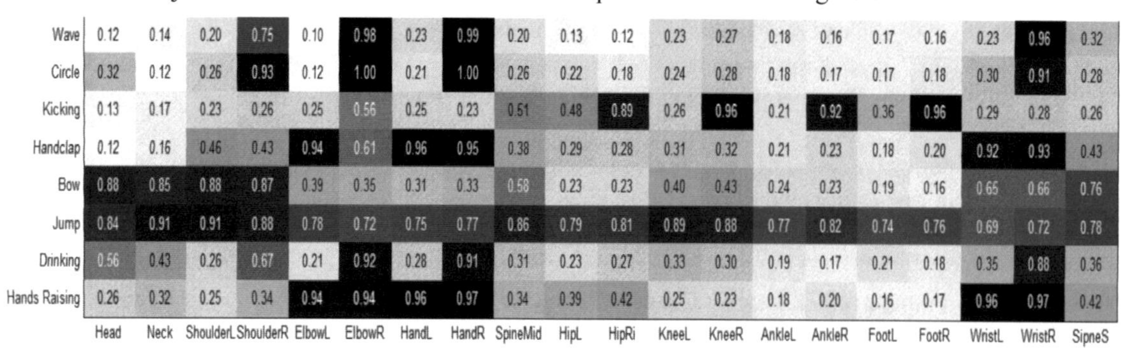

Figure 6. Representation of weight coefficients in squares.

4. Experiment

4.1. Experimental environment
The experiments in this paper were implemented in the Win10 operating system. The hardware environment includes a Kinect2.0 senor and a computer with the Intel i57300-Q CPU. We chose Visual Studio 2017 as the integrated development environment, and use SDK Browser v2.0 (Kinect for Windows) to provide API development interface. The action recognition interface is based on WPF, and the program is written in C#.

4.2. Experimental process
The experimental process is shown in figure 7. First, we use Kinect2.0 sensor to get human motion information to train action samples and build action libraries. Then we use quaternions to represent the sequence of actions and calculate the weight coefficient of joints. The traditional DTW method and the weighted DTW method are used to match the test action sequence and the template action sequence to obtain the comparison results.

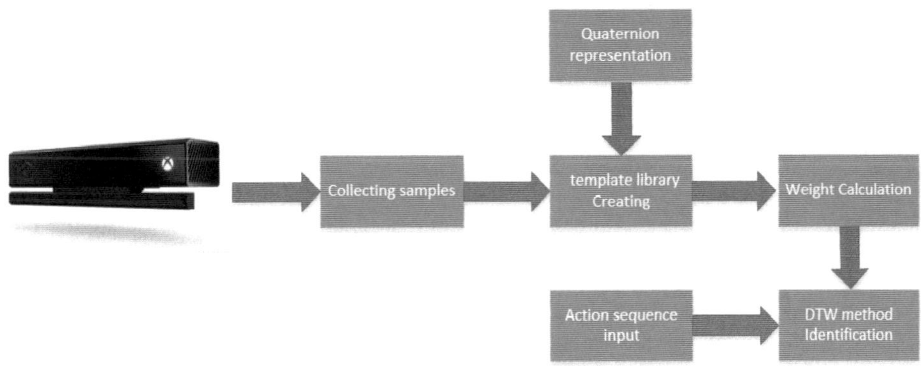

Figure 7. Experimental process using Weighted DTW method

4.3. Experimental operation

4.3.1. Template recording. We used Kinect2.0 to collect motions with a sampling frequency of 15 frames per second as shown in figure 8. 10 test subjects with different individual states were selected and each person performed 20 times on one action. Template recording is completed in WPF project. Click the Motion Capture button in the project to complete the corresponding action before the countdown ends. Repeat this process to complete the template library.

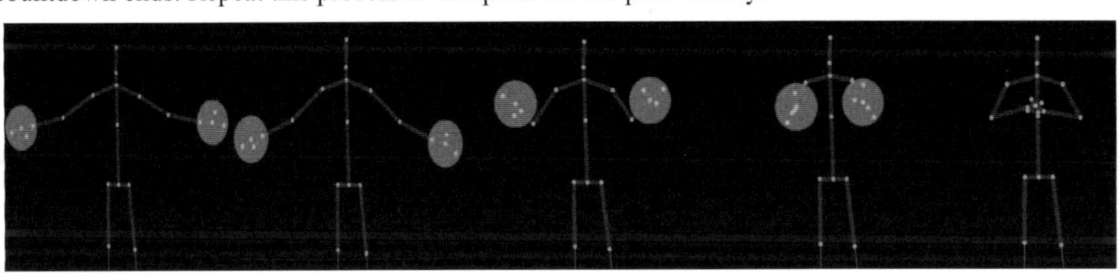

Figure 8. The action sequence of "Handclap".

4.3.2. Identification test. Click the Load Action Template button in the project to enter the recognition mode. As shown in figure 9-10,the tester performs the template action and then the system begins to identify the action sequence based on the Standard DTW algorithm, F-DTW algorithm and Weighted

DTW algorithm.

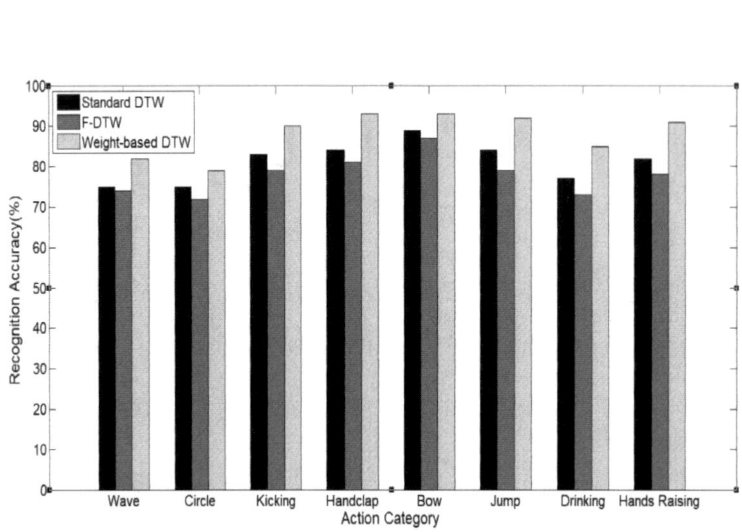

Figure 9. The recognition accuracy of Weighted DTW algorithm and other algorithms.

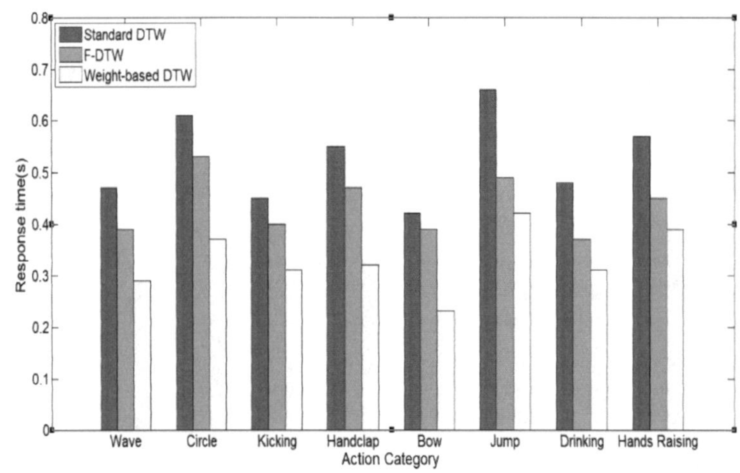

Figure 10. The response time of Weighted DTW algorithm and other algorithms.

4.4. Analysis of experimental results

In the recognition test of the eight action types, the Weighted DTW algorithm proposed in this paper shows a higher recognition rate and a shorter matching time than the Standard DTW Algorithm and the F-DTW Algorithm. Some of the actions in the experiment were not recognized because the Kinect sensor was not ideal for denoising when acquiring joints data. Overall, the Weighted DTW has shown satisfactory accuracy and timeliness in the recognition and action sequences matching.

5. Conclusion and future work

This paper proposes an action definition method using quaternion distances and a Weighted DTW algorithm based on the participation of joints. In the action description process, in order to reduce the calculation and the space occupied by the storage of calculation, we use quaternion distances to define action sequences. The Weighted DTW algorithm proposed in this paper solves the problem of the timeliness caused by the traversal calculation of the Standard DTW Algorithm and the low recognition accuracy of the F-DTW Algorithm. The experimental results show that the overall recognition result of the Weighted DTW algorithm has better accuracy and timeliness.

However, there are some deviations in the results of action recognition in the test when the experiment is disturbed by external conditions. Therefore, in the future work, we need to consider adding denoising and filtering methods to the improvement measures.

References

[1] V. Thanikachalam and K. K. Thyagharajan, "Human action recognition using accumulated motion and gradient of motion from video," 2012 Third International Conference on Computing, Communication and Networking Technologies (ICCCNT'12), Coimbatore, 2012, pp. 1-6.

[2] J. Wang, Z. Liu, Y. Wu and J. Yuan, "Mining actionlet ensemble for action recognition with depth cameras," 2012 IEEE Conference on Computer Vision and Pattern Recognition, Providence, RI, 2012, pp. 1290-1297.

[3] S. Cherla, K. Kulkarni, A. Kale and V. Ramasubramanian, "Towards fast, view-invariant human action recognition," 2008 IEEE Computer Society Conference on Computer Vision and Pattern Recognition Workshops, Anchorage, AK, 2008, pp. 1-8.

[4] R. Vemulapalli, F. Arrate and R. Chellappa, "Human Action Recognition by Representing 3D Skeletons as Points in a Lie Group," 2014 IEEE Conference on Computer Vision and Pattern Recognition, Columbus, OH, 2014, pp. 588-595.

[5] X. Tang, A. Song 2018 Human pose estimation and its implementation in scenario interaction system of rehabilitation training, *Chinese Journal of Scientific Instrument*, Vol.**39**, No.11, pp. 195-203

[6] S. Sempena, Nur Ulfa Maulidevi and Peb Ruswono Aryan, "Human action recognition using Dynamic Time Warping," Proceedings of the 2011 International Conference on Electrical Engineering and Informatics, Bandung, 2011, pp. 1-5.

[7] T. Vajda, "Action recognition based on Fast Dynamic-Time warping method," 2009 IEEE 5th International Conference on Intelligent Computer Communication and Processing, Cluj-Napoca, 2009, pp. 127-131.

[8] H. Alabbasi, A. Gradinaru, F. Moldoveanu and A. Moldoveanu, "Human motion tracking & evaluation using Kinect V2 sensor," 2015 E-Health and Bioengineering Conference (EHB), Iasi, 2015, pp. 1-4.

[9] B. Jablonski 2012 Quaternion Dynamic Time Warping, *Transactions on Signal Processing*, Vol. **60**, No. 3, pp. 1174-1183.

[10] X. Ruan and C. Tian, "Dynamic gesture recognition based on improved DTW algorithm," 2015 IEEE International Conference on Mechatronics and Automation (ICMA), Beijing, 2015, pp. 2134-2138.

[11] G. Zhang, H. Yu and L. Xiao, "CEW-DTW: A new time series model for text mining," 2018 International Conference on Information and Communications Technology (ICOIACT), Yogyakarta, 2018, pp. 158-162.

[12] S. S. Jambhale and A. Khaparde, "Gesture recognition using DTW & piecewise DTW," 2014 International Conference on Electronics and Communication Systems (ICECS), Coimbatore, 2014, pp. 1-5.

[13] Y. Jin and Z. Bi, "Power Load Curve Clustering Algorithm Using Fast Dynamic Time Warping and Affinity Propagation," 2018 5th International Conference on Systems and Informatics (ICSAI), Nanjing, 2018, pp. 1132-1137.

[14] Young-Seon Jeong, Myong K. Jeong, Olufemi A. Omitaomu 2011 Weighted dynamic time warping for time series classification, *Pattern Recognition*, Vol.**44**, No.9, pp.2231-2240.

[15] Y. Lee and J. Lee, "Kinematic Constraint Method for Human Gesture Recognition Based on Weighted Dynamic Time Warping," 2016 6th International Conference on IT Convergence and Security (ICITCS), Prague, 2016, pp. 1-5.

Multi-Agent Aviation Search Task Allocation Method

Wang Yijuan*, **Pan Weijun and Liu Kaiyuan**
Civil Aviation Flight University of China,Air Traffic Management College

*1420982601@qq.com

Abstract. The multi-agent technology is used to reasonably allocate the search area determined after the aircraft crash, and the research on the task assignment problem of aircraft crash area based on multi-agent system is proposed.In the simulation experiment, the model of multiple rescue points, single disaster points and multiple emergency resource allocations is shown. The results show that the method can optimize the allocation of the search task in the aircraft crash area and dynamically change with the change of the crash area. The aircraft's search-assisted decision-making system provides some support.

Keywords: aviation search; multi-agent system; resource allocation

1 Introduction

The search after the aircraft crash is the last guarantee to ensure the safety of the aviation industry. The establishment of an efficient aircraft search assistant decision system can greatly reduce the loss of life and property.The main tasks of the aircraft search aid decision system include determining the search area. [1]And the rules for the allocation of search resources.The search area is determined by calculating the initial search area by the aircraft's last report of the latitude, longitude and speed of the crash location. The search area is determined based on whether the aircraft is found to be wrecked and survivors. The allocation rule for aviation search resources is the most The optimal allocation model, in the time of the aircraft's wreckage search and rescue[2], develop a corresponding search plan, and complete the optimal search resource allocation in the shortest time to ensure the safety of life and property to the greatest extent.

In recent years, artificial intelligence technology has developed rapidly. At present, Multi-Agent System (MAS) technology has been derived. Multi-agent systems can take synergy between multiple agents.Each agent can be an independent individual and interact with other agents to establish a new development plan or solution.Multi-agent theory has been applied in task allocation in the early years. In 1980, Smith established a contract network protocol for task assignment, which differentiated multi-agent systems into managers and collaborators, enabling dynamic task assignment.[3].

Introducing multi-agent system into the aircraft's search-assisted decision-making system, the problem of resource allocation can be further solved from the interaction of multiple agents. At the same time, the task assignment method based on multi-agent system is studied, and the mathematical model of auction rules is established. The traditional auction rules are combined with the multi-agent system to achieve the optimal resource allocation method for aviation search and rescue.

It can be seen from the above research that the multi-agent task assignment method based on the auction rule mathematical model has a wide range of applications in the field of maritime search and rescue, but it is rarely used in aircraft search assistant decision-making systems.Therefore, this paper proposes to apply the auction rule algorithm to solve the task allocation problem of aircraft search area, and design a multi-agent air search task allocation method based on the mathematical model of auction rules.This method of allocation not only satisfies the needs of basic aircraft search, but also adapts to the aircraft search area to dynamically change with the identification of some aircraft wrecks and survivors, and can use the limited search resources of aircraft crashes to achieve optimal search resources. Distribution, has a strong practicality in practice.

2. Task assignment model

2.1 Auction Rules

Auction is a basic economic phenomenon in human society[3]The seller formulates the corresponding bidding rules, the buyer bids according to the rules set by the seller, and finally the highest bidder is the successful bidder.There are many forms of traditional auction rules. This article uses the rules of unilateral auction, that is, a single seller, multiple buyers, the seller is the air search and rescue mission after the crash, and the buyer is the search and rescue force around the search and rescue area.

Introduce the multi-agent system into the traditional auction rules. When starting the task assignment work, consider the factors such as the profit and cost of the bidder[4], and make the bidder's price as much as possible while ensuring the maximum profit of the bidder. Decrease, increase the difference between the bidder's return and the cost, that is, increase the net profit of the bidder, and get the best task allocation plan at this time. [5]

The auction rule a can be changed with the seller, and the auction plan under the multi-agent system should be dynamically changed with the change of the auction rules.During the adjustment process, it is necessary to continuously update the algorithm scheme, and strive to calculate the task arrangement with the most benefit and the least cost in a short time.

Mathematical model based on multi-agent system auction rules

Au=<A, B, C, D, E, F>, where:

a represents the seller'sauction task, A is a collection in the traditional auction rules, $a=\{A_1,A_2,......,A_n\}$ is a positive integer; B is a collection of auction agents, which is a collection of agents that auction the task, $B=\{B_1,B_2,......,B_n\}$, n is a positive integer; C is a collection of auction agents, $C=\{C_1,C_2,......,C_n\}$, n is a positive integer; D is the benefit collection of the buyer's successful auction, $D=\{D_1,D_2,......,D_n\}$, is a positive integer, wherein the auction agent in the auction agent collection B corresponds to the auction task in the traditional auction task collection A, the auction agent in the auction agent collection C, and the auction benefit in the benefit collection D. E is the collection of the price of the auction for the auction, which is the price for each auction agent to complete the auction task, $E=\{E_{B_j}^{A_i}\}$, i<n, j<n, and both are positive integers; F is the auction price collection of the bidders, $F=\{F_1,F_2,......,F_n\}$, n is a positive integer.

So in the mathematical model of the auction rules, when$\sum_{i=0}^{n}\sum_{j=0}^{n}(D_i - E_{B_j}^{A_i})$When the maximum value is obtained, the bidder at this time is the optimal task assignment method.In the formula, we can see that when the difference between the bidder's return and the cost is the maximum, then the bidder gets the maximum net benefit of the task, which is the optimal task allocation method.The auction process based on the multi-agent system is shown in Figure 1.

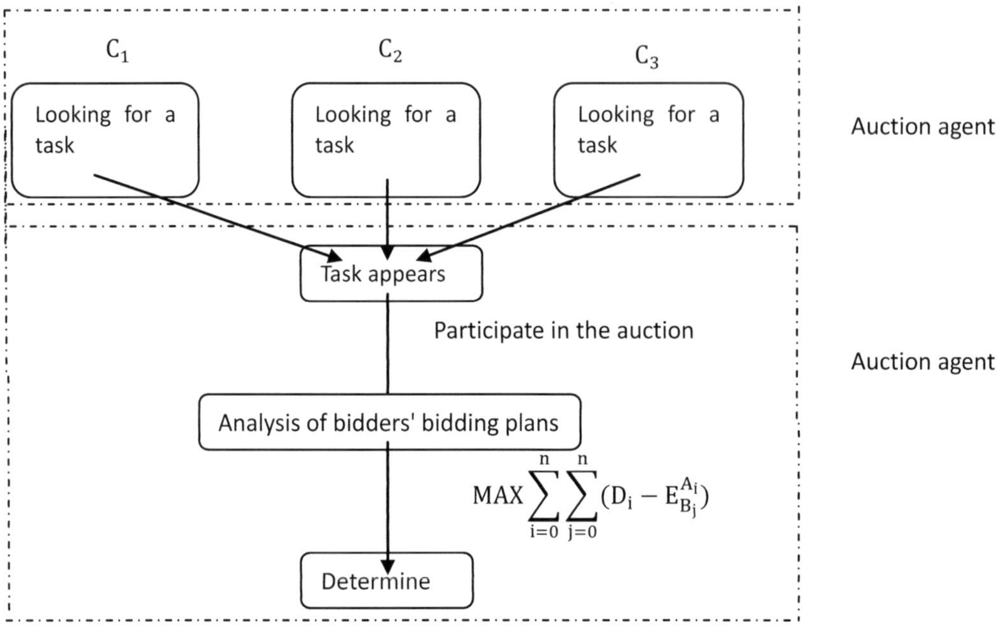

Figure 1 Auction process based on multi-agent system

2.2 Aircraft search task allocation principle
After the aircraft crashed, the Search and Rescue Coordination Center received the information and launched search and rescue work.After the Search and Rescue Coordination Center determines the search and rescue area, the search and rescue mission is released to the search and rescue forces within a certain range from the search and rescue area. The multiple search and rescue forces are their respective bidding agents, and the auction information is released to form a set of bidding agents.

2.3 Basic principles of auction assignment
The primary indicator of aviation search and rescue is fast. The time after the aircraft crash is life. Assigning the search and rescue mission to the professional forces around the crashed area within a limited time is the basic principle of the aircraft search task allocation method. [6]

Set the rescue response time after the aircraft is set to t, and the average moving speed of the surrounding rescue equipment is set to v. The center of the search and rescue area of the crashed aircraft is taken as the center, and the effective distance issued by the search and rescue mission can be obtained with the radius of v*t/2.Search and rescue forces within this distance can participate in the auction of aircraft search missions. If the search and rescue force is too small or the search and rescue requirements are not met within this distance, the scope of the auction information will be expanded to find a suitable bidder.

2.4 Aircraft search task allocation process introducing auction rules
In addition to the basic time to reach the search and rescue area, the task allocation method of the search area after the aircraft crash should also consider the comprehensive factors.The search and rescue capabilities, search and rescue speed, search and rescue costs of the bidding agent can be comprehensively considered into the benefits and costs of the bidding agent. Finally, the appropriate bidding agent is selected within the effective time to complete the search area after the aircraft crash. Initial task assignment.

With the introduction of multi-agent system auction rules into aircraft search task assignments, the new distribution process is shown in Figure 2.

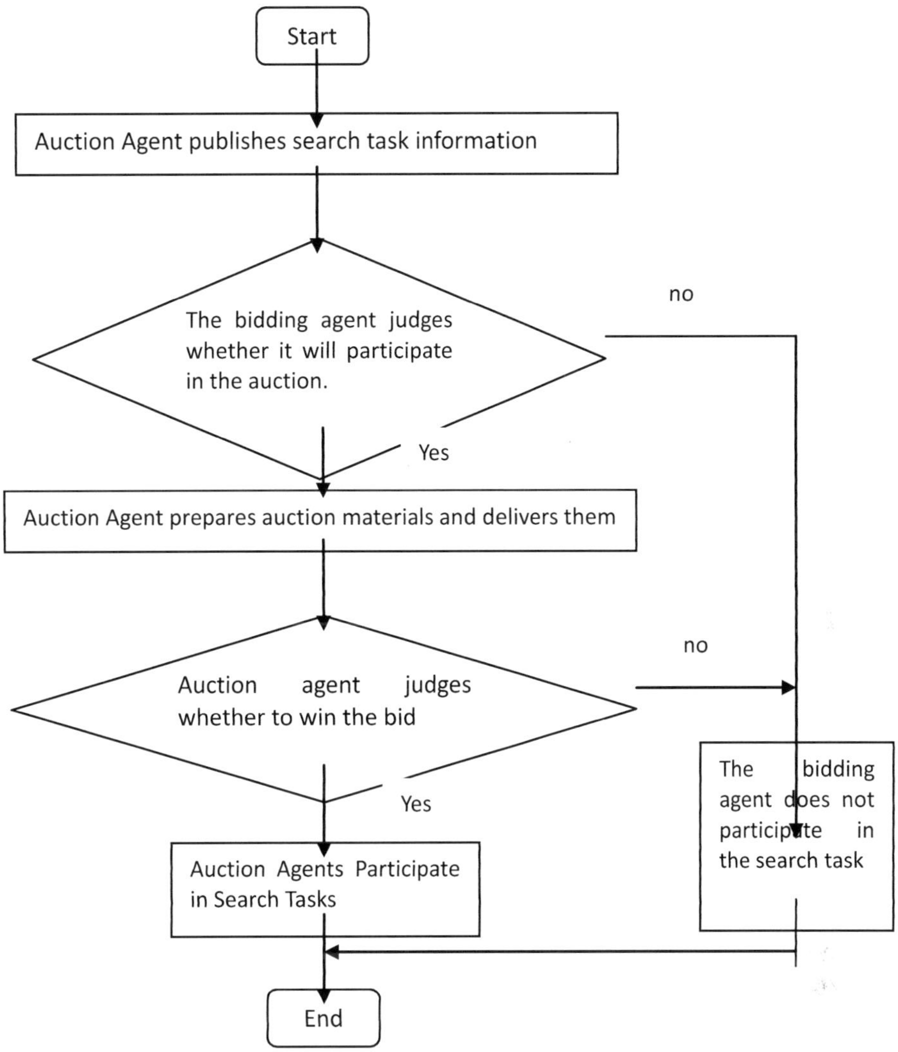

Figure 2 Aircraft search task allocation process introducing auction rules

2.5 Dynamic adjustment of search tasks

The search area will be dynamically adjusted according to whether or not the aircraft is found to be wrecked and survivors. At this time, the published search task needs to be adjusted to achieve the optimal allocation of search power.

The dynamic adjustment of the search task is divided into searching for the search area and part of the search area by searching for the wreckage and survivors of the plane to complete the search for the initial bid search area.If the search area is expanded, the expanded search area is released as a new task to the surrounding search forces, and the new tasks are distributed according to the auction rules to ensure the distribution of search resources.If the search force completes the search task, it can apply to the search and rescue coordination center for searching for the bid search area[7].

3. Multiple rescue points, single disaster points, and multiple emergency resource allocation models

In this question, there is a single disaster point and multiple rescue points are needed.For example, serious natural disasters in a local earthquake require various types of emergency resources. However, due to the destruction of local emergency reserve resources, it is necessary to mobilize emergency resources from neighboring cities for rescue.This is the typical multi-rescue point, single disaster point, and multiple emergency resource allocation problems. The road network structure of this problem is shown in Figure 3.

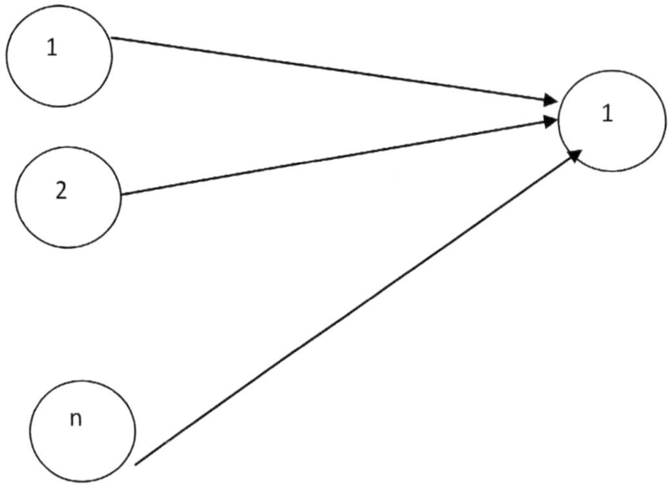

Figure 3 Road network structure of multiple rescue points, single disaster points, and multiple emergency resource allocation problems

4. Task assignment simulation experiment

In the simulation experiment of the search area task assignment after the aircraft crash based on the multi-agent system auction rule, multiple sets of search power comparisons are set, and the best search and rescue plan is obtained by sorting the final calculated bid price.

The influencing factors of aviation search and rescue forces are divided into: 1 the time required to reach the aircraft crash zone (determined by the ratio of distance to speed, in the simulation experiment, the assumed distance is 100km); 2 the maximum moving speed of the search rescue force; 3 the maximum range[8].According to the access data, the data in Table 1 is obtained:

Table 1 Data on the influencing factors of rescue forces

Influencing factor / Name, model	①	②	③
AW119Kx	0.35h	282km/h	954km
AW109SP	0.32h	311km/h	859km
AW139	0.32h	310km/h	1061km
Airbus helicopter h130	0.35h	287km/h	617km
Airbus helicopter h135	0.35h	287km/h	635km
Airbus helicopter h145	0.37h	268km/h	680km
total	2.06h	1745kn/h	4806km

The stratification method is used to sort these influencing factors as follows: 1) the speed of reaching the aircraft wrecked area; 2) searching for the moving speed of the rescue force under the premise of ensuring the efficiency of search and rescue; 3) the maximum voyage; 4) loading the rescue equipment.

Normalize the above parameters, and get Table 2:

Table 2 Auction price ranking of agents

Influencing factor / Name, model	①	②	③	Auction price
AW119Kx	0.170	0.161	0.198	0.529
AW109SP	0.155	0.177	0.179	0.511
AW139	0.155	0.176	0.221	0.552
Airbus helicopter h130	0.170	0.162	0.128	0.46
Airbus helicopter h135	0.170	0.162	0.133	0.465
Airbus helicopter h145	0.180	0.162	0.141	0.483

It can be seen from the above calculation results that after the bidding agent is sorted according to the auction price from high to low, the highest bidding price is selected for the search and rescue force, that is, the helicopter of the model aw139.

5. Conclusion

This paper proposes an aviation search task allocation method based on multi-agent system auction rules. This method can select the best search and rescue force to participate in the search action for the actual situation of the aircraft crash in a short time, and the dynamics occur in the search area. When changing, the task assignment method can be dynamically changed with the change of the search task, and the search efficiency after the aircraft crash is continuously improved. According to the simulation result, the assignment of the search and rescue task can be completed quickly after the actual aircraft accident occurs, which has strong practicability.

References

[1] Zhang Bingxiang. Method for determining the search area when an aircraft crashes [j]. Journal of Civil Aviation University of China, 2000, 18(4): 32-35.
[2] Liu Jinxi, ErLianjie. A review of the application of multi-agent technology [j]. Control and Decision, 2001 (2): 133-140+180.
[3] Ma Qiaoyun. Research on dynamic task assignment based on multi-agent system[D]. Huazhong University of Science and Technology, 2006.
[4] Jin Wei, Shi Chunyi. The Introduction of Auction Method into Multi-Agent System[J].Computer Science,2003(3):104-107.
[5] Yu Wentao, Peng Jun, Wu Min, Zhang Xiaoyong.Multi-agent task assignment algorithm based on auction[J].Computer Simulation,2008(12):184-188.
[6] Li Yan,XingShengwei,ZhangYingjun.Multi-agent maritime search task assignment method [j].China Navigation, 2018(3):91-94+100.
[7] Wang Wenjun, Jiang Qideng, Yang Quanwei, Li Zhirui. Indirect control method for aircraft landing sinking speed [j]. Aviation Science and Technology, 2018(9): 22-26.
[8] Zhu Yuzhu, Zhao Depeng, Huang Xiaoli.Optimization Principles and Methods for Rescue Ships in Maritime Relief Work[J].Journal of Dalian Maritime University,1999(4):48-51+64.

Design of Attitude Controller of Two-Dimensional Correction Projectile Based on UAS-SMC Method

Zhong Yang[1,a,*], Wang Liangming [1,b] and Fu Jian[1]

[1]School of Energy and Power Engineering, Nanjing University of Science and Technology, NanJing , China

[*]E-mail: [a]fishing0508@163.com; [b] lmwang802@163.com

Abstract. For the spin stabilized two-dimensional trajectory correction projectiles with utation and precession phenomenon, an attitude controller with sliding mode control method and unidiational auxilairy surfaces (UAS-SMC) is proposed in this paper. And the underactuated control allocation in two-dimensional trajectory correction projectile with UAS-SMC controller is also discussed here. Simulation results show that the proposed controller can effectively control the attitude of the projectile and has strong robustness, which provides reference for engineering applications.

1. Introduction

For the low-speed fin-stabilized projectiles, the relationship between the direction of the control moment and the direction of the projectile swing is simple and clear. For high-elastic bombs, the gyro effect causes the two-dimensional trajectory correction projectiles to produce nutation and precession under the action under control moments, which leading to the control relationship becomes very complicated. At the same time, the addition of correction components to the head of the projectile will cause mass eccentricity, installation errors, and complex aerodynamic characteristics, which cause model uncertainty and external disturbance. Therefore, the robustness of the controller algorithm is extremely high.

In practical engineering applications, traditional sliding mode control methods have unacceptable chattering phenomena. This chattering phenomenon may provoke vibrations in the unmodeled dynamics of the system, thereby reducing the performance of the control system. In order to solve this problem, experts and scholars from various countries have proposed control methods such as boundary layer sliding mode[1], high-order sliding mode[2], and double sliding mode[3], but these control methods require more or less different costs for the system.

Aiming at this problem, this paper proposes a new sliding mode control method, UAS-SMC by modifying the control structure of the traditional sliding mode by modifying the control system robustness and constraint ability. Under the premise of retaining the strong robustness of the traditional sliding mode, the method has no chattering of the controller, and the approaching speed near the switching surface is strengthened. No differential information is needed, and there is a positive invariant set in the system state space. The UAS-SMC improves the control performance while taking into account the application requirements in engineering practice, without the additional cost of the controller. Therefore, it has broad application prospects in practical engineering practice.

2. Attitude Controller of Two-dimensional Correction Projectile

The nonlinear attitude model [4-5] of trajectory correction projectiles is given as follows:

$$\begin{aligned}\dot{\boldsymbol{\Omega}} &= \boldsymbol{f}_s + \boldsymbol{g}_s\boldsymbol{\omega} \\ \dot{\boldsymbol{\omega}} &= \boldsymbol{f}_f + \boldsymbol{g}_f\boldsymbol{M}_c + \boldsymbol{d}(t)\end{aligned} \tag{1}$$

where

$$\boldsymbol{f}_s = \begin{bmatrix} -\dfrac{1}{mv\cos\beta}\left(F_x\sin\alpha + F_y\cos\alpha\right) + \dfrac{g}{v\cos\beta}\left(\sin\vartheta\sin\alpha + \cos\vartheta\cos\alpha\right) \\ \dfrac{1}{mv}\left(-F_x\cos\alpha\sin\beta + F_y\sin\alpha\sin\beta + F_z\cos\beta\right) + \dfrac{g\sin\beta}{v}\left(\cos\alpha\sin\vartheta - \sin\alpha\cos\vartheta\right)\end{bmatrix},$$

$$\boldsymbol{g}_s = \begin{bmatrix} \tan\beta\left(\sin\alpha - \cos\alpha\tan\vartheta\right) & 1 \\ 0 & \left(\cos\alpha + \sin\alpha\tan\vartheta\right)\end{bmatrix}, \quad \boldsymbol{g}_f = \begin{bmatrix} 1/I_y & 0 \\ 0 & 1/I_z \end{bmatrix}, \quad \boldsymbol{M}_c = \begin{bmatrix} M_{fy} \\ M_{fz}\end{bmatrix},$$

$$\boldsymbol{f}_f = \begin{bmatrix} \dfrac{M_y}{I_y} - \dfrac{(I_{xa}\omega_{xa} + I_{xf}\omega_{xf})\omega_z}{I_y} + \omega_z\omega_y\tan\vartheta \\ \dfrac{M_z}{I_y} + \dfrac{(I_{xa}\omega_{xa} + I_{xf}\omega_{xf})\omega_y}{I_y} - \omega_y^2\tan\vartheta \end{bmatrix},$$

and $\boldsymbol{\Omega} = \left[\alpha, \beta\right]^{\mathrm{T}}$ are the attitude angles , $\boldsymbol{\omega} = \left[\omega_y, \omega_z\right]^{\mathrm{T}}$ are the angular rates, $\boldsymbol{d}(t) = [d_1, d_2]^T$ are the unknown torque disturbances in attitude motion of the projectle. Attitude tracking command signal $\boldsymbol{\Omega}_c = \left[\alpha_c, \beta_c\right]^T$.

In order to design the UAS-SMC controller for trajectory correction projectiles (1), the following reasonable assumptions are needed.

a) Assume that the attitude tracking command signals $\boldsymbol{\Omega}_c = \left[\alpha_c, \beta_c\right]^T$ are always continuous.

b) Assuming that matrix $\boldsymbol{g}_s$, g_f , always reversible during system motion.

c) Assume that all elements $\boldsymbol{g}_s$, f_f , g_f in the matrix are continuous functions.

3. Design of Control System Based on UAS-SMC

The tracking errors of two-dimension trajectory correction projectile are defined as

$$\begin{aligned}\boldsymbol{e}_1 &= \boldsymbol{\Omega} - \boldsymbol{\Omega}_c \\ \boldsymbol{e}_2 &= \boldsymbol{\omega} - \boldsymbol{v}_1\end{aligned} \tag{2}$$

where $\boldsymbol{e}_1 = \left[e_{11}, e_{12}\right]^T \in R^2$ is the attitude angle error, $\boldsymbol{e}_2 = \left[e_{21}, e_{22}\right]^T \in R^2$ is the angular rate error, $\boldsymbol{v}_1$ is the virtual control law of the design. Then equation(1) can be converted into the form of error equation.

$$\dot{\mathbf{e}}_1 = \boldsymbol{f}_s + \boldsymbol{g}_s(\mathbf{e}_2 + \mathbf{v}_1) - \dot{\boldsymbol{\Omega}}_c,\ \dot{\mathbf{e}}_2 = \boldsymbol{f}_f + \boldsymbol{g}_f\boldsymbol{M}_c + \mathbf{d}(t) - \dot{\mathbf{v}}_1 \tag{3}$$

The virtual control law is designed for the attitude loop error in equation (3). Switching surface is

$$s_{11} = e_1 + \xi_{11}\int e_1 = 0,\ s_{12} = e_1 + \xi_{12}\int e_1 = 0 \tag{4}$$

where ξ_{11}= diag{ξ_{111}, ξ_{112}}, ξ_{12}= diag{ξ_{121}, ξ_{122}}, $\boldsymbol{s}_{11}$= $[s_{111}, s_{112}]^T$, $\boldsymbol{s}_{12}$= $[s_{121}, s_{122}]^T$, $\xi_{11i} > \xi_{12i} > 0$, $i \in \{1,2\}$.

The auxiliary surface h_{1i} is designed as follows with two auxiliary parameters ω_{10i2} , ω_{13i2} and two switching surface parameters ξ_{11i} , ξ_{12i} .

$$h_{1i} = \omega_{1i1}e_{1i} + \omega_{1i2}\int e_{1i} + m_{1i} \tag{5}$$

where $\omega_{1i1} \neq 0$ and

$$\omega_{1i1}=\begin{cases}1 & s_{11i}<0, s_{12i}<0\\ \omega_{11i1} & s_{11i}<0, s_{12i}\geq 0\\ \omega_{12i1} & s_{11i}\geq 0, s_{12i}<0\\ -1 & s_{11i}\geq 0, s_{12i}\geq 0\end{cases},\quad \omega_{1i2}=\begin{cases}\omega_{10i2} & s_{11i}<0, s_{12i}<0\\ \omega_{11i2} & s_{11i}<0, s_{12i}\geq 0\\ \omega_{12i2} & s_{11i}\geq 0, s_{12i}<0\\ -\omega_{13i2} & s_{11i}\geq 0, s_{12i}\geq 0\end{cases},$$

$$\omega_{11i1}=[-(\xi_{11i}+\xi_{12i})+(\omega_{10i2}+\omega_{13i2})]/(\xi_{11i}-\xi_{12i})$$
$$\omega_{11i2}=[-2\xi_{11i}\xi_{12i}+(\omega_{10i2}\xi_{11i}+\omega_{13i2}\xi_{12i})]/(\xi_{11i}-\xi_{12i})$$
$$\omega_{12i1}=[(\xi_{11i}+\xi_{12i})-(\omega_{10i2}+\omega_{13i2})]/(\xi_{11i}-\xi_{12i})$$
$$\omega_{12i2}=[2\xi_{11i}\xi_{12i}-(\omega_{10i2}\xi_{11i}+\omega_{13i2}\xi_{12i})]/(\xi_{11i}-\xi_{12i})$$

The attitude control loop virtual control law is shown below.

$$\boldsymbol{v}_1=\boldsymbol{g}_s^{-1}(-\boldsymbol{f}_s+\dot{\boldsymbol{\Omega}}_c-\boldsymbol{\Omega}_{11}^{-1}\boldsymbol{\Omega}_{12}\boldsymbol{e}_1+\boldsymbol{\Omega}_{11}^{-1}\boldsymbol{N}_1) \tag{6}$$

where $\boldsymbol{N}_1=[N_{11},N_{12}]^T$ is the design approach law and Satisfies $N_{11}>0, N_{12}>0$。The compacted form for the Unidirectional auxiliary surface is as follows.

$$h_1=\boldsymbol{\Omega}_{11}e_1+\boldsymbol{\Omega}_{12}\int e_1+m_1 \tag{7}$$

where $\boldsymbol{h}_1=[h_{11}, h_{12}]^T$,$\boldsymbol{\Omega}_{11}$=diag$\{\omega_{111},\omega_{121}\}$, $\boldsymbol{\Omega}_{12}$=diag$\{\omega_{112},\omega_{122}\}$, $\boldsymbol{m}_1=[m_{11}, m_{12}]^T$. From equation (5), it is noted that $\omega_{1i1}\neq 0$, and the matrix $\boldsymbol{\Omega}_{11}$ is a reversible matrix.

It can be known from equation (5), equation (6) and equation (7), there is the following relationship.

$$\dot{h}_1=N_1+\boldsymbol{\Omega}_{11}\cdot\boldsymbol{g}_s\cdot\boldsymbol{e}_2 \tag{8}$$

The controller is designed for the angular rate loop error equation in equation (3). Switching surface is

$$S_{21}=e_2+\xi_{21}\int e_2=0,\ S_{22}=e_2+\xi_{22}\int e_2=0 \tag{9}$$

where $\boldsymbol{\xi}_{21}$= diag$\{\xi_{211}, \xi_{212}\}$, $\boldsymbol{\xi}_{22}$= diag$\{\xi_{221}, \xi_{222}\}$, $\boldsymbol{s}_{21}=[s_{211}, s_{212}]^T$, $\boldsymbol{s}_{22}=[s_{221}, s_{222}]^T$, $\xi_{21i}>\xi_{22i}>0$, $i\in\{1,2\}$.

In the same way, design auxiliary parameters ω_{20i2}, ω_{23i2}, ξ_{21i}, ξ_{22i} can complete the design of the auxiliary surface.

$$h_{2i}=\omega_{2i1}e_{2i}+\omega_{2i2}\int e_{2i}+m_{2i} \tag{10}$$

where

$$\omega_{2i1}=\begin{cases}1 & s_{21i}<0, s_{22i}<0\\ \omega_{21i1} & s_{21i}<0, s_{22i}\geq 0\\ \omega_{22i1} & s_{21i}\geq 0, s_{22i}<0\\ -1 & s_{21i}\geq 0, s_{22i}\geq 0\end{cases},\quad \omega_{2i2}=\begin{cases}\omega_{20i2} & s_{21i}<0, s_{22i}<0\\ \omega_{21i2} & s_{21i}<0, s_{22i}\geq 0\\ \omega_{22i2} & s_{21i}\geq 0, s_{22i}<0\\ -\omega_{23i2} & s_{21i}\geq 0, s_{22i}\geq 0\end{cases},$$

$$\omega_{21i1}=[-(\xi_{21i}+\xi_{22i})+(\omega_{20i2}+\omega_{23i2})]/(\xi_{21i}-\xi_{22i})$$
$$\omega_{21i2}=[-2\xi_{21i}\xi_{22i}+(\omega_{20i2}\xi_{21i}+\omega_{23i2}\xi_{22i})]/(\xi_{21i}-\xi_{22i})$$
$$\omega_{22i1}=[(\xi_{21i}+\xi_{22i})-(\omega_{20i2}+\omega_{23i2})]/(\xi_{21i}-\xi_{22i})$$
$$\omega_{22i2}=[2\xi_{21i}\xi_{22i}-(\omega_{20i2}\xi_{21i}+\omega_{23i2}\xi_{22i})]/(\xi_{21i}-\xi_{22i})$$

Angular rate loop controller is

$$\boldsymbol{M}_c=\boldsymbol{g}_f^{-1}(-\boldsymbol{f}_f+\dot{\boldsymbol{v}}_1-\boldsymbol{\Omega}_{21}^{-1}\boldsymbol{\Omega}_{22}\boldsymbol{e}_2+\boldsymbol{\Omega}_{21}^{-1}\boldsymbol{N}_2) \tag{11}$$

where $N_2=[N_{21},N_{22}]^T$ is the reaching law and satisfies $N_{2i}>0$. The linear equation of the Unidirectional auxiliary surface is

$$H_{2ki}=\omega_{2ki1}e_{2i}+\omega_{1ki2}\int e_{2i}+m_{2i} \tag{12}$$

where $k \in \{0,1,2,3\}$, $\omega_{2ki1} \neq 0$ and $m_{2i} > 0$ are real numbers. The current unidirectional auxiliary sliding surface linear equation is

$$\boldsymbol{h}_2 = \boldsymbol{\Omega}_{21}\boldsymbol{e}_2 + \boldsymbol{\Omega}_{22}\int \boldsymbol{e}_2 + \boldsymbol{m}_2 \tag{13}$$

where $\boldsymbol{h}_2=[h_{21}, h_{22}]^T$, $\boldsymbol{\Omega}_{21}=\text{diag}\{\omega_{211},\omega_{221}\}$, $\boldsymbol{\Omega}_{22}=\text{diag}\{\omega_{212},\omega_{222}\}$, $\boldsymbol{m}_2=[m_{21}, m_{22}]^T$. It Can be seen by the equation (12) that $\omega_{2i1} \neq 0$ and Ω_{21} is reversible matrix.

Therorem 1 [6] Considering system (1), the closed -loop system is stable under control (11) if the switching surfaces $S_{11}=0,\ S_{12}=0, S_{21}=0,\ S_{22}=0$ are stable and the approaching laws $\boldsymbol{N}_1=[N_{11},N_{12}]^T$, $\boldsymbol{N}_2=[N_{21},N_{22}]^T$ satisfy $N_{11}>0, N_{12}>0, N_{21}>0, N_{22}>0$.

4. Control Allocation

As shown in Figure 1, the projectile correction component is mounted on the head of the projectile, consisting of four fixed rudders, two of which are in the same direction (control rudder) and two in opposite directions (differential rudder). The control rudder can generate control moment to change the attitude of projectile, and the differential rudder provides a torque moment which can change the direction of control aerodynamics. By the control rudder and differential rudder, two-dimensional trajectory correction projectiles can achieve the purpose of trajectory correction.

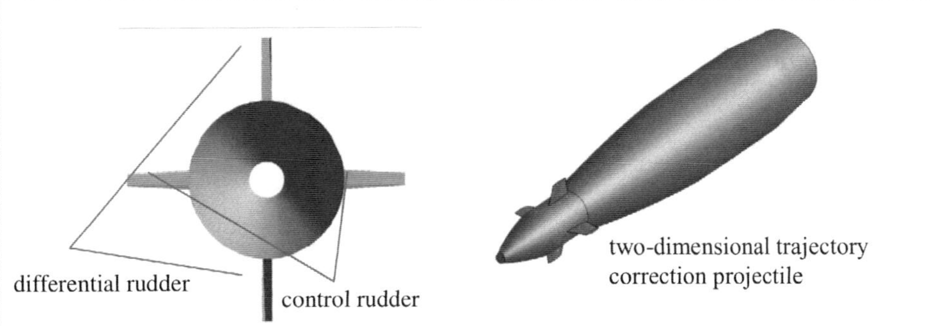

Figure 1. Schematic diagram of two-dimensional trajectory correction projectiles.

It is known from equation (1) that the desired control torque generated by the controller is $\boldsymbol{M}_c=[M_{fy},M_{fz}]^T$ and meet the following relationship.

$$M_{fy}=\frac{1}{2}\rho v_r^2 SlC_{M\delta}\delta_z(-\sin\gamma_f),\quad M_{fz}=\frac{1}{2}\rho v_r^2 SlC_{M\delta}\delta_z\cos\gamma_f \tag{14}$$

where γ_f is the roll angle of the correct component. The roll angle is expressed as

$$\gamma_f=\begin{cases} a\tan(|M_{fy}/M_{fz}|) & M_{fy}\le 0,\ M_{fz}>0 \\ \pi-a\tan(|M_{fy}/M_{fz}|) & M_{fy}\le 0,\ M_{fz}<0 \\ \pi+a\tan(|M_{fy}/M_{fz}|) & M_{fy}\ge 0,\ M_{fz}<0 \\ 2\pi-a\tan(|M_{fy}/M_{fz}|) & M_{fy}\ge 0,\ M_{fz}>0 \\ \pi/2 & M_{fy}<0,\ M_{fz}=0 \\ 3\pi/2 & M_{fy}>0,\ M_{fz}=0 \end{cases} \tag{15}$$

where $\arctan(\cdot)$ represents the arctangent function.

5. Simulation and Analysis of Trajectory Correction Projectile Controller

The actual control moment $\boldsymbol{M}_{creal}$ is shown below.

$$\boldsymbol{M}_{creal} = k_f \boldsymbol{M}_c \tag{16}$$

where $k_f = \sqrt{M_{fz}^{\ 2} + M_{fy}^{\ 2}} / (0.5 \rho v_r^2 S l C_{M\delta} \delta_z)$. First, substituting the desired control torque into equation (16), then substituting M_{creal} for $\boldsymbol{M}_c$ with equation (1). Finally, the closed loop control equation under the actual control torque is obtained. The specific simulation results are as follows.

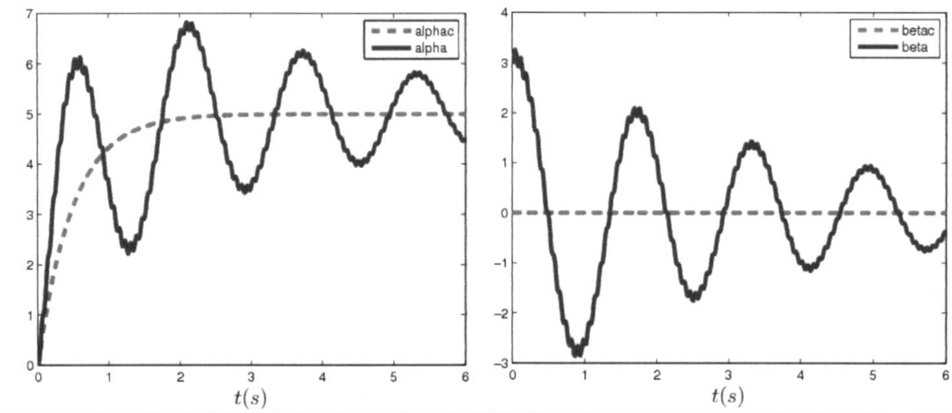

Figure 2. Dynamic inverse controller without control allocation.

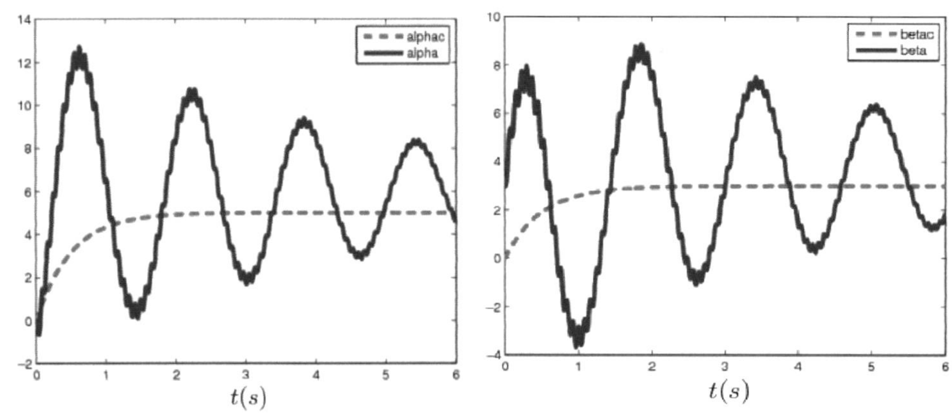

Figure 3. Dynamic inverse controller with control allocation.

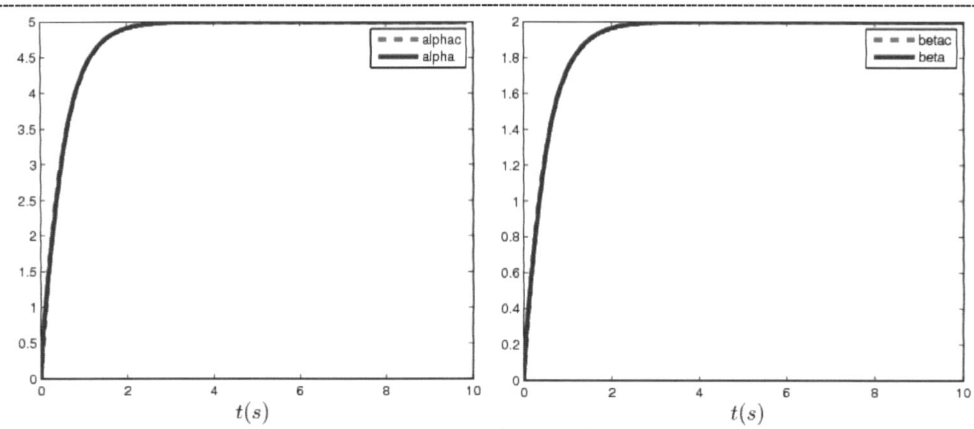

Figure 4. UAS-SMC controller without control allocation.

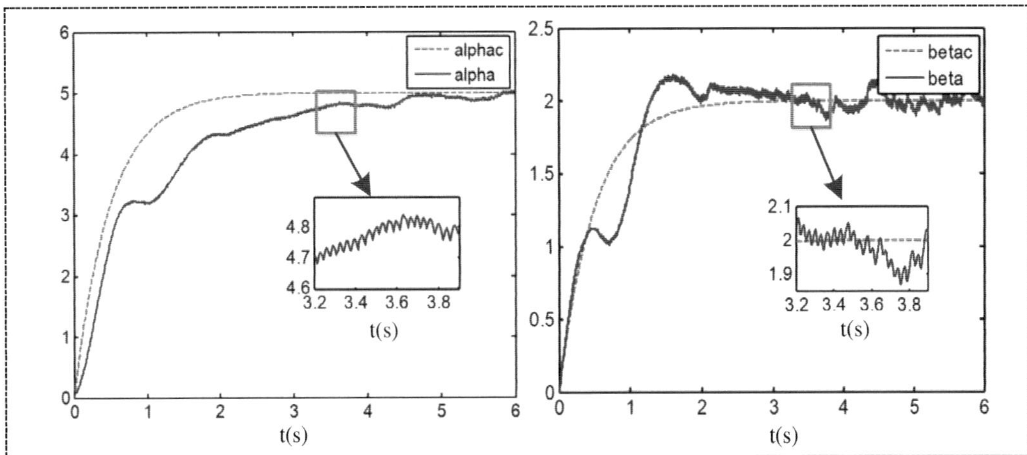

Figure 5. UAS-SMC controller with control allocation.

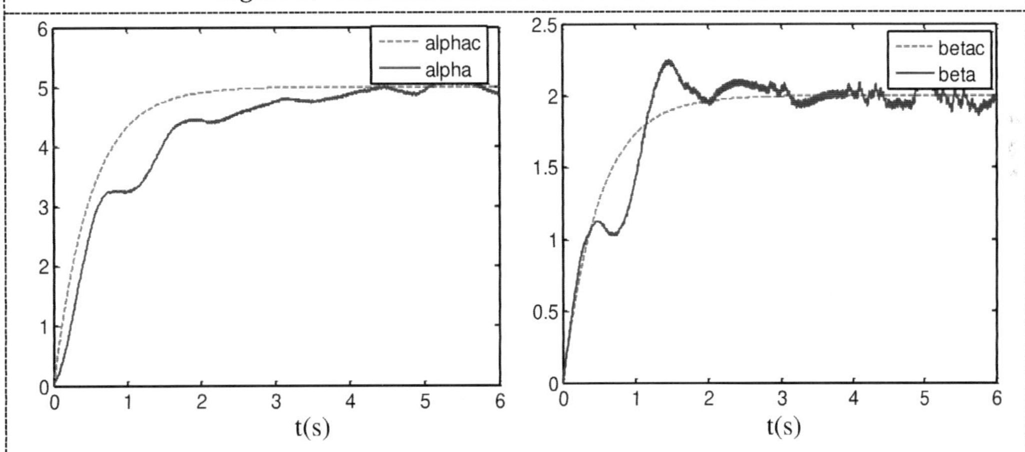

Figure 6. UAS-SMC controller with control allocation and disturbances.

Simulation results of attitude control system based on Dynamic inverse controller and UAS-SMC controller are shown in Figure 2 to Figure 6. As can be seen from Figure 2 and Figure 3, the open loop control based on the inverse system is less effective. Moreover, the control effect of controlling the magnitude and direction of the moment is better than the control effect of controlling only the moment direction. However, if the magnitude and direction of the control moment are to be achieved, it is necessary to achieve this through the equivalent periodic force control. This control method needs very large moment which is very difficult for the motor. Therefore, this paper adopts a control method that the direction of control moment can be variable and the magnitude is not variable. Figure 4 shows the UAS-SMC controller without control allocation, and it works well under ideal conditions. But in practice, you need to consider control allocation. Figure 5 gives a result of UAS-SMC controller with control allocation, and it can effectively control the attitude of the projectile. Figure 6 shows the result of UAS-SMC controller with disturbances of 5% aerodynamics error, and it proved strong robustness and the control effect can meet the system requirements.

6. Conclusions

Aiming at the nonlinear attitude control system of ballistic correction projectile, a backstepping unidirectional sliding mode control method is developed, and a nonlinear attitude control system is designed. Finally, verify its control effect through simulation. The simulation results show that open loop control based on inverse system is less effective, and the control effect of controlling the magnitude and direction of the moment is better than the control effect of controlling only the moment direction. In fact, in order to meet the low-cost design requirements in the design, it is recommended to use a control method with variable torque direction and variable magnitude. Results of UAS-SMC controller with control allocation and aerodynamic disturbance show that the method has strong robustness and the control effect can meet the system requirements.

References

[1] Chen Zhi and Wang Daobo 2016 Modeling and Sliding Mode Control with Boundary Layer for Unmanned Coaxial Rotor Ducted Fan Helicopter *Transactions of Nanjing University of Aeronautics and Astronautics* **33** 199

[2] Floquet T, Perruquetti P, Barbot J P 2000 Angular Velocity Stabilization of a Rigid Body Via VSS Control *Journal of Dynamic Systems, Measurement, and Control* **122** 669

[3] Chen Weidong, Chen Yangyang, Xue Zhixia 2013 Design of Control System for Cruise Missile Based on Double Sliding Mode Variable Structure *Aeronautical Computing Technique* **43** 4

[4] Fu Jian, Wu Qingxian, Chen Wenhua 2013 Chattering-free condition for sliding mode control with unidirectional auxiliary surfaces *Transactions of the Institute of Measurement and Control* **35** 593

[5] Levant A and Wu Qingxian 2011 Universal single-input-single-output sliding mode controller with finite-time convergence *IEEE Transactions on Automatic control* **46** 1447

[6] Fu Jian, Wu Qingxian, Jiang Changsheng, Cheng Lu 2011 Robust sliding mode control with unidirectional auxiliary surfaces for nonlinear system with state constraints *Control and Decision* **26** 1288

IOP Conf. Series: Materials Science and Engineering **646** (2019) 012060 doi:10.1088/1757-899X/646/1/012060

Aerial Automatic Initial Calibration Method for Geomagnetic Sensors Based on Spherical Harmonic Model

An Liangliang[1,*a] and Wang Liangming [1,*b]

[1]School of Energy and Power Engineering, Nanjing University of Science and Technology, Nanjing, China

[*]Corresponding author e-mail: [a]anliangno1@126.com, [b]lmwang802@163.com

Abstract. For the traditional calibration method, the problem of zero offset and amplitude variation caused by degaussing during the launching of the rotating projectile cannot be solved. Based on the practical engineering application, an automatic initial calibration method for the geomagnetic sensor combination is proposed. Based on the spherical harm onic model of the Earth's magnetic field, the output law of the non-orthogonal geomagnetic sensor combination and the instantaneous attitude of the projectile are analyzed, and the geomagnetic sensor combination is automatically calibrated. The experimental results show that after the automatic calibration in the air, the geomagnetic data error collected by the experiment is significantly reduced.

1. Introduction

With the development of geomagnetic measurement technology, the method of geomagnetic measurement to realize attitude test and guidance has become one of the important means of acquiring the attitude information of the projectile. Since the geomagnetic field is a weak magnetic field, the geomagnetic sensor signal is highly susceptible to the surrounding disturbing magnetic field. The projectile system is ferromagnetic due to the long-term placement in the earth's magnetic field [1]. Therefore, in order to obtain high-precision geomagnetic information, it is necessary to perform effective error calibration and compensation for the magnetic sensor. Commonly used compensation methods include Poisson formula compensation method, elliptical or ellipsoidal fitting compensation method and least squares compensation method [2-4], etc., but most of the research is based on ground experiments or simulations, without considering the changes of surrounding magnetic field environment during launching. However, a large number of experiments have shown that the demagnetization phenomenon occurs when the projectile is launched, and the geomagnetic intensity output signal will exhibit zero drift and amplitude change. Aiming at these problems, based on the spherical harmonic model of the geomagnetic field, an automatic initial calibration method for engineering is proposed.

2. Introduction to geomagnetic field and spherical harmonic model

2.1. Coordinate system to describe geomagnetic field

When describing the geomagnetic field, the geomagnetic field intensity M and its components are usually utilized. As shown in Figure 1, taking the northern hemisphere as an example, the vertical plane where the geomagnetic field intensity M is located is the magnetic meridian plane, the X axis is positive along the geographic meridian to the north, the Y axis is positive along the latitude to the east, and the Z axis is vertically downward.

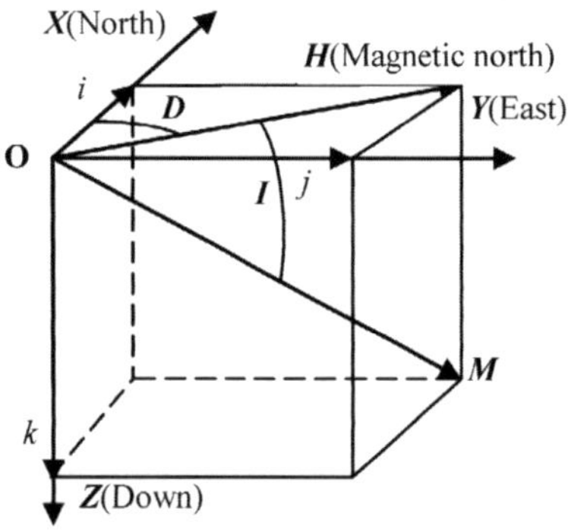

Figure 1. Coordinate system to describe geomagnetism.

The projections of M on the X , Y and Z axes, i , j , k , are called north, east, and vertical intensities, respectively. The projection of M on the XOY plane of the horizontal plane, H , is called the horizontal intensity. The angle between the magnetic meridian plane and the geographic meridian plane, D , is called the magnetic declination, and the deviation is positive to the east. The angle between M and the horizontal plane, I , is called the magnetic inclination. In the northern hemisphere, M points below the horizon and the magnetic inclination is positive.

2.2. Geomagnetic field model

Commonly used geomagnetic models include spherical harmonic models, dipole models, and regional magnetic field models. The regional magnetic field model has high precision, but missing measured data of the boundary and adjacent area will cause the distortion of the model field, that is, the boundary effect. The spherical harmonic model and the dipole model are global magnetic field models with lower accuracy than the regional magnetic field model. The dipole model needs to be iterated several times to obtain more satisfactory parameters, and the calculation process is cumbersome. Therefore, the spherical harmonic model is generally applicable internationally. The spherical harmonic function representing the earth's magnetic field is:

$$V = a\sum_{n=1}^{\infty}\sum_{m=0}^{n}\left(\frac{a}{r}\right)^{n+1}(g_n^m \cos m\lambda + h_n^m \sin m\lambda)P_n^m(\cos\theta) \tag{1}$$

Where a is the average radius of the Earth; $r = a + h$, the distance from the earth center to the point to be calculated; θ is the residual latitude from the Arctic, $\theta = 90° - \varphi$, φ is the magnetic latitude; λ is the longitude calculated from the Greenwich to the east; g_n^m and h_n^m are Schmidt's standardized spherical harmonic coefficients, also known as Gaussian coefficients.

Schmidt quasi-normalized n times and m-order Legendre function is defined as:

$$\begin{cases} P_n^m(\cos\theta) = \dfrac{1}{2^n n!}\left(\dfrac{C_m(n-m)!(1-\cos^2\theta)^m}{(n+m)!}\right)^{\frac{1}{2}} \dfrac{d^{n+m}(\cos^2\theta - 1)^n}{d\cos\theta^{n+m}} \\ C_m = \begin{cases} 1(m=0) \\ 2(m \le 1) \end{cases} \end{cases} \tag{2}$$

According to the theory of potential field transformation, the individual component expressions of the geomagnetic field are

$$\begin{cases} i = -\frac{1}{r}\frac{\partial V}{\partial \theta} = \sum_{n=1}^{\infty}\sum_{m=0}^{n}(\frac{a}{r})^{n+1}(g_n^m \cos m\lambda + h_n^m \sin m\lambda)\frac{d}{d\theta}P_n^m(\cos\theta) \\ j = \frac{1}{r\sin\theta}\frac{\partial V}{\partial \theta} = \sum_{n=1}^{\infty}\sum_{m=0}^{n}(\frac{a}{r})^{n+2}(\frac{m}{\sin\theta})(g_n^m \cos m\lambda - h_n^m \sin m\lambda)P_n^m(\cos\theta) \\ k = \frac{\partial V}{\partial r} = -\sum_{n=1}^{\infty}\sum_{m=0}^{n}(\frac{a}{r})^{n+2}(n+1)(g_n^m \cos m\lambda + h_n^m \sin m\lambda)P_n^m(\cos\theta) \end{cases} \quad (3)$$

The geomagnetic field intensity M is

$$M = \sqrt{i^2 + j^2 + k^2} \quad (4)$$

3. Automatic initial calibration method in the air

3.1. Three-axis non-orthogonal geomagnetic sensor combination

The three-axis non-orthogonal sensor combination consists of a magnetic sensor along the axis of the rotation and two magnetic sensors with an angle of 90° and 60°, respectively, as shown in Figure 2, denoted as axis x, y, z respectively. Regardless of the mounting error, the x-axis coincides with the axis of rotation of the projectile, the y-axis is perpendicular to the x-axis, and the z-axis is in the xoy plane and is at an angle of 60° to the x-axis.

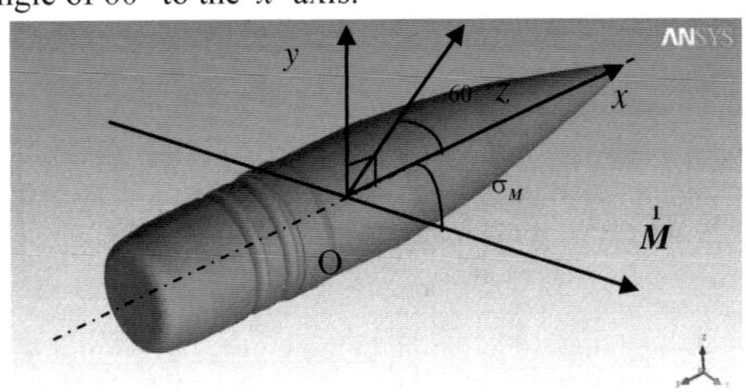

Figure 2. Geomagnetic sensor installation diagram.

3.2. Degaussing

When the magnetized material is affected by external energy, such as heating and impact, the magnetic moment direction of each magnetic domain will become inconsistent, and the magnetic properties will be weakened or disappeared. This process is called degaussing. Through a large number of shooting experiments, it is found that when the projectile is launched, due to the high overload and high temperature, the magnetic properties of the projectile will change. The ground calibration coefficient cannot guarantee the accuracy of the geomagnetic output signal and needs to be recalibrated.

3.3. Sensors output law analysis

As shown in Figure 2, when the geomagnetic sensor with an angle of β is rotated with the projectile in the geomagnetic field, the instantaneous field intensity along the sensor's sensitive axis is [5-6]:

$$M_S = \cos(\beta)\left|\vec{M}\right|\cos(\sigma_M) + \sin(\beta)\left|\vec{M}\right|\sin(\sigma_M)\sin(\phi_S) \quad (5)$$

Where σ_M is the magnetic azimuth, i.e. the angle between the spring axis and the earth's magnetic field; ϕ_S is the roll angle of the projectile.

If $\beta = 0^{\circ}$, the geomagnetic output of the x -axis is:

$$\boldsymbol{Mx} = \left|\vec{\boldsymbol{M}}\right| \cos(\sigma_M) \tag{6}$$

If $\beta = 90^{\circ}$, the geomagnetic output of the y -axis is:

$$\boldsymbol{My} = \left|\vec{\boldsymbol{M}}\right| \sin(\sigma_M)\sin(\phi_S) \tag{7}$$

If $\beta = 60^{\circ}$, the geomagnetic output of the z -axis is:

$$Mz = \frac{1}{2}\left|\overset{r}{\boldsymbol{M}}\right| \cos(\sigma_M) + \frac{\sqrt{3}}{2}\left|\overset{r}{\boldsymbol{M}}\right| \sin(\sigma_M)\sin(\phi_S) = \frac{1}{2}\boldsymbol{Mx} + \frac{\sqrt{3}}{2}\boldsymbol{My} \tag{8}$$

Due to the high-speed rotation of the projectile during flight, the magnetic azimuth is almost constant over several rotation cycles and can be regarded as a fixed value. Therefore, the geomagnetic output of the x-axis can be regarded as a fixed value related to the total geomagnetic intensity. The geomagnetic output of the y-axis is a sine wave with a zero offset, and the geomagnetic output of the z-axis is a sine wave signal with the same phase as y, as shown in Figure 3.

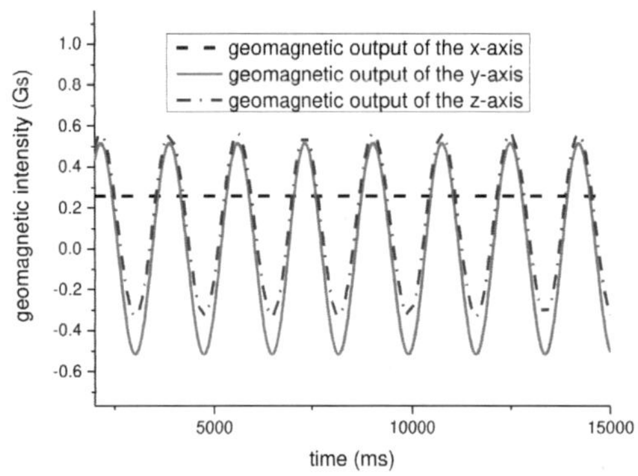

Figure 3. Three axes geomagnetic intensity output.

3.4. Automatic initial calibration method

3.4.1. Calibration parameter determination. When the projectile is flying in the air, the conventional method cannot perform the calibration work. Therefore, the airborne automatic calibration method is performed step by step, that is, the x -axis and y -axis geomagnetic signal output is first calibrated, the z -axis is compensated by the calibrated x -axis and y -axis.

Assume that the geomagnetic output of x -axis and y -axis after ground calibration is (Mx, My), and the high-temperature and high-load during launch causes the geomagnetic output signal to exhibit amplitude change and zero-bias drift, then the actual geomagnetic output is

$$\begin{bmatrix} Mx' \\ My' \end{bmatrix} = \begin{bmatrix} k_{11} & k_{12} \\ k_{21} & k_{22} \end{bmatrix} \begin{bmatrix} Mx \\ My \end{bmatrix} + \begin{bmatrix} b1 \\ b2 \end{bmatrix} \tag{9}$$

The coupling effect between the two axes can be regarded as a small amount, and equation (6) can be simplified as

$$\begin{bmatrix} Mx' \\ My' \end{bmatrix} = \begin{bmatrix} k_{11} & 0 \\ 0 & k_{22} \end{bmatrix} \begin{bmatrix} Mx \\ My \end{bmatrix} + \begin{bmatrix} b1 \\ b2 \end{bmatrix} \tag{10}$$

In equation (6), the matrix $K=\begin{bmatrix} k_{11} & 0 \\ 0 & k_{22} \end{bmatrix}$ is the amplitude change of the two-axis output, that is, the corrected scale factor, and $B=\begin{bmatrix} b1 \\ b2 \end{bmatrix}$ is zero-bias drift.

Since Mx is a fixed value, according to formula (10), the calibration equation is as follows

$$\begin{bmatrix} Mx \\ My \end{bmatrix}=\begin{bmatrix} 1 & 0 \\ 0 & k_{22} \end{bmatrix}^{-1}\left(\begin{bmatrix} Mx' \\ My' \end{bmatrix}-\begin{bmatrix} b1 \\ b2 \end{bmatrix}\right) \tag{11}$$

It can be seen from the above equation that the calibration of the x-axis and y-axis geomagnetic signals requires only three unknown parameters to be calibrated.

3.4.2. Calibration method. The initial calibration time point is selected after the launch of the projectile, just a short period of time from the muzzle, thus avoiding the influence of the launching device and the ground building on the geomagnetic sensor combination, and ensuring the accuracy of the local geomagnetic information is calculated by using the initial position information instead of the instantaneous position information of the projectile.

Before the experiment, the local geomagnetic field information is calculated by using the emission position information, which mainly includes the magnetic declination D, the magnetic inclination I and the total geomagnetic intensity M. Since the projectile has just emerged from the muzzle, it can be considered that the pitch angle of the projectile is the fire angle θ and the yaw angle is zero.

Regardless of the meridional convergence angle, the firing direction is denoted as α_N.According to the geometric relationship between the attitude of the projectile and the direction of the geomagnetism, the value of the magnetic azimuth σ_M can be calculated as follows

$$\cos(\sigma_M)=\cos(I)\cos(D-\alpha_N)\cos(\theta)-\sin(I)\sin(\theta) \tag{12}$$

Then the theoretical output of the geomagnetic intensity of the x-axis calculated by equation (6) is

$$Mx_t=\left|\vec{M}\right|\cos(\sigma_M)=\left|\vec{M}\right|[\cos(I)\cos(D-\alpha_N)\cos(\theta)-\sin(I)\sin(\theta)] \tag{13}$$

Then the theoretical output of the geomagnetic intensity of the y-axis calculated by equation (7) is

$$My_t=\left|\vec{M}\right|\sin(\sigma_M)\sin(\phi_S)=\left|\vec{M}\right|(1-[\cos(I)\cos(D-\alpha_N)\cos(\theta)-\sin(I)\sin(\theta)]^2)^{\frac{1}{2}}\sin(\phi_S) \tag{14}$$

Because the geomagnetic theory of the x-axis outputs a sine wave with zero offset, the theoretical amplitude is

$$Mx_{t\max}=\left|\vec{M}\right|(1-[\cos(I)\cos(D-\alpha_N)\cos(\theta)-\sin(I)\sin(\theta)]^2)^{\frac{1}{2}} \tag{15}$$

During the experiment, the y-axis output signal is detected, and two adjacent maximum value points p_1, p_n and a minimum value point q are acquired, and n is the number of acquisition points. The acquired x-axis geomagnetic output is the actual output $Mx'=\begin{bmatrix} x_1 & x_2 & \dots & x_n \end{bmatrix}$. The acquired y-axis geomagnetic output is the actual output $My'=\begin{bmatrix} y_1 & y_2 & \dots & y_n \end{bmatrix}$.

Then calculate the offset of x-axis geomagnetic output as follows

$$b1=\bar{M}x-Mx_t=\frac{1}{n-1}\sum Mx'-Mx_t \tag{16}$$

The y-axis geomagnetic output is a sine wave with zero offset, and its offset is

$$b2=\frac{1}{n-1}\sum My' \tag{17}$$

The amplitude of the actual output waveform of the y -axis is

$$My_{real} = \frac{1}{4}(p_1 + p_n - 2q) \tag{18}$$

The corrected scale factor of the y -axis is

$$k_{22} = \frac{My_{t\max}}{My_{real}} \tag{19}$$

In this way, the x -axis and the y -axis are calibrated, and the inclined axis z is compensated according to equation (8) by the calibrated x -axis and y -axis.

4. Experimental results and analysis

The position coordinates of the experiment for Verify: 124.4810°E, 47.5177°N, and 157m above sea level. The experimental time is June 20, 2016. The geomagnetic information of the experimental location is detected by the geomagnetic measuring equipment: the total geomagnetic intensity M = 56581.0nT, the magnetic declination D = -10°51', and the magnetic inclination I = 65°33'. The ground calibration experimental equipment is shown in Figure 4.

Figure 4. Non-magnetic turntable and related test equipment.

After the ground calibration, the shooting experiment was carried out. The projectile was launched at an angle of 10.2° and directed at 101.155°. In the experiment, the two sets of geomagnetic intensity data, not automatically calibrated and automatically calibrated, were acquired at the same time. The theoretical geomagnetic intensity output was simulated by the collected ballistic angle information to compare the calibration effects, as shown in Figure 5 and Figure 6.

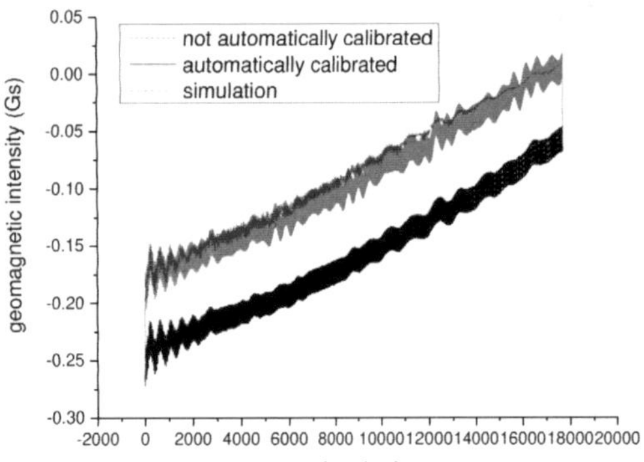

Figure 5. X-axis geomagnetic output calibration effect

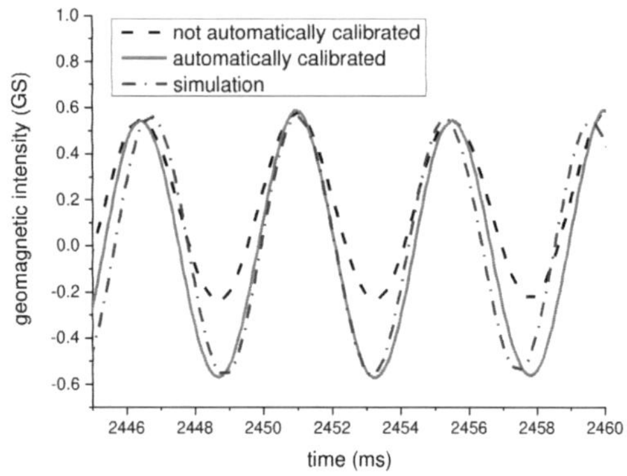

Figure 6. Y-axis geomagnetic output calibration effect.

Figure 5 and Figure 6 compare the geomagnetic intensity output not automatically calibrated, after automatic calibration, and the simulated. Figure 5 is the x -axis geomagnetic intensity output. It can be seen that the geomagnetic intensity output that is not automatically calibrated deviates significantly from the simulated output signal. After the automatic calibration, the geomagnetic intensity output signal is pulled back to a position that is consistent with simulated output signal. At the same time, compared with the analog output signal, the x -axis geomagnetic output collected in the experiment has a swing phenomenon due to the existence of installation errors. However, the presence of installation errors does not affect the results of the verification experiment. Figure 6 is a comparison of the geomagnetic intensity output of the y -axis. Since the rolling speed of the projectile is not constant during the flight, the rolling speed of the simulated projectile cannot be exactly the same as the rolling speed in the experiment, but the geomagnetic intensity output after calibration is more consistent with the simulated intensity output, regardless of amplitude or offset. The z -axis geomagnetic intensity output signal is similar to the y -axis geomagnetic intensity output signal and will not be described again. The comparative effects shown in Figures 5 and 6 demonstrate the effectiveness of the automatic initial calibration method in the air.

5. Conclusion

When the projectile is launched, due to the influence of high temperature and high overload, the remanence of the projectile changes. Only the ground calibration cannot accurately compensate the geomagnetic output of the projectile. It must be initially calibrated in the air. In the calculation of the geomagnetic information required for calibration, since the position of the automatic calibration is close to the emission position, the geographical information of the calibration position can be replaced by the geographic information of the emission location, and the geomagnetic information of the geomagnetic field is calculated by the spherical harmonic model, thereby calculating the automatic calibration coefficient. Then calculate the air calibration coefficient of the bullet. The experimental results show that after the initial automatic calibration in the air, the accuracy of the sensor output is greatly improved, which is very helpful for practical engineering applications.

References

[1] Brunotte X, Meunier G and Bongiraud J 1993 Ship magnetizations modelling by the finite element method *IEEE Trans Magn* **29** 1970-75.

[2] Zhu Jianliang, Wang Xingquan, Wu Panlong, Bo Yuming and Zhang Jie 2012 Three-dimensional magnetic compass error compensation algorithm based on ellipsoid surface fitting *J. Chinese Inertial Technol.* **20** 562-6.

[3] Fei Jie, Yang Yingdong, Wang Weidong and Qian Feng 2013 A Study of Magnetic Calibration in Navigation System Based on LS and Genetic Algorithm *Microcomputer Applications* **29** 17-19.

[4] Rogers J, Costello M and Harkins T 2011 Effective use of magnetometer feedback for smart projectile applications *Navigation* **58** 203-19.

[5] Thomas H and David H 2000 *Magsonde(patent pending): A Device for Making Angular Measurements on Spinning Projectiles using Magnetic Sensors* Proceedings of SPIE Orlando FL USA **4025** 60-67.

[6] Thompson A A 2002 *A procedure for calibrating magnetic sensors* ARL-MR-524 1-2.

IOP Conf. Series: Materials Science and Engineering **646** (2019) 012061 doi:10.1088/1757-899X/646/1/012061

A Case Study on Three-Dimensional Single-Beacon Localization for Autonomous Underwater Vehicles

Huapeng Yu[a] and Chao Han[b]

National Innovation Institute of Defense Technology, Academy of Military Sciences, Beijing 100010, China

[a]hpyu_qtxy@163.com; [b]fatal_han@163.com

Abstract. Circular trajectory is most frequently used when an autonomous underwater vehicle calibrates positions of underwater acoustic beacons or corrects its own navigation errors using beacons. However, the space-time effect on control and navigation data processing and the impact of the angle of attack on three-dimensional localization performance of AUVs have rarely been mentioned in previous studies. To improve the three dimensional localization performance of AUVs with a single-beacon aiding, methods for space-time trade-offs from a control perspective and determination of the optimal angle of attack are investigated in this study. Through theoretical deduction aided with numerical simulation, concerned quantitative values with their corresponding preconditions are given, and some general conclusions which can greatly improve the navigation capability of AUVs have been made.

1. Introduction

For long-duration underwater navigation or reliable control and guidance of the booming autonomous underwater vehicles (AUVs), there is a great demand for acoustic beacon aiding [1, 2]. Meanwhile, homing and navigating using a single-beacon is always the most cost-effective and robust method in complex underwater environments [1, 3-7]. For the geometry of a single-beacon, circular trajectory is easy to realize and be controlled for AUVs. There are lots of published works to show the various applications of circular trajectories [5-10]. However, it should be pointed out that few published studies have discussed in-depth the three-dimensional single-beacon localization issue for AUVs using circular trajectories. Most researchers always project the three-dimensional localization problem to the planar coordinates for simplification. Despite the aid of depth sensors, ignoring the impact of angle of attack on localization performance yet to be possible. Our research is motivated by the fact that, the space-time effect on control and navigation data processing in the planar systems and the impact of the angle of attack on three-dimensional localization performance of AUVs under the circular trajectory case need to be studies sufficiently.

The remaining of this paper is arranged as follows. In Section II, we give some detailed preliminaries and problem statements for AUV navigation with single-beacon aiding. Theoretical analysis about the circular trajectory case study on three-dimensional localization for AUVs is comprehensively presented in Section III. Numerical simulation and theoretical deduction are performed and discussed in Section IV. Finally, Section V summarizes the analysis of the preceding sections.

2. Preliminaries and problem setup

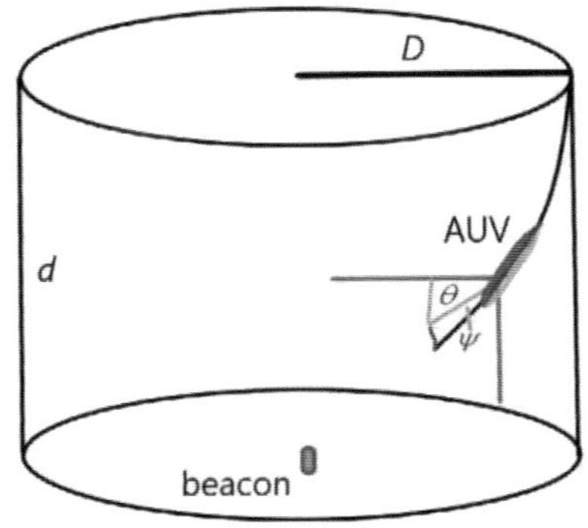

Figure 1. AUV localization in a circular trajectory using a single-beacon.

Figure 1 illustrates the system configuration. Using a single beacon, the AUV maneuvers relatively to accomplish certain tasks, such as homing, position correcting, etc. In Figure 1, we denote the radius of rotation of the AUV in circular trajectory as *D*, the speed of the AUV as *V*, the depth of the AUV as *d*, the angle-of-attack of the AUV as θ, and the sideslip angle of the AUV as ψ.

Take positioning of an unknown beacon using an AUV as an example. Let the position of the single beacon be $P_b = [x_b \quad y_b \quad z_b]^{\mathrm{T}}$ and the real-time position of the AUV be $P_{AUV} = [x_{AUV} \quad y_{AUV} \quad z_{AUV}]^{\mathrm{T}}$, we can obtain the following expression with the acoustic range measurement *r* [3, 4]:

$$2A_P P_b = \begin{bmatrix} (r_1)^2 - (r_2)^2 + \|P_{AUV,2}\| - \|P_{AUV,1}\| \\ (r_2)^2 - (r_3)^2 + \|P_{AUV,3}\| - \|P_{AUV,2}\| \\ (r_3)^2 - (r_4)^2 + \|P_{AUV,4}\| - \|P_{AUV,3}\| \\ (r_4)^2 - (r_1)^2 + \|P_{AUV,1}\| - \|P_{AUV,4}\| \end{bmatrix}, \quad A_P = \begin{bmatrix} (P_{AUV,2} - P_{AUV,1})^{\mathrm{T}} \\ (P_{AUV,3} - P_{AUV,2})^{\mathrm{T}} \\ (P_{AUV,4} - P_{AUV,3})^{\mathrm{T}} \\ (P_{AUV,1} - P_{AUV,4})^{\mathrm{T}} \end{bmatrix} \tag{1}$$

To obtain a unique solution of equation (1), the rank of the left coefficient matrix A_P of P_b matrix must be 3, that is $|A_P| = (P_{AUV,2} - P_{AUV,1})^{\mathrm{T}} \bullet ((P_{AUV,3} - P_{AUV,2}) \times (P_{AUV,4} - P_{AUV,3})) \neq 0$. Therefore, it is illustrated that, only when the changes of angle of attack and sideslip angle of the AUV all occur and these changes are not synchronized, we can obtain the unique solution of the beacon position, which is the basis of the analysis in this paper.

By linearizing the range measurement about a certain point $P_{AUV,0}$, we can describe the system model for correcting navigation error of the AUV with a single beacon [4, 8-10]:

$$r = \|(P_{AUV,0} - P_b)\| + C(P_{AUV,0}, P_b)\delta P_{AUV} + w_r, \quad w_r \text{-} \mathrm{N}(0, \sigma_r) \tag{2}$$

where, δP_{AUV} represents the three-dimensional incremental travel displacement, w_r denotes range measurement noise, and

$$C(P_{AUV,0}, P_b) = 2\left[(x_{AUV,0} - x_b)/r \quad (y_{AUV,0} - y_b)/r \quad (z_{AUV,0} - z_b)/r\right] \tag{3}$$

Thus, the localization performance can be evaluated in the Cramér Rao lower bound (CRLB) form through assembling multiple measurements estimation performance in the Cramér Rao lower bound (CRLB) form:

$$\mathbf{E} = \left[\mathbf{C}(P_{AUV,0}, P_b)^{\mathrm{T}} \boldsymbol{\sigma}_r^{-1} \mathbf{C}(P_{AUV,0}, P_b)\right]^{-1} \tag{4}$$

This tool would be used and discussed in-depth later.

3. Three-dimensional localization solution

3.1. Sampling period from a control perspective

First, to obtain the optimal numerical solution of the sampling period, we only consider the navigation and control of an AUV in the planar rectangular coordinate system.

Assuming the sampling period is T, the Jacobian can be then evaluated at the fixed position of the single beacon. Without losing generality, the initial projection is considered to be located on the x axis (Figure 1) of the rectangular coordinate system centered on the beacon. Thus, we have:

$$\mathbf{C}\left(P_{AUV,0}, P_b\right) = -\frac{2}{D}\begin{bmatrix} D & 0 \\ D\cos\left(VT/D\right) & -D\sin\left(VT/D\right) \\ D\cos\left(2VT/D\right) & -D\sin\left(2VT/D\right) \end{bmatrix} \tag{5}$$

The CRLB representing the geometry of the single beacon can be written as follows:

$$\begin{aligned} \mathbf{E}^{2\mathrm{D}} &= \sigma_r\left[\mathbf{C}\left(P_{AUV,0}, P_b\right)^{\mathrm{T}}\mathbf{C}\left(P_{AUV,0}, P_b\right)\right]^{-1} \\ &= \frac{\sigma_r\begin{bmatrix} \sin\left(VT/D\right)^2 + \sin\left(2VT/D\right)^2 & \sin\left(2VT/D\right)/2 + \sin\left(4VT/D\right)/2 \\ \sin\left(2VT/D\right)/2 + \sin\left(4VT/D\right)/2 & 1 + \cos\left(VT/D\right)^2 + \cos\left(2VT/D\right)^2 \end{bmatrix}}{4\mathrm{DET}\left(\begin{bmatrix} \sin\left(VT/D\right)^2 + \sin\left(2VT/D\right)^2 & \sin\left(2VT/D\right)/2 + \sin\left(4VT/D\right)/2 \\ \sin\left(2VT/D\right)/2 + \sin\left(4VT/D\right)/2 & 1 + \cos\left(VT/D\right)^2 + \cos\left(2VT/D\right)^2 \end{bmatrix}\right)} \end{aligned} \tag{6}$$

where, the mathematical symbol "DET" denotes solving the determinant for a matrix; σ_r represents the range measurement noise variances between the AUV and the single beacon. Apparently, σ_r on both axes have been assumed to equal in the planar rectangular coordinate system for simplification.

A scalar metric commonly known in the field of Global Positioning System (GPS) applications, the dilution of position (*DOP*), is used here to further predict the navigation performance. From equation (6), a navigation performance index in *DOP* form can be obtained. And, the resulting performance index equation can be expressed as:

$$\boldsymbol{DOP} = \frac{3\sigma_r}{4\left\{\left(\sin^2\frac{VT}{D} + \sin^2\frac{2VT}{D}\right)\left[3 - \left(\sin^2\frac{VT}{D} + \sin^2\frac{2VT}{D}\right)\right] - \left(\frac{1}{2}\sin\frac{2VT}{D} + \frac{1}{2}\sin\frac{4VT}{D}\right)^2\right\}} \tag{7}$$

3.2. Determination of the optimal angle of attack

In order to analyze the impact of the angle of attack on three-dimensional localization performance, the depth of the AUV (d) needs to be considered. Besides, to intuitively investigate the system performance, the Jacobian $\mathbf{C}\left(P_{AUV,0}, P_b\right)$ is expressed in the single-beacon centered polar coordinate system as follows:

$$\mathbf{C}^{3\mathrm{D}}\left(P_{AUV,0}, P_b\right) = 2\begin{bmatrix} e_1^x & e_1^y & e_1^z \\ e_2^x & e_2^y & e_2^z \\ e_3^x & e_3^y & e_3^z \\ e_4^x & e_4^y & e_4^z \end{bmatrix} = 2\begin{bmatrix} \sin\theta_1\cos\psi_1 & \sin\theta_1\sin\psi_1 & \cos\theta_1 \\ \sin\theta_2\cos\psi_2 & \sin\theta_2\sin\psi_2 & \cos\theta_2 \\ \sin\theta_3\cos\psi_3 & \sin\theta_3\sin\psi_3 & \cos\theta_3 \\ \sin\theta_4\cos\psi_4 & \sin\theta_4\sin\psi_4 & \cos\theta_4 \end{bmatrix} \tag{8}$$

where, the subscripts of e, θ and ψ denote continuous multiple measurements.

Similarly, the range measurement noise level projected on the three polar coordinate axes are assumed to be the same, and their variances are denoted as σ_r. Then, we can derive the CRLB estimation.

$$\mathbf{\Lambda}^{3D} = \sigma_r \begin{bmatrix} M_1 & M_2 \\ M_3 & M_4 \end{bmatrix}^{-1} = \sigma_r \begin{bmatrix} M_1 & M_2 \\ M_3 & M_4 \end{bmatrix}^{*} \Bigg/ \mathrm{DET}\begin{bmatrix} M_1 & M_2 \\ M_3 & M_4 \end{bmatrix} \tag{9}$$

where, the superscript of a matrix represents its adjoint matrix, and

$$M_1 = \sum_{i=1}^{4} \begin{bmatrix} \sin^2\theta_i \cos^2\psi_i & \sin^2\theta_i \cos\psi_i \sin\psi_i \\ \sin^2\theta_i \cos\psi_i \sin\psi_i & \sin^2\theta_i \sin^2\psi_i \end{bmatrix} \tag{10}$$

$$M_2 = \sum_{i=1}^{4} \begin{bmatrix} \frac{1}{2}\sin 2\theta_i \cos\psi_i \\ \frac{1}{2}\sin 2\theta_i \sin\psi_i \end{bmatrix} \tag{11}$$

$$M_3 = \sum_{i=1}^{4} \begin{bmatrix} \frac{1}{2}\sin 2\theta_i \cos\psi_i & \frac{1}{2}\sin 2\theta_i \sin\psi_i \end{bmatrix} \tag{12}$$

$$M_4 = \sum_{i=1}^{4} \cos^2\theta_i \tag{13}$$

Using the general matrix operating formula $\mathrm{DET}\left(\begin{bmatrix} A & B \\ C & D \end{bmatrix}\right) = \mathrm{DET}(D)\mathrm{DET}\left(A - BD^{-1}C\right)$, a conclusion can be made that the determinant of the matrix $\begin{bmatrix} M_1 & M_2 \\ M_3 & M_4 \end{bmatrix}$ is the largest when the following conditions hold:

$$\begin{cases} \sum_{i=1}^{4} \sin^2\theta_i \sin 2\psi_i = 0, \quad \sum_{i=1}^{4} \sin^2\theta_i \cos 2\psi_i = 0 \\ \arg\min \left(\dfrac{\sum_{i=1}^{4} \begin{bmatrix} \sin 2\theta_i \cos\psi_i \\ \sin 2\theta_i \sin\psi_i \end{bmatrix} \sum_{i=1}^{4} \begin{bmatrix} \sin 2\theta_i \cos\psi_i & \sin 2\theta_i \sin\psi_i \end{bmatrix}}{\sum_{i=1}^{4} \cos^2\theta_i} \right) \end{cases} \tag{14}$$

Thus,

$$\mathrm{DET}\begin{bmatrix} M_1 & M_2 \\ M_3 & M_4 \end{bmatrix} = \sum_{i=1}^{4} \cos^2\theta_i \left(2 - \frac{1}{2}\sum_{i=1}^{4} \cos^2\theta_i \right)^2 \tag{15}$$

Under the predefined condition (14), the CRLB estimation is the lowest, and thus the localization performance is the best-case estimation.

4. Results and discussions

4.1. Sampling period

To analytically solve the derived navigation performance index equation (7) in *DOP* form is rather complicated. In this subsection, we will determine the sampling period by synthesizing theoretical deduction and simulation.

As mentioned before, the control of the AUV and data processing of navigation are key issues for autonomous underwater travel. From a control perspective, the incremental rudder angle for a circular trajectory can be expressed as $\omega = VT/D$. Therefore, through numerical simulation, we can derive the relative *DOP* value with incremental rudder angle variation. From the simulation results, Figure 2 is plotted for intuitive investigation on characteristics of rudder control scheme, and Figure 3 is plotted for comparative analysis.

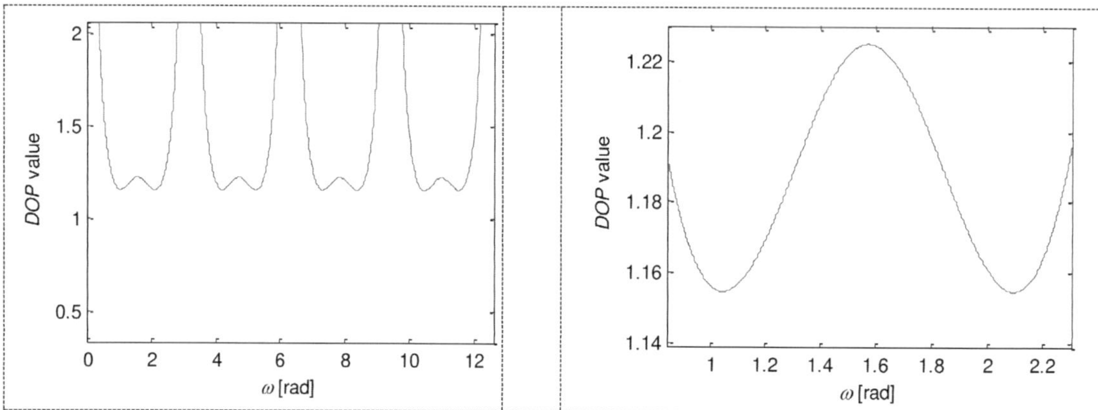

Figure 2. DOP value versus incremental rudder angle and locally enlarged drawing.

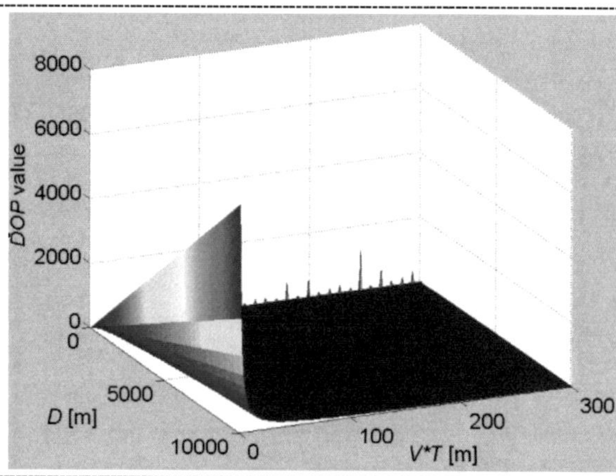

Figure 3. A commonly used navigation and correction Configuration.

In Figure 2, it can be seen that, the DOP values reaches their local extreme points when the incremental rudder angle accumulates up to integer multiple of $\pi/3$ or $2\pi/3$. Apparently, the AUV localization performance in a circular trajectory of the single-beacon geometry are strongly relative with control maneuver scheme, which are determined by the speed *V*, the radius of rotation *D* and the sampling period *T* of the AUV. Compared to Figure 3, it is obviously that the *DOP* values decrease rapidly with the distance the AUV travelled during the sampling period and increase linearly with the radius of rotation. Subsequently, the AUV can determine its data processing period for optimal navigation and control using real-time online measurements about *D* and *V*, which has provided the AUV with a degree of decision and control intelligence.

4.2. The optimal angle of attack

By getting the maximum value of the determinant $\mathrm{DET}\left[\begin{array}{c|c} M_1 & M_2 \\ \hline M_3 & M_4 \end{array}\right]$, the lowest CRLB estimation of Λ^{3D} can be obtained, which means that the localization error of the AUV would be restrained to the greatest extent.

Substituting $\sum_{i=1}^{4}\cos^2\theta_i = \xi$ into the derive determinant, we have:

$$\mathrm{DET}\left[\begin{array}{c|c} M_1 & M_2 \\ \hline M_3 & M_4 \end{array}\right] = \xi\left(2-\frac{1}{2}\xi\right)^2 \tag{16}$$

Differentiating equation (16), it can be easily concluded that the derived determinant reaches its maximum value when $\xi = 4/3$. Then, when a constant angle of attack is chosen, the optimal θ value is determined:

$$\cos^2\theta = 1/3 \quad \Rightarrow \quad \theta = 54.74^{\circ} \tag{17}$$

5. Conclusions

This study makes a thorough analysis on the three-dimensional single-beacon localization problems for AUVs in circular trajectory. Theoretical and numerical methods for space-time trade-offs from a control perspective and determination of the optimal angle of attack of the AUV are proposed and verified. Our future work will focus on how to apply these methods to in-motion usage of the AUV with a moving single beacon and to evaluate the sea current influences on three-dimensional localization.

Acknowledgments

This research was supported by the project of the National Natural Science Foundation of China (Grant No. 61803381).

References

[1] J. Sliwka, B. Clement, and I. Probst. Sea glider guidance around a circle using distance measurements to a drifting acoustic source. In *Proceedings of 2012 IEEE/RSJ International Conference on Intelligent Robots and Systems*, pages 94-99, Vilamoura, Algarve, Portugal, 7-12 October 2012.

[2] H. Yu and X. Zhou. Observability Degree-Based AUV Single Beacon Navigation Trajectory Optimization with Range-Only Measurements. *Complexity*, 2018: 1-10, 2018, https://doi.org/10.1155/2018/4056870.

[3] M. Liu, T. Dong, and L. Zhang. Underwater SLAM navigation algorithm based on random beacons. *Systems Engineering and Electronics*, 37(12): 2830-2834, 2015.

[4] A. Rahman, V. Muthukkumarasamy, and X. Wu. Coordinates and Bearing of Submerged Sensors Using a Single Mobile Beacon (CSMB). *Journal of Networks*, 10(8): 462-469, 2015.

[5] P. Baccou, B. Jouvencel, and V. Creuze. Homing and navigation using one transponder. In *Proceedings of IEEE International Conference on Robotics and Automation*, Washington DV, USA, 2002.

[6] A. Pavin, A. Inzartsev, and Y. Matvienko. Experience of AUV Automatic Homing to Hydroacoustic Transponder. In *Proceedings of Sixth International Symposium on Underwater Technology*, pages 201-206, Wuxi, China, April 2009.

[7] G. Vallicrosa, P. Ridao, D. Ribas, and A. Palomer. Active Range-only Beacon Localization for AUV Homing. In *Proceedings of 2013 IEEE/RSJ International Conference on Intelligent Robots and Systems*, pages 2286-2291, Chicago, IL, USA, 14-18 September 2014.

[8] I. N. Burdinsky, and S. A. Otcheskiy. AUV Localization Using a Single Transponder Acoustic Positioning System. *JOURNAL OF ATMOSPHERIC AND OCEANIC TECHNOLOGY*, 19:1860-1872, 2002.

[9] M. Pedro, D. Moreno-Salinas, N. Crasta, and A. Pascoal. Underwater Single-Beacon Localization: Optimal Trajectory Planning and Minimum-Energy Estimation. In *Proceedings of IFAC Workshop on Navigation, Guidance and Control of Underwater Vehicles*, Girona, Spain, 28-30 April 2015.

[10] P. Baccou, and B. Jouvencel. Simulation results, Post-processing experimentations and Comparisons results for navigation, homing and multiple vehicles operations with a new Positioning method using one transponder. In *Proceedings of 2003 IEEE/RSJ International Conference on Intelligent Robots and Systems,* pages 811-817, Las Vegas, NV, USA, 27-31 October 2003.

IOP Conf. Series: Materials Science and Engineering **646** (2019) 012062 doi:10.1088/1757-899X/646/1/012062

Vision-based Autonomous Landing of UAV on Moving Platform using a New Marker

Huang Xiaoyun[1,*], **Xu Qing**[1] **and Wang Jianqiang**[1]

[1]School of Vehicle and Mobility, Tsinghua University, Hai Dian, Beijing, 100084

[*]hxly0820@126.com

Abstract. The research of Unmanned Aerial Vehicle (UAV) flight autonomous landing on a moving platform based on computer vision has important significance, especially for the rescue and spray UAV in the future. A vision-based relative position and attitude estimation algorithm with a novel color marker and its recognition method was proposed for the UAV autonomous landing system. The new marker is designed with a pattern which make it is as visible as possible in the whole landing process. The proposed relative position and attitude estimation algorithm is based on the four corner points on the marker which are detected by the combined approach of vanishing point of parallel lines and Levenberg-Marquardt (L-M) optimization method. Indoor and outdoor experiments were carried out. The results indicate that the method of marker detection has real-time and robust performance and the position and attitude estimation accuracy attain cm-level, which shows the feasibility of landings on moving platform automatically.

1. Introduction

As an indispensable part of mini UAV, vision system has been paid great attention by colleges and research organizations in the world. UAV is increasingly replacing manned systems in situations that are dangerous, remote, or difficult for manned aircraft to access. UAV stabilization by pose estimation from visual sensors is an active research area. It can be used where Global positioning sensor (GPS) or Inertial Navigation Sensor (INS) system are not available or with other sensors to obtain better estimations. Vision-based techniques are proving their effectiveness and robustness in handling this problem [1, 2].

Vision-based methods were first introduced by Ettinger [3]. He proposed to install a perspective camera on a Micro AIR Vehicle (MAV) to have vision-guided flight stability. After that, Omnidirectional sensors (such as fisheye and catadioptric cameras) attitude estimation was also arisen [4, 5]. Omnidirectional sensors have the advantage of wide surrounding of the UAV, which make horizon is always present in the image. However, these images contain significant deformations due to the geometry of the mirror and to the sampling of the camera. So they are usually used in high altitude.

There are basically three group methods of position and attitude estimation with monocular camera. The first group tries to detect the skyline in the world to estimate the up direction and hence the attitude of the vehicle [6]. Nevertheless, the skyline might not be visible in urban environments. The second group tries to detect a horizontal reference from artificial structures to estimate the pose of UAV [7, 8], such as structures and runway. However, artificial structures are not existed everywhere. Most of them could not obtain the position estimation.

The third group tries to use an artificial marker which could be recognized on a platform by UAV vision system [9]. Then, the position and altitude information could be estimated from the functional relationship between the coordination of image and camera. Sharp, C.S. et al [10] designed a marker which consists of 6 rectangles. They used the corners' information of those rectangles to estimate UAV pose. Lei et al [11] developed a color marker which consists of 6 solid circles. In [12], the marker consists of one hollow and solid circle and ten points on them were detected. The centres of them were extracted based on joint method of projection and wavelet transform. Na Meng et al [13] used four corners information to calculate UAV pose based on a rectangular marker made up of a red and blue square. All of those methods are based on point information on marker. Some researchers used lines information from markers, such as Liu Shiqing [14] and Guili X [15]. They estimated the UAV pose using the vanishing point of parallel lines on the marker. Comparing with point-based method, line-based method is more stable and reliable due to more ample information is used. Additionally, there are other markers which are designed to use the surface information under some assumptions, such as area [16] and image moment [17]. However, surface-based method is difficult to obtain all of the pose information with complex calculation. Furthermore, those markers hardly meet the requirement of landing on a moving platform.

Landing on a moving platform becomes to be a hot and difficult issue in recent years [18-21]. Borowczyk [22] at el used the marker of AprilTag to estimate relative position and velocity with kalman filter. Borshchova [23] at el achieved relative pose estimation using color-based object detection. Abdullah M. Almeshal [24] designed a vision-based neural network controller to track the target for autonomous Landing. Overall, most of them are under the stage of simulation. Usually, the marker or target will be out of the field of camera view when the UAV land at a lower altitude as a big size of it is needed for high altitude.

In this paper, in order to make the marker is as visible as possible in the whole landing process, a new color marker is designed as well as its detection method with the single perspective camera. A combining pose estimation algorithm based on vanishing point of parallel lines on the marker and L-M with lightweight computation and strong robustness is proposed.

This paper is organized as follows: in Section 2, a system and equipment description is developed; Section 3 describes how to obtain the relative position and attitude estimation based on vision system; Section 4 presents a designed landing marker and its detection approach; experimental results of the whole system are described in Section 5; finally, Section 6 concludes the paper.

2. System Description

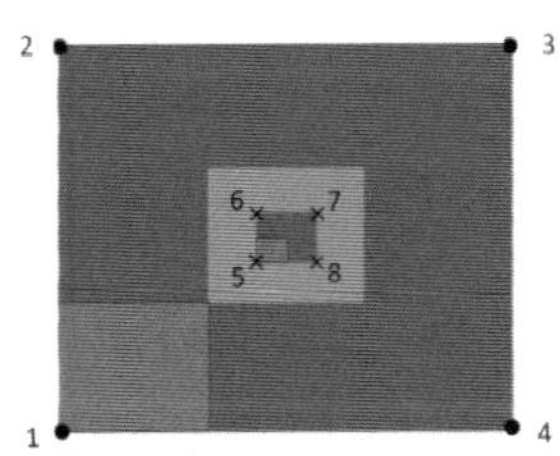

Figure 1. The marker designed in our application.

In our study, the vision-based autonomous landing system is consisted of a UAV with monocular camera and a landing marker which could be put on the moving or un-moving platform. There also have some conventional on-board sensors on the UAV, such as inertial measure unit (IMU) and GPS.

As we assume that the each part size of the marker is known, the 3-D reconstruction could be calculated using the projecting model and the camera calibration parameters by the vision system. We also assume the UAV computer system can receive the real-time GPS location of the landing platform. Therefore, UAV could fly to the around area to prepare for landing. In the light of the following four factors, a RGB tri-phosphor based marker is designed as shown in figure 1. a) Compared to black-and-white marker, color marker helps to distinguish from the surrounding environment with rich color information; b) Red, green and blue are basic three primary colors, which helps to detect points or lines from the marker; c) Because of landing on a moving platform, the location of the marker may changes at any time. The relative position and attitude should be calculated throughout the whole landing processing; d) As the platform may be moving, the direction is also necessary.

So, the new marker consists of two blue parts as undertone, two red parts as directional reference and a green part as an isolation belt. It is painted on the spot landing surface. And the centre point is the landing spot. The bigger blue and red blocks on the outer circle of the marker constitute a pattern that will be used when the UAV flies at high altitude. On the other hand, when UAV is close to the landing spot at low altitude, this pattern will be out of the camera view. The smaller blue and red blocks in the isolation belt will be detected as an inner pattern to support the autonomous landing of UAVs on a moving platform. Flow diagram of landing program is shown in figure 2.

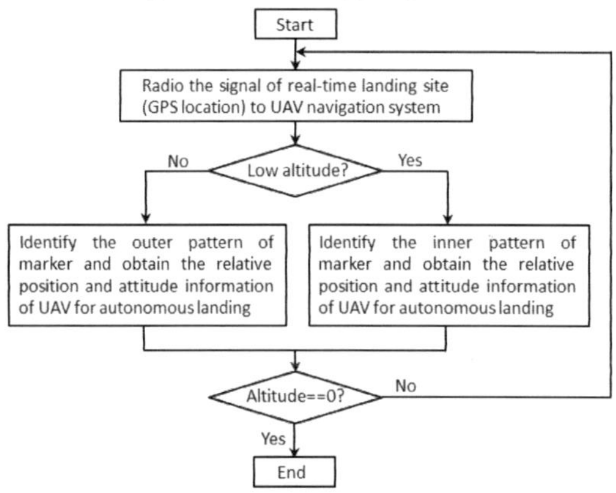

Figure 2. Flow diagram of landing program.

3. Vision-based Position and Attitude Estimation for UAV

3.1. Vision-based Position and Attitude Estimation Principle

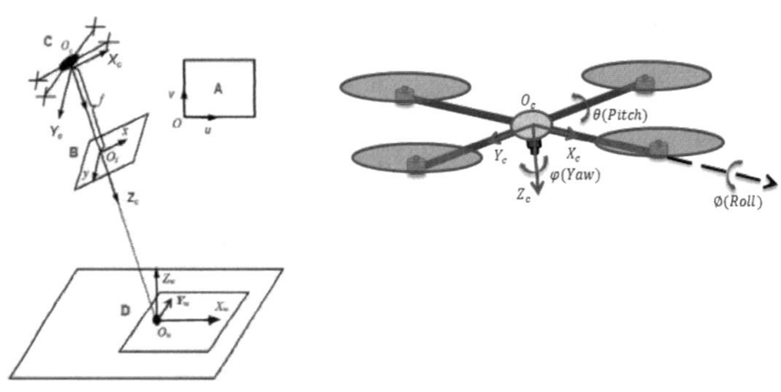

Figure 3. Relationship between different coordinate system.

Shown in figure 3 are the coordinate systems adopted in the UAV vision systems. There are four following coordinate in the system:

- WCS (X_W, Y_W, Z_W): World coordinate system, x_w–y_w plane is parallel to ground;
- CCS (X_C, Y_C, Z_C): Camera coordinate system, which represents the camera mounted on the UAV;
- IPCS (x, y): Image plane coordinate system, which means the coordinate system of the sensor;
- ICS (u, v): Image coordinate system, which is a two-dimensional coordinate system. The origin is set at the lower left point.

The aim of this research is to estimate the relative position (x_W, y_W, z_W) and attitude $(\emptyset, \theta, \varphi)$ of UAV in WCS from ICS. The relationship between ICS and WCS coordinate systems could be derived from the relationship of ICS, IPCS and CCS [10].

The relationship between CCS and WCS coordinate systems could be descripted as follow:

$$\begin{bmatrix} X_C \\ Y_C \\ Z_C \end{bmatrix} = R \begin{bmatrix} X_W \\ Y_W \\ Z_W \end{bmatrix} + T \tag{1}$$

Where, R is the rotation matrix of the camera coordinate system relative to the world coordinate system. T is the three-dimensional translation vector from the world coordinate to the camera coordinate system.

The rotation matrix is defined as follow:

$$R = \begin{bmatrix} r_{11} & r_{12} & r_{13} \\ r_{21} & r_{22} & r_{23} \\ r_{31} & r_{32} & r_{33} \end{bmatrix} =$$

$$\begin{bmatrix} \cos\theta\cos\varphi & -\cos\theta\sin\varphi & -\sin\theta \\ \cos\emptyset\sin\varphi - \sin\emptyset\sin\theta\cos\varphi & \cos\emptyset\cos\varphi + \sin\emptyset\sin\theta\sin\varphi & -\sin\emptyset\cos\theta \\ \sin\emptyset\sin\varphi + \cos\emptyset\sin\theta\cos\varphi & \sin\emptyset\cos\varphi - \cos\emptyset\sin\theta\sin\varphi & \cos\emptyset\cos\theta \end{bmatrix} \tag{2}$$

The three-dimensional translation vector is as follow:

$$T = \begin{bmatrix} t_x \\ t_y \\ t_z \end{bmatrix} \tag{3}$$

Combining the relationship between ICS and CCS, the relationship between ICS and WCS coordinate systems is:

$$Z_C \begin{bmatrix} u \\ v \\ 1 \end{bmatrix} = \begin{bmatrix} f_u & 0 & u_0 \\ 0 & f_v & v_0 \\ 0 & 0 & 1 \end{bmatrix} \begin{bmatrix} r_{11}x_w + r_{12}y_w + r_{13}z_w + t_x \\ r_{21}x_w + r_{22}y_w + r_{23}z_w + t_y \\ r_{31}x_w + r_{32}y_w + r_{33}z_w + t_z \end{bmatrix} \tag{4}$$

equal to:

$$\begin{cases} u = f_u \dfrac{r_{11}x_w + r_{12}y_w + r_{13}z_w + t_x}{r_{31}x_w + r_{32}y_w + r_{33}z_w + t_z} + u_0 \\ v = f_v \dfrac{r_{21}x_w + r_{22}y_w + r_{23}z_w + t_y}{r_{31}x_w + r_{32}y_w + r_{33}z_w + t_z} + v_0 \end{cases} \tag{5}$$

Where, assume f is the focal length of camera, $f_u = f/d_u$, $f_v = f/d_v$. (d_u, d_v) is the pixel size of the camera sensor. (u_0, v_0) are the pixel value of the IPCS origin in ICS. Assume that we have coordinates in ICS and WCS of n points, a set of equation by equation (5) will be shown as follow:

$$\begin{cases} (u_1 - u_0)(r_{31}x_{w1} + r_{32}y_{w1} + r_{33}z_{w1} + t_z) - f_u(r_{11}x_{w1} + r_{12}y_{w1} + r_{13}z_{w1} + t_x) = 0 \\ (v_1 - v_0)(r_{31}x_{w1} + r_{32}y_{w1} + r_{33}z_{w1} + t_z) - f_v(r_{21}x_{w1} + r_{22}y_{w1} + r_{23}z_{w1} + t_y) = 0 \\ \vdots \\ (u_n - u_0)(r_{31}x_{wn} + r_{32}y_{wn} + r_{33}z_{wn} + t_z) - f_u(r_{11}x_{wn} + r_{12}y_{wn} + r_{13}z_{wn} + t_x) = 0 \\ (v_n - v_0)(r_{31}x_{wn} + r_{32}y_{wn} + r_{33}z_{wn} + t_z) - f_v(r_{21}x_{wn} + r_{22}y_{wn} + r_{23}z_{wn} + t_y) = 0 \end{cases} \tag{6}$$

In this set of equation, f_u, f_v, u_0, v_0 can be obtained from camera calibration. Theoretically, if we have the coordinates in ICS and WCS of 3 points at least, t_x, t_y, t_z, $\emptyset$, θ and φ could be calculated employing only the equations set by combining equation (2) and (6). However, due to the existing of error from calibration and point detection, the exact solution of the equations probably cannot be obtained. So, a optimization method of L-M is employed in this work.

3.2. Pose Estimation using L-M and Vanishing Point Method

Setting $\boldsymbol{x} = (\emptyset, \theta, \varphi, t_x, t_y, t_z)$ and $\boldsymbol{F}(\boldsymbol{x}) = (\boldsymbol{f}_1(\boldsymbol{x}), \boldsymbol{f}_2(\boldsymbol{x}), \cdots, \boldsymbol{f}_{2n}(\boldsymbol{x}))^T$ (means that $\boldsymbol{F}(\boldsymbol{x})$ represents the left side of equation (6)), the solution of equation (6) could be described as an optimization problem as equation (7).

$$\min \boldsymbol{F}(\boldsymbol{x}) = \min \sum_{i=1}^{2n} \|\boldsymbol{f}_i(\boldsymbol{x})\|^2 \tag{7}$$

Levenberg-Marquardt (L-M) is a widely used least square method to calculate the minimum value of multidimensional nonlinear equations. For this issue, the objective function is $\boldsymbol{F(x)=0}$. The steps of L-M method are shown as follows:

Step1: Set termination control parameter $\boldsymbol{\varepsilon}$, damping variable $\boldsymbol{\lambda_0}$ and $\boldsymbol{s}$. Initializing count of number $\boldsymbol{k=0}$, $\boldsymbol{x=x_0}$, get $\boldsymbol{\varepsilon_k=\|F(x_0)\|}$;

Step2: Calculate Jacobi matrix $\boldsymbol{J_k}$ and $\boldsymbol{\bar{N}_k={J_k}^T J_k+\lambda_k I}$;

Step3: Calculate adjustment coefficient $\boldsymbol{\delta_k}$ from incremental normal equation $\boldsymbol{\bar{N}_k\delta_k={J_k}^T\varepsilon_k}$;

Step4:

a) If $\boldsymbol{\|F(x_k+\delta_k)\|<\varepsilon_k}$ and $\boldsymbol{\|\delta_k\|<\varepsilon}$, stop the loop. Export the result $\boldsymbol{x_{k+1}=x_k+\delta_k}$;

b) If $\boldsymbol{\|F(x_k+\delta_k)\|<\varepsilon_k}$ and $\boldsymbol{\|\delta_k\|\geq\varepsilon}$, set $\boldsymbol{x_{k+1}=x_k+\delta_k}$, $\boldsymbol{\lambda_{k+1}=\lambda_k/s}$, $\boldsymbol{k=k+1}$, then go to step 2;

c) If $\boldsymbol{\|F(x_k+\delta_k)\|\geq\varepsilon_k}$, set $\boldsymbol{\lambda_{k+1}=s\times\lambda_k}$, then go to step 2.

Initial value $\boldsymbol{x_0}$ is important for L-M method. Otherwise, the result will be diverging with an improper value. When UAV vision system has been under tracking processing, the estimation result of this moment could be used as $\boldsymbol{x_0}$ for next moment. However, at the start of this system, we have to use another way to get $\boldsymbol{x_0}$, which as explained as followed.

A vanishing point is a point on the image plane of a perspective drawing where the two-dimensional perspective projections (or drawings) of mutually parallel lines in three-dimensional space appear to converge. Assume that there is a line on the two-dimensional plant (X_W, Y_W) and the equation is $\boldsymbol{y=ax+b}$ ($\boldsymbol{z=0}$, $\boldsymbol{a}$ is the slope of this line), combining equation (5) the vanishing point $P(u_p, v_p)$ will be as follows:

$$\begin{aligned}(u_p, v_p) &= (\lim_{x\to\infty} f_u\frac{(r_{11}+r_{12}a)x+r_{12}b+t_x}{(r_{31}+r_{32}a)x+r_{32}b+t_z}+u_0,\ \lim_{x\to\infty} f_v\frac{(r_{21}+r_{22}a)x+r_{22}b+t_y}{(r_{31}+r_{32}a)x+r_{32}b+t_z}+v_0)\\ &= (f_u\frac{r_{11}+r_{12}a}{r_{31}+r_{32}a}+u_0,\ f_v\frac{r_{21}+r_{22}a}{r_{31}+r_{32}a}+v_0)\end{aligned} \tag{8}$$

From equation (8), we can see that the coordinates of vanishing point in ICS is only related to the slope $\boldsymbol{a}$ except the internal parameters $\boldsymbol{f_u, f_v, u_0}$ **and** $\boldsymbol{v_0}$. These internal parameters could be calculated by camera calibration. Therefore, if we set $\boldsymbol{a=0}$, the vanishing point $\boldsymbol{P_0(u_{p\_0}, v_{p\_0})}$ of the lines parallel to X_W-axis will be as follow:

$$(u_{p\_0}, v_{p\_0}) = (f_u\frac{r_{11}}{r_{31}}+u_0,\ f_v\frac{r_{21}}{r_{31}}+v_0) \tag{9}$$

In addition, the roll angle $\emptyset$ is usually very small in landing. In order to get the initial values which close to the true value, we also set $\emptyset_0=0$, combining equation (2) the vanishing point $\boldsymbol{P_0(u_{p\_0}, v_{p\_0})}$ will be converted to:

$$(u_{p\_0}, v_{p\_0}) = (f_u\frac{1}{\tan\theta_0}+u_0,\ f_v\frac{\tan\varphi_0}{\sin\theta_0}+v_0) \tag{10}$$

When $\boldsymbol{(u_{p\_0}, v_{p\_0})}$ is obtained from image process, the $\boldsymbol{\theta_0}$ and $\boldsymbol{\varphi_0}$ could be calculated from Eqs.13. Then, combining equation (5) the initial value of translation vector $\boldsymbol{T_0(t_{x\_0}, t_{y\_0}, t_{z\_0})}$ could be obtained too. Lastly, using all of the initial values into the steps of L-M method, the final result of relative position and attitude will be calculated.

4. Marker Detection Algorithm

As mentioned in section 3.1, theoretically, only 3 points are enough to calculate t_x, t_y, t_z, $\emptyset$, θ and φ. In this paper, we use the corners of outer or inner pattern. See figure 1, when the outer pattern is used, the four black corner points 1~4 are detected for calculating relative position and attitude. When the inner pattern is used, the four corner points 6~8 are detected. On the other hand, once the four corner points are detected, the sidelines of marker rectangle are also determined. Then vanishing point will be calculated by these sidelines which parallel to X_W-axis in WCS.

4.1. Detection of outer marker pattern

The flow chart of outer marker pattern detection method is illustrated in figure 4. This method includes two main stages: marker region detection and target points detection. Marker region detection is to segment marker region from whole image. Target points detection is to detect corner points and location from segmented region. Then the needed vanishing point coordinate could be calculated.

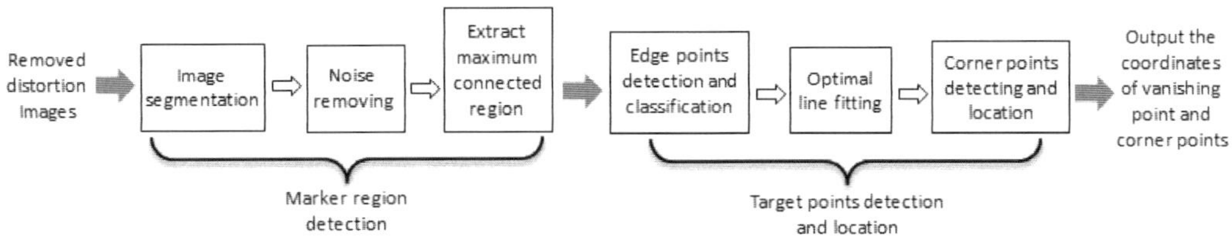

Figure 4. The flow chart of outer pattern detection algorithm.

This new marker is designed based on color. So color information will be used in the detection method. For RGB model, the color image could be divided into R, G and B three gray images. In order to keep blue and red pixels and wipe off background, the calculation between three gray images will be employed. First, the R, G and B images are divided from color image; then, different-images of B-G and R-G are calculated; lastly, the different-images are merged into a binaryzation image by union. However, small noises and holes often remain in the binary images. So, open and close processes are employed to remove them (see figure 5).

Target edge points detection and location. In order to detect the four corner points which are shown in figure 6, we had tried to use common corner detection method such as Harris. However, because of the noise, the edges of marker region are generally jagged and irregular. Some ruleless points are detected, which lead to more difficult to identify the true corners. On the other hand, line features are more reliable than point features in real environment. So, line edge detection algorithm is employed to identify the four marker edges first. Then, the coordinate values of corner points will be calculated from these linear equations.

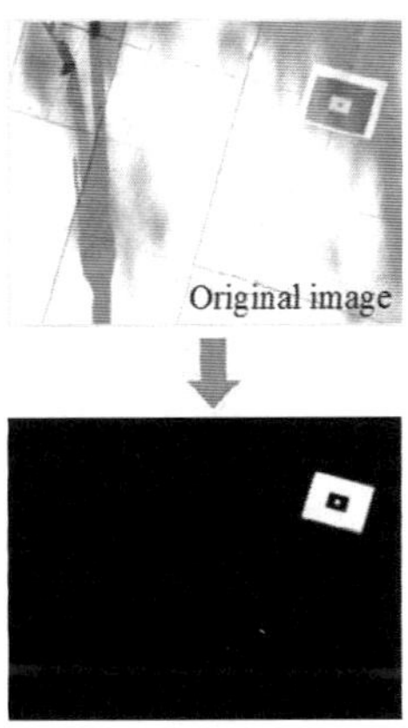

Figure 5. Result of marker region detection

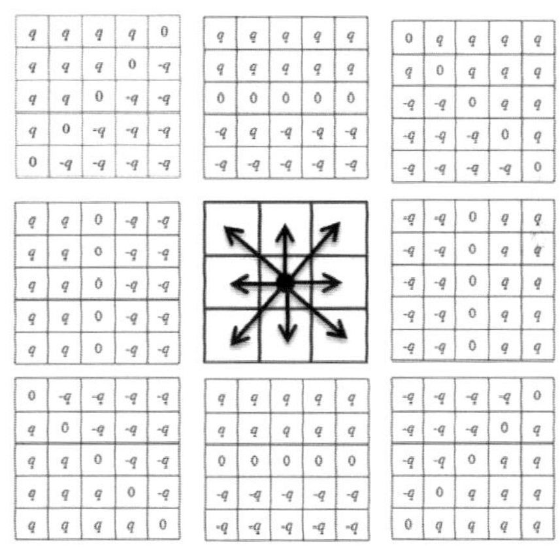

Figure 6. Gradient operators for 8 directions

(1) Edge points classification

Given that the points on the same edge have same or similar gradient characteristic including gradient magnitude and direction. The point gradient characteristic could be the classification judgment to identify whether in the same edge. For this work, we designed 8 gradient operators to calculate the gradient magnitude of the 8 directions for every pixel (see figure 5, $q = 0.1$). The pixel value means the gradient magnitude in this direction. So, the larger the gradient of the pixel, the value in result image is closer to white. So, the points on the same edges present max magnitude in the same direction. Because the outer marker has the maximum connected region, the inner pixels could be removed.

(2) Linear fitting

Although most of edge points have correct classification, error points will cause false linear fitting if the fitting method is weak robustness such as least-squares estimation. So, we adopt Huber fitting method which is insensitivity to outliers. The fitting principle is to minimize the sum of distance function $\sum \rho(r)$. The distance function $\rho(r)$ is defined in equation (11).

$$\rho(r) = \begin{cases} \frac{r^2}{2}, & \text{if } r \leq \delta \\ \delta\left(r - \frac{\delta}{2}\right), & \text{otherwise} \end{cases} \tag{11}$$

where, r is the difference between the fitting value and true value: $|y - f(x)|$ ($f(x)$ is the equation of the line currently fitted). δ is a threshold. When r is bigger than δ, the distance function of this point will be decreased to cut down the impact of outliers.

(3) Corner and vanishing points calculation

When the edge lines are obtained after linear fitting process, corner points are the intersections of two lines. Corner points coordinate value could be calculated by the equations of edge lines. Similarly, the vanishing point coordinate value we needed in the L-M method will be calculated by the intersection of the two edge lines that parallel to X_W-axis.

4.2. Detection of inner marker pattern

Under whole landing process, the outer marker pattern will be out of the field of camera view when the UAV fly blow a certain height. At this moment, the aim of this algorithm is to detect the points 5~8 of inner marker pattern in figure 1.

Inner marker pattern is surrounded by a green isolation belt, which is a trait we will use to detect the image area of it. So, the basic flow chart of inner marker pattern detection method is the same as figure 3. The different are the processing details in the stages of 'Image segmentation' and 'Extract maximum connected region'.

In the image segmentation stage, the R, G and B images are also divided from color image. Then, different-images of G-B and G-R are calculated. Lastly, the different-images are merged into a binaryzation image by and operation. In the extract maximum connected region stage, conversely, the inner edge points are preserved. After this stage, the detection method for inner marker pattern is the same as for outer marker pattern.

5. Experiments

In the marker design and its detection method research process, we did experiments for different verification goals. One is to test the marker detection algorithm. Another one is to test the accuracy of the whole method proposed in this paper. In those tests, moreover, algorithms are programmed by Visual Studio C++ platform with OpenCV. All of results are calculated by offline processing.

5.1. Marker detection algorithm test

Given the security and operability, we did indoor test for this algorithm. The equipment and tools are shown in figure 7. UAV-M100 from DJI company is used for this test. The camera calibration is also finished before test.

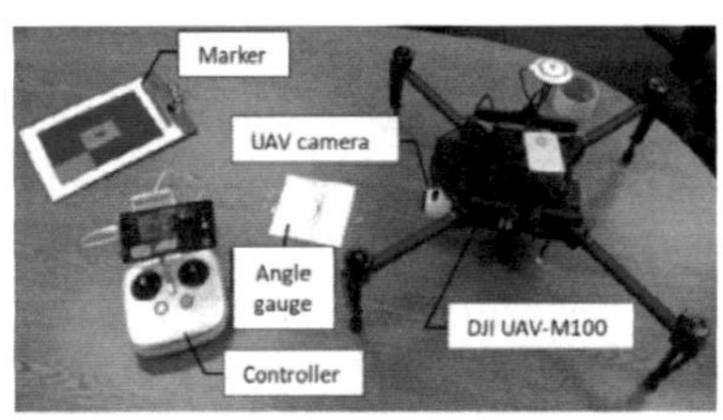

Figure 7. The equipment and tools for marker detection algorithm test

In order to test the robustness of this this algorithm, a scene with complex background is selected. There are not only lots of objects with different colors, but also have interference from strong light. Figure 8 shows the mainly processing procedures and results for 5 images. 63 images were tested and the average error is 10 pixels with 4000 × 3000 camera resolution.

Moreover, analysing the intermediate processing image which has large line fitting error, we found that most of them have hackly edge after marker region detection process. The hackly edge is formed from binaryzation for different-images of B-G which is introduced in Section 4.1.1. So, we changed the threshold of binaryzation and observe the accuracy of line fitting as shown in figure 9. Therefore, binaryzation is sensitive to the threshold value as well as the final detection results.

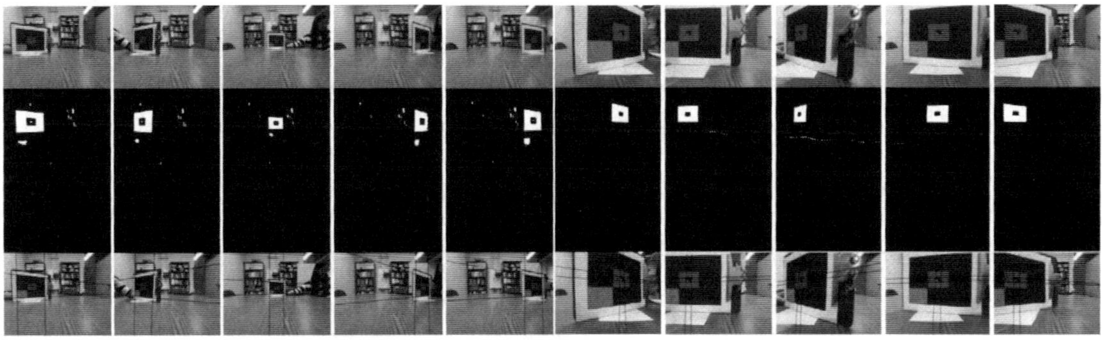

Figure 8. Partial marker detection algorithm test results

Threshold = 50 Threshold = 40 Threshold = 30

Figure 9. Partial line fitting results with different binarization thresholds

5.2. Whole method test

An outdoor experiment is implemented for whole pose estimation method. A printed marker with the size of 600×500 mm is fixed on the ground. At the same time as the UAV test, two other cameras record the flight track from X_W-axis and Y_W-axis directions respectively. To obtain the reference ground truth, we draw the measure unit of distance on the ground for X_W-axis and Y_W-axis. For Z_W-axis direction, a rod is used as the measure unit. It is set on the centre of the marker and perpendicular to the ground. The cameras in the X_W-axis and Y_W-axis directions take the pictures which will be the standard to measure the height of UAV and its camera (see figure 10).

a) From X_W-axis direction

b) From Y_W-axis direction

Figure 10. Z_W-axis standard measure images

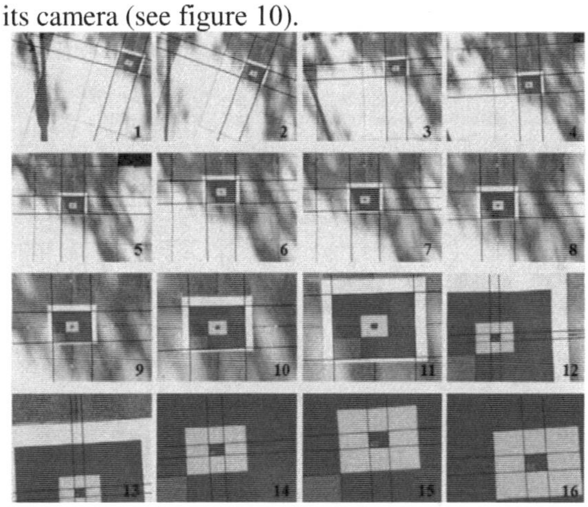

Figure 11. A set of line fitting result images captured from UAV camera

Figure 11 is a set of line fitting result images captured from UAV camera on this scene. From these images we can see that abundant shadows provided by the profusion of the woods. Even so, this algorithm still works. Using those fitting lines, the coordinate value of corner and vanishing points will be obtained. Then, import them into the position and attitude estimation program. The comparison from estimation results and reference ground truth is shown in figure 12. The numbers in figure 12 correspond to the numbers in figure 11. The average error of this test is 22 mm.

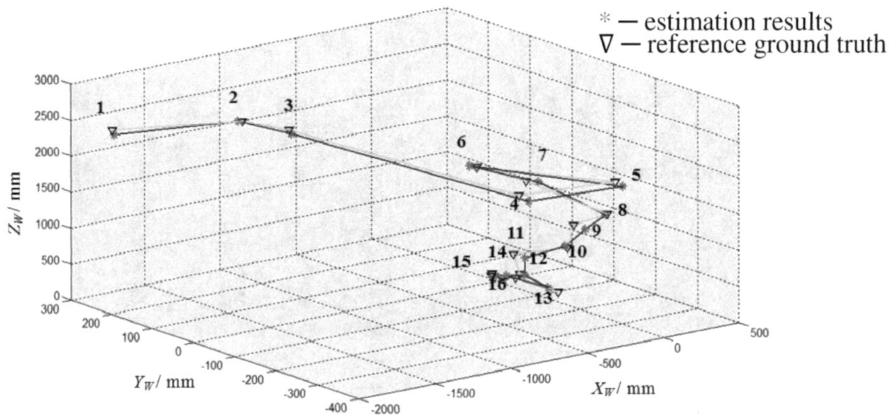

Figure 12. Comparison from pose estimation results and reference ground truth

6. Conclusion

After analysing the key and difficult points of autonomous landing technology of UAV on unmoving or moving platform, a new inner and outer combined color-based fixed point landing marker is presented, as well as its vision detection algorithm. Position and attitude could be estimated accurately and quickly using L-M and vanishing point combined method. On the other hand, irregular quadrilateral edge points are successfully classified using maximum gradient magnitude and direction feature, which is an important foundation to the edge lines fitting using Huber loss function and final corner point calculation. The indoor and outdoor experiments in various scenarios are implemented. From these results we can see that the proposed method fulfils the requirements of being accurate, whole covered and robust for vision-based landing system.

In future work, we plan to research the track algorithm to improve the real timing. Combining with the UAV control, real tests landing on a removing platform are required to update the marker and its detection algorithm.

Acknowledgments

The research is sponsored by the National Science Fund for Distinguished Young Scholars (51625503). The authors would like to thank all partners for their cooperation and valuable contribution.

References

[1] Shabayek, A E R, Cédric Demonceaux, Morel, O and Fofi, D 2012 Vision based UAV attitude estimation: progress and insights *Journal of Intelligent & Robotic Systems* **65** pp 295–308

[2] Angelino C V, Baraniello V R and Cicala L 2013 High altitude UAV navigation using IMU, GPS and camera *Int. Conf. on Information Fusion IEEE*

[3] Ettinger S M, Nechyba M C, Ifju P G and Waszak M 2003 Vision-guided flight stability and control for micro air vehicles *Advanced Robotics* **17** pp 617–640

[4] Tarhan M and Altug, E 2011 EKF based attitude estimation and stabilization of a quadrotor UAV using vanishing points in catadioptric images *Journal of Intelligent & Robotic Systems* **62** pp 587–607

[5] Natraj A, Cédric Demonceaux, Vasseur P and Sturm P F 2011 A geometrical approach for

vision based attitude and altitude estimation for UAVs in dark environments 2011 *Int. Conf. on Intelligent Robots and Systems IEEE/RSJ (San Francisco)*
[6] Moore R J D, Thurrowgood S, Bland D, Soccol D and Srinivasan M V 2011 A fast and adaptive method for estimating UAV attitude from the visual horizon *Int. Conf. on Intelligent Robots & Systems IEEE*
[7] Hwangbo M and Kanade T 2011 Visual-inertial UAV attitude estimation using urban scene regularities *IEEE Int. Conf. on Robotics and Automation (shanghai)* pp 2451–2458
[8] Laiacker M, Kondak K, Schwarzbach M and Muskardin T 2014 Vision aided automatic landing system for fixed wing UAV *IEEE/RSJ Int. Conf. on Intelligent Robots and Systems*
[9] Skoczylas M, Gadomer L and Walendziuk W 2017 Multirotor micro air vehicle autonomous landing system based on image markers recognition *Society of Photo-optical Instrumentation Engineers* Society of Photo-Optical Instrumentation Engineers (SPIE) Conference Series
[10] Sharp C S, Shakernia O and Sastry S S 2003 A vision system for landing an unmanned aerial vehicle *IEEE Int. Conf. on Robotics & Automation* vol 2 pp 1720–1727
[11] Li Xiaojie 2009 *Study and Implementation of Position and Attitude Estimation for Four-rotor Mini Rotorcraft Based on Vision* (Changchun:Jilin University)
[12] Wan Li 2014 *Target Location and Landing Based on Unmanned Aerial Vehicle* (Xi'an: Xidian University)
[13] Na Meng and Jia Peifa 2006 A vision system for autonomous flight of micro helicopter *Computer Engineering and Applications* **30** pp 220–22
[14] Liu Shiqing, Hu Chunhua and Zhu Jihong 2004 A method for estimating position and orientation of an unmanned helicopter based on vanishing line information *Computer Engineering and Applications* **40** pp 50–54
[15] Guili X, Xiaopeng Q, Qinghua Z, Yupeng T, Ruipeng G and Biao W 2013 Use of land's cooperative object to estimate UAV's pose for autonomous landing *Chinese Journal of Aeronautics 26* pp 1498–1505
[16] Lange, Sunderhauf and Protzel 2009 A vision based onboard approach for landing and position control of an autonomous multirotor UAV in GPS-denied environments *IEEE Int. Conf. on Advanced Robotics*
[17] Yang S W, Sebastian AS and Zell A 2013 An onboard monocular vision system for autonomous takeoff, hovering and landing of a micro aerial vehicle *J. Intell. Robot Syst.* **69** pp 499–515
[18] Muskardin T, Balmer G, Wlach S, Kondak K, Laiacker M and Ollero A 2016 Landing of a fixedwing UAV on a mobile ground vehicle *IEEE Int. Conf. on Robotics and Automation* pp 1237–1242
[19] Chen X, Phang S K, Shan M and Chen M B 2016 System integration of a vision-guided UAV for autonomous landing on moving platform *IEEE Int. Conf. on Control and Automation*
[20] Polvara R, Sharma S, Wan J, Manning A and Sutton R 2017 Towards autonomous landing on a moving vessel through fiducial markers *European Conference on Mobile Robots* pp 1-6
[21] Francisco A, Manuel G and Ivan M 2019 A precise and gnss-free landing system on moving platforms for rotary-wing UAVs *Sensors* **19** 886
[22] Skoczylas M, Gadomer L and Walendziuk W 2017 Multirotor micro air vehicle autonomous landing system based on image markers recognition *Society of Photo-optical Instrumentation Engineers* Society of Photo-Optical Instrumentation Engineers (SPIE) Conference Series
[23] Borshchova I and O"Young S 2017 Marker-guided auto-landing on a moving platform *International Journal of Intelligent Unmanned Systems* **5**
[24] Abdullah M. Almeshal and Mohammad R. Alenezi 2018 A vision-based neural network controller for the autonomous landing of a quadrotor on moving targets *Robotics* **7** 71

IOP Conf. Series: Materials Science and Engineering **646** (2019) 012063 doi:10.1088/1757-899X/646/1/012063

Trajectory Estimation of Flexible Needle Using PVA Tissue Material

Shijian Zhao[1], Dedong Gao[1,*], Mengxiao Zhao[1] and Jiajie Fu[1]

[1]School of Mechanical Engineering, Qinghai University, Xining 810016, China

*Corresponding author: gaodd@qhu.edu.cn.

Abstract. In order to determine the movement trajectory and characteristics of puncture needle in tissue, a series of puncture is selected for the puncture of PVA prosthesis. The image captured by CCD camera is preprocessed, and the starting coordinate point and target in the image are extracted. A fitting model based on least squares method is proposed to fit the radius of puncture needle. Comparing the simulation trajectory of forward kinematics and experimental trajectory, the experimental results show that the fitting model has a high success rate over 93% in the trajectory radius of the prosthesis puncture needle, and has a high success rate, stability and accuracy. It has a certain reference significance for the future path planning and guidance process.

1. Introduction

Recently, needle insertion has been widely used in clinic, but the traditional rigid needle cannot plan its trajectory well to get around the obstacles such as blood vessel and nerve[1]. Also for its low accuracy, it may cause secondary injury to patients during insertion. The flexible needle is made by nickel titanium, which let it have good elasticity, for the tip of the needle has an asymmetric diagonal angle, when insert into the tissue, subjected to asymmetric force the flexible needle deflects along the direction of force. There for, he bevel-tip flexible needle has good maneuverability and can bypass obstacles to reach the target[2].As a minimally invasive surgical technique, needle insertion is widely used in clinical practice such as biopsy, epidural injection of spinal, near-range radiotherapy, anesthesia and ablation[3,4]. Needle insertion different situations have different precision requirements, usually, such as proximity radiotherapy, anesthesia and biopsy require mm-grade needle puncture accuracy, when the operation involves infants, eyes and ears, etc., the accuracy of the needle puncture is required to reach the um-grade[5].

How to control the insertion needle to carry on the specified route puncture has become the current difficulty, use the robot assisted needle insertion control technology has become a hot research topic in recent years for it has ability to accurately control the injection volume and needle tip direction[6,7]. The main point in robot assisted needle insertion controls technology is path planning and barrier avoidance technology. Recently, although a large number of relevant literature on the needle path planning problem has been studied, but because the flexible puncture needle is an incomplete system and the existence of tissue deformation leads to target migration and complex organizational structure and many other factors[8]. Most flexible needle model is divided into kinetic model and kinematics model. The kinetic model establishes the relationship between the force of needle tip and the deflection of needle tip. There are two ways to solve this problem. First, install the force sensors in the needle body. The force data is using building the friction model[9,10], and second, modeling and analyzing the force of the needle tip directly[7,11]. Most of the dynamic models are based on static analysis, which cannot deal with the multiple rotation of flexible needles effectively, and the model is more complex and stays in the theoretical analysis stage. In 2005 Webster and other experiments show that in ideal conditions the sharp of flexible needle is insertion along the ARC path of the fixed radius, and the azimuth of the arc curve can be changed by changing the angle of the bevel-tip[8,12].

Then a large number of relevant literature is based on Webster et al.[8,13-16] to study kinematics, path planning and obstacle avoidance techniques of inclined tip flexible puncture needle. In 2005, Okazaw et al. designed a handheld maneuver[17], combined with software to puncture through the needle tip direction of the control needle, and in 2006 proposed a segmentation algorithm based on Ough transform, and the deflection of the needle body linear fitting[18], However, linear fitting of the needle body is not suitable for practical application. In 2007, Davneet et al.[19] analyzed the bending characteristics of the flexible needle, and proposed the use of rotating needle to control the trajectory of the needle body in the tissue, the purpose of controlling the trajectory of the needle body is achieved by adjusting the duty-free ratio of the needle body during rotation. In addition, HuoBenyan et al.[20] aiming at the relationship between control quantity and path, an algorithm based on multi-objective particle swarm is proposed, the model is simplified, and the efficiency is improved. However, the characteristics of biological tissue will change over time, resulting in path failure, and because the difficulty to accurately locate the pinpoint position within the biological tissue, how to solve the feedback problem has become a problem to be solved. In 2007 Gao et al.[21] established the kinematics model of puncture needle by using parameter method. By simulating and validating the model of the recording shows that, it has relatively accurate accuracy. However, it has not been verified experimentally.

In this paper, a model is designed to extract the coordinates of the starting point, the target point and the radius of the needle. In the experiment, the image is collected by using CCD camera. After preprocessing the image, the parameters are extracted and the needle trajectory is simulated. The experimental results show that the model has a high accuracy, which has a certain reference significance for the future needle trajectory planning.

2. Forward kinematics model

When bevel-tip flexible needle insert the tissue, many factors determine the puncture trajectory of the needle. In the model of geometric approximation, we make the following assumptions about the insertion system: First, there is no delay between the needle tip and the needle seat. Secondly, when the needle inserts the tissue, the tissue does not cause deformation of the needle body that affects its trajectory. The last, the position of the trajectory is entirely determined by the needle tip shape, and the radius value of the needle trajectory is a certain value in the tissue.

Forward kinematics is used to calculate the motion trajectory of the needle body. It mainly includes a few control variables: needle pitch angle (α), needle rotation angle (β), injection volume (L) and the bending radius value of needle body (r). In the process of injection, the transformation of coordinates is divided into two main steps. First, before insert the needle adjust the value of β to select the insertion angel. Secondly, translation coordinate system Om-XmYmZm to coordinate system On-XnYnZn . The coordinate transformation matrix is as follows:

$$
{}^{m}_{n}T(\beta, L) = Rot(Z, \beta)Trans(h, 0, d)
= \begin{bmatrix} c\alpha & -s\alpha & 0 & 0 \\ s\alpha & c\alpha & 0 & 0 \\ 0 & 0 & 1 & 0 \\ 0 & 0 & 0 & 1 \end{bmatrix} \begin{bmatrix} 1 & 0 & 0 & h \\ 0 & 1 & 0 & 0 \\ 0 & 0 & 1 & d \\ 0 & 0 & 0 & 1 \end{bmatrix} = \begin{bmatrix} c\alpha & -s\alpha & 0 & c\alpha h \\ s\alpha & c\alpha & 0 & s\alpha h \\ 0 & 0 & 1 & d \\ 0 & 0 & 0 & 1 \end{bmatrix} \quad (1)
$$

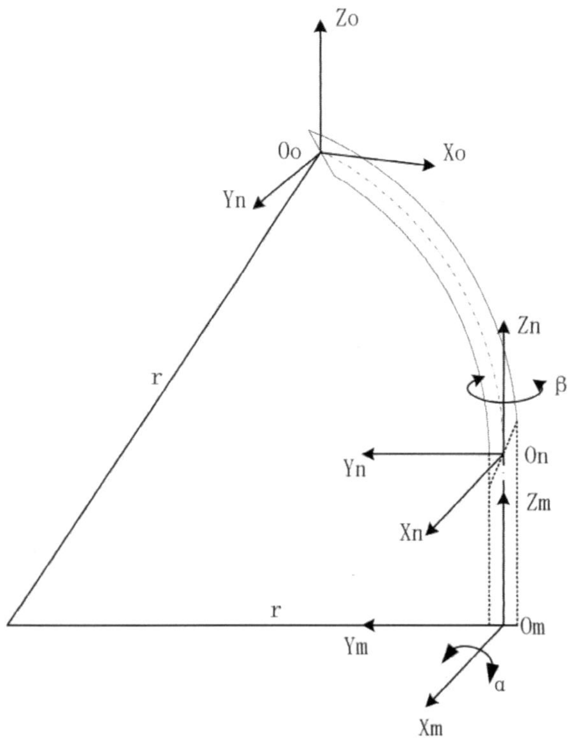

Figure 1. Schematic diagram of coordinate transformation of needle insertion

Third, change the value of α, rotate coordinate On-XnYnZn around Zn axis to get coordinate On-XoYoZo. Therefore, the coordinate transformation matrix is as follows:

$$
\begin{aligned}
{}^{m}_{o}T(\beta, L) &= {}^{m}_{n}T(\beta, L)Rot(Y, \beta) \\
&= \begin{bmatrix} c\alpha & -s\alpha & 0 & c\alpha h \\ s\alpha & c\alpha & 0 & s\alpha h \\ 0 & 0 & 1 & d \\ 0 & 0 & 0 & 1 \end{bmatrix} \begin{bmatrix} c\beta & 0 & s\beta & 0 \\ 0 & 1 & 0 & 0 \\ -s\beta & 0 & c\beta & 0 \\ 0 & 0 & 0 & 1 \end{bmatrix} \\
&= \begin{bmatrix} c\alpha c\beta & -s\beta & c\alpha s\beta & c\alpha h \\ s\alpha c\beta & c\alpha & s\alpha s\beta & s\alpha h \\ -s\beta & 0 & c\beta & d \\ 0 & 0 & 0 & 1 \end{bmatrix}
\end{aligned} \tag{2}
$$

Where $\beta = L/r$, $h = r(1-\cos\beta)$, $d = r\sin\beta$, $s\alpha = \sin\alpha$, $s\beta = \sin\beta$, $c\alpha = \cos\alpha$, $c\beta = \cos\beta$.

By repeating the above two steps, the coordinate transformation in the whole puncture process is established. In addition, the coordinate transformation of needle insertion is shown in Figure 1.

When the needle tip is from the initial moment to the puncture to paragraph I, the position of the needle tip relative to the global coordinate system can be expressed as follows:

$$
{}^{g}_{i}T = {}^{g}_{0}T\,{}^{0}_{1}T(\beta, L) \cdots {}^{I-1}_{I}T(\beta, L) = {}^{g}_{0}T \prod_{k=1}^{I} {}^{k-1}_{k}T(\beta, L) \tag{3}
$$

Where $I = 1, 2, \cdots, n$. ${}^{g}_{0}T$ is the homogeneous transformation matrix in needle-tip coordinate system relative to global coordinate system. Therefore, the position coordinates of the needle tip relative to the global coordinate system can be described in any period.

3. Experiments setup

To verify the accuracy of the model and the extracted data, a puncture needle was used to insert the PVA tissue. Where the puncture experimental platform shown in Figure 2.

In this experiment, polyvinyl alcohol (PVA) is used as a soft tissue phantom. The ratio of polyvinyl alcohol, dimethyl sulfoxide and distilled water was 8:60:40. After heating and stirring at 90 degrees for 6 hours, then standing at 20 degrees below zero for 10 hours and thawing for 3 hours, the PVA tissue was made.

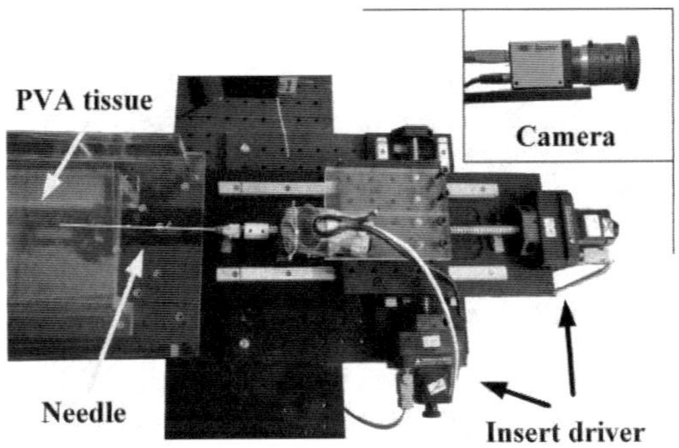

Figure 2. Needle insertion system

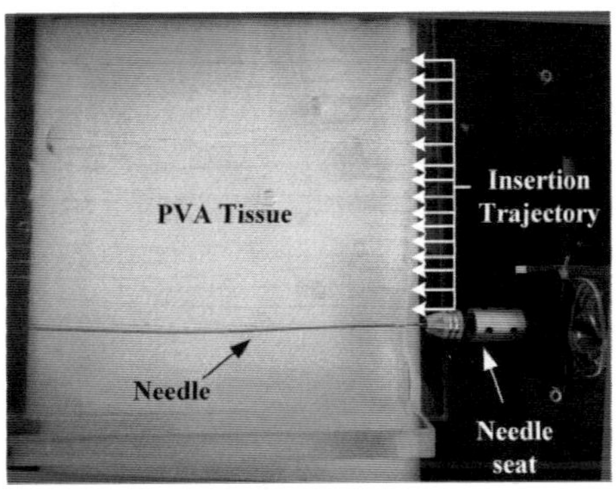

Figure 3. PVA insertion experiment

The system has 3 degree of freedom (DOF) to that the needle can move, feed and rotate relative to the soft tissue during the insertion. A CCD camera is mounted above the PVA tissue to collect images. The CCD camera is used to take pictures of the whole needle body, which is as shown in Figure 3. The Figure 3 shows a series of insertion trajectories of the needle in soft tissue.The type of needle we are using is 18~23g hakko puncture needle. The bevel of needle is 20 degrees and the length is 200mm. The size of tissue is $100\times100\times20mm^3$. Both type of needle is insert the PVA tissue in 2mm/s. In all experiments, the needle without deflection.

4. Data extraction

4.1. Extraction of needle trajectory

In the process of extracting the above basic parameters, the image taken by CCD should be pre-processed first. Firstly, the image is processed by binarization. Using binarization to image can make needle body more prominent in image and filter out needle-independent tissue structure. The value we choose here is 0.85, as shown in Figure 4 a). Thinning the needle in the image by using the skeleton processing of the image, as shown in Figure 4 b). The trajectory of the needle can be extracted by thinning the image. In addition, the trajectory of the needle which is shown in Figure 4 c). As can be obtained from Figure 4 c) the trajectory of a needle body consists of several coordinate points.

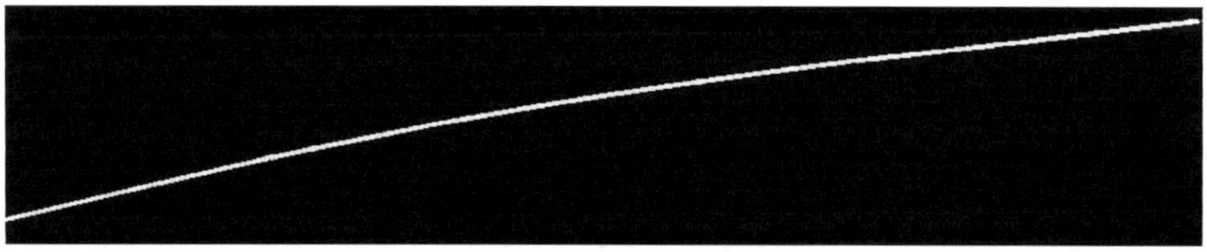

a) Image binarization results

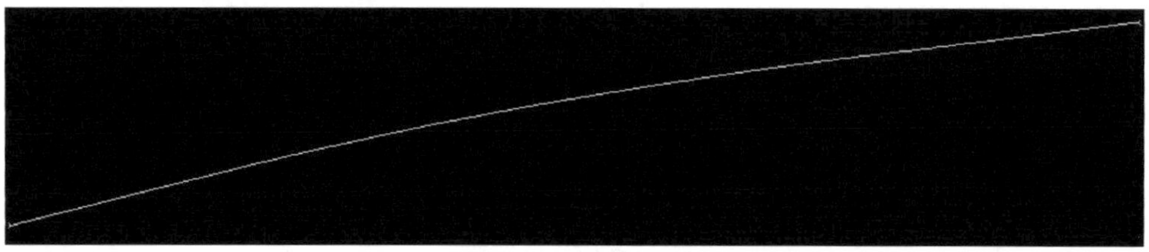

b) Image skeletonization results

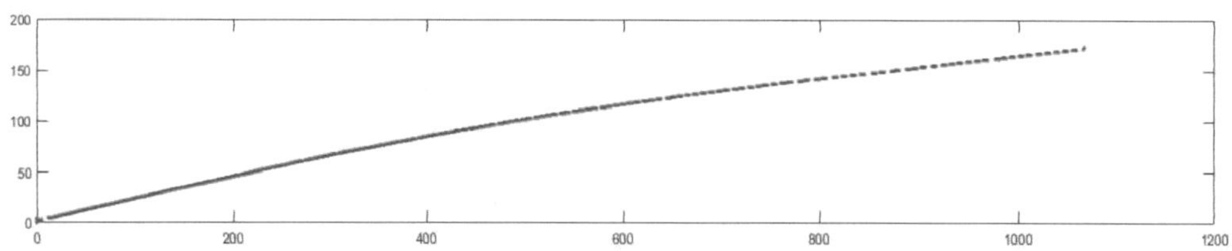

c) Needle-body trajectory

Figure 4. The process and results of image processing

4.2. Radius fitting

By using the coordinate points, the radius of the needle in tissue could be determined. For the Hough circles transformation is inefficient and computational complexity is large. We here use the ordinary least squares to fitting the radius of puncture needle. The function formula of the ordinary least squares is as follows:

$$f = \sum\left(\sqrt{\left(x_i - x\right)^2 + \left(y_i - y\right)^2} - R\right)^2 \tag{4}$$

Where x_i and y_i is the coordinate points of the needle. x, y is the center coordinate points of the needle body trajectory, R is the radius of the needle.

For the ordinary, least squares the quality of sitting punctuation is very important. In order to ensure the accuracy of the fitting process, the generation of outliers in the original data should be reduced. Therefore, the generation of interference points should be reduced in image denoising, thresholding and filtering.

5. Verification

By using the trajectory coordinate points, the radius of the needle body is determined. The simulation trajectory of the needle body is obtained by bringing the needle coordinate points, target coordinate points and radius values into the forward kinematics model. The simulation and experimental trajectory shown in Figure 5.

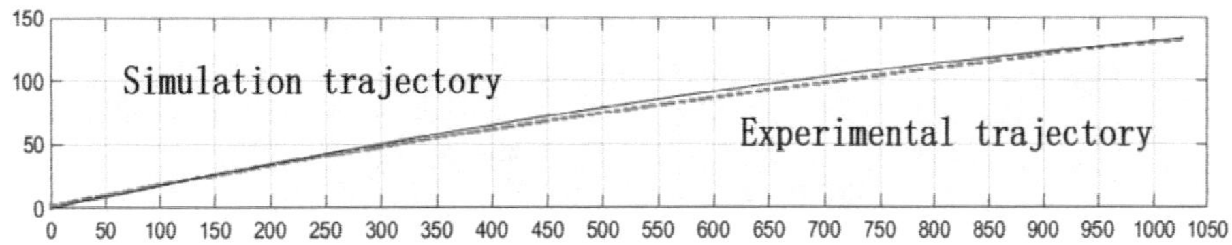

Figure 5. Contrast of trajectory curves

In the needle insertion experiments, the type of needle included with inner core and without inner core. The simulation and experimental error result are shown in Table 1. Table 1 shows that the error between simulation and experiment is less than 7%, which the deviation ranges are all below 4%.

Table 1. Non-deflection simulation and experimental error results

Needle type	With inner core		Without inner core	
	Average Error (%)	Deviation Range	Average Error (%)	Deviation Range
18g	1.41103	1.28339	2.49968	1.28339
19g	2.31193	1.5396	5.09432	1.5396
20g	0.43969	1.44891	2.09591	1.44891
21g	3.33652	3.38101	3.53084	3.38101
22g	2.55718	0.273	1.60456	0.273
23g	2.47023	0.93764	1.95708	0.93764

6. Conclusions and future work

In this paper, a fitting model is established to combine the trajectory of the puncture needle with the experimental verification of the established forward kinematics. The CCD camera is using to take image in the experiments. In order to improve the accuracy of the fitting process, the image is pre-processed by binarization and noise reduction. After thinning the image, the radius of the needle in the image is fitted by the ordinary least squares. And the key coordinate points in the needle body puncture are extracted. By compared the experimental trajectory and the simulation trajectory of forward kinematics, the results shown that the model of fitting needle trajectory has high accuracy and stability which provides some reference significance for the future path planning.

In the future work, we will use the camera to record the result of deflecting one and more times during puncture needle puncture. By analyzing these results, we will be able to understand the puncture trajectory of puncture needle in more detail, and have some reference significance for the future puncture needle path planning. In addition, the radius measurement of the puncture needle should be extended to the animal tissue, and the trajectory and characteristics of the puncture needle in animal tissue should be studied by using B-ultrasonography to collect the image. This paper will take these issues as the content of future research.

Acknowledgments

This research work is sponsored by the National Natural Science Foundation of China (No. 51665049) and Qinghai Provincial Natural Science Foundation of China (Grant No.2015-ZJ-906).

References

[1] Reed K B, Majewicz A, Kallem V, et al. 2011 Robot-Assisted Needle Steering *IEEE Robotics & Automation Magazine* **18** 35-46
[2] Alterovitz R, Goldberg K, Okamura A. Planning for Steerable Bevel-tip Needle Insertion Through 2D Soft Tissue with Obstacles[C]. IEEE International Conference on Robotics & Automation, 2006.

[3] Asadian A, Patel R V, Kermani M R 2014 Dynamics of translational friction in needle-tissue interaction during needle insertion *Annals of Biomedical Engineering* **42** 73-85

[4] Heverly M, Dupont P, Triedman J. Trajectory Optimization for Dynamic Needle Insertion[C]. IEEE International Conference on Robotics & Automation, 2006.

[5] Abolhassani N, Patel R, Ayazi F. Effects of different insertion methods on reducing needle deflection[C]. International Conference of the IEEE Engineering in Medicine & Biology Society, 2007.

[6] Glozman D, Shoham M 2007 Image-Guided Robotic Flexible Needle Steering *IEEE Transactions on Robotics* **23** 459-467

[7] Misra S, Reed K B, Schafer B W, et al. 2010 Mechanics of Flexible Needles Robotically Steered through Soft Tissue *International Journal of Robotics Research* **29** 1640-1660

[8] Webster R J, Cowan N J, Chirikjian G, et al. 2006 Nonholonomic Modeling of Needle Steering *International Journal of Robotics Research* **25** 509-525

[9] Roesthuis R J, Veen Y R J V, Jahya A, et al. Mechanics of needle-tissue interaction[C]. IEEE/RSJ International Conference on Intelligent Robots & Systems, 2011.

[10] Asadian A, Patel R V, Kermani M R. A distributed model for needle-tissue friction in percutaneous interventions[C]. IEEE International Conference on Robotics & Automation, 2011.

[11] Wei D, Han H, Du Z. The tip interface mechanics modeling of a bevel-tip flexible needle insertion[C]. International Conference on Mechatronics & Automation, 2012.

[12] Iii R J W, Memisevic J, Okamura A M. Design Considerations for Robotic Needle Steering[C]. IEEE International Conference on Robotics & Automation, 2005.

[13] Duindam V, Xu J, Alterovitz R, et al. 2010 Three-dimensional Motion Planning Algorithms for Steerable Needles Using Inverse Kinematics *International Journal of Robotics Research* **29** 789-800

[14] Yanjiang Z, Yongde Z, Junpeng S 2010 Kinematic model and experimental study of flexible needle *ROBOT* **32** 666-673

[15] Alterovitz R, Branicky M, Goldberg K 2008 Motion Planning Under Uncertainty for Image-guided Medical Needle Steering *Int J Rob Res* **27** 1361-1374

[16] Park W, Wang Y, Chirikjian G S 2010 The Path-of-Probability Algorithm for Steering and Feedback Control of Flexible Needles *International Journal of Robotics Research* **29** 813

[17] Ebrahimi R, Okazawa S, Rohling R, et al. 2003 Hand-Held Steerable Needle Device *IEEE/ASME Transactions on Mechatronics* **10** 285-296

[18] Okazawa S H, Ebrahimi R, Chuang J, et al. 2006 Methods for segmenting curved needles in ultrasound images *Medical Image Analysis* **10** 330-342

[19] Minhas D S, Engh J A, Fenske M M, et al. Modeling of needle steering via duty-cycled spinning[C]. Engineering in Medicine and Biology Society, 2007. EMBS 2007. 29th Annual International Conference of the IEEE, 2007: 2756-2759.

[20] Benyan H, Xingang Z, Jianda H, et al. 2015 Puncture Path Planning for Bevel-tip Flexible Needle Based on Multi-objective Particle Swarm Optimization Algorithm *ROBOT* **37** 385-394

[21] De-Dong G, Qiang L, Yong L, et al. 2017 Geometric approximation approach based research on kinematics of bevel-tip flexible needles *Joural of Zhejiang University (English Science)* **51** 706-713

Research on Intelligent Speech Guide Robot Control Method Based on Machine Vision

Zhengbo Wang, Xing Ma* and Chunyang Mu

North Minzu University, Yinchuan, Ningxia, China

*Corresponding author: maxingsky@126.com

Abstract. Robotic navigation via speech guidance gain potential attention on the field artificial intelligent and it stat-of-art. This paper highlights speech guide for robot toward machine vision were designed. The robot navigation can determine and identify the objects around the blind area. The process of determining the direction and distance of the blind people to consider for deciding and predicating the next step. The information of the sensors is fused together, and a path information suitable for the blind person is transmitted to the user via speech voice broadcast are the key elements for identification part. Moreover, deep learning algorithm are involve for accurate classifiers target into categories, to archive the high est recognition accuracy. The direction and distanceare judgment part toadoptinclude the improvements binocular ranging, angle measurement algorithm with a consideration of errors marge less than 2%. The novelsystem can -autocorrect and, guide blind people to move froward ,advance, stop, turn, which greatly compensates for the visual defects of the blind, with respect of constraint ofimproving the indoor safety of the blind people.

1. Introduction

According to 2010 data released by the world health organization, there are nearly 45 million blind peoples in the world, and yearly around7 million become blind[1]. In our daily life, most of the information are obtained by visual sense. However, the physical defects of the blind and the influence of the complex social environment have brought great inconvenience to the life of the blindpeople [2]. Some countries has given more support to this category of people, moreover it has more institutions for training centers. The integration of this category into social demands creation more job opportunities, the emergence of guide sticks andassistance guide such as dogs were has greatimpact for helping in trips. But unfortunate these have not really helped to solveall the inconveniences of lifeissues .

Carnegie Mellon university of Pittsburgh haddeveloped a robot suitable for guiding the blind[3]. The robot emphasizes the stability of movement, providing support for the blind or elderly, but requires a handle. Zhang Zhimei et al. developed a crawler guide robot, which uses various sensors to track the black trajectory preset on the ground to realize navigation and obstacle avoidance, so it cannot work in an environment without black trajectory[4]. In this paper, present conditions and constraints ofprocess (the guide robot and find trajectory) however is designed a novel approach of intelligence for navigat ion andguide robot based on voice with machine vision. The robot is mainly used indoor (household), office environment, to detect and identify the blind around objects, perception object distance of the blind angle and distance, in dynamic area, variety of sensor information fusion together, passed to the user in the form of speech.

2. Design of intelligent speech guide robot

In terms of approach applied, intelligent speech guide robot used to design anobject detection and recognition[5] based on deep learning, binocular calibration includingranging, motion control, ultrasonic and short infrared ranging, wireless communication and speech broadcast system[6]. Amongvariety and complex elements , the design is mainly considerthe robot's perception of the external environment.(indoor) The detection and recognition of external objects is hold by machine vision, as with perception of the distance and angle between objects and users, to calculate the optimal path suitable for blind people.

In terms of physical component, four dc motors are used to control McNamee wheels, and the wheeled chassis as the carrier to realize 360 degree rotation of the car, optimize obstacle avoidance effect, and equipped with binocular camera, controller, ultrasonic and short infrared sensor, wireless transmission equipment, to complete information acquisition, information processing, communication and other functions.

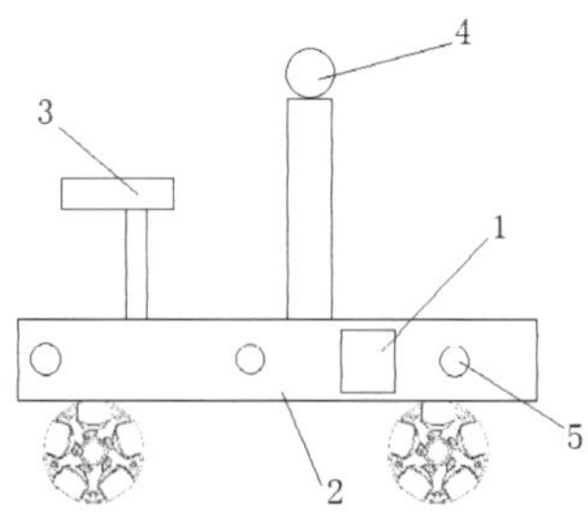

Figure 1. Schematic diagram of hardware structure

As shown in figure 1, this design includes 1 as the main controller, 2 as the robot chassis, 3 as the voice broadcaster, 4 as the binocular camera, and 5 as the infrared sensor.

3. Intelligent guide robot

3.1 Object detection and recognition based on neural network

The robot's perception of the surrounding environment with accurate detection and recognition of objects are the focus of this system. Therefore, it is necessary to detect multiple objects in the image signals received by the camera and accurately identify object categories. Joseph Redmon[7] mentioned in literature that YOLO algorithm can ensure the accuracy and detection rate of detection. Standard YOLO algorithm can detect a frequencies of45 frames/s, while Fast YOLO algorithm can detect 155 frames/s. Author Wang yuning analyzed in literature[8]. By comparing several different target detection algorithms, it is concluded that YOLO algorithm has the fastest detection speed.

In addition, YOLO algorithm is most suitable for the requirements of simultaneous detection of multiple obstacles in this paper, so the object detection part of this paper adopts YOLO V2[9] detection system. YOLO algorithm belongs to CNN, which also consists of convolution layer, pooling layer and full connection layer. However, the difference is that in output layer is a tensor, and its training samples need not be specially cut out from the sample image, but the whole image is trained and detected, which improves the stability of the system and reduces the false detection on the background. Input image is divided into S * S small grid, each grid is responsible for the inspection is to be in the center of the object on the grid, the grid inspection when produce B a frame, each frame contains 5 D information (x, y, w, h, C_{obj}), respectively, represents the abscissa and ordinate of border center, width and height of the whole image, boxes of confidence, confidence level as shown in formula (1), (2):

$$C_{Obj} = Pr(Object) * IOU_{pred}^{truth} \tag{1}$$

$$\mathrm{IOU}=\frac{\text{area}\ (BB_{\mathrm{dt}}\cap BB_{\mathrm{gt}})}{\text{area}\ (BB_{\mathrm{dt}}\cup BB_{\mathrm{gt}})} \tag{2}$$

In equation (1), Pr (Object) is the probability of the Object in the grid, and if there is, it is 1, otherwise, it is 0. IOU represents the ratio of the intersection area between the reference standard box and the detection box.

YOLO algorithm extracts features in the convolution layer, and target prediction is carried out in the full connection layer. When Pr(Object)=1, the confidence of some kind in the whole picture is shown in formula (3).

$$\Pr(Class\mid Object)*\Pr(Object)*IOU_{pred}^{truth}=\Pr(Class_i)*IOU_{pred}^{truth} \tag{3}$$

Where $\Pr(Class \mid \mathrm{Object})$ is the classification conditional probability of the target to be detected. $\Pr(Class_i)$ is predicting the probability of a certain class. After the above calculation, set the threshold value of the detection border, filter out the surrounding border whose score is lower than the threshold value, and conduct non-maximum suppression treatment on the reserved border to obtain the detection result.

3.2 Stereo matching and ranging based on binocular vision

3.2.1 Camera calibration. The calibration and balance of objects in this paper are two logitech C270 cameras with fixed relative positions. A 9*6 checkerboard plane was used as the calibration template, and the side length of the positive square was 19mm. In the experiment, the template plane was placed in the field of view of two cameras, and images with pixel size of 640×480 were collected. Then adjust the angle and position of the template, so that the left and right cameras can successfully detect the corner coordinates of 20 pictures, and then implement calibration program to solve projection matrix, internal parameter matrix and external parameter matrix.

3.2.2 Stereo correction. In real binocular stereo vision, there is no complete coplanar alignment between two camera planes. Therefore, we need to carry out stereo correction to correct the two images that are actually aligned with non-coplanar rows into coplanar rows. OpenCV provides us with a function "cvStereoRectify" for the left and right image alignments. Figure 2 shows the left and right image pairs without stereo correction. Obviously, the same points in the left and right images are not on the same line. Figure 3 shows the image pair after stereo correction to eliminate the distortion.

Figure 2. Left and right images before stereo correction

Figure 3. Left and right images after stereo correction

3.2.3 Stereo matching and ranging. SGBM stereo matching is performed on the corrected left and right images to find the corresponding coordinates of the same object in the two images for obtaining the distance between them -- parallax. In fact, stereo matching is to match the same points in the left and right camera image pairs to get a parallax map.

The main steps are as follows:

- preprocess and filter to normalize the brightness and enhance the image texture.
- conduct matching search.
- remove the false matching points.

Analysis of the above table shows that when the target object to be measured is relatively close to the camera, the depth information accuracy of the target measured by the system is relatively high, and the measurement accuracy decreases as the distance gets farther. In this paper, we do not pursue the accuracy of distance, and the experimental results meet the requirements of broadcasting environmental information for blind people.

3.2.4 Calculation of migration angle. According to the actual needs, we only need to guide the blind people in the general direction and distance, so it is very important to choose a simple and convenient path with fewer obstacles. We propose a method to judge the number of obstacles (whether the path is complex or not) in front based on the depth map information. In the depth map, the area where obstacles may appear. And split into left and right parts on average, and the pixel points with known depth information are counted. The side with more points is identified as the side with more obstacles and more complex paths, so the other side is selected as the area where the robot passes.

If the selected area on the left side of the image, to detect objects with the latest information of the pixel area of the leftmost pixel position, the position and the center of the image pixels, using trigonometric function, calculate the robot to offset angle: according to the edge of the object in the position of image pixels, and according to the pixel location and image pixel location x direction difference among pixel values divided by the x direction to take the perspective of the whole image (image represents the view angle), obtains the real object and robot angle deviation. The angle is obtained and then transferred to the robot chassis control. The motor controls the universal wheel rotation angle, which is the angle that the blind person should move to the left. Similarly, if the selected area is on the right half of the image, that is, the angle that the blind person should move to the right.

3.3 Voice broadcasting system

In this experiment, espeak speech synthesis tool under Linux system is used to convert the text information collected from images into speech information, so that the robot can broadcast the distance between the current obstacle and the robot and the angle formed directly in front of the person.

Espeak speech synthesizer is simple, fast, small, can read text from standard input, speech output can be saved as .wav. For the voice broadcasting system, there are three parts: the type of obstacles, the distance of obstacles and pedestrians' obstacle avoidance.

3.4 Robot motion control and sensor signal receiving system

Robot control program with raspberry PI, python programming. Figure 4 shows the flow chart of the control program. The control of the chassis walking device is carried out by judging which direction sensor has signal.

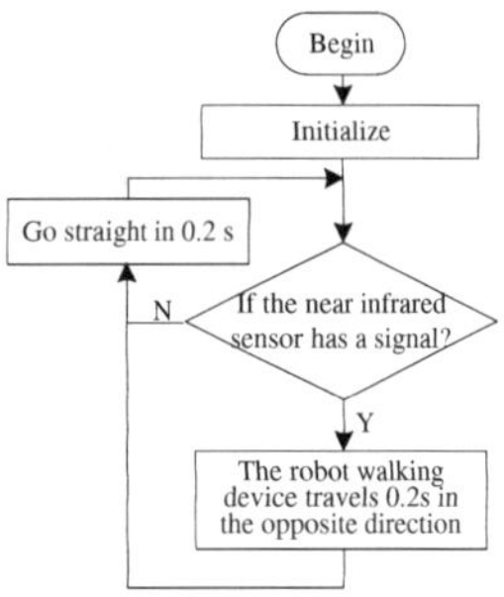

Figure 4. Robot control flow chart

As shown in Figure 4, when the signal is received by the sensor in a certain direction, the robot will walk 0.2s in the opposite direction, with a speed of 0.5m/s and a distance of 10cm under control. The motion time and speed can be modified according to the situation to change the travel distance.

4. Experimental analysis

4.1 Analysis of experimental results of hardware system

IOP Conf. Series: Materials Science and Engineering **646** (2019) 012064 doi:10.1088/1757-899X/646/1/012064

This novel approach can control the robot chassis motion speed and eight direction motion, but also can achieve autonomous obstacle avoidance function. In the design, 8 near-infrared sensors are mainly used to realize obstacle avoidance of the robot, which are respectively installed one 45 degrees apart, to detect whether the object in front is less than 20cm from the robot.

As shown in Figure 5, the actual obstacle avoidance effect is shown as follows: (a) there is an obstacle within a distance of 20cm behind the right, and the robot drives ahead to the left for 0.2s, (b) where there are obstacles within a distance of 20cm behind, the robot drives ahead for 0.2s. (c), (d) have the same effect as (a) and (b).

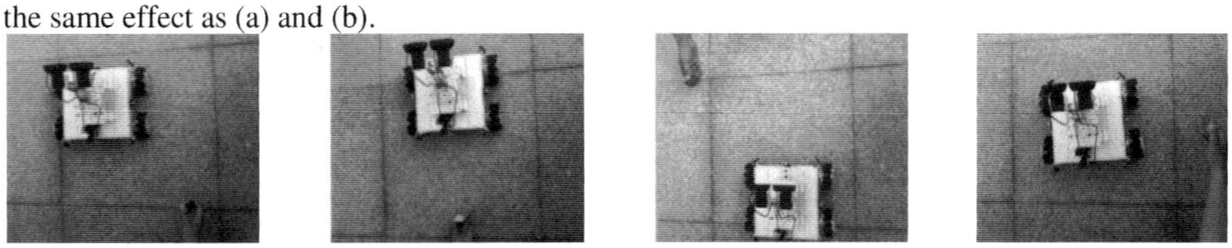

(a) drive ahead to the left (b) drive to the front (c) drive rear to the right (d) drive to the left

Figure 5. Hardware experimental results

According to the analysis in Figure 5, when the sensor in a certain direction has a signal, the robot will drive 0.2s in the opposite direction to avoid obstacles.

4.2 Analysis of experimental results of the software system

4.2.1 Analysis of stereo matching and ranging results. According to the system design, we selected the object closest to the robot in the original picture, and displayed the distance between it and the robot and the angle of the object's most right end to the robot. The actual nearest distance is 32.3cm with an angle of 21 degrees to the left, as shown in Figure 6. The system can automatically find out the nearest object and display the distance and angle information, and the values are basically consistent with the actual values. SGBM algorithm is used to realize the matching of right and left corrected images, and the parallax map is obtained. Object depth can be implemented by the function cvReprojectImageTo3D, and the experimental results are shown in table 1.

Table 1 Ranging results

number	actual distance/mm	20 mean measurements/mm	relative error/%
1	400	404.55	1.138
2	500	506.55	1.310
3	600	609.65	1.608
4	700	711.87	1.696
5	800	814.20	1.775
6	900	918.28	2.031
7	1000	1021.57	2.157
8	1100	1125.77	2.343

Figure 6. binocular ranging results

4.2.2 Object detection and recognition result analysis. Object detection and recognition based on neural network can detect multiple objects in the same image and display the object category and recognition accuracy. As shown in Figure 7, the network accurately identifies cups, laptops, potted plants and chairs in the real scene.

Figure 7. object detection and recognition

4.2.3 Result analysis of voice broadcasting system. The voice broadcast system collects the output information of binocular ranging, object detection and recognition, and can carry out intelligent voice broadcast. The robot will broadcast "there are obstacles in front of you -- laptop, potted plant, chair and cup. Please pay attention to the distance of the nearest obstacle in front is 32 centimeters, please walk 21 degrees to the left".

5. Conclusion

This paper mainly integrates motion control technology, sensor technology and image processing technology to complete the design of the entire intelligent speech guide robot system, including hardware circuit design, car control code writing, image recognition code writing and so on. According to the experimental results, the guide robot basically realizes obstacle avoidance, accurate object recognition, accurate and reasonable ranging and azimuth measurement, voice broadcast and other functions.

Acknowledgments

This work was financially supported by Ningxia natural science foundation project (2019AAC03121), University Science and Technology Innovation Platform of Ningxia Province (Industry-University Cooperative Research Base for Key parts of advanced equipment and system innovation), Open project of Ningxia key laboratory of intelligent information and big data processing(2019KLBD002), Scientific research project of North Minzu university(2019XYZDX01), the Master graduate student innovation project of North Minzu University(YCX19087).

References

[1] Z Li, G Xu, M Guo. Design and research of audio-visual fusion guide robot [J]. Progress in laser and optoelectronics, 2017, 54(12).
[2] Monteiro-Junior R S, Figueiredo L F D S, Maciel-Pinheiro P D T, et al. Acute effects of exergames on cognitive function of institutionalized older persons: a single-blinded, randomized and controlled pilot study[J]. Aging Clinical & Experimental Research, 2016, 29(3):1-8.
[3] Morris A, Donamukkala R, Kapuria A, et al. A robotic walker that provides guidance[C]// IEEE International Conference on Robotics & Automation. 2003.
[4] Z Zhang, L Cheng , Y Zhao, et al. Fuzzy-PID arithmetic for blind guiding robots.Journal of Shenyang Normal University(Natural Science Edition) 2015,33(1):81-8.
[5] He K, Zhang X, Ren S, et al. Deep Residual Learning for Image Recognition[J]. 2015:770-778.
[6] Kastrati R, Hamiti M, Abazi L. The opportunity of using eSpeak as text-to-speech synthesizer for Albanian language[J]. 2014.
[7] Redmon J, Divvala S, Girshick R, et al. You Only Look Once: Unified, real-time Object Detection[C]// Computer Vision and Pattern Recognition. IEEE, 2016:779-788.
[8] Y Wang,Z Pang, D Yuan. Real-time vehicle detection based on YOLO algorithm [J]. Journal of wuhan university of technology, 2016, 38(10):41-46.
[9] C Cai, K Wu, Y Yan. Airport runway target detection based on optimized YOLO method [J]. Command information system and technology, 2018(3).

Bayesian Sparse Linear Prediction with Pearson Type VII Distribution

Hongjie Wan[1*] and Haiyun Zhang[2]

[1]Information Engineering Dept, Beijing University of Chemical Technology, China
[2]College of Information Science and Engineering, Fujian University of Technology, China

[*]Corresponding author: Wanhj@mail.buct.edu.cn

Abstract. The speech signal can be modelled as AR models with an innovation noise model. The Pearson type VII distribution is used to model the real excitation. Variational Bayesian framework is used to estimate the posteriors of the AR coefficients and noise model parameters. The model is not conjugate, so MCMC is embed into VB framework to estimate the degree of freedom (DOF) parameter of Pearson type VII distribution. The model order selection is carried out by setting ARD priors on the coefficients. Simulation is carried out on synthetic and real data, the results show that the algorithm performs well for linear prediction both for synthetic data and speech signal, and the results are better than using least square method.

1. Introduction

For speech signal analysis, the autoregressive (AR) model is a typical model. The AR model is used widely for speech coding, speech recognition and speech synthesis. Traditionally, the innovation model of the AR model is assumed to be Gaussian model, and the usually methods for estimating the model parameters are least squares (LS), Durbin, Burg, and Lattice [1]. In this paper, the LS method be compared with.

Variational Bayesian (VB) is a common tool to estimate the posterior of the parameters for a model that is hard to compute the integral. It has great advantage than Maximum likelihood method and Maximum a posterior method in that it can have a posterior estimation other than merely point estimation. In addition, his framework can prevent over-fitting [2]. For the convenience of this method, Penny and Roberts provide a method based on VB to learn the AR parameter and the model order [3]. Their model takes Gaussian as the innovation noise model. The model order selection is based on the lower bound of the VB framework. Gaussian noise model is not robust to outliers, and it can't model long tailed noise.

In this paper, a Bayesian model is built for AR model parameter estimation with Pearson type VII distribution [4][5] as the noise model. The VB based learning algorithm is derived to estimate the posterior of the parameters. ARD priors are set on each coefficient, other than cast isotropic priors on the coefficient as in [3]. The ARD prior can determine the model order by pruning the coefficients during VB iterations. When the coefficients are very near to zero, then they can be pruned automatically. This type of prior and pruning has been used in RVM [6], and the algorithm of this paper can be seen as the modified version of RVM. The EM based algorithm and the VB framework based algorithm has been given in [6] and [7] respectively. The VB framework requires computing the expected statistics of the parameters on the posteriors, then calculate these statistics round by round until convergence. To have this type of computation, the posteriors should have closed form. It is unfortunate that closed form distribution for the DOF parameter of Pearson type 7 distribution cannot be reached. To avoid this, the MCMC algorithm is embedded in VB to compute the expected statistics. As long as the sample number tends to large enough number, closely estimation of the expected statistics of the true posterior can be obtained. In VB framework, the lower bound is used to monitor the convergence of the algorithm. The lower bound is also derived to control the algorithm. In the embedded MCMC algorithm, a normal distribution is adopted to approximate the true posterior, for which the mean and covariance can be achieved. By this assumption, the approximated lower bound can be achieved and can be used to control the convergence.

This paper is organized as follows. In Section II, a Bayesian model is built for AR model using Pearson type 7 distribution and the ARD priors on every coefficients. In section III, the posteriors of the parameters except for the DOF parameter is firstly derived; Secondly, the embedded MCMC are described to approximate the posterior of the DOF parameter; in the last part of this section, the detailed approximated lower bound is given out. Experiments on synthetic and speech signal are performed in section IV. In section V, conclusions are given.

2. Bayesian AR Model

Given the signal with length $T\mathbf{s}=(s_1,s_2,K\ ,s_T)$, and the known model order p, the autoregressive model can be described as

$$s_t=\sum_{i=1}^{p}w_i s_{t-i}+n_t \tag{1}$$

Where $(w_1,L\ ,w_p)$ are the coefficients of the AR model, n_t denotes the additive noise at time t. Instead of using normal distribution as the additive noise model, the additive noise is modelled using the Pearson type VII distribution. The formulation of this distribution is as follows

$$p(n_t\,|\,0,\lambda,\eta)=\frac{\sqrt{\lambda}}{B(\eta-1/2,1/2)}\left(1+\frac{n_t^2}{\lambda^{-1}}\right)^{-\eta} \tag{2}$$

Where $B(gg)$ is the Beta function. It has been proved that the Pearson type VII distribution can be formulated as the convolution of a normal and a Gamma distribution as following

$$p(n_t\,|\,0,\lambda,\eta)=\int_0^\infty N(n_t\,|\,0,\lambda^{-1}x_t^{-1})\mathcal{G}(x_t\,|\,\eta-\frac{1}{2},\frac{1}{2})dx_t \tag{3}$$

Where $\mathcal{G}(g)$ is the Gamma distribution and η is the degree of freedom parameter with the constraint condition $\eta>\frac{1}{2}$. From above equation, after the complex form (2) is decomposed into two simple distributions, a hidden variable x_t is introduced at every time t. All the hidden variables are collected into a vector $\mathbf{x}=(x_1,x_2,K\ ,x_T)$ .To form Bayesian model, the Gamma distributions is cast on the parameters λ and η as priors

$$\mathrm{p}(\lambda)=\mathcal{G}(\lambda\,|\,a_\lambda,b_\lambda) \tag{4}$$

For the DOF parameter η, note that $\eta-\frac{1}{2}>0$, so $\upsilon=\eta-\frac{1}{2}$ is used as the parameter and cast Gamma distributed prior on it

$$\mathrm{p}(\upsilon)=\mathcal{G}(\eta\,|\,a_\upsilon,b_\upsilon) \tag{5}$$

Then similarly as sparse learning, the AR model is transformed as the following form

$$\mathbf{s} = \mathbf{\Phi w} + \mathbf{n} \tag{6}$$

This equation is given by writing all the observance in to one vector **s**, and all additive noise into one vector $\mathbf{n} = (n_1, n_2, \ldots, n_T)$, and $\mathbf{w} = (w_1, \ldots, w_p)$ are the collection of all the coefficients. To give sparse learning and determine the model order in the VB iteration, the length of the coefficient vector **w** is changed as a large number *M*, which is larger than the maximum probable model order *p*. Then the coefficient vector is $\mathbf{w} = (w_1, \ldots, w_M)$.The matrix $\mathbf{\Phi}$ is formed the observance that each row contains the current observance and its *M*-1 previous observed values, such as $(s_0, s_{-1}, \ldots, s_{-M+1})$. Then priors should be placed on the coefficients to form a Bayesian model. Here ARD priors are adopted, which can introduce automatic model selection. The prior on each coefficient is set with a normal distribution with zero mean and γ_m variance as follows

$$\mathrm{p}(w_m \mid \gamma_m) = \mathrm{N}(w_m \mid 0, \gamma_m^{-1}) \tag{7}$$

Where γ_m determines the amplitude of the coefficient w_m in VB iteration. Then each γ_m is set with a Gamma distributed prior with the same shape and rate parameters as follows

$$\mathrm{p}(\gamma_n) = \mathcal{G}(\gamma_n \mid a_\gamma, b_\gamma) \tag{8}$$

Also, all the γ_m parameters are written into one vector $\boldsymbol{\gamma} = (\gamma_1, \ldots, \gamma_p, \ldots, \gamma_N)^T$.

3. Parameter Estimation using Variational Bayesian

All the parameters and hidden variables can be written into a set as $\boldsymbol{\theta} = \{\mathbf{w}, \boldsymbol{\gamma}, \mathbf{x}, \lambda, \upsilon\}$. VB is used to approximate the true posterior p(**θ**|**s**), thus reaching a closest approximation to p(**s**). The following factorized product is used to approximate the posterior of all the items of **θ.**

$$\mathrm{q}(\boldsymbol{\theta}) = \mathrm{q}(\mathbf{w})\mathrm{q}(\boldsymbol{\gamma})\mathrm{q}(\mathbf{x})\mathrm{q}(\lambda)\mathrm{q}(\upsilon) = \prod_{i=1}^{I} \mathrm{q}(\boldsymbol{\theta}_i) \tag{9}$$

Where $\boldsymbol{\theta}_i$ is the *i*-th item of the parameter set **θ**, and I is the total number of parameter items, here I equals to 5.The posterior approximation q($\boldsymbol{\theta}_i$) can be obtained by Variational EM algorithm [8]

$$\mathrm{q}(\boldsymbol{\theta}_i) \propto \exp\langle \mathrm{logp}(\mathbf{s}, \boldsymbol{\theta}) \rangle_{j \neq i} \tag{10}$$

Where $\langle \bullet \rangle_{j \neq i}$ denotes the expected statistics with respect to distributions other than q($\boldsymbol{\theta}_i$). The VB framework is to compute the expected statistics of each $\boldsymbol{\theta}_i$ round by round until convergence. The joint probability of all the parameters and the observed values p(**θ**|**s**) is

$$\mathrm{p}(\mathbf{s}, \boldsymbol{\theta}) = \mathrm{p}(\mathbf{s} \mid \mathbf{w}, \lambda, \mathbf{x})\,\mathrm{p}(\mathbf{w} \mid \boldsymbol{\gamma})\,\mathrm{p}(\boldsymbol{\gamma})\,\mathrm{p}(\mathbf{x} \mid \upsilon)\,\mathrm{p}(\upsilon)\,\mathrm{p}(\lambda) \tag{11}$$

3.1. Closed form Posteriors

Applying (10) to **w**, it can be seen that q(**w**) is a multivariate normal distribution, which has the covariance and mean as follows

$$\mathbf{\Sigma} = \left(\langle \lambda \rangle \mathbf{\Phi}^T diag(\langle \mathbf{x} \rangle)\mathbf{\Phi} + diag(\langle \boldsymbol{\gamma} \rangle)\right)^{-1} \tag{12a}$$

$$\boldsymbol{\mu} = \mathbf{\Sigma} \langle \lambda \rangle \mathbf{\Phi}^T diag(\langle \mathbf{x} \rangle)\mathbf{s} \tag{12b}$$

Notice that the subscript denoting the expected statistics with respect to which distribution is omitted here and it will be omitted in the following chapters. Then taking the same procedure, it can be derived that each γ_m is Gamma distributed with shape and rate parameters as follows

$$\alpha_{\gamma} = a_{\gamma} + \frac{1}{2} \tag{13a}$$

$$\beta_{\gamma_m} = b_{\gamma} + \frac{1}{2}\left\langle w_m^2 \right\rangle \tag{13b}$$

With the same steps, each hidden variable x_t is obtained as gamma distributed with the following shape and rate parameters

$$\alpha_x = \frac{\langle \upsilon \rangle + 1}{2} \tag{14a}$$

$$\boldsymbol{\beta}_{\mathbf{x}} = \frac{\langle \upsilon \rangle}{2} + \frac{\langle \lambda \rangle}{2}\left[\mathbf{s} \circ (\mathbf{s}\text{-}2\boldsymbol{\Phi}\langle \mathbf{w} \rangle + \text{diag}\left(\boldsymbol{\Phi}\left\langle \mathbf{w}\mathbf{w}^T \right\rangle \boldsymbol{\Phi}^T \right) \right] \tag{14b}$$

Where $\boldsymbol{\beta}_{\mathbf{x}} = \left(\beta_{x_1}, \text{K}, \beta_{x_T} \right)$ is the collection of all rate parameters of **x,** and $\circ$ is the item by item product of two vectors; and $\left\langle \mathbf{w}\mathbf{w}^T \right\rangle = \boldsymbol{\Sigma} + \langle \mathbf{w} \rangle \langle \mathbf{w} \rangle^T$ .Then the posterior of λ is also a gamma distribution, with the following parameters

$$\alpha_{\lambda} = a_{\lambda} + \frac{T}{2} \tag{15a}$$

$$\beta_{\lambda} = b_{\lambda} + \frac{1}{2}(\mathbf{s} \circ \langle \mathbf{x} \rangle)^T (\mathbf{s} - 2\boldsymbol{\Phi}\langle \mathbf{w} \rangle) + \frac{1}{2} tr\left(diag(\langle \mathbf{x} \rangle) \boldsymbol{\Phi} \left\langle \mathbf{w}\mathbf{w}^T \right\rangle \boldsymbol{\Phi}^T \right) \tag{15b}$$

3.2. Embedded MCMC for Unclosed form Posterior

Based on (10), the posterior of the freedom parameter of Pearson type VII distribution will be

$$\text{q}(\upsilon) \propto \exp\left\{ -T \log \Gamma(\upsilon) + \left[-T \log 2 + \sum_{t=1}^{T} \langle \log x_t \rangle - b_{\upsilon} \right] \upsilon + (a_{\upsilon} - 1) \log \upsilon \right\} \tag{16}$$

The above equation contains Gamma function of the variable υ, so it can't be formed by a closed form distribution. From (14), it can seen that only the estimation of the mean of q(υ) should be obtained. Markov Chain Mote Carlo method can be used to get this estimation. By MCMC method, a sequence of samples can be got until the samples can be seen from a stationary distribution which is exactly the true distribution. Then the mean $\bar{\upsilon}$ of the samples can be reached. Metropolis-Hasting (MH) [9] sampler is used to get samples from the distribution q(υ). Normal distribution denoting g is adopted to be the sampler. Let the sample to be generated be $\tilde{\upsilon}$, and the last sample be υ_{t-1} . The MH acceptance probability is

$$\alpha = \min\left(1, \frac{\text{q}(\tilde{\upsilon})}{\text{q}(\upsilon_{t-1})} \frac{g(\upsilon_{t-1} \mid \tilde{\upsilon})}{g(\tilde{\upsilon} \mid \upsilon_{t-1})} \right) \tag{17}$$

Then a rand number r is generated from a uniform distribution, if $r \le \alpha$ the proposed sample is accepted, otherwise rejected. To avoid different initialization affections, burnin in the sampling process is considered. The variance σ_{υ}^2 of the sampled samples can also be estimated. To form complete VB framework, q(υ) is assumed to be a normal distribution $N(\upsilon \mid \bar{\upsilon}, \sigma_{\upsilon}^2)$.

In above equations, the mean of the variable with Gamma distribution, and the expected value of the logarithm of the variable with Gamma distribution should be computed, which can be the following equations

$$\langle x \rangle_{\text{p(x)}} = \alpha / \beta \tag{18}$$

$$\langle \log x \rangle_{p(x)} = \Psi(\alpha) - \log(\beta) \tag{19}$$

Where x is the Gamma distributed variable, α and β is the shape and rate parameters respectively.

4. Experiments

In this section, experiments are performed to evaluate the proposed algorithm. Firstly, the algorithm is run on simulation data to prove the efficiency of the algorithm. Then, the algorithm is applied to speech signal, and compares the result with the least square method. In the following, the proposed algorithm is denoted as VBMC and the traditional least square method as LS.

4.1. Experiment on Synthetic Data

In this experiment, synthesized data are generated by AR systems with non-Gaussian distributed innovations. Then the algorithm is run to estimate the parameters of AR coefficients and the innovation distributions. The AR coefficients are generated by polynomials whose roots are random selected within the unit circle of complex plane. The order of AR is set to be 10, and the maximum order in the algorithm to be 20. With known AR order and defined values υ and λ, the model can be defined. Then the observed sequences can be produced following (1) with 10 initial values. In the algorithm, υ and λ are initialized randomly and the sample number is set to be 4000.

Firstly, the noise model parameters υ and λ are set to be 1.5 and 1/3 respectively, and then the algorithm is verified whether it can estimate the noise model parameters. The AR coefficients are generated by the above methods, and the sample number is 800. It should be noted that the parameters of the noise distribution can't be estimated by LS approach, and only the mean and variance of the normal distribution can be estimated by the prediction errors. The noise distribution is plotted in figure 1. From the figure, it can be seen that the proposed algorithm can accurately estimate the parameter of the noise model.

Secondly, with the noise parameters given above, AR observations with random initial values are generated. The true value of AR coefficients and the coefficients estimated by VBMC and LS are all plotted in figure 2. The figure shows that the VBMC approach can estimate the large coefficients accurately, but some of the small coefficients not very accurate. It can be also seen that the model order by VBMC is 10, which shows that it can estimate the model order automatically. Although the estimated coefficients are not exactly the same as the original, they are closer than that estimated by LS method.

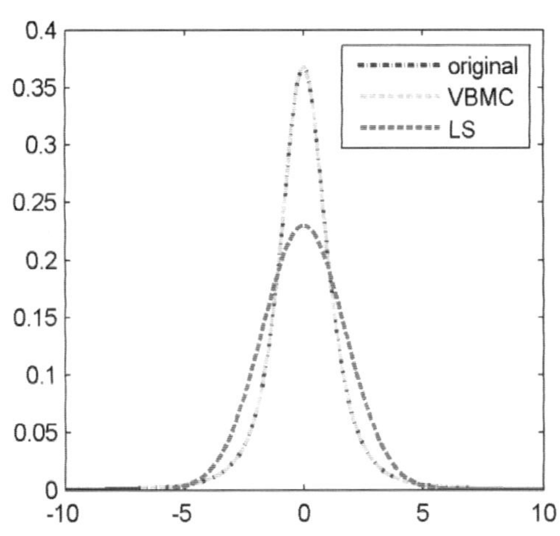

Figure 1. Original noise model, estimated by the proposed algorithm and by LS method.

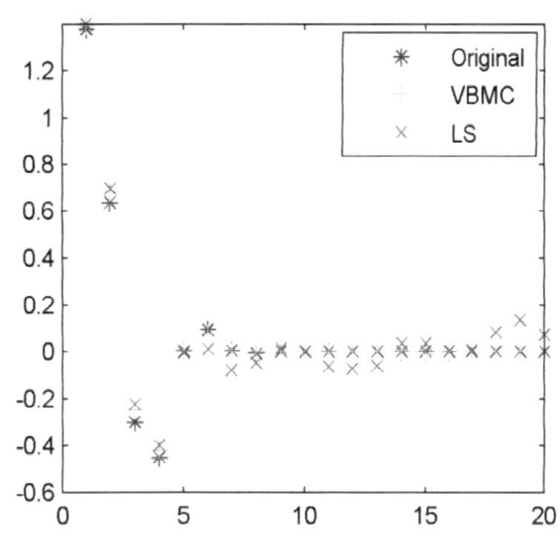

Figure 2. AR coefficients of the original, estimated by VBMC and LS method.

4.2. Experiment on speech signals

Speech signal can be modelled by AR model and several methods have been presented for AR coefficients estimation [1]. Here, experiment is used to prove that the speech signal is not generated by normal distributed noise model. If the noise model is normal distribution, then the observation should be normal distributed. A section of speech signal is selected with duration of 20ms and 16 KHz sampling rate, then plot the quantiles of the data versus the normal distribution in figure 3. If the speech signal can be modelled by AR model with Gaussian noise model, then the quantile will be a line. Figure 3 shows that it is not a line thus proved that speech signal is not generated by Gaussian noise model.

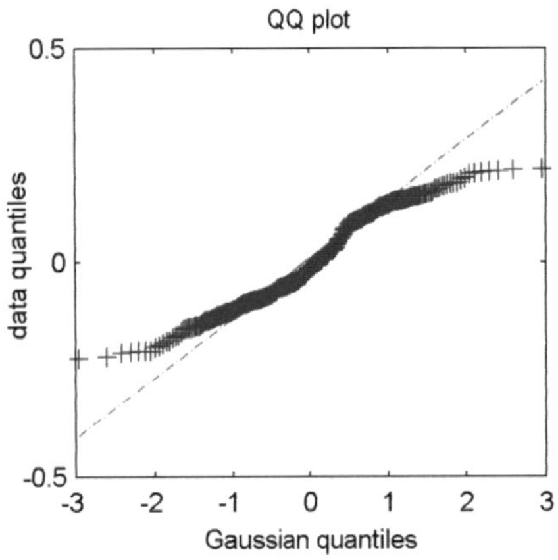

Figure 3. Quantiles of speech signal vs. standard normal distribution

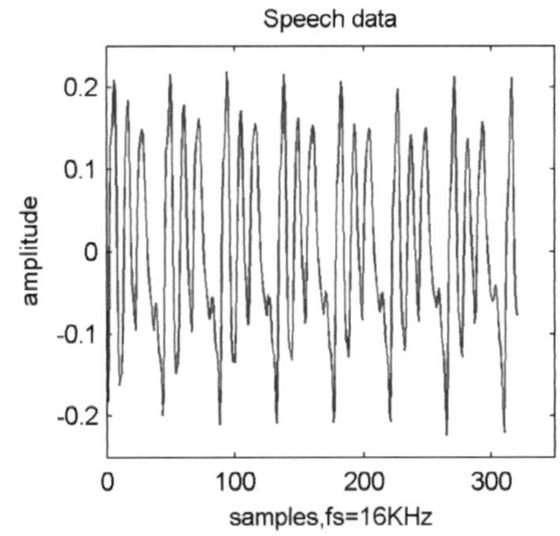

Figure 4. The original speech signal

The original signal is shown in figure 4, which is a periodic signal and the sharp peak and valley shows that there are outliers in the signal. Note that the speech signal is normalized, so the values are small. For signal with outliers, the noise model with long tail is suitable for innovation model of AR modelling. The prediction error and the estimated noise model are plotted in figure 5. The result shows that there really exist outliers in prediction error which correspond to the innovation model of AR model. The estimated noise model has long tail, thus the model can process such outliers. While for Gaussian noise model, it doesn't have long tail and can't process such data points.

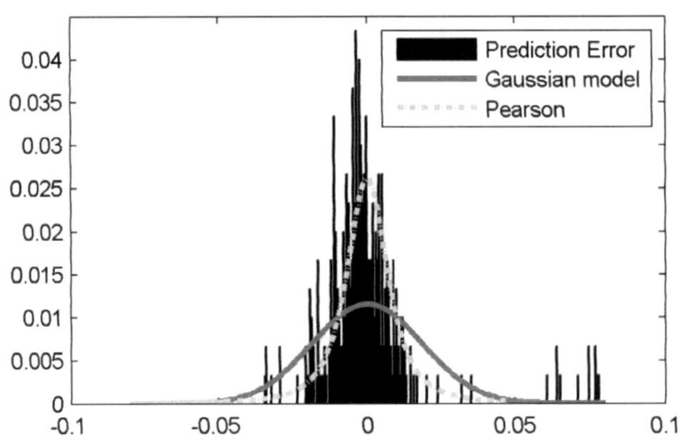

Figure 5. The prediction error and the estimated noise model

The AR model order is set to be from 10 to 20, and the RMSE between the original signal and the one time prediction signal are computed. The results are listed in Table 1 and the AR coefficients estimated by VBMC and that estimated by LS are plotted in figure 6. In table 1, the RMSE for order 17 to 20 is the same as 16, so only the results from 10 to 16 are listed. From table 1 and figure 6, it can be seen that LS can't automatically decide which the best model order is and it can only give result under given model order. The model order is 16 and VBMC can automatically give out this model order by automatically setting the irrelevant coefficients to be zero or almost zero.

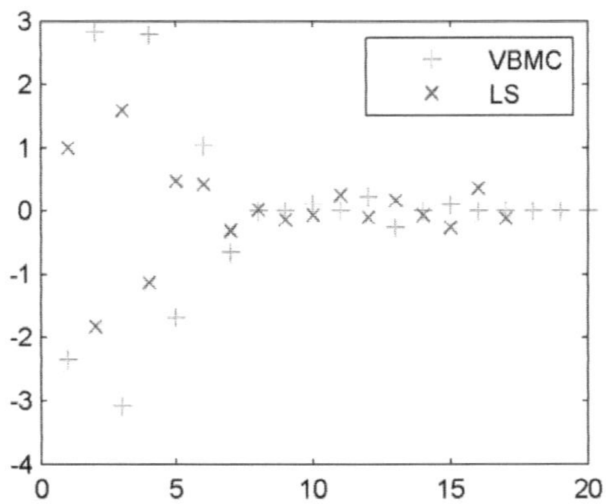

Figure 6. The estimated coefficients of speech signal

Table 1. RMSE by LS approach for different order and VBMC for automatically decision of model order

LS		VBMC	
10	0.0198		
11	0.0196		
12	0.0194		
13	0.0194	15	0.0168
14	0.0186		
15	0.0187		
16	0.0184		

5. Conclusion

In this paper, Bayesian method is proposed to estimate the AR coefficients of speech signal. Speech signal can be modelled by AR model and long tailed Pearson type 7 distributions. Variational Bayesian (VB) framework is used to get the posterior estimate of the coefficients and noise model parameters. To eliminate the non-conjugate problem, MCMC is embedded into the VB framework. The model order can be computed automatically by ARD. Experiments on synthetic data and speech signal are given for the proposed method and LS approach. Results show that the proposed algorithm is suitable for speech linear prediction.

References

[1] C Ben G, Nelson M and Dan E 2011 *Speech and Audio Signal Processing: Processing and Perception of Speech and Music, 2nd edition* (New York: Wiley–Interscience).
[2] Tzikas D G , Likas A C and Galatsanos N P The variational approximation for Bayesian inference 2008 *IEEE Signal Processing Magazine* **35(6)** 131-46.
[3] Penny W D and Roberts S J Bayesian methods for autoregressive models 2002 *Neural Networks for Signal Processing X, IEEE Signal Processing Society Workshop.*
[4] Pearson K Mathematical contributions to the theory of evolution, xix: Second supplement to a memoir on skew variation 1916 *Philos. Trans. Roy. Soc. Of London, Ser. A* **216** 429–57.
[5] Sun J Ata K and Garibaldi J M Robust mixture clustering using pearson type vii distribution 2010 *Pattern Recognition Letters* **31(16)** 2447-54.
[6] Tipping M E The relevance vector machine 2000 *Advances in Neural Information Processing Systems*.
[7] Bishop C M and Tipping M E Variational relevance vector machines 2000 *Proc. 16th Conf. Uncertainty in Artificial Intelligence* p 46–53.
[8] Wan H , Ma X and Li X Variational Bayesian learning for removal of sparse impulsive noise from speech signal 2018 *Digital Signal Processing* **73** 106–16.
[9] Gilks W R, Richardson S and Spiegelhalter D J 1996 *Markov Chain Monte Carlo in Practice* (London: Chapman and Hall).

An Implementation Method of SVPWM Modulation Algorithm

Wang Chao[1,*], Gao Qiang[1,2] and Tang Jing[2]

[1]Faculty of Information Technology, College of Microelectronics, Beijing University of Technology, Beijing, 100124, People's Republic of China
[2]Jiangsu Simand Electronics Co., Ltd., JiangSu province, People's Republic of China

[*]E-mail: 240333056@qq.com

Abstract: SVPWM algorithm is widely used in the field of power electronic frequency conversion power supply, motor frequency conversion control, power quality optimization and other technical fields. In this paper, the principle of SVPWM modulation is analyzed, in order to make the over-modulation algorithm more simplistic, the adverse effects on the system are smaller, and a phase and amplitude are used to change the modulation method to optimize the algorithm of the cross modulation II zone[1,2]. And using Matlab software to build SVPWM model for simulation verification. Explains the correctness of this method.

1. Introduction

With the development of power electronics technology, frequency conversion power supply, motor control, AC frequency conversion speed regulation and other technologies have been greatly improved. Space vector pulse width (SVPWM) technology is widely used in permanent magnet synchronous motor (PMSM). At present, there are two main ways to deal with SVPWM modulation[3,4]: One is the over-modulation algorithm that only changes the amplitude, [5]and the other is the over-modulation algorithm which changes the phase angle and amplitude at the same time. The former implementation method is simple, but the output base wave voltage amplitude can only be reached, the latter existing modulation algorithm is complex, and cause frequency distortion, so that the output voltage harmonic components increase, but the output base wave voltage amplitude can be achieved[6]. Therefore, in order to realize the over-modulation algorithm, the adverse effects on the system are smaller, the minimum phase error over modulation is used in the over modulation I area, and the phase and amplitude are used to change the modulation mode simultaneously in the II region.Using Matlab for simulation verification. Simulation results show the feasibility of the algorithm[7] .

2. SVPWM Fundamentals

The basic principle of SVPWM is to regroup the basic vector of voltage in a switching period, so that the actual voltage vector and reference voltage vector approximation, by fixing the opening source shutdown of a three-phase inverter in a time period to achieve this control effect, that is, through the superposition of the voltage vector of a certain period, Then control the action time of the three-phase inverter and make the magnetic chain circle of the voltage synthesis closer to a complete circle, the difference between the two after the coordinate transformation into the PWM modulation module, and finally form the PWM wave output to the inverter.

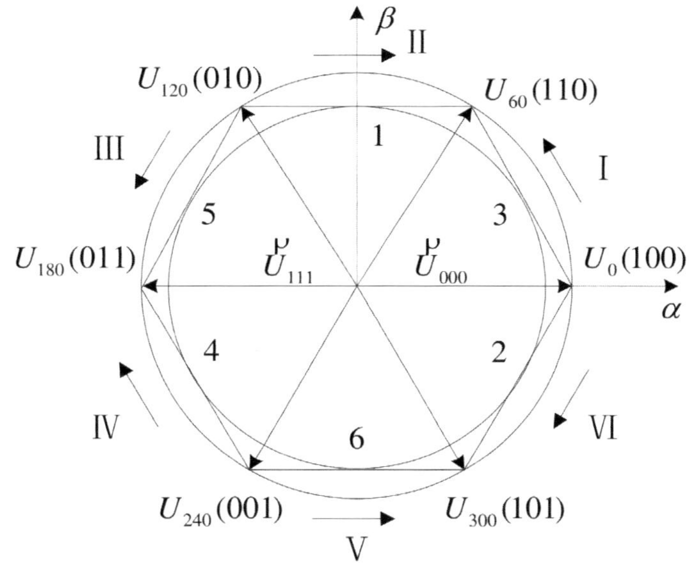

Figure 1. Voltage Space vector diagram

3. SVPWM over-modulation principle

In order to measure the over-modulation depth of the inverter, the modulation coefficient is defined: $MI = |V_{ref}| / \frac{2}{\pi} V_{dc}$. The $\frac{2}{\pi} V_{dc}$ is the phase voltage base amplitude value that the inverter outputs under the working state of six beat ladder wave, and when the modulation coefficient MI reaches 0.9069, the inverter begins to enter the modulation region, and the greater the modulation coefficient, indicates that the deeper the modulation depth, the less the reverse. Linear modulation zone when MI≤0.9069, when MI>0.9069 is a modulation zone. According to the different modulation coefficients of modulation coefficient, the over-modulation I area and the over-modulation II zone are divided.

3.1 Liner Modulation Zone

As shown in Figure 1, in a linear modulation region, any reference vector can be combined with two adjacent basic voltage vectors and zero vectors. At this point, the trajectory of the voltage vector is circular. By calculating the action time of two adjacent non-zero basic voltage vectors and zero vectors in the sector, then the voltage vector is synthesized.

Take the first sector, for example, $T_1 = \left(-\frac{3}{2}U_\alpha + \frac{\sqrt{3}}{2}U_\beta\right)T_s / U_d$; $T_2 = \sqrt{3}U_\beta T_s / U_d$; $T_0 = T_s - T_1 - T_2$; When the reference voltage angle is Π/6, the maximum amplitude of the output voltage vector is $V_{dc} / \sqrt{3}$. That is,the inner tangent circle radius of the hexagon.At this point $MI = 0.9069$. The output base wave reaches the maximum value of the linear zone. This modulation zone is called a linear modulation zone.

3.2 Overmodulation Area I

At that time, $0.9069 < MI \leq 0.9517$, in the modulation I area. The trajectory part of the reference voltage vector V_{ref} in the Overmodulated I region is inside the regular hexagon. and partly outside the positive hexagon. The Overvoltage I area limits the voltage vector in part beyond the hexagonal boundary to the hexagonal boundary without changing the phase of the voltage vector, and the voltage vector amplitude of this loss is compensated by the compensating voltage vector with a amplitude greater than the reference voltage vector amplitude value. The action phase of the compensating voltage vector is determined by reference angle α_r, and the trajectory of the whole region voltage vector is shown by the black trajectory in the Figure 2.

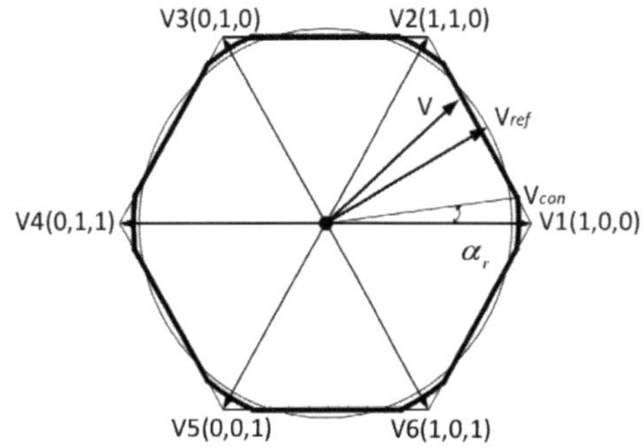

Figure 2. Voltage vector trajectory of dual-mode overmodulation I region

3.3 Over modulation II area

At that time, $0.9517 < MI \leq 1$, it was in the modulation II area. The reference voltage vector trajectory in the over-modulation II area is located outside the positive hexagon. Because the reference voltage vectors in the over-modulation II area are outside the positive hexagonal region, linear modulation cannot be implemented. It is necessary to pass the modulation algorithm to correct the base amplitude value of the reference voltage. Although the reference voltage of the over-modulated II area exceeds the positive hexagonal excision circle, the base wave of its actual output is still in the positive hexagonal excision circle.

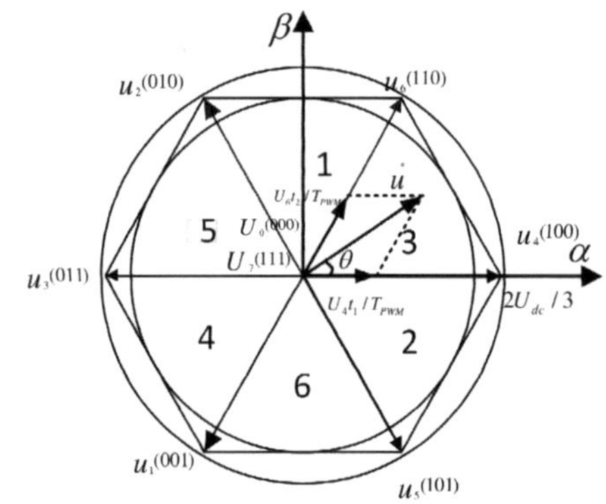

Figure 3. Overvoltage vector trajectory of over-modulation II zone

4. Over-modulation algorithm

4.1 Over modulation I area

In the over-modulation I area, the phase error minimization method is used, which is not to correct the phase of the reference voltage, only the amplitude is modified accordingly. Take sector 3 As an example, when the reference voltage vector $\vec{u}^*$ moves from point A to synchronous rotational angular frequency ω_e along the arc ABCD, $\vec{u}_1^*$ (fixed voltage phase) moves along the arc segment AB, linear segment BC, arc segment CD trajectory with the same rotation angle frequency, the dotted circle trajectory and the thick solid line trajectory are the motion tracks of $\vec{u}^*$ and $\vec{u}_1^*$ in a sinusoidal period, respectively The specific solution of this interval to the modified voltage vector $\vec{u}_1^*$ is as follows: Calculate t_1, t_2 and then modify the action time of two adjacent basic voltage vectors to t_1', t_2' by formula (1), and the action time of 0 vectors is 0. According to this correction method, when the reference voltage vector amplitude gradient in the $(U_{dc}/\sqrt{3} \sim 2U_{dc}/3)$ range, the actual output voltage base amplitude value changes within $(U_{dc}/\sqrt{3} \sim 0.9517*2U_{dc}/3)$ range.

$$t_1^{'} = \frac{t_1}{t_1 + t_2} T_{PWM}$$
$$t_2^{'} = \frac{t_2}{t_1 + t_2} T_{PWM} \quad (1)$$

4.2 Over modulation II area

The amplitude and phase in the over modulation II area are corrected at the same time, still in Sector 3 example, as shown in Figure 3, when $\vec{u}^*$ moves along the arc and the phase changes from zero to the retention angle θ_h (reference formula(2)), the $\vec{u}_1^*$ vector remains motionless at point A; in the process of $\vec{u}^*$ moving along the arc to synchronize frequency ω_e from B to C, $\vec{u}_1^*$ moves from A to D along the hexagonal boundary at A rotational frequency faster than ω_e , and when $\vec{u}^*$ moves along the arc from C point to phase $\pi/6$, the correction voltage vector $\vec{u}_1^*$ remains unchanged at D Point, and other sectors are modified in a similar way. When the reference voltage vector amplitude in $(2U_{dc}/3 \sim 2.2U_{dc}/\pi)$ gradient, the retention angle θ_h in $0 \sim \pi/6$ according to the modulation ratio of linear operation, the correction angle θ_1 in the retention angle is also calculated in a linear manner, when the actual output pressure vector $\vec{u}_1^*$'s base amplitude value in the $(0.9517*2U_{dc}/3 \sim 2U_{dc}/3)$ range, θ_h is 0 that corresponds to the upper modulation 2 zone output voltage value limit, When θ_h is $\pi/6$, it meant the beginning of a six-shot ladder wave.

The implementation algorithm of the over-modulation II zone is as follows: first select the Standard one Datum u_{base} (in this article, $u_{base} = U_{dc}/\sqrt{3}$) $U_{dc}/\sqrt{3}$, $2U_{dc}/3$, $2.2U_{dc}/\pi$ the corresponding standard value is u_{1b}, u_{2b}, u_{3b} , the voltage vector $\vec{u}_1^*$ $\vec{u}^*$, $\vec{u}_1^*$ standard homogeneous amplitude are recorded as u^*_{PU}, u^*_{1PU} Interval 3 θ_h and u^*_{PU} the relationship available formula (2) is indicated, In the whole $\alpha - \beta$ two-dimensional static coordinate system plane, The relationship expression between the amplitude u^*_{1PU} and phase θ_1 of the actual output voltage vector $\vec{u}_1^*$ and the phase angle θ of the reference voltage vector $\vec{u}^*$ is as as (3), (4). when located in the 3,1,5,4,6,2 sector in the figure above, the value θ_s is 0, $\pi/3$, $2\pi/3$, π , $4\pi/3$, $5\pi/3$,respectively.

$$\theta_h = \frac{(u^*_{PU} - u_{2b})}{u_{3b} - u_{2b}} * \pi/6 \quad (2)$$

$$\theta_1 = \begin{cases} \theta_s & (0 \le \theta - \theta_s \le \theta_h) \\ \dfrac{(\theta - \theta_h)\pi/6}{\pi/6 - \theta_h} + \theta_s & (\theta_h < \theta - \theta_s < \dfrac{\pi}{3} - \theta_h) \\ \dfrac{\pi}{3} + \theta_s & (\dfrac{\pi}{3} - \theta_h \le \theta - \theta_s \le \dfrac{\pi}{3}) \end{cases} \quad (3)$$

$$u^*_{1PU} = \begin{cases} u_{2b} & (0 \le \theta - \theta_s \le \theta_h) \\ \dfrac{u_{1b}}{\cos(\theta_1 - \theta_s - \pi/6)} & (\theta_h < \theta - \theta_s < \dfrac{\pi}{3} - \theta_h) \\ u_{2b} & (\dfrac{\pi}{3} - \theta_h \le \theta - \theta_s \le \dfrac{\pi}{3}) \end{cases} \quad (4)$$

5. SVPWM over-modulation simulation and experimental results

The SVPWM-modulated simulation model is built under the Matlab/Simulink software platform as shown in the Figure 4:

IOP Conf. Series: Materials Science and Engineering **646** (2019) 012066 doi:10.1088/1757-899X/646/1/012066

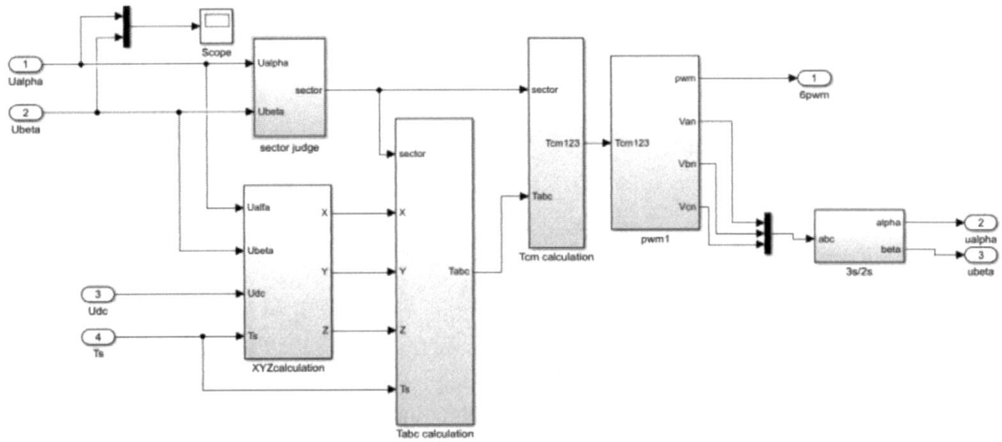

Figure 4. SVPWM simulation model of cross modulation

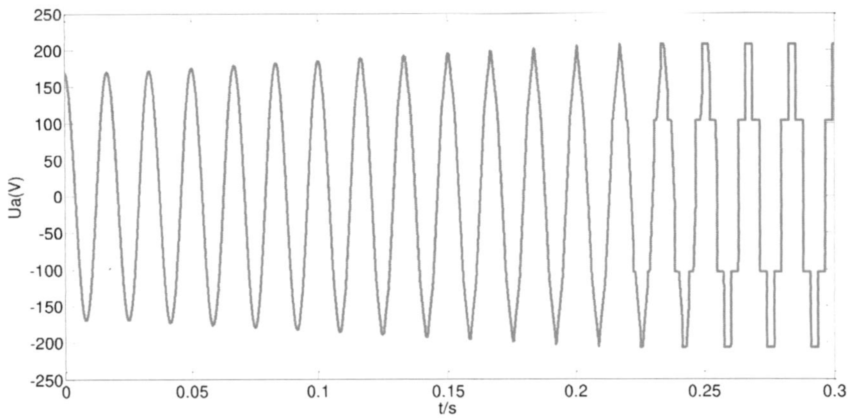

Figure 5. Phase voltage waveform

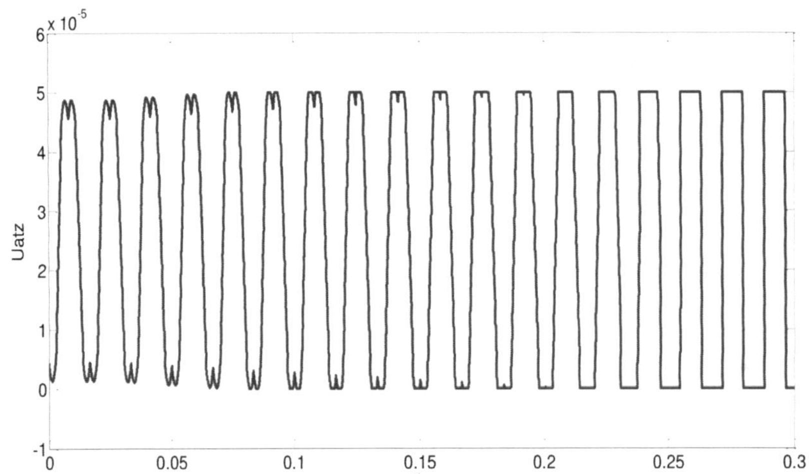

Figure 6. Modulation wave

The waveform Figure 5 of phase voltage can be obtained by voltage reconstruction technology, and the phase voltage amplitude transitions smoothly from linear modulation region to six beat ladder wave. SVPWM modulation wave in linear modulation region is saddle Wave Figure 6, and then gradually transition to square wave modulation mode.

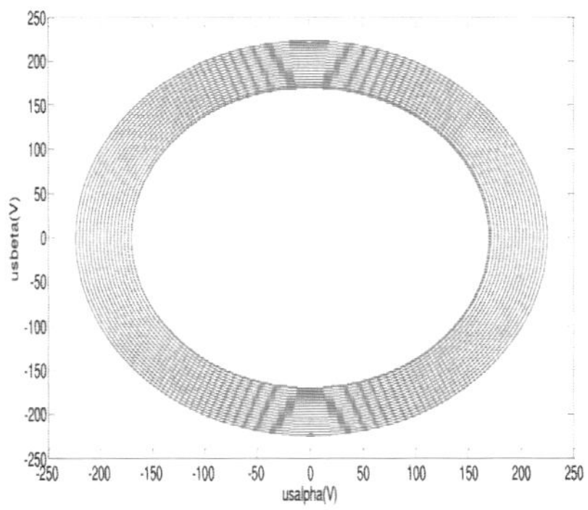

Figure 7 Reference voltage Vector

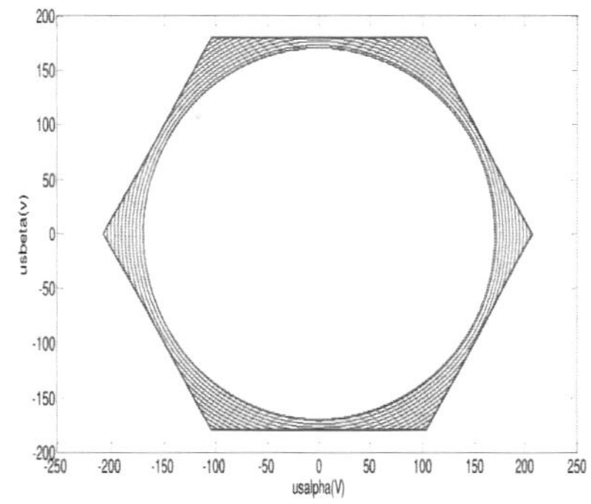

Figure 8 Fixed voltage vector

By comparing Figure 7 and Figure 8, it is found that when the reference voltage is less than $U_{dc}/\sqrt{3}$, the waveforms are exactly the same, and when greater than $U_{dc}/\sqrt{3}$, the exterior of the inner tangent circle is limited to the hexagonal.

6. Conclusion

In this paper, the SVPWM algorithm is simplified, so that its algorithm is more simplistic, the adverse effects on the system are smaller, and it is easier to engineer and realize. And the Matlab/simulink software is used to verify the SVPWM modulation algorithm. The results show that the modulation algorithm can realize the transition from linear modulation state to six beat step wave running state.

References

[1] S. Venugopal, G.Narayanan.An overmodulation scheme for vector controlled induction motor drives [J].IEEE,2006.

[2] Jul Ki Seok, Seung Ki Sul. A new overmodulation strategy for induction motor drive using space vector PWM [C]. APEC'95,1995.10(1):211-216.

[3] Lee D C, Lee G M.Linear control of inverter output voltage in overmodulation [J]. IEEETransaction on Industrial Electronics,1997,44 (4) :590-592.

[4] Lee Dong-Choon. [J].IEEE Trans. on PE, 1998,13 (6): 1144-1150.

[5] Holtz J.On Continuous Control of PWM Inverters in the Over-modulation Range Including the Six-Step Mode[J].IEEE Trans. on PE, 1993, 8 (4): 546-553.

[6] Holtz J., Lotzkat W., Khambadkone A.M., "On Continuous Control of PWM Inverters in the Overmodulation Range Including the Six-StepMode", IEEE Trans. On Power Electronics, vol.8,no.4, October 1993.

[7] Jul Ki Seok, Jhoon Sheok Kim, Jong Woo Choi. Overmodulation strategy for high Performance torque control[C].PESC,96,1996,2(23):1549-1554.

IOP Conf. Series: Materials Science and Engineering **646** (2019) 012067 doi:10.1088/1757-899X/646/1/012067

Human Motion Trajectory Prediction in Human-Robot Collaborative Tasks

Shiqi Li, Haipeng Wang*, Shuai Zhang, Shuze Wang and Ke Han

School of Mechanical Science and Engineering, Huazhong University of Science and Technology, Wuhan, China

*wanghai463100@126.com

Abstract. A method is introduced to predict human motion trajectory in the process of human-robot collaboration (HRC). In the method, the human-robot distances are assumed to be a Gaussian Process (GP). To achieve this, a human-robot handover task is conducted by a human and a collaborative robot, while the positions of the human hand and the robot end-effector are recorded. Some of the recorded data are used for the Gaussian Process Regression (GPR), a GP and a 95% confidence convince about the GP are obtained by the GPR. Experimental results show that about 80% of the testing data are included in the 95% confidence convince. The method and results here are useful to other human-robot collaborative tasks where existing human-robot relative motions, especially, the method is able to predict the human motion trajectory with varying initial position of the human hand and varying locations of the robot end-effector.

1. Introduction

A natural human-robot collaboration (HRC) flow is important to reduce human's waiting time, increase efficiency and enhance naturalness [1, 2, 3]. To achieve that, it is critical for the robot to comprehend their human peers' on-going actions and predict their behaviours in the near future, to start motion planning and execution early enough to smooth the turn-taking transition. So, not only the human motion target, but also the human motion trajectory over time is required.

The robots make prediction of human motion mainly for the purpose of collision avoidance [4, 5] or adapting to the human [6, 7]. The realm of human motion prediction can be divided into two categories [8]: the unsupervised approaches and the supervised approaches. In terms of the unsupervised approaches, the Hidden Markov Model (HMM) [9], the Support Vector Machine (SVM) [10], the Partially Observable Markov Decision Processes (POMDPs) [11] and the social force model [12] are always used, in recent years, the brain-machine interface (BMI) have demonstrated success by means of the sparse linear regression to predict hand trajectory [13, 14]. In terms of the supervised approaches, the inverse optimal control [15, 16] and the neural network, especially the Long Short-Term Memory network [17, 18, 19] are always used, in which a series of labelled demonstrations of specific tasks are required, the labelled demonstrations may be a series of motions of a specific task [20], or a motion library of basic human motion [21, 22].

However, it is found that most of the researches are focused on the human motion target prediction or the human behaviour recognition, the problem with this is that the robot is usually able to predict what the human is about to do, but the motion trajectories in a future time is still unknown.

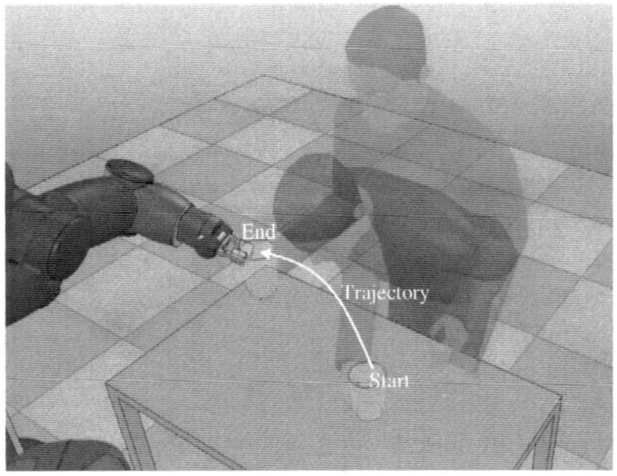

Figure 1. Illustration of the human motion trajectory: in these scenarios, the starting and ending points are known, while the human motion trajectory is unknown. The knowledge of the human motion trajectory can help the robot make early motion planning and perform more natural and smooth actions.

The human always moves with considerable uncertainty, even to a specific task, the human motion trajectories varies, based on this issue, a method of human trajectory prediction in HRC is proposed in this paper, in a human-robot handover task (see Figure 1), the distances between the human hand and the robot end-effector alone three axis of coordinates are assumed to be Gaussian Processes as time goes on, the human motion trajectories can be described by the combination of the three Gaussian Processes, experimental results show that the proposed method can effectively predict the human trajectories.

2. Proposed Approach

2.1. Task description

In many collaborative tasks, such as the human-robot object handover, the human and the robot should move close to each other and make physical contact, the task can be divided into three phase:

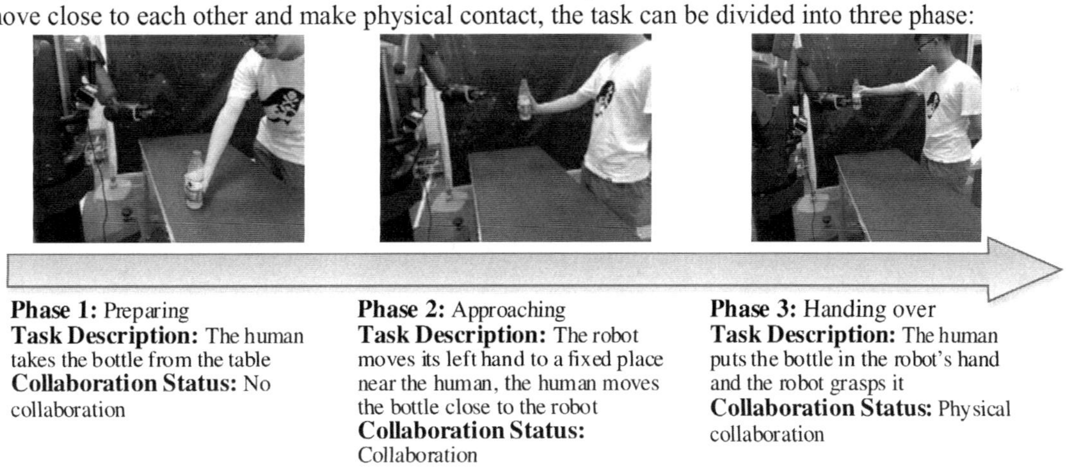

Figure 2. The overall task flow

In the first phase (see Figure 2), the human holds the bottle from the table, the initial position of the bottle on the table varies in each demonstration. In the second phase, the robot moves its hand to a fixed place which is in the scope of the human hand's movement space, the human moves the bottle close to the robot end-effector, a Microsoft Kinect is used to track the motion trajectory of the human during the entire task. In the last step, the robot grasps the bottle.

IOP Conf. Series: Materials Science and Engineering **646** (2019) 012067 doi:10.1088/1757-899X/646/1/012067

2.2. The Gaussian Process(GP)

A series of pre-experiments show that the human moves with considerable random, even for a same task—the human moves a bottle from a fixed position to another fixed position, the trajectories usually vary. the Gaussian Process is used in this paper to predict the human motion trajectories, in which the human hand position probability, not the accurate position, is predicted. It is assumed that at time t, the distance between the human hand and the robot end-effector $f(t)$ obey the Gaussian distribution, and the change of human-robot distance with time obey the Gaussian Process:

$$\begin{bmatrix} f(t_1) \\ \vdots \\ f(t_n) \end{bmatrix} \sim \mathbf{N}\left(\begin{bmatrix} m(t_1) \\ \vdots \\ m(t_n) \end{bmatrix}, \begin{pmatrix} k(t_1,t_1) & \cdots & k(t_1,t_n) \\ \vdots & \ddots & \vdots \\ k(t_n,t_n) & \cdots & k(t_n,t_n) \end{pmatrix}\right) \quad (1)$$

Where $m(x_i)$ is the mean value of the human-robot distance at time t_i and $k(\cdot,\cdot)$ is the covariance matrix. At time t_i, the human-robot distance $f(t_i)$ obey a Gaussian distribution with an expectation of $m(t_i)$ and a variance of $k(t_i,t_i)$. And in different time t_i and t_j, the $f(t_i)$ and $f(t_j)$ are not statistically independent, the $k(t_i,t_j)$ (not zero) is used to represent the mutual relationship.

3. Algorithms

Our approach to predicting human motion trajectories consists of three phases (see Figure 3). In the first phase we provide demonstrations of human-robot handover and gather data (the position trajectories of the human and the robot), the data is processed in the second phase, including the coordinate transformation and the data segmentation, in the third phase the training data is used for the Gaussian process regression, and then the GP is used to predict the human trajectories.

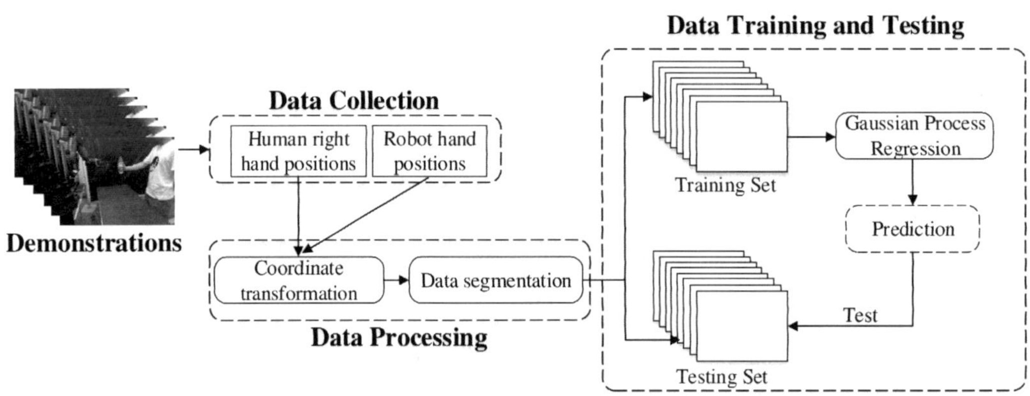

Figure 3. Diagram of the proposed framework

3.1. Data Collection

The human-robot handover demonstrations are performed by a human and a collaborative robot, A Microsoft Kinect was used to track the motion trajectory of the human (the data acquisition cycle is 0.03s). The bottle is placed on the table, the human holds the bottle and moves it to the robot, 30 demonstrations were made, the initial positions of the bottle on the table, and the positions of the robot end-effector are different in each demonstration.

3.2. Data Process

Data segmentation: A total of 30 groups of experiments were conducted, 20 of which were used for training, and the others were used for testing. In this paper, the relative distance between the human and the robot along three coordinate axes were studied separately.

3.3. Data training and testing

Gaussian Process Regression: At any time step, the mean distance between the human hand and the robot is calculated:

$$(m_1,,m_i,,m_n) = \left(\sum_{j=1}^{n_1}\frac{d_{1j}}{n_1},,\sum_{j=1}^{n_i}\frac{d_{ij}}{n_2},,\sum_{j=1}^{n_n}\frac{d_{nj}}{n_n}\right) \tag{2}$$

At time t_i, n_i is the total number of data collected, d_{ij} is the human-robot distance of each point.

There is a strong correlation between the human-robot distance and time, so the covariance at different time is not zero, a Radial Basis Function (RBF) is used to represent the covariance matrix, the first reason is that the covariance matrix is required to be semi-positive definite matrixes, the RBF meets that need, the other is that the RBF providing a fast approximation of the covariance matrix [23], has good generalization ability and a fast learning convergence speed:

$$k(t_i, t_j) = \partial^2 e^{-\frac{(t_i-t_j)^2}{2*l*l}} \tag{3}$$

Where $k(t_i, t_j)$ are the covariance of human-robot distance between time t_i and t_j. ∂ and l are the parameters to be iterated optimization, the constraint condition is to maximize $\mathrm{p}(Y|X)$ while $\mathrm{f(x)}\sim\mathrm{N}(\mathrm{m(x)}, \mathrm{k}(\mathrm{x}, x^T))$, the object function is set to:

$$log\mathrm{p}(\mathrm{Y}|\mathrm{X}) = logN(\mu, K) \tag{4}$$

Prediction: the mean value and the covariance matrix are calculated based on the algorithms above and the training data, and the distance between the human and the robot at the other time also subordinate to the Gaussian Process, so:

$$\begin{bmatrix} f \\ f* \end{bmatrix} \sim \mathrm{N}\left(\begin{bmatrix} m \\ m* \end{bmatrix}, \begin{pmatrix} k & k* \\ k*^T & k** \end{pmatrix}\right) \tag{5}$$

Where $f*$ are the distance at the other time, $k*$ and $k**$ are the covariance matrix which can be calculated by (5).

$$f(t_i)^{upper} = m(t_i) + 1.96 \cdot \sqrt{k(t_i, t_i)} \tag{6}$$

$$f(t_i)^{lower} = m(t_i) - 1.96 \cdot \sqrt{k(t_i, t_i)} \tag{7}$$

Where $f(t_i)^{upper}$ and $f(t_i)^{lower}$ are the upper and lower limits of the 95% confidence interval respectively.

4. Experiments and Results

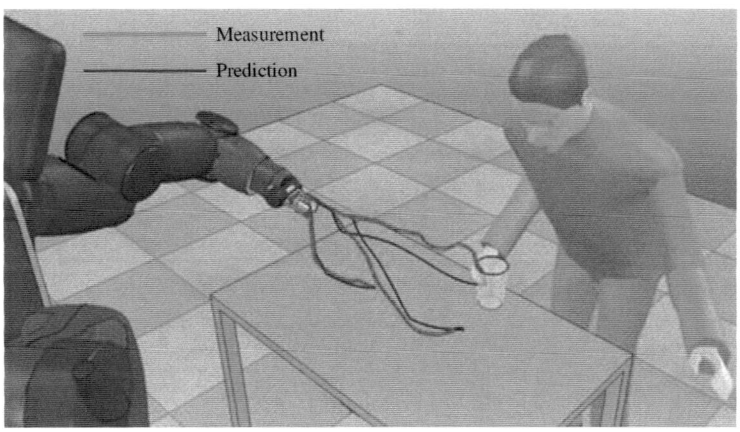

Figure 4. some of the measured trajectories and the predicted trajectories. The pink and thick curves represent the measured trajectories and the blue and thin curves represent the predicted trajectories.

Three of the tested trajectories is shown in a virtual 3D scenario (see Figure 4), as can be shown, two of the prediction trajectories coincidence with the measured trajectory, the other shows a relative big difference.

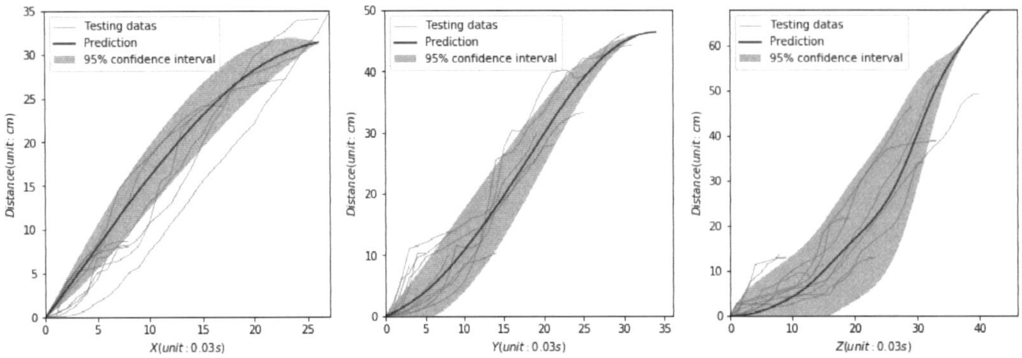

Figure 5. the testing data and the GP

In all 10 testing trajectories (see Figure 5), 8 of the trajectories (about 80%) are included in the 95% confidence interval, the remaining shows a relatively bigger difference. On the whole, the 95% confidence interval include most of the trajectories, which can be used as a prediction interval of human trajectories in the human-robot handover.

5. Conclusion

An approach of human trajectories prediction is proposed in this paper, in which the human motion trajectories alone along three coordinate axes are fitted into Gaussian Process respectively, and the Gaussian Processes are then used for prediction, Experimental results show that the method can make prediction of human motion trajectories in a relatively high accuracy. The method and results presented in this paper could be extended to other human-robot collaborative tasks when the human and the robot move closer to each other.

Acknowledgments

We acknowledge the support received for the Manned Aerospace Research Project of China (Grant No. 060601).

References

[1] Mainprice J, Berenson D. Human-robot collaborative manipulation planning using early prediction of human motion[C]//2013 IEEE/RSJ International Conference on Intelligent Robots and Systems. IEEE, 2013: 299-306.

[2] Zhou T, Wachs J P. Early prediction for physical human robot collaboration in the operating room[J]. Autonomous Robots, 2018, 42(5): 977-995.

[3] Ajoudani A, Zanchettin A M, Ivaldi S, et al. Progress and prospects of the human–robot collaboration[J]. Autonomous Robots, 2018, 42(5): 957-975.

[4] Ragaglia M, Zanchettin A M, Rocco P. Safety-aware trajectory scaling for human-robot collaboration with prediction of human occupancy[C]//2015 International Conference on Advanced Robotics (ICAR). IEEE, 2015: 85-90.

[5] Luo R, Hayne R, Berenson D. Unsupervised early prediction of human reaching for human–robot collaboration in shared workspaces[J]. Autonomous Robots, 2018, 42(3): 631-648.

[6] Lawitzky M, Medina J R, Lee D, et al. Feedback motion planning and learning from demonstration in physical robotic assistance: differences and synergies[C]//2012 IEEE/RSJ International Conference on Intelligent Robots and Systems. IEEE, 2012: 3646-3652.

[7] Nikolaidis S, Kuznetsov A, Hsu D, et al. Formalizing human-robot mutual adaptation: A bounded memory model[C]//The Eleventh ACM/IEEE International Conference on Human Robot Interaction. IEEE Press, 2016: 75-82.

[8] Kong Y, Fu Y. Human action recognition and prediction: A survey[J]. arXiv preprint arXiv:1806.11230, 2018.
[9] Liu H, Wang L. Human motion prediction for human-robot collaboration[J]. Journal of Manufacturing Systems, 2017, 44: 287-294.
[10] Huang C M, Andrist S, Sauppé A, et al. Using gaze patterns to predict task intent in collaboration[J]. Frontiers in psychology, 2015, 6: 1049.
[11] Pellegrinelli S, Admoni H, Javdani S, et al. Human-robot shared workspace collaboration via hindsight optimization[C]//2016 IEEE/RSJ International Conference on Intelligent Robots and Systems (IROS). IEEE, 2016: 831-838.
[12] Luber M, Stork J A, Tipaldi G D, et al. People tracking with human motion predictions from social forces[C]//2010 IEEE International Conference on Robotics and Automation. IEEE, 2010: 464-469.
[13] Nakanishi Y, Yanagisawa T, Shin D, et al. Mapping ECoG channel contributions to trajectory and muscle activity prediction in human sensorimotor cortex[J]. Scientific reports, 2017, 7: 45486.
[14] Nakanishi Y, Yanagisawa T, Shin D, et al. Prediction of three-dimensional arm trajectories based on ECoG signals recorded from human sensorimotor cortex[J]. PloS one, 2013, 8(8): e72085.
[15] Mainprice J, Hayne R, Berenson D. Goal set inverse optimal control and iterative replanning for predicting human reaching motions in shared workspaces[J]. IEEE Transactions on Robotics, 2016, 32(4): 897-908.
[16] Mainprice J, Hayne R, Berenson D. Predicting human reaching motion in collaborative tasks using inverse optimal control and iterative re-planning[C]//2015 IEEE International Conference on Robotics and Automation (ICRA). IEEE, 2015: 885-892.
[17] Alahi A, Goel K, Ramanathan V, et al. Social lstm: Human trajectory prediction in crowded spaces[C]//Proceedings of the IEEE conference on computer vision and pattern recognition. 2016: 961-971.
[18] Cheng Y, Zhao W, Liu C, et al. Human motion prediction using adaptable neural networks[J]. arXiv preprint arXiv:1810.00781, 2018.
[19] Fernando T, Denman S, Sridharan S, et al. Soft+ hardwired attention: An lstm framework for human trajectory prediction and abnormal event detection[J]. Neural networks, 2018, 108: 466-478.
[20] Vogt D, Stepputtis S, Grehl S, et al. A system for learning continuous human-robot interactions from human-human demonstrations[C]//2017 IEEE International Conference on Robotics and Automation (ICRA). IEEE, 2017: 2882-2889.
[21] Luo R, Berenson D. A framework for unsupervised online human reaching motion recognition and early prediction[C]//2015 IEEE/RSJ International Conference on Intelligent Robots and Systems (IROS). IEEE, 2015: 2426-2433.
[22] Pérez-D'Arpino C, Shah J A. Fast target prediction of human reaching motion for cooperative human-robot manipulation tasks using time series classification[C]//2015 IEEE international conference on robotics and automation (ICRA). IEEE, 2015: 6175-6182.
[23] Curtó J D, Zarza H C, Kim T. McKernel: Approximate Kernel Expansions in Log-linear Time through Randomization[J]. arXiv preprint arXiv:1702.08159, 2017.